Lecture Notes in
Computer Science

Lecture Notes in Computer Science

Lecture Notes in Computer Science

Edited by G. Goos and J. Hartmanis

473

I.B. Damgård (Ed.)

Advances in Cryptology — EUROCRYPT '90

Workshop on the Theory and Application
of Cryptographic Techniques
Aarhus, Denmark, May 21–24, 1990
Proceedings

Springer-Verlag
Berlin Heidelberg New York London
Paris Tokyo Hong Kong Barcelona

Volume Editor

Ivan Bjerre Damgård
Matematisk Institut, Århus Universitet
Ny Munkegade, DK-8000 Århus C, Denmark

CR Subject Classification (1987): D.4.6, E.3, H.2.0

ISBN 978-3-540-53587-4 Springer-Verlag Berlin Heidelberg New York
Springer-Verlag New York Berlin Heidelberg

Printing and binding: Druckhaus Beltz, Hemsbach/Bergstr.
2145/3140-543210 — Printed on acid-free paper

Preface

EUROCRYPT is a conference devoted to all aspects of cryptologic research, both theoretical and practical. In the last 7 years, the meeting has taken place once a year at various places in Europe. Both these meetings and the annual Crypto meetings in California are sponsored by The International Association for Cryptologic Research (IACR). Most of the proceedings from these meetings are, like this one, published in Springer-Verlag's *Lecture Notes in Computer Science* series.

EuroCrypt 90 took place on May 21-24 at conference center Scanticon, situated in Århus, Denmark. There were more than 250 participants from all over the world. It is a pleasure to take this opportunity to thank the general chairman Peter Landrock, Århus Congress Bureau, Scanticon, and the organizing committee, who all contributed with hard work and dedication to make a well organized and successful conference.

A total of 85 papers from all over the world were submitted to the conference. This number marks a continuation of the steady growth of interest in the EuroCrypt meetings. Out of the papers submitted, 41 were rejected, 1 was withdrawn, and 2 papers were asked to merge. This resulted in a set of 42 papers presented at the conference. The submissions were in the form of extended abstracts. All program committee members received a full set of submissions, and each submission was refereed independently by at least two members of the program committee (not including the program chairman). The experiment from Crypto 89 with blind refereeing was continued at this conference, and has now become standard policy at IACR conferences. The final papers appearing in these proceedings were not refereed, and the authors retain, of course, full responsibility for the contents. Several of the papers can be expected to appear in various journals in more polished form. There will a special issue of the Journal of Cryptology containing selected papers from the conference.

In addition to the formal contributions, a number of informal talks were given at the traditional rump session. These proceedings include short abstracts of some of these impromptu talks.

Finally, it is a pleasure to acknowledge all those who contributed to putting together the program of EuroCrypt 90 and making these proceedings a reality.

First of all, thanks to the program committee. All of its members put a tremendous amount of hard work into the refereeing, and many of them even took the time to make detailed comment on other papers than the 20 they were asked to read carefully. Also some of my colleagues at Århus University kindly offered their help on various technical questions; among these were Torben Pedersen and Jørgen Brandt.

Of course, no conference could have taken place without the authors' contribution. I would like to thank all those who submitted papers, also those whose submissions could not be accepted because of the large number of high quality submissions we received. Many of the authors have been extremely cooperative in changing the format of their papers to fit into the proceedings. Were it not for this attitude, these proceedings would have been significantly delayed.

Århus, September 1990   Ivan Bjerre Damgård

EUROCRYPT 90

A conference on the theory and application of cryptology

Sponsored by The International Association for Cryptologic Research (IACR)

and

CRYPTOMAT_HIC AS, DATACO AS, Den Danske Bank AS,

Jutland Telephone Company AS

General Chairman: Peter Landrock (Aarhus University)
Organizing Committee:
Jørgen Brandt (Aarhus University)
Palle Brandt Jensen (Jutland Telephone Company)
Torben Pedersen (Aarhus University)
Århus Congress Bureau

Program Chairman: Ivan Damgård (Aarhus University)
Program Committee:
Ueli Maurer (ETH, Zürich)
Andrew J. Clark (Computer Security Ltd., Brighton)
Claude Crépeau (LRI, Paris)
Thomas Siegenthaler (AWK, Zürich)
Joan Boyar (Aarhus University)
Stig Frode Mjølsnes (ELAB, Trondheim)
Marc Girault (SEPT, Caen)
Walter Fumy (Siemens AG, Erlangen)
Othmar Staffelbach (Gretag, Regensdorf)

Contents

Session 1: Protocols

Session 2: Number-Theoretic Algorithms

Session 3: Boolean Functions

Session 4: Binary Sequences

Session 5: Implementations

Session 6: Combinatorial Schemes

Session 7: Cryptanalysis

Session 8: New Cryptosystems

Session 9: Signatures and Authentication

Rump Session: Impromptu Talks

ALL LANGUAGES IN **NP** HAVE DIVERTIBLE ZERO-KNOWLEDGE PROOFS AND ARGUMENTS UNDER CRYPTOGRAPHIC ASSUMPTIONS*

(Extended Abstract)

Mike V. D. Burmester [†]
Dept. of Mathematics
RHBNC - University of London
Egham, Surrey TW20 0EX
U.K.

Yvo Desmedt [‡]
Dept. EE & CS
Univ. of Wisconsin – Milwaukee
P.O. Box 784
WI 53201 Milwaukee
U.S.A.

Abstract

We present a divertible zero-knowledge proof (argument) for SAT under the assumption that probabilistic encryption homomorphisms exist. Our protocol uses a simple 'swapping' technique which can be applied to many zero knowledge proofs (arguments). In particular we obtain a divertible zero-knowledge proof for graph isomorphism. The consequences for abuse-free zero-knowledge proofs are also considered.

I. Introduction

Okamoto-Ohta defined divertible zero-knowledge proofs in [OO89] and showed that commutative random self-reducible relations have such proofs, provided certain conditions are satisfied. The first divertible zero-knowledge proof was given in [DGB88, pp. 37–38] in the context of an abuse-free zero-knowledge proof.

In this paper we generalize this result to *all* problems in NP under cryptographic assumptions and consider the consequences for abuse-free proofs. We also remark that most divertible zero-knowledge proofs of membership presented here will not convince unconditionally two (independent) verifiers simultaneously. So the framework of divertible zero-knowledge has to be modified if it is to be used for this purpose.

This paper is organized as follows. We first state our results. Then we present the protocol and finally we sketch the proofs.

---

*Research done while visiting the EISS, University of Karlsruhe, West Germany.
[†]Research partially supported by SERC Grant GR/F 5700.
[‡]Research is being supported by NSF Grant NCR-9004879.

II. Main results

II.1. Notation and Definitions

(A, B, C) is a divertible interactive triple of Turing machines [OO89]. For the definition of divertible proofs and abuse-free systems see [OO89,Des90]; for the SAT proof (argument) see [BCC88,BC89]. A probabilistic encryption function $f.(\cdot)$ satisfies the properties that $f_r(b)$ can be computed in polynomial time when r, b are given, and that $f_r(b) = f_{r'}(b') \Rightarrow b = b'$. Here r, r' are any random bit strings and b, b' are bits. f is a probabilistic *homomorphism* if $f_r(b) \cdot f_{r'}(b') = f_{r''}(b \oplus b')$, where r'' can be computed from r, r', b and b' in polynomial time, and $\oplus$ is exclusive-or. A well-known example of an encryption homomorphism [GM84] is given by $f_r(b) \equiv s^b r^2 (\bmod\, n)$, where n is a Blum integer and s is an appropriate quadratic non-residue. (It is instructive to compute r'' in this case, given s, n, and r, r', $b = b' = 1$.) The modulus n and s parameterize f. We shall assume that all the probabilistic encryption functions considered in this paper are parameterized, but for simplicity we ignore this in our notation.

We denote by $\{z\}$ a string which is a concatenation of strings of type z with delimiters.

II.2. Theorems and implications for abuse-free proofs

Theorem 1 *If probabilistic encryption homomorphisms exist and are provided by an oracle, then all languages in NP have divertible zero-knowledge proofs.*

Corollary 1 *If probabilistic encryption homomorphisms exist then all languages in NP have conditional abuse-free zero-knowledge proofs.*

Theorem 2 *If probabilistic encryption functions exist then all languages in NP have unconditional abuse-free zero-knowledge proofs.*

Theorem 3 *Given an oracle similar to the one in Theorem 1: If factoring is hard then all languages in NP have divertible statistical zero-knowledge arguments.*

Corollary 2 *If probabilistic blob functions exist then all languages in NP have abuse-free zero-knowledge arguments.*

Theorem 4 *There exists an 'unconditional'* [1] *divertible zero-knowledge proof for graph isomorphism.*

Remarks: We will describe a protocol which can be used for many zero-knowledge proofs with slight modifications. This protocol does not require that the structures involved are commutative. Furthermore it can easily be adapted to make the authentication system [Des88] unconditionally divertible (so that two or more independent wardens can be used).

---

[1]The quotation marks are due to the unnatural condition (iii) of Definition 1 in [OO89], which implies that the protocol is only divertible when graph isomorphism is not decidable in probabilistic polynomial time. In the final paper we will restate this definition but without this property.

III. Main approach

Many interactive zero-knowledge proofs, as, [Blu87,CEvdG88,GMW86,GMR89,BCC88] (and arguments [BCC88,BC89]) have protocols with a loop in which:

Step 1 the prover sends a 'commitment' (blob),

Step 2 the verifier asks a one bit question,

Step 3 the prover replies to this,

Step 4 the verifier checks the reply.

These steps are repeated t times independently. In this paper we are only interested in such protocols.

To prove the theorems in Section II. we will first adapt such a protocol and show that the resulting protocol is also a zero-knowledge proof (argument). We then apply this procedure to the SAT protocol(s). Finally we transform the adapted SAT protocol(s) and obtain a divertible zero-knowledge proof (argument). This transformation uses a 'swapping' technique.

III.1. Adapting a zero-knowledge protocol

In this section A is the prover and B the verifier. Consider a general protocol P of the type described above.

Protocol P: input x.
B checks that x has the appropriate form. Then the following steps are repeated t times independently:

Step 1 A sends B: $Z \in \mathcal{H}$,

Step 2 B sends A: $q \in_R \{0,1\}$,

Step 3 A sends B: $Y \in \mathcal{G}$,

Step 4 B verifies that $p(x, Z, q, Y) = 1$, where p is an appropriate polynomial time predicate.

(Here '$\in_R$' means 'selected randomly with uniform distribution'). This protocol is adapted as follows:

Protocol P': input x.
B checks that x has the appropriate form. Then the following steps are repeated t times independently:

Step 1 A sends B: $(Z_0, Z_1) \in \mathcal{H} \times \mathcal{H}$,

Step 2 B sends A: $q \in_R \{0,1\}$,

Step 3 A sends B: $(Y_0, Y_1) \in \mathcal{G} \times \mathcal{G}$,

Step 4 B verifies that $p(x, Z_0, q, Y_0) = 1$ and that $p(x, Z_1, \bar{q}, Y_1) = 1$.

We assume that the honest prover chooses the $Z_0 \in \mathcal{H}$ with the same distribution as the Z in the protocol P, and similarly for Z_1. Let us study the relation between the protocols P and P'. Hereto let us consider the query of B in P' as a pair of queries $(q, \bar{q})$. It is then easy to verify that Y_0 corresponds with an answer which would have been given in protocol P when Z_0 would have been the cover and q the query. A similar observation is valid for Y_1, Z_1, $\bar{q}$.

Theorem 5 *If for appropriate conditions P is a zero-knowledge proof (argument) then for the same conditions P' is also a zero-knowledge proof (argument).*

Proof. The completeness and soundness conditions are obvious. To prove that P' is zero-knowledge we describe a simulator $M'_{B'}$ for any (possibly cheating) verifier B'. $M'_{B'}$ uses the simulator M_B of P, where B is the honest verifier (of P), as an oracle to obtain valid conversations $T = (Z, q, Y)$. Clearly P and P' define the same language. When $x \in L$, M_B outputs valid conversations T with a distribution which is identical to (indistinguishable from) the actual distribution. Suppose that $M'_{B'}$ receives from M_B the valid conversations $T_0 = (Z_0, q_0, Y_0)$ and $T_1 = (Z_1, q_1, Y_1)$. $M'_{B'}$ checks until it gets $q_0 \neq q_1$. When this is so, $M'_{B'}$ 'submits' $T' = ((Z_0, Z_1), q, (Y_0, Y_1))$, $q = q_0$, to the verifier B'. If the query of B' is q then $M'_{B'}$ outputs T'. Otherwise it resets B' and tries again with another pair T_0, T_1. Because the prover and the verifier of P are honest and because the distribution of M_B is equal to (indistinguishable from) the actual distribution, the conversations (Z, q, Y) are independent and have the appropriate distribution. □

IV. The divertible zero-knowledge protocols

To show how the adaptation and swapping technique is used we will first apply it to the graph isomorphism protocol, making it divertible. Then we extend this and obtain a divertible protocol for SAT. A sketch of the proofs is given in the following section.

IV.1. Graph isomorphism

An introduction: The [GMW86] protocol

Let Γ_0 and Γ_1 be graphs with vertex set V and $\sigma : \Gamma_1 \rightarrow \Gamma_0$ be an isomorphism (σ is a permutation of the vertex set V). In the [GMW86] graph isomorphism protocol the verifier B first checks that the input (Γ_0, Γ_1) is a proper description of two graphs. Then, in Step 1, the prover A chooses a random permutation π and sends B the graph $Z = \pi(\Gamma_0)$. In Step 2, B asks the random bit-question q. In Step 3, A sends the permutation $Y = \pi\sigma^q$. In Step 4, B checks that $Z = Y(\Gamma_q)$. These steps are repeated t times (t is the length of the input).

A divertible protocol for graph isomorphism

First B checks that the input (Γ_0, Γ_1) is a proper description of two graphs. Repeat t times, where t is the length of the input:

Prover A	**B**	**Verifier C**
$\pi_k \in_R \operatorname{Sym} V,$		
$Z_k := \pi_k(\Gamma_0),$	$e \in_R \{0,1\},$	
$k = 0,1.$	$\pi'_k \in_R \operatorname{Sym} V,$	
$\xrightarrow{\quad Z_0, Z_1 \quad}$		
	$Z'_e := \pi'_0(Z_0),$	
	$Z'_{\bar{e}} := \pi'_1(Z_1).$	$\xrightarrow{\quad Z'_0, Z'_1 \quad}$
		$\xleftarrow{\quad q \quad}$ $\quad q \in_R \{0,1\}.$
$\xleftarrow{\quad q_1 \quad}$	$q_1 := q \oplus e.$	
$Y_0 := \pi_0 \sigma^{q_1},$		
$Y_1 := \pi_1 \sigma^{\bar{q}_1}.$	$\xrightarrow{\quad Y_0, Y_1 \quad}$	
	$Y'_e := \pi'_0 Y_0,$	
	$Y'_{\bar{e}} := \pi'_1 Y_1,$	
	B checks as C. $\xrightarrow{\quad Y'_0, Y'_1 \quad}$	
		C checks that:
		$Z'_0 = Y'_0(\Gamma_q),$
		$Z'_1 = Y'_1(\Gamma_{\bar{q}}).$

Observe that when $e = 1$, B 'swaps' the Z_k and the Y_k to obtain the Z'_k and the Y'_k.

IV.2. SAT

An introduction: The [BCC88] protocol for SAT

The [BCC88] protocol is a zero-knowledge proof (argument) for a satisfying assignment of a Boolean circuit. This circuit consists of h logic gates with truth tables T_m, $1 \leq m \leq h$, and the connecting lines (wires). A satisfying assignment can be regarded as a collection of pointers, one for each truth table, which point to the computation rows of the T_m. In Step 1 of the [BCC88] protocol, the prover, for *each* m:

- complements some of the columns of $T_m = (b_{i,j})_m$ using bits c_j (one for each line),

- permutes the rows i of $T'_m = (b_{i,j} \oplus c_j)_m$ using a permutation π (one for each truth table),

- 'commits' to each bit of $T''_m = (b_{\pi(i),j} \oplus c_j)_m$ using a probabilistic encryption function.

In Step 2 the verifier asks the bit-question q. In Step 3 the prover reveals to the verifier $Y = Y(q)$ which, when $q = 0$ consists of opening all the commitments, and when $q = 1$ consists of opening the commitments of the computation rows with the corresponding row pointers. In Step 4 the verifier checks if the corresponding commitments are appropriate (results of encryptions and content of tables). Therefore in Step 1 the prover sends to the verifier $Z = \{f_{r_{\pi(i),j}}(b_{\pi(i),j} \oplus c_j)\}$. Observe that if f is a probabilistic encryption homomorphism then

$$\{f_{r'}(c'_j)\} \cdot \{f_r(b_{\pi(i),j} \oplus c_j)\} = \{f_{r''}(b_{\pi(i),j} \oplus c_j \oplus c'_j)\} , \tag{1}$$

and r'' can be computed given r, r', $b_{\pi(i),j} \oplus c_j$, and c'_j.

We denote by $X = \{(c_j, r_{i,j}, \pi)\}$ the strings which contain the complementation bits c_j, the random strings $r_{i,j}$ and the permutations π. These form a direct product group $\mathcal{G}$.

A divertible protocol for SAT

The protocol is described in Figure 1. In this protocol

$$u_{l,j} \cdot z_{\pi'(l),j} = f_{r''}(b_{\pi'\pi(i),j} \oplus c_j \oplus c'_j)$$

by (1), since $u_{l,j} = f_{r'_{l,j}}(c'_j)$, $z_{\pi'(l),j} = z_{\pi'\pi(i),j} = f_{r_{\pi'\pi(i),j}}(b_{\pi'\pi(i),j} \oplus c_j)$, and since we are assuming that f is a probabilistic encryption homomorphism. Furthermore the Y_k consist of all, or part of the

$$(r_{\pi(i),j} , b_{\pi(i),j} \oplus c_j) , \tag{2}$$

and the Y'_k consist of all, or part of the

$$(r'', b_{\pi'\pi(i),j} \oplus c_j \oplus c'_j) . \tag{3}$$

Therefore the encryptions of Y'_k produce all, or part of the Z'_k. The product $\{(c', r', \pi')\}_k \circ Y_k$ is obtained by applying the operator (c', r', π') to the parts (2) of Y_k to give strings of type (3).

V.   Sketch of proofs

Proof of Theorem 1: In the final paper we shall show that the above protocol satisfies the conditions of Theorem 1. □

First B checks that x (the input) is a proper description of a Boolean circuit. Then the protocol starts. Repeat $t = \Theta(|x|)$ times:

Prover A	B	**Verifier C**
$\{(c, r, \pi)\}_k \in_R \mathcal{G},$		
$z_{i,j} := f_{r_{i,j}}(b_{i,j} \oplus c_j),$		
$l := \pi(i),$	$e \in_R \{0, 1\},$	
$Z_k := \{z_{l,j}\}_k,$	$\{(c', r', \pi')\}_k \in_R \mathcal{G},$	
$k = 0, 1.$	$u_{l,j} := f_{r'_{l,j}}(c'_j),$	
$\xrightarrow{\quad Z_0, Z_1 \quad}$		
	$z'_{l,j} := u_{l,j} \cdot z_{\pi'(l),j},$	
	$Z'_e := \{z'_{l,j}\}_0,$	
	$Z'_{\bar{e}} := \{z'_{l,j}\}_1.$	
	$\xrightarrow{\quad Z'_0, Z'_1 \quad}$	
		$q \in_R \{0, 1\}.$
	$\xleftarrow{\quad q \quad}$	
	$q_1 := q \oplus e.$	
$\xleftarrow{\quad q_1 \quad}$		
$Y_0 := Y_0(q_1),$		
$Y_1 := Y_1(\bar{q}_1).$		
$\xrightarrow{\quad Y_0, Y_1 \quad}$		
	$Y'_e :=$	
	$\{(c', r', \pi')\}_0 \circ Y_0,$	
	$Y'_{\bar{e}} :=$	Let p be a
	$\{(c', r', \pi')\}_1 \circ Y_1,$	predicate as in
	B checks as C.	[BCC88].
	$\xrightarrow{\quad Y'_0, Y'_1 \quad}$	

C checks that:
$$p(x, Z'_0, q, Y'_0),$$
$$p(x, Z'_1, \bar{q}, Y'_1),$$
are satisfied.

Figure 1: A divertible protocol for SAT

Observe that Theorem 1 does not imply Theorem 2 since our protocol is only conditionally abuse-free. Indeed suppose that only once during the execution of the protocol A decides to replace one row (e.g. the row $(1\,1\,1)$ by $(0\,0\,0)$). The probability that B will detect this is only $1/2$. If the encryption is insecure then the verifier C will find out that this has happened.

Proof of Theorem 2: The prover A first commits to X_0, X_1 by sending Z_0, Z_1. B then sends A: X_0', X_1'. After having combined X_0 with X_0' and X_1 with X_1', A commits to those two combinations. So B obtains from A: Z_0', Z_1'. Then A proves *only* to B (using another proof) that X_0', X_1' have been used in Z_0', Z_1' appropriately. B checks this proof. The prover A does not reveal X_0, X_1 to (the warden) B, and B does not reveal e to A. Then the protocol continues as previously. So when $e=1$, B switches components to get the Z_k'' in Step 1, etc. Observe that the verifier C does *not* have to commit to his question. So the proof is unconditionally sound. $\square$

Proof of Theorem 3: The proof is identical to that of Theorem 1 with the only difference that the encryption function is replaced by a blob function. $\square$

VI.   Conclusion and remarks

For all so far proposed divertible zero-knowledge proofs, the question that B asks A is the exclusive-or of the question that C asks B and B's random bit. This may give one the impression that A can convince independently two verifiers simultaneously (B and C). However after careful analysis it is clear that when A and C collaborate the soundness related to B is conditional for many proofs of membership.

To illustrate let us consider the graph isomorphism case. Let us assume that dishonest $\tilde{A}$ and dishonest $\tilde{C}$ have infinite computer power and that the graphs Γ_0 Γ_1 are *not* isomorphic. $\tilde{A}$ now sends Z_0 isomorphic to Γ_0 and Z_1 isomorphic to Γ_1. $\tilde{C}$ is now able to calculate e (using exponential computer power). Then $\tilde{C}$ can manipulate q_1.

The same remark is valid for some of the schemes presented earlier [DGB88,OO89]. For some it is sufficient that C knows some trapdoor information to perform above fraud. This problem implies that the [OO89] formal definition of divertible zero-knowledge has to be revised in this context.

By analyzing Theorem 1 and Theorem 2 we see that even though divertibility and abuse-freeness have common aspects they are essentially different concepts.

Acknowledgement

We wish to thank Joan Boyar for suggesting, at Eurocrypt '89, that we investigate the divertible zero-knowledge aspects of the [BDPW89] protocol. We also thank an anonymous referee for pointing out an error in an earlier version of this paper.

REFERENCES

[BC89]   G. Brassard and C. Crépeau. Sorting out zero-knowledge. Presented at Eurocrypt'89, Houthalen, Belgium, to appear in: Advances in Cryptology. Proc. of Eurocrypt'89 (Lecture Notes in Computer Science), Springer-Verlag, April 1989.

[BCC88]   G. Brassard, D. Chaum, and C. Crépeau. Minimum disclosure proofs of knowledge. *Journal of Computer and System Sciences*, 37(2), pp. 156–189, October 1988.

[BDPW89]   M. V. D. Burmester, Y. G. Desmedt, F. Piper, and M. Walker. A general zero-knowledge scheme. Presented at Eurocrypt '89, Houthalen, Belgium, to appear in: Advances in Cryptology. Proc. of Eurocrypt '89 (Lecture Notes in Computer Science), Springer-Verlag, April 1989.

[Blu87]   M. Blum. How to prove a theorem so no one else can claim it. In *Proceedings of the International Congress of Mathematicians*, pp. 1444–1451, August 3–11, 1987. Berkeley, California, U.S.A., 1986.

[CEvdG88] D. Chaum, J.-H. Evertse, and J. van de Graaf. An improved protocol for demonstrating possession of discrete logarithms and some generalizations. In D. Chaum and W. L. Price, editors, *Advances in Cryptology — Eurocrypt'87 (Lecture Notes in Computer Science 304)*, pp. 127–141. Springer-Verlag, Berlin, 1988. Amsterdam, The Netherlands, April 13–15, 1987.

[Des88]   Y. Desmedt. Subliminal-free authentication and signature. In C. G. Günther, editor, *Advances in Cryptology, Proc. of Eurocrypt '88 (Lecture Notes in Computer Science 330)*, pp. 23–33. Springer-Verlag, May 1988. Davos, Switzerland.

[Des90]   Y. Desmedt. Making conditionally secure cryptosystems unconditionally abuse-free in a general context. In G. Brassard, editor, *Advances in Cryptology — Crypto '89, Proceedings (Lecture Notes in Computer Science 435)*, pp. 6–16. Springer-Verlag, 1990. Santa Barbara, California, U.S.A., August 20–24.

[DGB88]   Y. Desmedt, C. Goutier, and S. Bengio. Special uses and abuses of the Fiat-Shamir passport protocol. In C. Pomerance, editor, *Advances in Cryptology, Proc. of Crypto '87 (Lecture Notes in Computer Science 293)*, pp. 21–39. Springer-Verlag, 1988. Santa Barbara, California, U.S.A., August 16–20.

[GM84]   S. Goldwasser and S. Micali. Probabilistic encryption. *Journal of Computer and System Sciences*, 28(2), pp. 270–299, April 1984.

[GMR89]   S. Goldwasser, S. Micali, and C. Rackoff. The knowledge complexity of interactive proof systems. *Siam J. Comput.*, 18(1), pp. 186–208, February 1989.

[GMW86] O. Goldreich, S. Micali, and A. Wigderson. Proofs that yield nothing but their validity and a methodology of cryptographic protocol design. In *The Computer Society of IEEE, 27th Annual Symp. on Foundations of Computer Science (FOCS)*, pp. 174–187. IEEE Computer Society Press, 1986. Toronto, Ontario, Canada, October 27–29, 1986.

[OO89] T. Okamoto and K. Ohta. Divertible zero knowledge interactive proofs and commutative random self-reducibility. Presented at Eurocrypt'89, Houthalen, Belgium, to appear in: Advances in Cryptology. Proc. of Eurocrypt'89 (Lecture Notes in Computer Science), Springer-Verlag, April 1989.

On the Importance of Memory Resources in the Security of Key Exchange Protocols

(Extended Abstract)

George Davida   Yvo Desmedt   René Peralta*

Dept. EE & CS,
Univ. of Wisconsin – Milwaukee
P.O. Box 784
WI 53201 Milwaukee
U.S.A.

Abstract

We present a protocol for key exchange which relies on the existence of permutations which are not necessarily trap-door, and which are one-way in a weaker sense than that usually assumed in the literature. Our main result is that, under this assumption, two players can exchange a secret key over an open channel in such a way that an eavesdropper must spend time proportional to $TIME \cdot SPACE$, where $TIME$ is the time spent by the two players and $SPACE$ is the amount of information which can be stored and transmitted by the two players. Hence the importance of storage technology for security.

1 Introduction

It is not known whether or not one-way trap-door functions exist. Moreover, proving (from a complexity theory point of view) that these functions do exist implies proving $P \neq NP$, and therefore such a proof is not likely to be found in the near future. In fact, every year a number of researchers claim they have proven $P = NP$ (even though their proofs are invariably incorrect or incomprehensible). Given this state of affairs, it is reasonable to explore the possibility of solving the main cryptographic problems under weaker assumptions.

*Supported in part by NSF Grant Number CCR-8909657.

In this paper we present a protocol for key exchange which relies on the existence of permutations (bijections) which are not necessarily trap-door, and which are one-way in a weaker sense than that usually assumed in the literature. Our main result is that, under this assumption, two players can exchange a secret key over an open channel in such a way that an eavesdropper must spend time proportional to $TIME \cdot SPACE$, where $TIME$ is the time spent by the two players and $SPACE$ is the amount of information which can be stored and transmitted by the two players. Hence the importance of storage technology for security. Using current optical-disk technology both for storage and transfer of information, we can think of $SPACE$ as being in the gigabytes range. Therefore, if the players are willing to invest one week in computation time each, then an eavesdropper will have to spend gigaweeks to obtain the secret. This scenario is reasonable, for example, in the case of embassies exchanging keys with their governments on a weekly basis.

Our protocol combines techniques appearing in [Mer78, DDP90] for key exchange without trap-door functions and uses Carter-Wegman universal hashing [CW79] to implement ideas similar to Hellman's time-memory tradeoff [Hel80]. The security achieved is similar to that of the protocols in [Mer78, DDP90] but our assumptions are weaker. In particular we do not assume, as is done in [DDP90], that (weakly) one-way functions exist which have arbitrarily low rates of encryption.

2 The assumptions

Let F_α be a family of bijections parametrized by α and with domain $\{1...K\}$. We suppose F_α is implemented by a specific circuit. In our protocol, players A and B will use F_α to exchange a secret key k over an open channel. Player E (the eavesdropper) will have access to the whole communication. Player E's goal is to compute k given A and B's communication. We make the following assumptions:

- $|\alpha| < \sqrt{K}$.

- the fastest algorithm to compute k given α and $F_\alpha(k)$ uses exhaustive search on a set of expected size $O(K)$.

- We assume the existence of an authenticated channel.

- We assume that E's technology is comparable to A and B's technology (E, however, may spend much more resources computing k than A and B do).

Note that the first and second assumptions do not imply that calculating k from $\alpha, F_\alpha(k)$ takes exponential time, since $|\alpha|$ itself is allowed to be exponential in $|k|$.

3 The protocol

In the following protocol, players A and B will agree on a common secret key $k \in \{1...K\}$. A set T of size h, is defined as follows:

- a random hashing function $H : \{1...K\} \rightarrow \{1...K/h\}$ is chosen from a *universal$_2$* family of hashing functions (see [CW79] for the definition of universal hashing functions).

- we let $T = \{x \mid H(x) = 1\}$

The use of universal hashing is for the purpose of making T behave like a randomly chosen subset of the key space (a truly random subset cannot be described in polynomial time in $|h|$).

Let $F_\alpha^1(x) = F_\alpha(x)$ and $F_\alpha^i(x) = F_\alpha^{i-1}(F_\alpha(x))$ for $i > 1$. Given T, we define $G_{\alpha,T}(x) = F_\alpha^{u_x}(x)$ where u_x is the minimum positive integer such that $F_\alpha^{u_x}(x) \in T$. If no such integer exists, then $G_{\alpha,T}(x)$ is undefined. Note that if u_x is defined, then it has expected value $\leq K/h$, under the assumption that T is a truly random subset of K.

The protocol is as follows:

precomputation:

Step 1 Player A chooses α and H at random.

Step 2 Player A computes and stores $(x, G_{\alpha,T}(x), u_x)$ for n distinct randomly chosen $x \in \{1...K\}$.

Step 3 Player A sends α and a description of H to player B.

communication: Steps 4-5 are repeated until an agreement is achieved.

Step 4 Player B chooses h/n distinct random $z \in \{1...K\}$ and sends $(G_{\alpha,T}(z), u_z)$ to player A.

Step 5 Player A checks whether $G_{\alpha,T}(z) = G_{\alpha,T}(x)$ for some x in the table computed at Step 2 and some z sent at Step 4. If this is the case, then A sends u_x and $G_{\alpha,T}(z)$ to player B. The secret key is x if $u_x < u_z$ and z otherwise.

Note that if $u_x < u_z$ then B can calculate x by computing $F_\alpha^{u_z-u_x}(z)$. If $u_z < u_x$ then A can calculate z by computing $F_\alpha^{u_x-u_z}(x)$.

It is not hard to show that each iteration of this protocol has a chance of about $1 - e^{-1}$ of reaching agreement on a secret key. Alternatively, B may send, at Step 4, sufficiently many random $G_{\alpha,T}(x)$'s so that the probability of at least one $G_{\alpha,T}(x)$ being in A's table is exponentially high. This has the desirable effect of reducing the number of rounds in the protocol to, essentially, one.

4 Analysis

The security of the protocol follows from the fact that the key agreed on is randomly chosen from the key space (in this version of the protocol a slight deviation from the uniform distribution is caused by the fact that, in Step 5, x is favored over z

if $u_x < u_z$). The information available to the eavesdropper is, essentially, a pair $(i, F_\alpha^i(x))$ with $i > 1$. Recovering x from this information can be no easier than recovering x from $F_\alpha(x)$. By assumption, the fastest way to recover x from $F_\alpha(x)$ is by exhaustive search.

The costs of the protocol depend on the parameters K, n,and h. Let $ATIME$ and $BTIME$ be the cost of the protocol, in number of computations of F, to A and B respectively. We assume $ATIME \geq \sqrt{K} > |\alpha|$ so that we may ignore the time incurred in transmitting α.

Let $AMEM$ be the memory costs of A, in terms of triples $(x, G_{\alpha,T}(x), u_x)$ stored at Step 2. Let C be the communication cost of the protocol in terms of pairs $(G_{\alpha,T}(z), u_z)$ sent by B in Step 4.

Under the heuristic assumption that, given a random x, the sequence $\{F_\alpha^i(x)\}_i$ behaves (until it loops) as a random walk in the key space, it is easy to derive the following:

- $ATIME \approx \frac{nK}{h}$.

- $BTIME \approx \frac{K}{n}$.

- $AMEM \approx n$.

- $C \approx \frac{h}{n}$.

Let $\overline{ATIME}$ be the maximum value of $ATIME$ acceptable to player A. Similarly define $\overline{BTIME}$, $\overline{AMEM}$, and $\overline{C}$.

Thus, ignoring logarithmic and constant factors, we have the following constraints:

- $K \geq h \geq n$.

- $\overline{ATIME} \geq \frac{nK}{h}$.

- $\overline{BTIME} \geq \frac{K}{n}$.

- $\overline{AMEM} \geq n$.

- $\overline{C} \geq \frac{h}{n}$.

Since the security of the protocol is proportional to K, we must maximize K subject to these constraints. Under the assumption that $\overline{C} \leq \overline{BTIME}$ and $\overline{AMEM} \leq \overline{ATIME}$, the solution to this optimization problem is

$$n = \overline{AMEM}; h = \overline{AMEM} \cdot \overline{C}$$

and

$$K = min(\overline{BTIME} \cdot \overline{AMEM}, \overline{ATIME} \cdot \overline{C}).$$

Thus, the security of our protocol is proportional to

$$min(\overline{BTIME} \cdot \overline{AMEM}, \overline{ATIME} \cdot \overline{C}).$$

From this we can derive the impact of future technology on the security of this protocol. It turns out that faster chips do not help, since the effect of this is to increase both K and the eavesdropper's speed by the same factor. On the other hand, if both $\overline{C}$ and $\overline{AMEM}$ increase, then security increases by a proportional amount. This would be the effect of technology which increases the capacity of storage devices.

5   An open problem

We have assumed the existence of families of bijections F_α of a space of size N which require exhaustive search to invert. This assumption implies the existence of one-way-functions as usually defined in the literature, unless the size of the key α is large (i.e. more than polylogarithmic in N). To our knowledge, all bijections which have been proposed in the literature and which remain one-way after the key is made public have a key-size which is $O(\log N)$. On the other hand, if we could truly choose random permutations of a space of size N, then it would take $O(N \log N)$ bits to describe these permutations. The problem we propose is finding a family of permutations F_α on a space of size N such that it seems plausible that exhaustive search is the fastest way to invert F_α and α has length more than $poly(\log N)$. Note that the difficulty in achieving this is because of the condition that α is public. Otherwise, DES-like functions with the required property can be easily constructed.

6   Acknowledgement

Several useful comments from referees are gratefully acknowledged.

References

[CW79]   J.L. Carter and M.N. Wegman. Universal classes of hash functions. *Journal of Computer and System Sciences*, 18(2):143–154, 1979.

[DDP90] G. Davida, Y. Desmedt, and R. Peralta. A key distribution system based on any one-way function. In *Advances in Cryptology - proceedings of EU-ROCRYPT '89*, Lecture Notes in Computer Science. Springer-Verlag, 1990. to appear.

[Hel80] M. E. Hellman. A cryptanalytic time-memory tradeoff. *IEEE Tr. Inform. Theory*, 26(4):401–406, July 1980.

[Mer78] Ralph Merkle. Secure communications over insecure channels. *Communications of the ACM*, 21(4):294 – 299, 1978.

[QD88]   J.-J. Quisquater and J.-P. Delescaille. Other cycling tests for DES. In C. Pomerance, editor, *Advances in Cryptology, Proc. of Crypto '87 (Lecture Notes in Computer Science 293)*, pages 255–256. Springer-Verlag, 1988. Santa Barbara, California, U.S.A., August 16–20.

Provably Secure Key-Updating Schemes in Identity-Based Systems

S. Shinozaki †   T. Itoh ‡   A. Fujioka †   S. Tsujii †

† Department of Electrical and Electronic Engineering
Faculty of Engineering
Tokyo Institute of Technology
O-okayama, Meguro-ku, Tokyo 152, Japan

‡ Department of Information Processing
The Graduate School at Nagatsuta
Tokyo Institute of Technology
4259 Nagatsuta, Midori-ku, Yokohama 227, Japan
e-mail: titoh@cc.titech.ac.jp

Abstract:

In this paper, we present *Key-Updating Schemes* in identity-based (identification or signature) systems, and consider the security of the schemes. We propose two kinds of key-updating schemes, i.e., one is sequential type and the other is parallel type, and show that both schemes are equivalent to each other in a polynomial time sense, i.e., there exists a deterministic polynomial time algorithm that transforms the sequential key-updating scheme to the parallel one, and vice versa. We also show that even if any polynomially many entities conspire to find a secret-key of any other entities, both key-updating schemes are provably secure against polynomially many times key-updating if decrypting RSA is hard.

1 Introduction

In identity-based systems, each entity i has his(her) own identity number ID_i, and a trusted center needs to generate a pair of a public information P (known to all entities) and a secret information S (known to only the trusted center), and a pair of public-key PK_i and secret-key SK_i for entity i. Let a probabilistic polynomial time algorithm CKG be a *center-key generator* that, on input 1^k, outputs a pair of the public information P ($|P| = O(k^c)$ for some constant $c > 0$) and the secret information S ($|S| = O(k^d)$ for some constant $d > 0$), i.e., $CKG(1^k) = \langle P, S \rangle$, and let a probabilistic polynomial time algorithm EKG be a *entity-key generator* that, on input 1^k, P, S, and ID_i, outputs a pair of public-key PK_i and secret-key SK_i for entity i, i.e., $EKG(\langle 1^k, P, S, ID_i \rangle) = \langle PK_i, SK_i \rangle$. Note that k is the security parameter.

When a foolish entity j carelessly loses his secret-key SK_j or reveals it and asks the trusted center again to generate a new pair of public-key PK_j' and secret-key SK_j' for him, what should the trusted center do? If the system is provably secure (see, e.g., [FS], [FFS], [GQ], [OO].), i.e., there exist no efficient algorithms for entity j to derive the secret information S from P, ID_j, and a single pair of $\langle PK_j, SK_j \rangle$, then (presumably) the simplest and secure way to update the secret-key SK_j to SK_j' is to make the trusted center run CKG on input 1^k in order to regenerate a new pair of public information P' and secret information S' and to make the trusted center regenerate a new pair of public-key PK_j' and secret-key SK_j' for the entity j by running EKG on input 1^k, P', S', and ID_j (or SK_j). This scheme, however, imposes cumbersome procedures on the trusted center and all entities, because the trusted center must regenerate not only a new pair of public-key PK_j' and secret-key SK_j' for the foolish entity j but a new pair of public-key PK_i' and secret-key SK_i' for every entity i ($\neq j$).

Another way to update the secret-key SK_j to SK_j' is to make the trusted center run only EKG on input 1^k, P, S, and ID_j (or SK_j) and to regenerate a new pair of public-key PK_j' and secret-key SK_j' only for the foolish entity j, while those for the other entities i ($\neq j$) are unchanged. This scheme is much simpler than before, but unfortunately there might be a possibility that the entity j can derive the secret information S efficiently from P, ID_j, PK_j, SK_j, PK_j', and SK_j'.

Thus this provokes us to construct efficient and *provably* secure key-updating schemes in identity-based systems in the above sense. To do this, we take the extended Fiat-Shamir scheme [GQ], [OO] as an identity-based system, and apply two kinds of key-updating schemes, one is sequential and the other is parallel, to the extended Fiat-Shamir scheme. (The details will be discussed in Section 2.) We also show that our key-updating schemes are *provably* secure against polynomially many times key-updating even if any polynomially many entities conspire to find a secret-key of any other entities.

The organization of this paper is as follows: Section 2 presents a brief description of *key-generation* and *key-distribution* in the extended Fiat-Shamir scheme [GQ], [OO], and proposes two kinds of *key-updating* schemes, sequential one and parallel one; Section 3 shows that both schemes are equivalent to each other in a polynomial time sense, i.e., there exists a polynomial time algorithm that transforms the sequential key-updating scheme to the parallel one, and vise versa; Section 4 gives a main result that both key-updating schemes are provably secure against polynomially many times key-updating, i.e., any polynomially many conspiring entities can not find a secret-key of any other entities under the assumption that decrypting RSA is hard; and Section 5 finally gives conclusion and remarks, and refers to extensions of our results to more general settings and the security of the schemes against conspiracy of entities.

2 Key-Updating Schemes

2.1 Extended Fiat-Shamir Scheme

This subsection presents a brief description of key-generation and key-distribution in the extended Fiat-Shamir scheme [GQ], [OO]. The extended Fiat-Shamir scheme is an extension of the Fiat-Shamir scheme [FS], [FFS], and is shown, under the assumption that factoring is hard, to be zero-knowledge in the sequential execution (of the protocol) and to be non-transferable in the parallel execution (of the protocol). This scheme is an identity-based system, and thus the trusted center needs to generate a pair of public information P (known to all entities) and secret information S (known to only trusted center) and to distribute a pair of public-key PK_i and secret-key SK_i for each entity i with his identity number ID_i, in the following way:

The trusted center has two probabilistic polynomial time algorithms, i.e., *center-key generator* CKG and *entity-key generator* EKG; On input 1^k, the center-key generator CKG outputs a pair of public information $n\ (= p \cdot q)$ and secret information $\langle p, q \rangle$, where $p, q \in \mathcal{OP}$ and $|p| = |q| = k$, and on input 1^k, n, $\langle p, q \rangle$, and ID_i, the entity-key generator EKG outputs a pair of public-key e_i and secret-key S_i for entity i such that $S_i^{e_i} \equiv ID_i \pmod{n}$. Note that $\mathcal{OP}$ denotes a set of odd primes and $|a|$ denotes the length of binary encoding of a. For details of identification and signature protocols in the extended Fiat-Shamir scheme, see [GQ], [OO].

2.2 Key-Updating Schemes

In this subsection, we propose two kinds of key-updating schemes, sequential one [FT] and parallel one, in the extended Fiat-Shamir scheme. Consider the case where some entity i asks the trusted center to issue a new pair of public-key e_i' and secret-key S_i' for the entity i in some reason, e.g., losing or revealing his original secret-key S_i.

Informally, our key-updating schemes are as follows: (1) Sequential Key-Updating Scheme (**SKU**) is a key-updating scheme in which the trusted center runs the entity-key generator EKG on input 1^k, n, $\langle p, q \rangle$, e_i, and S_i (instead of ID_i), and generates a new pair of $\langle e_i', S_i' \rangle$ such that $S_i'^{e_i'} \equiv S_i \pmod{n}$ and $e_i \neq e_i'$, and (2) Parallel Key-Updating Scheme (**PKU**) is a key-updating scheme in which the trusted center runs the entity-key generator EKG on input 1^k, n, $\langle p, q \rangle$, e_i, and ID_i, and generates a new pair of $\langle e_i', S_i' \rangle$ such that $S_i'^{e_i'} \equiv ID_i \pmod{n}$ and $e_i \neq e_i'$. Note that for entity i, a pair of public-key and secret-key will be $\langle e_i e_i', S_i' \rangle$ in **SKU**, while will be $\langle e_i', S_i' \rangle$ in **PKU**.

This formulation, however, does not necessarily match our desire, because a malicious entity j might ask the trusted center to issue new pairs of $\langle e_j', S_j' \rangle$ many times for compromising the secret information $\langle p, q \rangle$. Then we formally define our key-updating schemes in more general settings.

Let $U(|n|)$ be any fixed polynomial in $|n|$, and let $\mathcal{OP}(\ell)$ denote a set of odd primes less than ℓ. Here we assume that each entity i is allowed to ask the trusted center to issue new pairs of $\langle e_i', S_i' \rangle$ at most $U(|n|)$ times.

Sequential Key-Updating Scheme (SKU):

Initial Key-Setting Stage: For each entity i (with $ID_i \in \mathcal{Z}_n^*$), the trusted center distributes a pair of his public-key $e_i^{(0)}$ and his secret-key $S_i^{(0)}$ such that $ID_i \equiv \left\{ S_i^{(0)} \right\}^{e_i^{(0)}} \pmod{n}$, where $e_i^{(0)} \in \mathcal{OP}(\lfloor \sqrt{n}/4 \rfloor)$, $S_i^{(0)} \not\equiv ID_i \pmod{n}$, and $ID_i^2 \not\equiv 1 \pmod{n}$.

Key-Updating Stage: For entity i in the r_i-th ($1 \leq r_i \leq U(|n|)$) key-updating, the trusted center distributes a new pair of $\langle e_i^{(r_i)}, S_i^{(r_i)} \rangle$ such that $S_i^{(r_i-1)} \equiv \left\{ S_i^{(r_i)} \right\}^{e_i^{(r_i)}} \pmod{n}$, where $e_i^{(r_i)} \in \mathcal{OP}(\lfloor \sqrt{n}/4 \rfloor)$, $e_i^{(j)} \neq e_i^{(r_i)}$ ($0 \leq j < r_i$), $S_i^{(r_i)} \not\equiv ID_i \pmod{n}$, and $S_i^{(j)} \not\equiv S_i^{(r_i)} \pmod{n}$ ($0 \leq j < r_i$).

Remark 2.1: In the r_i-th key-updating of **SKU**, a pair of the public-key and the secret-key will be $\langle e_i^{(0)} e_i^{(1)} \cdots e_i^{(r_i)}, S_i^{(r_i)} \rangle$. The condition $S_i^{(j)} \not\equiv ID_i \pmod{n}$ ($0 \leq j \leq r_i$) shows that the trusted center avoids distributing trivial secret-key $S_i^{(j)}$, and the condition $S_i^{(j)} \not\equiv S_i^{(r_i)} \pmod{n}$ ($0 \leq j < r_i$) implies that the trusted center does not distribute the same secret-key $S_i^{(r_i)}$ again, because old secret-keys might be known to someone else. The trusted center does not care about collisions of secret-keys among entities.

Parallel Key-Updating Scheme (PKU):

Initial Key-Setting Stage: For each entity i (with $ID_i \in Z_n^*$), the trusted center distributes a pair of his public-key $f_i^{(0)}$ and his secret-key $T_i^{(0)}$ such that $ID_i \equiv \left\{T_i^{(0)}\right\}^{f_i^{(0)}}$ (mod n), where $f_i^{(0)} \in \mathcal{OP}(\lfloor \sqrt{n}/4 \rfloor)$, $T_i^{(0)} \not\equiv ID_i$ (mod n), and $ID_i^2 \not\equiv 1$ (mod n).

Key-Updating Stage: For entity i in the r_i-th ($1 \leq r_i \leq U(|n|)$) key-updating, the trusted center distributes a new pair of $\langle f_i^{(r_i)}, T_i^{(r_i)} \rangle$ such that $ID_i \equiv \left\{T_i^{(r_i)}\right\}^{f_i^{(r_i)}}$ (mod n), where $f_i^{(r_i)} \in \mathcal{OP}(\lfloor \sqrt{n}/4 \rfloor)$, $f_i^{(j)} \neq f_i^{(r_i)}$ ($0 \leq j < r_i$), $T_i^{(r_i)} \not\equiv ID_i$ (mod n), $T_i^{(j)} \not\equiv T_i^{(r_i)}$ (mod n) ($0 \leq j < r_i$), and $ID_i^{f_i^{(j)} \cdots f_i^{(r_i)}} \not\equiv ID_i$ (mod n) ($0 \leq j < r_i$).

Remark 2.2: In the r_i-th key-updating of **PKU**, a pair of the public-key and the secret-key will be $\langle f_i^{(r_i)}, T_i^{(r_i)} \rangle$. The meaning of conditions $T_i^{(j)} \not\equiv ID_i$ (mod n) ($0 \leq j \leq r_i$) and $T_i^{(j)} \not\equiv T_i^{(r_i)}$ (mod n) ($0 \leq j < r_i$) is similar to the one in the Remark 2.1. The trusted center does not care about collisions of secret-keys among entities.

3   Transforms Between SKU and PKU

This section shows that key-updating schemes **SKU** and **PKU** are equivalent to each other in a polynomial time sense, i.e., there exists a deterministic polynomial time algorithm that transforms **SKU** to **PKU**, and vice versa.

Let $\mathcal{SC}_k$ denote a set of strong composites with the security parameter k, i.e.,

$$\mathcal{SC}_k = \{n \mid n = p \cdot q, \ p \neq q, \ |p| = |q| = k,$$
$$p = 2p' + 1, \ q = 2q' + 1, \ p, q, p', q' \in \mathcal{OP}\}.$$

To prove that for $n \in \mathcal{SC}_k$, key-updating schemes **SKU** and **PKU** are deterministic polynomial time transformable to each other, we need to show the following lemmas:

Lemma 3.1: Let $n \in \mathcal{SC}_k$. Then for any odd e less than $\lfloor \sqrt{n}/4 \rfloor$, $e < \min\{p', q'\}$ and $\gcd(e, \lambda(n)) = 1$, where $\lambda(n)$ is the Carmichael function [Kr] of n.

Proof: From the definition of $\mathcal{SC}_k$, it follows that

$$\lambda(n) = \mathrm{lcm}(p - 1, q - 1) = \mathrm{lcm}(2p', 2q') = 2p'q'.$$

Note that $n \in \mathcal{SC}_k$, i.e., $|p| = |q| = k$, then $2 \cdot \min\{p, q\} > \max\{p, q\}$. Hence,

$$\begin{aligned}
\lfloor \sqrt{n}/4 \rfloor &\leq \lfloor \max\{p, q\}/4 \rfloor \\
&\leq \lfloor \min\{p, q\}/2 \rfloor \\
&= \lfloor (2 \cdot \min\{p', q'\} + 1)/2 \rfloor \\
&= \lfloor \min\{p', q'\} + 1/2 \rfloor = \min\{p', q'\},
\end{aligned}$$

and thus $e < \min\{p', q'\}$. It immediately follows, from the fact that $\lambda(n) = 2p'q'$, that $\gcd(e, \lambda(n)) = \gcd(e, 2p'q') = 1$, because $e < \min\{p', q'\}$, e is odd, and $p', q' \in \mathcal{OP}$. $\square$

Lemma 3.2: Let $n \in \mathcal{SC}_k$, and $x \in Z_n^*$ such that $x^2 \not\equiv 1$ (mod n). For any distinct odd numbers a_1 and a_2, $x^{a_2 - a_1} \not\equiv 1$ (mod n), where $a_1, a_2 < \lfloor \sqrt{n}/4 \rfloor$.

Proof: By Contradiction. Without loss of generality, we assume that $a_1 < a_2$. Assume that $x^{a_2 - a_1} \equiv 1$ (mod n). This implies that the order of x modulo n divides both $\lambda(n)$ and $a_2 - a_1$. Since $n \in \mathcal{SC}_k$, $\lambda(n) = 2p'q'$ and $0 < a_2 - a_1 < \min\{p', q'\}$ (see Lemma 3.1.), and thus the order of x modulo n is equal to either 1 or 2. This, however, contradicts the assumption that $x^2 \not\equiv 1$ (mod n). Hence $x^{a_2 - a_1} \not\equiv 1$ (mod n). $\square$

Lemma 3.3: Let r be any positive integer and let $n \in \mathcal{SC}_k$. Let $ID \in Z_n^*$, $S^{(i)} \in Z_n^*$ $(0 \le i \le r)$, and $e^{(i)} \in \mathcal{OP}(\lfloor \sqrt{n}/4 \rfloor)$ $(0 \le i \le r)$ satisfy the relation that $ID \equiv \left\{ S^{(0)} \right\}^{e^{(0)}}$ (mod n), $S^{(i-1)} \equiv \left\{ S^{(i)} \right\}^{e^{(i)}}$ (mod n) $(1 \le i \le r)$, where $ID^2 \not\equiv 1$ (mod n) and $e^{(i)} \ne e^{(j)}$ $(0 \le i < j \le r)$. Then $ID \not\equiv S^{(i)}$ (mod n) $(0 \le i \le r)$ and $S^{(i)} \not\equiv S^{(j)}$ (mod n) $(0 \le i < j \le r)$ iff $ID^{e^{(i)} \cdots e^{(j)}} \not\equiv ID$ (mod n) $(0 \le i < j \le r)$.

Proof: Since $n \in \mathcal{SC}_k$ and ID $(\in Z_n^*)$ satisfies $ID^2 \not\equiv 1$ (mod n), it follows, from Lemma 3.2, that for $e^{(i)} \in \mathcal{OP}(\lfloor \sqrt{n}/4 \rfloor)$ $(0 \le i \le r)$, $ID^{e^{(i)} - 1} \not\equiv 1$ (mod n) $(0 \le i \le r)$, and hence $ID^{e^{(i)}} \not\equiv ID$ (mod n) $(0 \le i \le r)$. Note that $x^e \equiv y^e$ (mod n) iff $x \equiv y$ (mod n) for e such that $\gcd(e, \lambda(n)) = 1$. (see Lemma 3.1.) Then for all i $(0 \le i \le r)$,

$$ ID \equiv S^{(i)} \text{ (mod } n) \iff ID^{e^{(0)} \cdots e^{(i)}} \equiv \left\{ S^{(i)} \right\}^{e^{(0)} \cdots e^{(i)}} \text{ (mod } n) $$
$$ \iff ID^{e^{(0)} \cdots e^{(i)}} \equiv ID \text{ (mod } n). $$

On the other hand, for any i, j $(0 \le i < j \le r)$, we have

$$ S^{(i)} \equiv S^{(j)} \text{ (mod } n) \iff \left\{ S^{(i)} \right\}^{e^{(0)} \cdots e^{(i)} e^{(i+1)} \cdots e^{(j)}} \equiv \left\{ S^{(j)} \right\}^{e^{(0)} \cdots e^{(j)}} \text{ (mod } n) $$
$$ \iff ID^{e^{(i+1)} \cdots e^{(j)}} \equiv ID \text{ (mod } n). $$

Hence $ID \not\equiv S^{(i)}$ (mod n) $(0 \le i \le r)$ and $S^{(i)} \not\equiv S^{(j)}$ (mod n) $(0 \le i < j \le r)$ iff $ID^{e^{(i)} \cdots e^{(j)}} \not\equiv ID$ (mod n) $(0 \le i < j \le r)$. $\square$

Lemma 3.4: Let r be any positive integer and let $n \in \mathcal{SC}_k$. Let $ID \in Z_n^*$, $T^{(i)} \in Z_n^*$ $(0 \le i \le r)$, and $f^{(i)} \in \mathcal{OP}(\lfloor \sqrt{n}/4 \rfloor)$ $(0 \le i \le r)$ satisfy the relation that $ID \equiv \left\{ T^{(i)} \right\}^{f^{(i)}}$ (mod n) $(0 \le i \le r)$, where $ID^2 \not\equiv 1$ (mod n), and $f^{(i)} \ne f^{(j)}$ $(0 \le i < j \le r)$. Then $ID \not\equiv T^{(i)}$ (mod n) $(0 \le i \le r)$ and $T^{(i)} \not\equiv T^{(j)}$ (mod n) $(0 \le i < j \le r)$.

Proof: Since $n \in SC_k$ and $ID(\in Z_n^*)$ satisfies $ID^2 \not\equiv 1 \pmod{n}$, it follows, from Lemma 3.2, that for $f^{(i)} \in \mathcal{OP}(\lfloor \sqrt{n}/4 \rfloor)$ $(0 \leq i \leq r)$, $ID^{f^{(j)}-f^{(i)}} \not\equiv 1 \pmod{n}$ $(0 \leq i < j \leq r)$ and $ID^{f^{(i)}-1} \not\equiv 1 \pmod{n}$ $(0 \leq i \leq r)$. Then for all i $(0 \leq i \leq r)$, we have

$$ID \equiv T^{(i)} \pmod{n} \iff ID^{f^{(i)}} \equiv \left\{T^{(i)}\right\}^{f^{(i)}} \pmod{n}$$
$$\iff ID^{f^{(i)}} \equiv ID \pmod{n}$$
$$\iff ID^{f^{(i)}-1} \equiv 1 \pmod{n}.$$

On the other hand, for any i, j $(0 \leq i < j \leq r)$, we also have

$$T^{(i)} \equiv T^{(j)} \pmod{n} \iff \left\{\left\{T^{(i)}\right\}^{f^{(i)}}\right\}^{f^{(j)}} \equiv \left\{\left\{T^{(j)}\right\}^{f^{(j)}}\right\}^{f^{(i)}} \pmod{n}$$
$$\iff ID^{f^{(j)}} \equiv ID^{f^{(i)}} \pmod{n}$$
$$\iff ID^{f^{(j)}-f^{(i)}} \equiv 1 \pmod{n},$$

hence $ID \not\equiv T^{(i)} \pmod{n}$ $(0 \leq i \leq r)$ and $T^{(i)} \not\equiv T^{(j)} \pmod{n}$ $(0 \leq i < j \leq r)$. $\square$

Let $U(|n|)$ be any fixed polynomial in $|n|$ and let r be any positive integer not greater than $U(|n|)$. Here we define $\mathcal{C}_{SKU}$ to be a set of tuples $\langle n, ID, \boldsymbol{S}^{(r)}, \boldsymbol{e}^{(r)} \rangle$ that satisfy

$$ID \equiv \left\{S^{(0)}\right\}^{e^{(0)}} \pmod{n};$$
$$S^{(i-1)} \equiv \left\{S^{(i)}\right\}^{e^{(i)}} \pmod{n} \ (1 \leq i \leq r);$$
$$ID^2 \not\equiv 1 \pmod{n};$$
$$e^{(i)} \in \mathcal{OP}(\lfloor \sqrt{n}/4 \rfloor) \ (0 \leq i \leq r);$$
$$e^{(i)} \neq e^{(j)} \ (0 \leq i < j \leq r);$$
$$ID^{e^{(i)} \cdots e^{(j)}} \not\equiv ID \pmod{n} \ (0 \leq i < j \leq r),$$

in the r-th key-updating of **SKU** (see Lemma 3.3.), where $n \in SC_k$, $ID \in Z_n^*$, $\boldsymbol{S}^{(r)} = (S^{(0)}, S^{(1)}, \ldots, S^{(r)})$, and $\boldsymbol{e}^{(r)} = (e^{(0)}, e^{(1)}, \ldots, e^{(r)})$. In a way similar to the above, we define $\mathcal{C}_{PKU}$ to be a set of tuples $\langle n, ID, \boldsymbol{T}^{(r)}, \boldsymbol{f}^{(r)} \rangle$ that satisfy

$$ID \equiv \left\{T^{(i)}\right\}^{f^{(i)}} \pmod{n} \ (0 \leq i \leq r);$$
$$ID^2 \not\equiv 1 \pmod{n};$$
$$f^{(i)} \in \mathcal{OP}(\lfloor \sqrt{n}/4 \rfloor) \ (0 \leq i \leq r);$$
$$f^{(i)} \neq f^{(j)} \ (0 \leq i < j \leq r);$$
$$ID^{f^{(i)} \cdots f^{(j)}} \not\equiv ID \pmod{n} \ (0 \leq i < j \leq r),$$

in the r-th key-updating of **PKU** (see Lemma 3.4.), where $n \in SC_k$, $ID \in Z_n^*$, $\boldsymbol{T}^{(r)} = (T^{(0)}, T^{(1)}, \ldots, T^{(r)})$, and $\boldsymbol{f}^{(r)} = (f^{(0)}, f^{(1)}, \ldots, f^{(r)})$.

We use $A_{\text{SKU} \to \text{PKU}}$ to denote any algorithm that, on input $\langle n, ID, \boldsymbol{S}^{(r)}, \boldsymbol{e}^{(r)} \rangle \in \mathcal{C}_{\text{SKU}}$, outputs $\langle n, ID, \boldsymbol{T}^{(r)}, \boldsymbol{f}^{(r)} \rangle \in \mathcal{C}_{\text{PKU}}$, and $A_{\text{PKU} \to \text{SKU}}$ to denote any algorithm that, on input $\langle n, ID, \boldsymbol{T}^{(r)}, \boldsymbol{f}^{(r)} \rangle \in \mathcal{C}_{\text{PKU}}$, outputs $\langle n, ID, \boldsymbol{S}^{(r)}, \boldsymbol{e}^{(r)} \rangle \in \mathcal{C}_{\text{SKU}}$. Then we have the following theorems on deterministic polynomial time transformability between **SKU** and **PKU**.

Theorem 3.5:   *There exists a deterministic polynomial time algorithm $A_{\text{SKU} \to \text{PKU}}$.*

Sketch of Proof: On input $\langle n, ID, \boldsymbol{S}^{(r)}, \boldsymbol{e}^{(r)} \rangle \in \mathcal{C}_{\text{SKU}}$, the algorithm $A_{\text{SKU} \to \text{PKU}}$ sets $T^{(0)} := S^{(0)}$, $f^{(i)} := e^{(i)}$ ($0 \leq i \leq r$), and computes $T^{(i)} \equiv \left\{ S^{(i)} \right\}^{e^{(0)} \cdots e^{(i-1)}}$ (mod n) ($1 \leq i \leq r$). Then it outputs $\langle n, ID, \boldsymbol{T}^{(r)}, \boldsymbol{f}^{(r)} \rangle$, where $\boldsymbol{T}^{(r)} = (T^{(0)}, T^{(1)}, \ldots, T^{(r)})$, $\boldsymbol{f}^{(r)} = (f^{(0)}, f^{(1)}, \ldots, f^{(r)})$. It is easy to see that the algorithm $A_{\text{SKU} \to \text{PKU}}$ runs in deterministic polynomial time and $\langle n, ID, \boldsymbol{T}^{(r)}, \boldsymbol{f}^{(r)} \rangle \in \mathcal{C}_{\text{PKU}}$. $\square$

Theorem 3.6:   *There exists a deterministic polynomial time algorithm $A_{\text{PKU} \to \text{SKU}}$.*

Sketch of Proof: Let n be an odd composite. We assume here that x_1 is the a_1-th root of y modulo n, and x_2 is the a_2-th root of y modulo n, where $\gcd(a_1, a_2) = 1$ and $y \in Z_n^*$. Then we can compute x, the $a_1 a_2$-th root of y modulo n, by algorithm E (see below.) in deterministic polynomial time without knowing prime factors of n. On input n, y, x_1, a_1, x_2, and a_2, the algorithm E computes two integers s and t such that $t a_1 + s a_2 = 1$ by Euclidean algorithm, and outputs $x \equiv x_1^s \cdot x_2^t$ (mod n). It is easy to see that the algorithm E runs in deterministic polynomial time and x is the $a_1 a_2$-th root of y modulo n. The algorithm $A_{\text{PKU} \to \text{SKU}}$ runs in the following way:

On input $\langle n, ID, \boldsymbol{T}^{(r)}, \boldsymbol{f}^{(r)} \rangle \in \mathcal{C}_{\text{PKU}}$, the algorithm $A_{\text{PKU} \to \text{SKU}}$ sets $T^{(0)} := S^{(0)}$, $f^{(0)} := e^{(0)}$, and computes x_i ($1 \leq i \leq r$) by running the deterministic polynomial time algorithm E on input $\langle n, ID, S^{(i-1)}, e^{(0)} e^{(1)} \cdots e^{(i-1)}, T^{(i)}, f^{(i)} \rangle$. Then the algorithm $A_{\text{PKU} \to \text{SKU}}$ substitutes x_i to $S^{(i)}$ and $f^{(i)}$ to $e^{(i)}$ ($1 \leq i \leq r$), and outputs $\langle n, ID, \boldsymbol{S}^{(r)}, \boldsymbol{e}^{(r)} \rangle$, where $\boldsymbol{S}^{(r)} = (S^{(0)}, S^{(1)}, \ldots, S^{(r)})$ and $\boldsymbol{e}^{(r)} = (e^{(0)}, e^{(1)}, \ldots, e^{(r)})$.

It is not difficult to see that the algorithm $A_{\text{PKU} \to \text{SKU}}$ runs in deterministic polynomial time and $\langle n, ID, \boldsymbol{S}^{(r)}, \boldsymbol{e}^{(r)} \rangle \in \mathcal{C}_{\text{SKU}}$. $\square$

4   SKU and PKU are Provably Secure

This section shows that key-updating schemes **SKU** and **PKU** are provably secure against polynomially many times key-updating under the assumption that decrypting RSA is hard for $n \in SC_k$, i.e., even if any polynomially many entities conspire, they can not find a secret-key of any other entity in polynomially many times key-updating.

To show this, we provide several lemmas in the following:

Lemma 4.1: Let $n \in SC_k$ and let $U(|n|)$ be any fixed polynomial in $|n|$. Let r be any positive integer not greater than $U(|n|)$ and let $e^{(i)} < \lfloor \sqrt{n}/4 \rfloor$ $(0 \leq i \leq r-1)$ be distinct r odd primes. Then the probability P that for any $d < \lfloor \sqrt{n}/4 \rfloor$, $d \in OP(\lfloor \sqrt{n}/4 \rfloor)$ and $d \neq e^{(i)}$ $(0 \leq i \leq r-1)$ is greater than $C/|n|$ for some $C > 0$ and sufficiently large n.

Proof: Let $\pi(x)$ denote the number of primes not greater than x $(x \geq 2)$. From prime number theorem [HW], it follows that

$$\pi(x) > C_0 \frac{x}{\log_2 x},$$

for some constant C_0. Then the probability P is

$$
\begin{aligned}
P &> \frac{C_0 \dfrac{\lfloor \sqrt{n}/4 \rfloor - 1}{\log_2(\lfloor \sqrt{n}/4 \rfloor - 1)} - (r+1)}{\lfloor \sqrt{n}/4 \rfloor - 1} > \frac{C_1}{\log_2 \lfloor \sqrt{n}/4 \rfloor} - \frac{r+1}{\lfloor \sqrt{n}/4 \rfloor - 1} \\
&> \frac{C_1}{\log_2 \lfloor \sqrt{n}/4 \rfloor} - \frac{2U(|n|)}{\lfloor \sqrt{n}/4 \rfloor - 1} > \frac{C_2}{\log_2 \lfloor \sqrt{n}/4 \rfloor} \\
&> \frac{C_3}{\log_2(\sqrt{n}/4)} > \frac{C_4}{\lfloor \log_2 n \rfloor + 1},
\end{aligned}
$$

thus $P > C/|n|$ for some constant C and sufficiently large n. $\square$

Lemma 4.2: Let $n \in SC_k$ and let $U(|n|)$ be any fixed polynomial in $|n|$. Let r be any positive integer not greater than $U(|n|)$ and let $e^{(r-1)} = (e^{(0)}, e^{(1)}, \ldots, e^{(r-1)})$, where $e^{(i)} < \lfloor \sqrt{n}/4 \rfloor$ $(0 \leq i \leq r-1)$ are distinct r odd primes. Define $\hat{e}^{(r)} = (\hat{e}^{(0)}, \hat{e}^{(1)}, \ldots, \hat{e}^{(r)})$ to be $\hat{e}^{(r)} = (e^{(0)}, e^{(1)}, \ldots, e^{(r-1)}, d)$ for any $d \in OP(\lfloor \sqrt{n}/4 \rfloor)$ such that $d \neq e^{(i)}$ $(0 \leq i \leq r-1)$. Then for any $g \in Z_n^*$ such that $g^2 \not\equiv 1 \pmod{n}$, $g^{\hat{e}^{(i)} \cdots \hat{e}^{(j)}} \not\equiv g \pmod{n}$ $(0 \leq i < j \leq r)$ iff $g^{e^{(s)} \cdots e^{(t)}} \not\equiv g \pmod{n}$ $(0 \leq s < t \leq r-1)$ and $\left\{ \prod_{\ell=i}^{r-1} e^{(\ell)} \right\} \cdot d \not\equiv 1 \pmod{L}$ $(0 \leq i \leq r-1)$, where L is the order of g modulo n.

Proof: Let L denote the order of g modulo n. Then it is clear that

$$g^{e^{(s)} \cdots e^{(t)}} \not\equiv g \pmod{n} \iff \prod_{\ell=s}^{t} e^{(\ell)} \not\equiv 1 \pmod{L},$$

for all s, t $(0 \leq s < t \leq r-1)$, and

$$g^{\hat{e}^{(i)} \cdots \hat{e}^{(j)}} \not\equiv g \pmod{n} \iff \prod_{\ell=i}^{j} \hat{e}^{(\ell)} \not\equiv 1 \pmod{L},$$

for all i, j $(0 \leq i < j \leq r)$. Thus it suffices to show that

$$\prod_{\ell=i}^{j} \hat{e}^{(\ell)} \not\equiv 1 \pmod{L} \ (0 \leq i < j \leq r)$$

$$\Longleftrightarrow \prod_{\ell=s}^{t} e^{(\ell)} \not\equiv 1 \pmod{L} \ (0 \leq s < t \leq r - 1)$$

$$\wedge \left\{ \prod_{\ell=i}^{r-1} e^{(\ell)} \right\} \cdot d \not\equiv 1 \pmod{L} \ (0 \leq i \leq r - 1).$$

When $j < r$, $\prod_{\ell=i}^{j} \hat{e}^{(\ell)} \not\equiv 1 \pmod{L}$ $(0 \leq i < j \leq r - 1)$ iff $\prod_{\ell=s}^{t} e^{(\ell)} \not\equiv 1 \pmod{L}$ $(0 \leq s < t \leq r - 1)$, and when $j = r$, $\prod_{\ell=i}^{j} \hat{e}^{(\ell)} \not\equiv 1 \pmod{L}$ $(0 \leq i \leq r - 1)$ iff $\left\{ \prod_{\ell=i}^{r-1} e^{(\ell)} \right\} \cdot d \not\equiv 1$ $\pmod{L}$ $(0 \leq i \leq r - 1)$. Thus it is immediate to see that $g^{\hat{e}^{(i)} \cdots \hat{e}^{(j)}} \not\equiv g \pmod{n}$ $(0 \leq i < j \leq r)$ iff $g^{e^{(s)} \cdots e^{(t)}} \not\equiv g \pmod{n}$ $(0 \leq s < t \leq r - 1)$ and $\left\{ \prod_{\ell=i}^{r-1} e^{(\ell)} \right\} \cdot d \not\equiv 1 \pmod{L}$ $(0 \leq i \leq r - 1)$. $\square$

For $n \in \mathcal{SC}_k$ and $e^{(i)} \in \mathcal{OP}(\lfloor \sqrt{n}/4 \rfloor)$ $(0 \leq i \leq r - 1)$ such that $e^{(i)} \neq e^{(j)}$ $(0 \leq i < j \leq r - 1)$, we define $\mathcal{D}_{r+1}$ to be a set of $(r + 1)$ distinct $d_j \in \mathcal{OP}(\lfloor \sqrt{n}/4 \rfloor)$ $(1 \leq j \leq r + 1)$ such that $d_j \neq e^{(i)}$ $(1 \leq j \leq r + 1, 0 \leq i \leq r - 1)$.

Lemma 4.3: Let $n \in \mathcal{SC}_k$ and let $U(|n|)$ be any fixed polynomial in $|n|$. Let r be any positive integer not greater than $U(|n|)$ and let $e^{(r-1)} = (e^{(0)}, e^{(1)}, \ldots, e^{(r-1)})$, where $e^{(i)} < \lfloor \sqrt{n}/4 \rfloor$ $(0 \leq i \leq r - 1)$ are distinct r odd primes. Then for any $g \in Z_n^*$ such that $g^2 \not\equiv 1 \pmod{n}$, there exists at least one $d \in \mathcal{D}_{r+1}$ such that $\left\{ \prod_{\ell=i}^{r-1} e^{(\ell)} \right\} \cdot d \not\equiv 1 \pmod{L}$ for any i $(0 \leq i \leq r - 1)$, where L is the order of g modulo n.

Proof: Let L denote the order of g modulo n. Since $n \in \mathcal{SC}_k$ and g $(\in Z_n^*)$ satisfies $g^2 \not\equiv 1$ $\pmod{n}$, $L \geq \min\{p', q'\}$. From Lemma 3.1, it follows that $\lfloor \sqrt{n}/4 \rfloor \leq \min\{p', q'\}$, and thus for any $d < \lfloor \sqrt{n}/4 \rfloor$, $d < L$. For some i $(0 \leq i \leq r - 1)$, there exist at most r distinct $d_j < \lfloor \sqrt{n}/4 \rfloor$ that satisfy $\left\{ \prod_{\ell=i}^{r-1} e^{(\ell)} \right\} \cdot d_j \equiv 1 \pmod{L}$, hence at least one $d \in \mathcal{D}_{r+1}$ must satisfy $\left\{ \prod_{\ell=i}^{r-1} e^{(\ell)} \right\} \cdot d \not\equiv 1 \pmod{L}$ for any i $(0 \leq i \leq r - 1)$. $\square$

For simplicity, we assume that every entity i is numbered as $1, 2, \ldots$. Let $E(|n|)$ and $U(|n|)$ be any fixed polynomials in $|n|$. When m $(< E(|n|))$ entities, each of which is in the r_i-th $(1 \leq r_i \leq U(|n|))$ key-updating, conspire to find a secret-key of any other entity u $(> m)$, they can use the following information in **SKU**.

$$n \in \mathcal{SC}_k; \ e_u^{(r_u)} = e_0' e_1' \cdots e_{r_u}', \ e_i' \in \mathcal{OP}(\lfloor \sqrt{n}/4 \rfloor) \ (0 \leq i \leq r_u);$$

$$ID_u \in Z_n^* \ (ID_u^2 \not\equiv 1 \pmod{n}); \ \langle n, ID_i, S_i^{(r_i)}, e_i^{(r_i)} \rangle \in \mathcal{C}_{SKU} \ (1 \leq i \leq m),$$

where $S_i^{(r_i)} = (S_i^{(0)}, S_i^{(1)}, \ldots, S_i^{(r_i)})$, $e_i^{(r_i)} = (e_i^{(0)}, e_i^{(1)}, \ldots, e_i^{(r_i)})$ $(1 \le i \le m)$. Let R be m tuple of integers, $R = (r_1, r_2, \ldots, r_m)$, and each r_i $(1 \le i \le m)$ is not greater than $U(|n|)$. Then we use $INV_{\text{SKU}}^{(m,R)}$ to denote any algorithm that, on input

$$n \in \mathcal{SC}_k; \; e_u^{(r_u)} = e_0' e_1' \cdots e_{r_u}', \; e_i' \in \mathcal{OP}(\lfloor \sqrt{n}/4 \rfloor) \; (0 \le i \le r_u);$$
$$x \in \mathcal{Z}_n^* \; (x^2 \not\equiv 1 \pmod n)); \; \langle n, ID_i, S_i^{(r_i)}, e_i^{(r_i)} \rangle \in \mathcal{C}_{\text{SKU}} \; (1 \le i \le m),$$

outputs $y \in \mathcal{Z}_n^*$ such that $x \equiv y^e \pmod n$ for a non-negligible fraction of $x \in \mathcal{Z}_n^*$, and we use $INV_{\text{PKU}}^{(m,R)}$ to denote any algorithm that, on input

$$n \in \mathcal{SC}_k; \; f \in \mathcal{OP}(\lfloor \sqrt{n}/4 \rfloor); \; x \in \mathcal{Z}_n^* \; (x^2 \not\equiv 1 \pmod n));$$
$$\langle n, ID_i, T_i^{(r_i)}, f_i^{(r_i)} \rangle \in \mathcal{C}_{\text{PKU}} \; (1 \le i \le m),$$

outputs $y \in \mathcal{Z}_n^*$ such that $x \equiv y^f \pmod n$ for a non-negligible fraction of $x \in \mathcal{Z}_n^*$, where $T = (T_i^{(0)}, T_i^{(1)}, \ldots, T_i^{(r_i)})$, $f_i^{(r_i)} = (f_i^{(0)}, f_i^{(1)}, \ldots, f_i^{(r_i)})$ $(1 \le i \le m)$. In addition, we use INV to denote any algorithm that, on input $n \in \mathcal{SC}_k$, $e \in \mathcal{OP}(\lfloor \sqrt{n}/4 \rfloor)$, and $x \in \mathcal{Z}_n^*$, outputs $y \in \mathcal{Z}_n^*$ such that $x \equiv y^e \pmod n$ for a non-negligible fraction of $x \in \mathcal{Z}_n^*$.

From technical reasons, we assume, throughout the rest of this paper, that each ID_i such that $ID_i^2 \not\equiv 1 \pmod n$ is randomly chosen (by the trusted center) with uniform probability over $\mathcal{Z}_n^*$, but once assigned it is unchanged forever.

Theorem 4.4: *Given an expected polynomial time algorithm $INV_{\text{SKU}}^{(m,R)}$, there exists an expected polynomial time algorithm INV using $INV_{\text{SKU}}^{(m,R)}$ as an oracle.*

Proof: It suffices to show that, given $n \in \mathcal{SC}_k$, $e = e_0' e_1' \cdots e_{r_u}'$, and $x \in \mathcal{Z}_n^*$ such that $x^2 \not\equiv 1 \pmod n$, there exists an expected polynomial time $INV\left(INV_{\text{SKU}}^{(m,R)}\right)$ using an expected polynomial time algorithm $INV_{\text{SKU}}^{(m,R)}$ as an oracle.

Let $E(|n|)$ and $U(|n|)$ be any fixed polynomials in $|n|$. Let $m < E(|n|)$ and let $R = (r_1, r_2, \ldots, r_m)$, where $r_i \le U(|n|)$ $(1 \le i \le m)$. Note that for $\langle n, ID_i, S_i^{(r_i)}, e_i^{(r_i)} \rangle$ $(1 \le i \le m)$, the oracle $INV_{\text{SKU}}^{(m,R)}$ returns a correct answer if $\langle n, ID_i, S_i^{(r_i)}, e_i^{(r_i)} \rangle \in \mathcal{C}_{\text{SKU}}$ for all i $(1 \le i \le m)$; it might return garbage or something otherwise.

Algorithm $INV\left(INV_{\text{SKU}}^{(m,R)}\right)$:

Input. $n \in \mathcal{SC}_k; \; e = e_0' e_1' \cdots e_{r_u}'$, where $e_i' \in \mathcal{OP}(\lfloor \sqrt{n}/4 \rfloor)$ $(0 \le i \le r_u)$;
$x \in \mathcal{Z}_n^*$ such that $x^2 \not\equiv 1 \pmod n$.

Step 1.   Set $\mathcal{R} := \phi$ and $i := 1$.

Step 2.   Set $\mathcal{S}_i := \phi$ and $\ell_i := r_i$ and choose $S_i^{(r_i)} \in \mathcal{Z}_n^*$ such that $\left\{ S_i^{(r_i)} \right\}^2 \not\equiv 1 \pmod n$.

Step 3.   Choose $(r_i - \ell_i + 1)$ distinct $d_j \in \mathcal{OP}(\lfloor \sqrt{n}/4 \rfloor)$ such that $d_j \notin \mathcal{S}_i$, using primality testing. (see, e.g., [AH], [Ra], [SS].)

Step 4. For each $d_j \notin S_i$ ($1 \leq j \leq r_i - \ell_i + 1$), compute $S_i^{(\ell_i - 1)} \equiv \left\{ S_i^{(\ell_i)} \right\}^{d_j}$ (mod n)

until $S_i^{(\ell_i - 1)} \not\equiv S_i^{(\ell)}$ (mod n) for all ℓ ($\ell_i \leq \ell \leq r_i$).

Step 5. Set $e_i^{(\ell_i)} := d_j$ and $S_i := S_i \cup e_i^{(\ell_i)}$.

Step 6. If $\ell_i > 0$, then $\ell_i := \ell_i - 1$ and go to **Step 3**.

Step 7. Set $ID_i := S_i^{(-1)}$.

Step 7-1. If $ID_i \in \mathcal{R}$, then go to **Step 2**; otherwise set $\boldsymbol{S}_i^{(r_i)} = (S_i^{(0)}, S_i^{(1)}, \ldots, S_i^{(r_i)})$ and $\boldsymbol{e}_i^{(r_i)} = (e_i^{(0)}, e_i^{(1)}, \ldots, e_i^{(r_i)})$.

Step 7-2. If $i < m$, then $\mathcal{R} := \mathcal{R} \cup ID_i$, $i := i + 1$ and go to **Step 2**.

Step 8. Run the algorithm (or oracle) $INV_{\text{SKU}}^{(m,R)}$ on input $n \in SC_k$, $e = e_0' e_1' \cdots e_{r_u}'$, $x \in \mathcal{Z}_n^*$ such that $x^2 \not\equiv 1$ (mod n), and $\langle n, ID_i, \boldsymbol{S}_i^{(r_i)}, \boldsymbol{e}_i^{(r_i)} \rangle$ ($1 \leq i \leq m$).

Output. $y \in \mathcal{Z}_n^*$ such that $x \equiv y^e$ (mod n).

Trivially, **Step 1** (resp. **Step 2**) runs in deterministic (resp. expected) polynomial time. From Lemma 4.1 and the facts that deciding primality is in $\mathcal{ZPP}$ (see [AH], [Ra], [SS].) and $(r_i - \ell_i + 1) \leq U(|n|) + 1$, it follows that **Step 3** runs in expected polynomial time. From Lemmas 4.2 and 4.3, there must exist at least one $d_j \notin S_i$ such that for all ℓ ($\ell_i \leq \ell \leq r_i$), $S_i^{(\ell_i - 1)} \not\equiv S_i^{(\ell)}$ (mod n), thus **Step 4** runs in deterministic polynomial time. Since **Step 5** runs in deterministic polynomial time and the iteration times of a loop from **Step 3** to 6 is $r_i + 1 \leq U(|n|) + 1$, then the total running cost of the loop from **Step 3** to 6 is expected polynomial time.

In **Step 7-1**, the probability that $ID_i \in \mathcal{R}$ is negligibly small, because possible candidates of ID_i is *exponentially* many, while $\|\mathcal{R}\|$ is *polynomially* bounded, i.e., $\|\mathcal{R}\| \leq m < E(|n|)$, and then the expected iteration times from **Step 2** to 7-1 or 7-2 is $O(E(|n|))$. Hence the algorithm $INV \left(INV_{\text{SKU}}^{(m,R)} \right)$ runs in expected polynomial time and outputs $y \in \mathcal{Z}_n^*$ such that $y \equiv x^e$ (mod n) for a non-negligible fraction of $x \in \mathcal{Z}_n^*$. $\square$

Informally, Theorem 4.4 shows that when any polynomially many entities conspire in **SKU** even in polynomially many times key-updating, they can not invert $x \in \mathcal{Z}_n^*$ for a non-negligible fraction of $x \in \mathcal{Z}_n^*$. From the definition of *soundness* [FFS], [TW], this implies that in **SKU** any polynomially many conspiring entities can not misrepresent themselves for a non-negligible fraction of (possible) other entities, even in polynomially many times key-updating. A result similar to this holds for **PKU**.

Theorem 4.5: *Given an expected polynomial time algorithm $INV_{\text{PKU}}^{(m,R)}$, there exists an expected polynomial time algorithm INV using $INV_{\text{PKU}}^{(m,R)}$ as an oracle.*

Proof: It suffices to show that, given $n \in SC_k$, $f \in \mathcal{OP}(\lfloor \sqrt{n}/4 \rfloor)$, and $x \in \mathcal{Z}_n^*$ such that $x^2 \not\equiv 1 \pmod{n}$, there exists an expected polynomial time $INV\left(INV_{\mathrm{PKU}}^{(m,R)}\right)$ using an expected polynomial time algorithm $INV_{\mathrm{PKU}}^{(m,R)}$ as an oracle.

Let $E(|n|)$, $U(|n|)$, m, and $R = (r_1, r_2 \ldots, r_m)$ be defined in the same way as the proof of Theorem 4.4. It should be noted that for $\langle n, ID_i, T_i^{(r_i)}, f_i^{(r_i)} \rangle$ $(1 \le i \le m)$, the oracle $INV_{\mathrm{PKU}}^{(m,R)}$ returns a correct answer if $\langle n, ID_i, T_i^{(r_i)}, f_i^{(r_i)} \rangle \in \mathcal{C}_{\mathrm{PKU}}$ for all i $(1 \le i \le m)$; it might return garbage or something otherwise.

Algorithm $INV\left(INV_{\mathrm{PKU}}^{(m,R)}\right)$:

Input. $n \in \mathcal{SC}_k$; $f \in \mathcal{OP}(\lfloor \sqrt{n}/4 \rfloor)$; $x \in \mathcal{Z}_n^*$ such that $x^2 \not\equiv 1 \pmod{n}$.

Step 1. Run each **Step** from 1 to 7 in the algorithm $INV\left(INV_{\mathrm{SKU}}^{(m,R)}\right)$.

Step 2. Run the algorithm $A_{\mathrm{SKU} \to \mathrm{PKU}}$ on input $\langle n, ID_i, S_i^{(r_i)}, e_i^{(r_i)} \rangle \in \mathcal{C}_{\mathrm{SKU}}$ for each i $(1 \le i \le m)$, and output $\langle n, ID_i, T_i^{(r_i)}, f_i^{(r_i)} \rangle \in \mathcal{C}_{\mathrm{PKU}}$.

Step 3. Run the algorithm (oracle) $INV_{\mathrm{PKU}}^{(m,R)}$ on input $n \in \mathcal{SC}_k$, $f \in \mathcal{OP}(\lfloor \sqrt{n}/4 \rfloor)$, $x \in \mathcal{Z}_n^*$ such that $x^2 \not\equiv 1 \pmod{n}$, and $\langle n, ID_i, T_i^{(r_i)}, f_i^{(r_i)} \rangle$ $(1 \le i \le m)$.

Output. $y \in \mathcal{Z}_n^*$ such that $x \equiv y^f \pmod{n}$.

From the proof of Theorem 4.4, it follows that **Step 1** runs in expected polynomial time. Since $m < E(|n|)$ and Theorem 3.5 guarantees that the algorithm $A_{\mathrm{SKU} \to \mathrm{PKU}}$ runs in deterministic polynomial time, **Step 2** runs in deterministic polynomial time.

Hence the algorithm $INV\left(INV_{\mathrm{SKU}}^{(m,R)}\right)$ runs in expected polynomial time and outputs $y \in \mathcal{Z}_n^*$ such that $y \equiv x^f \pmod{n}$ for a non-negligible fraction of $x \in \mathcal{Z}_n^*$. □

5 Conclusion and Remarks

In this paper, we showed two kinds of secure key-updating schemes **SKU** and **PKU** in the extended Fiat-Shamir scheme. Here we define more general schemes **SKU'** and **PKU'**:

Let $n \in \mathcal{SC}_k$ and let $E(|n|)$ and $U(|n|)$ be any fixed polynomial in $|n|$. Then the key-updating scheme **SKU'** is completely the same as **SKU** except that for each entity i, $e_i^{(j)} < \lfloor \sqrt{n}/4 \rfloor$ $(0 \le j \le r_i \le U(|n|))$ is an odd number and is not necessarily distinct from each other, and the key-updating scheme **PKU'** is also completely the same as **PKU** except that for each entity i, $f_i^{(j)} \nmid f_i^{(k)}$ $(0 \le k < j \le r_i)$. Using a technique similar to the proofs of Theorems 4.4 and 4.5, we can show that both **SKU'** and **PKU'** are provably secure if decrypting RSA is hard for $n \in \mathcal{SC}_k$.

Observing the results in this paper, we can say that **SKU** and **PKU** have the same security with each other in a polynomial time sense, and seemingly so do **SKU'** and **PKU'**. The scheme **PKU**, however, seems to be better one than **SKU** in the light of efficiency, because in the r_i-th $(1 \leq r_i \leq U(|n|))$ key-updating of **PKU**, a public-key of each entity i is only a prime $f_i^{(r_i)}$, while in the r_i-th $(1 \leq r_i \leq U(|n|))$ key-updating of **SKU**, a public-key of each entity i is $\prod_{j=0}^{r_i} e_i^{(j)}$. This is also the case for **PKU'** and **SKU'**.

Our results can be generalized to more theoretical form — For any transitive trapdoor random self-reducible uniform relation (see [IST].), there exists a perfect zero-knowledge (identity-based) identification system with provably secure key-updating schemes, i.e., if any polynomially many entities conspire in polynomially many times key-updating, they can not find a secret-key of a non-negligible fraction of (possible) other entities, or they can not misrepresent themselves for a non-negligible fraction of (possible) other entities.

Acknowledgments

The authors wish to thank Mitsunori Ogiwara, Yuliang Zheng, Kouichi Sakurai, Kenji Koyama, Tsutomu Matsumoto, and Hiroki Shizuya for their valuable comments and suggestions. The authors also would like to thank Osamu Watanabe for his cooperative and suggestive discussion on the early version of this work.

References

[AH] Adleman, L.M. and Huang, M.D.A., "Recognizing Primes in Random Polynomial Time," Proc. of 19*th* Annual ACM Symposium on Theory of Computing, pp.462-469 (May, 1987).

[AL] Angluin, D. and Lichtenstein, D., "Provable Security of Cryptosystems: a Survey," Technical Report TR-288, Yale University (October, 1983).

[FFS] Feige, U., Fiat, A., and Shamir, A., "Zero Knowledge Proofs of Identity," *Journal of Cryptology*, Vol.1, No.1, pp.77-94 (1988).

[FS] Fiat, A. and Shamir, A., "How to Prove Yourself: Practical Solutions to Identification and Signature Problems," in *Advances in Cryptology* — Crypto'86, Lecture Notes in Computer Science 263, Springer-Verlag, Berlin, pp.186-194 (1987).

[FT] Fujioka, A. and Tsujii, S., "An ID-Based Identification System with Simple Key-Updating," Technical Report of IEICE, ISEC89-25 (November, 1989).

[GQ] Guillou, L.C. and Quisquater, J.J., "A Practical Zero-Knowledge Protocol Fitted to Security Microprocessors Minimizing both Transmission and Memory," in *Advances*

[GQ] Guillou, L.C. and Quisquater, J.J., "A Practical Zero-Knowledge Protocol Fitted to Security Microprocessors Minimizing both Transmission and Memory," in *Advances in Cryptology* — Eurocrypt'88, Lecture Notes in Computer Science 330, Springer-Verlag, Berlin, pp.123-128 (1988).

[HW] Hardy, G.H. and Wright, E.M., *An Introduction to the Theory of Numbers*, Oxford University Press, 5*th* Edition (1979).

[IST] Itoh, T., Shinozaki, S., and Tsujii, S., "Secure Key-Updating Schemes in Identification Protocols," manuscript (March, 1990).

[Kr] Kranakis, E., *Primality and Cryptography*, Wiley-Teubner Series in Computer Science, John Wiley & Sons, Chishester (1986).

[OO] Ohta, K. and Okamoto, T., "A Modification of the Fiat-Shamir Scheme," in *Advances in Cryptology* — Crypto'88, Lecture Notes in Computer Science 403, Springer-Verlag, Berlin, pp.232-243 (1989).

[Ra] Rabin, M.O., "Probabilistic Algorithm for Primality Testing," *Journal of Number Theory*, Vol.12, pp.128-138 (1980).

[RSA] Rivest, R.L., Shamir, A., and Adleman, L.M., "A Method for Obtaining Digital Signatures and Public-Key Cryptosystems," *Communication of the ACM*, Vol.21, No.2, pp.120-126 (February, 1978).

[SI] Shizuya, H. and Itoh, T., "A Group-Theoretic Interface to Random Self-Reducibility," to appear in The Transactions of the IEICE, Vol.E 73, No.7 (July, 1990).

[SS] Solovay, R. and Strassen, V., "A Fast Monte Calro Test for Primality," *SIAM Journal on Computing*, Vol.6, No.1, pp.84-85 (March, 1977).

[TW] Tompa, M. and Woll, H., "Random Self-Reducibility and Zero Knowledge Interactive Proofs of Possession of Information," Proc. of 28*th* Annual IEEE Symposium on Foundations of Computer Science, pp.472-482 (October, 1987).

Oblivious transfer protecting secrecy

An implementation for oblivious transfer protecting secrecy almost unconditionally and a bitcommitment based on factoring protecting secrecy unconditionally.

Bert den Boer

Philips Crypto B.V.
Postbus 218
5600 MD Eindhoven
The Netherlands

Abstract

We present a scheme for oblivious transfer in which contrary to earlier proposals the secrecy is protected (almost) unconditional and not by a cryptographic assumption. We also present a bitcommitment scheme based on factoring where the secrecy is unconditional.

Introduction

An Oblivious Transfer protocol as introduced in [Ra] and with one of the first mathematically based implementations in [Bl] is roughly a protocol where a sender puts a message in a communication channel where the message has a chance of (usually) one half to arrive at the intended recipient. The protocol has two important aspects; one is secrecy: the recipient cannot by deviating from the protocol increase the chance that eventually he gets hold of the message. The other aspect is authentication : the sender cannot find out whether the message reached the recipient. Based on heuristic arguments it is generally believed that both aspects cannot be met unconditionally in a mathematically-oriented protocol. So we need a cryptographic assumption to protect at least one of the two aspects. In the literature implementations for oblivious transfer based on a cryptographic assumption protected authenticity unconditionally, i.e. the sender could even with infinite computing power never find out what message reached the recipient.

In those protocols the cryptographic assumption is that factoring is a hard problem. So in those implementations we have to believe that the recipient cannot find a particular square root otherwise the aspect of secrecy is not fulfilled.

In our proposal we reverse which aspect is protected unconditionally and which aspect is protected by a cryptographic assumption. In our proposal the cryptographic assumption is the Quadratic Residuosity Assumption and this assumption must protect the authenticity. Our proposal protects secrecy almost unconditionally, i.e. the recipient could even with infinite computer power not find out what the message was if he could not find out what that message was by following the protocol; the word almost reflects the fact that the recipient could cheat in an initializing protocol, but the chance that this goes undetected can be made extremely small and of course if the sender detects this cheating he refuses to do the main protocol.

We also present a bitcommitment scheme. Such a scheme can be derived from our proposal for Oblivious Transfer but our bitcommitment scheme is with completely unconditional secrecy. In the presented form the underlying cryptographic assumption is the factoring problem. An earlier proposal for bit-commitment based on factoring is more efficient but guarantees secrecy only almost unconditionally while our bitcommitment guarantees secrecy unconditionally.

Oblivious Transfer, three flavors

Roughly speaking oblivious transfer is a protocol where a message usually has a chance of $1/2$ of reaching the recipient Bob and where the sender Alice can not find out whether this message reached Bob. For a single message this is a adequate definition. Further we have One out of Two Messages Oblivious Transfer and One Chosen out of Two Messages Oblivious Transfer.

In the first case Alice has two messages, one message will reach Bob but Alice does not know which one and Bob will know that he got say the second message but can not a priori decide which of the two messages he will get. In the second case Bob also gets only one message but this time he is able to decide which of the two he will get while Alice cannot find out his choice.

Description of the Protocol without mathematics

Alice the sender has a one bit message and encodes this message with a coin, say heads up for one. Before she comes to that she takes two coins and two identical wrapping papers and gives them to Bob sitting at the other end of the table. Bob wraps both coins in paper and gives them back to Alice. She unwraps one of the coins and puts the wrapping paper in a low tray. Secondly she puts clearly the still wrapped coin in the tray. And finally she puts the unwrapped coin with the side according to her message bit above ,her choice protected from Bobs sight, in the tray. The tray is filled with a liquid completely dissolving the papers. Alice puts the lid on the tray, shifts the tray long enough for both the dissolving process and the shuffling of the coins. The tray is that low that the coins cannot flip. Then the tray is handed to Bob and he opens the tray and exams the coins, this time such that Alice cannot see which sides are up. If Bob sees two heads he knows that Alice's bit was one but if he sees one head and one tail he does not know which of the two was Alice's choice and which was still wrapped when Alice put it in the tray.

Need for an Initializing Protocol

Basically our protocol works with two coins for a one bit message. Alice chooses the upper side of one coin according to her message and the other is thrown such that the Alice, the sender cannot see which side is up and such that Bob, the recipient, sees both coins but cannot see which was thrown and which was chosen. We approach this situation under the Quadratic Residuosity Assumption. Given a number N which is not a square and which among its set of divisors has exactly two prime numbers we know from elementary number theory that the group of residues with Jacoby symbol one is twice as big as the subgroup of square residues. Given two arbitrary residues J and Res with Jacoby symbol plus one Alice start with encoding her message bit by a square if it is zero and and by J times a square if her message bit is one. This is her first residue. She throws a coin by flipping a coin first and decoding zero by Res and one by Res times J. This is her second residue. If Bob made N and choosed J and Res then Alice does not know whether the second residue is a square or not. So for Alice the second residue is like throwing a coin. Bob cannot influence this coin unless J is a square. But then he cannot decrypt the first residue. So Bob better chooses J to be a non square. Up to now Bob can easily distinguish the (first) residue Alice made according to her choice and the (second) residue made as an implementation of a thrown coin even if Alice arbitrarily interchanges their order or not. The final tuning for the main algorithm is to divide the first residue on the second and sending this quotient residue and the first residue in arbitrary order to Bob. Bob checks that their product is one of the two admitted cases and decrypt a residue into a one if and only if it is not a square. If and only if both bits are the same he knows the message. This two-steps decoding only works if J is not a square. It is easy to analyze that if J is a square Alice message bit remains unknown.

So the only way to cheat for Bob is making N a square with two prime divisors or make a modulus with three or more prime divisors. The first trick is easy to recognize but to avoid the second way of cheating Bob needs to convince Alice that N has only two prime divisors.

As far as the author knows there is no direct proof for this (in the litterature there is even a claim that deciding whether a number has two or three prime divisors is a hard problem) without giving away the factorization and thereby the authenticity of the main protocol and so we need a interactive protocol(or a mutually trusted random source like in [GP] or [SP]) for convincing Alice that the number N chosen by Bob has only two prime factors.

Initializing Protocol

Bob ,the recipient in the main protocol, produces a R.S.A. modulus N and a non-square residue J with Jacobi-symbol 1. He sends them to Alice, the sender in the main protocol and Alice and Bob participate in a zero-knowledge protocol like the protocol described in [GP] (in their protocol there is no interaction because there the existence of a mutually trusted random source is assumed) where Alice challenges Bob to prove that the number of prime factors of N is equal to two and J is a non-square. Alice does not need a protocol to find out that N has at least two different prime factors because we assume that the order of the multiplicative group modulo a good R.S.A. modulus has only a factor two or perhaps another small factor in common with $N-1$. The other claimed properties are not feasible to check for Alice on her own so she needs this protocol. We will briefly sketch one round of this protocol. Both produce a residue with Jacobi-symbol 1, first Alice sends her residue to Bob and then Bob sends his to Alice, Alice makes a choice challenging Bob to write either his residue or the product of her residue and his residue as $J^b R^2$. Bob sends the bit b and the residue R to Alice and Alice checks for correctness.

If the number of different prime factors is three or more Bob has a chance of $3/4$ of being able to produce a adequate pair b and R. In the case that N contains only two different prime factors but J is a square Bob also has a chance of $3/4$ of being able to produce b and R. By using n rounds of this protocol with new choices for residues the chance of successfully cheating by Bob becomes $(3/4)^n$

In other words at the expense of linear cost Alice can get a "exponential confidence" in Bob's claim about N and J.

Main Protocol

The first protocol is for Single Message Oblivious Transfer.

Before Alice starts sending messages, she gets a set of residues, called *Res* , with Jacobi-symbol 1 from Bob . A message is written in binary form and let us say c is the next bit to send. To send this bit of the message Alice produces a random residue S and a random bit d and computes $A = J^c S^2$ and $B = Res \, J^d /$ A where *Res* is the lowest numbered element of the set of residues Alice got from Bob which is not used for other messages. Then A and B are interchanged precisely if their values were not in a natural order. After this the two residues are send in their natural ordering to Bob. Bob checks that the product is either an element of the list of residues he send Alice at the start of the main protocol or of the form J times such an element (and identical to products of pairs for sending other bits of the same message).

With high probability we may assume that N and J are what they are supposed to be and then Bob will find out what the message bit c is precisely when the product $Res \, J^d$ of the two residues A and B is a square. The chance that this occurs is exactly $1/2$ under Q.R.A.. This chance has this value because Alice is under Q.R.A. not supposed to see the quadratic characteristic of *Res* and by choosing d independently with a chance $1/2$ from her point of view there is a chance $1/2$ that the product $Res \, J^d$ of the two residues is a square. The reason that Bob can "encrypt" c in the case that the product of the residues is a square is first : Bob can easily see that the product is a square because he knows the factorisation of N. And the second reason for Bob's ability of encrypting is that in that case both residues he got from Alice have to be or both a square or both a non-square. In that case, so although Bob does not know which of the two residues is the message A containing the message bit c, he certainly knows whether A is a square (because both residues are squares) or not and therefore what the message bit c is. In case the product is a non square and of the form $Res\ J^d$

(this form disables Alice to know whether or not A and B have the same quadratic characteristic) A and B have a different quadratic characteristic and as Bob can not tell which of the two residues is A, from his point of view it is equally likely that c equals zero as well equals one. This dilemma means that the message does not reach Bob. In this case Bob can study A and B for years but he will never know (without the cooperation of Alice) so if according to the protocol Bob did not receive the message he will not find it by any deviation of the protocol. So Bob can, after committing to a good modulus and non-square residue, never find out the message bit if the product of the two residues is a non-square. So indeed secrecy is almost unconditionally protected.

On the other hand if Alice can distinguish between squares and non-squares (thus violating Q.R.A. in other words breaking the cryptographic assumption) she can for example always sends a message which never arrives. She does this by observing the quadratic characteristic of Res and choosing d zero if Res is a non-square and choosing d one if Res is a square. But as long as Alice does not know the quadratic characteristic of Res she does not know whether she "doubles" her message bit or sends a non-descriptive (unordered!) pair zero and one to Bob. So indeed the authenticity is protected by the cryptographic assumption.

The extra factor J which Alice can use (she can take $Res\ J$ or Res as product for the residues A and B) is to protect her against an attempt to cheat by Bob by making all *Res* a square.

For each message another element *Res* of the list is used otherwise authentication in a higher sense is violated, i.e., for example assume that for two message bits the probably different pairs A and B have the same product *Res* then Alice does not know whether the first message reaches Bob but she knows that either both message reach Bob or none of the two message bits. This dependency is exploited by using the same product *Res* or *Res J* for sending the bits of a single message and also in the One out of Two Messages Oblivious transfer.

The protocols for One out of Two Messages Oblivious Transfer are done by using the same residue *Res* twice but once with *Res* as product of two residues and once *Res J* as product of two residues. Bob can choose which message he gets if Alice has to use *Res* as the first product and *Res J* as the second product. If Alice is allowed to interchange the products, i.e., she is free to choose *Res* respectively as well as *Res J* respectively for the product of the first pair A and B but then she is obliged to make the second B such that the product of the second pair A and B is *Res J*, respectively *Res*. In this case there is no choice for Bob. Exactly one of the products will be a square and the secret bit put in one of the factors of that product will be revealed. It is also clear that the other product is a non-square, therefore one factor is a square and the other is a non-square. It is equally likely that the square or the non-square contains the message bit so this secret bit will never be known by Bob.

Efficiency Considerations

We can decrease message-lengths in the Single Message protocol by sending only the smallest of the pair A, B and the bit d and if necessary a pointer to *Res*.

We can further avoid modulo divisions by Alice the sender if Bob, who is the recipient and the maker of the modulus, is willing to compute inverses of J and his residues *Res* and send those numbers in corresponding pairs to Alice.

There is also a way to avoid the sending of the batch with the residues *Res* (or one residue *Res* per oblivious transferred message) by letting Alice use outputs of one-way functions. This last is not possible with the One Chosen out of two Oblivious Transfer because there Bob has to ensure that the product of the first two residues A and B is a square and therefore the product of the last two residues A and B is a non-square if he wants to know the first secret bit.

After the next chapter we will indicate how an initializing protocol for our Oblivious Transfer can be done in two and a half round.

Unconditionally secure bitcommitment

Under this heading we present a bitcommitment scheme for which the secrecy of the encrypted bit is unconditionally that is each well-formed encryption of zero is also a possible encryption of one and vice-versa. Before we present this we first make the association between the oblivious transfer above and the bitcommitment presented at the end of this section. In [Cr] a method is presented to derive a bit commitment scheme out of oblivious transfer, especially out of One out of Two Messages Oblivious Transfer. For the general method one has to do several transfers to commit to one bit. By using one of the One out of Two

messages Oblivious Transfer presented above it is clear that the Exclusive-Or

sum of the two secret bits is almost unconditionally secret for the recipient Bob.

In this case one (instead of several as in [Cr] for the general case) One out of

Two messages Oblivious Transfer is enough to commit to a bit.

To clarify this let *Res* be a residue made by Bob. Alice want to commit to bit *a*

. She produces a random bit *c* and two random residues *R* and *S*. She com-

putes the bit $b=c \oplus a$ and $A = J^b R^2$, $B = Res/A$, $C = J^c S^2$, $D = Res\, J/C$.

She interchanges the first pair if their values are not in their natural ordering and

does the same with the second pair. In a straightforward version this fourtuple is

a commitment to the bit *a*. In fact by analyzing this and thereby reducing the

message length we get an earlier proposal for a bitcommitment scheme [BCC]

based on factoring which is almost unconditionally secure with respect to secre-

cy. To reduce the message length Bob and Alice extend the initializing protocol

by a zero knowledge proof from Bob to Alice that this special *Res* is a square

[like in BCC]. The residues *A* and *B* are no longer necessary so the bit *c* is

made equal to the secret bit and only the smallest of the residues *C* and *D* (the

third element of the special fourtupel) is given to Bob. Alice can "open this bit-

commitment" (terminology of [BCC]) in two ways if she can compute a square

root of *Res*. So suddenly the authenticity of the bitcommitment is protected by

the difficulty of computing this square root (whereas the authenticity of the

oblivious transfer generating this bitcommitment was protected by Q.R.A.).

Now we will present our bitcommitment scheme. Like the bitcommitment scheme of [BCC] it is based on factoring. The secrecy of the encrypted bits in the [BCC] scheme is almost unconditional (it depends like our oblivious tranfer on luck in an initializing protocol). We change this protocol at the sacrifice of more computations to get unconditional secrecy of the encrypted bits. We get complete unconditionality for the secrecy of the committed bit by using repeated squaring.

In order to commit to a bit Alice asks Bob to produce an odd number N between M, a fixed power of two, and $2M$ and a residue K with Jacobi-symbol -1. Bob computes $L = K^M \bmod N$ and sends N, K, L to Alice. First Alice checks this triple. In order to commit to a bit c Alice takes a random residue S and computes $T = L^c S^{2M}$ and sends T to Bob. Bob, who better knows the factorization of N, checks that T has indeed odd multiplicative order. Bob can open T in two ways and therefore the secrecy is unconditional and Alice cannot find an alternative opening of T unless she can factor N. (The power M is taken to ensure that the gcd of $2M$ with the multiplicative order of the group is at most M and of course in practice this gcd is much lower.)

Because every residue which can be written as a a residue in the power M also can be written as a residue in the power $2M$ Bob gets no information about the secret bit c. But Alice can only change her commitment if she can write L as a residue in the power $2M$ and that is equivalent to factoring N.

So here we have a bitcommitment scheme with unconditional secrecy and authenticity based on the difficulty of factoring.

Initializing protocol in a bounded number of rounds

In [SP] a One out of Two Oblivious Transfer is presented where the initializing protocol is a single message (half a round) under the assumption of Trusted Randomness. Their protocol differs in the sense that even if N has two prime factors the recipient can cheat if their y is a square. Such a cheating is conceivable if the Public Random bits are known on beforehand and a recipient with unbounded computing power can thereafter select the prime divisors for his public modulus N. Without the assumption of mutually trusted random bits we can make our initializing protocol in a bounded number of rounds using the footsteps of [BCY] as follows:

Bob sends his modulus.

Alice makes random residues with Jacoby symbol one and raises them to the power 2^M (where N has $M+1$ bits) and sends those outputs to Bob.

Bob makes and sends random residues with Jacoby symbol one.

Alice opens the roots of the residues she send.

Bob computes the products of one residue of his own and one root of Alice (following the ordering) and shows that those products can be divided in two sets, where in each set the quotient of two members is a square.

This is a protocol with two and a half round which need an extra polishing if N-1 has Jacoby symbol one and Bob is not willing to disclose whether this number is a square or not. In that case Alice also has for each root to commit to whether twice this root is bigger or smaller than the modulus.

Conclusions

We have made an implementation for three flavors of Oblivious Transfer build from nothing in common by the participants.

References

[BCC] Brassard, G.,Chaum, D. and Crepeau, C., "Minimum disclosure proofs of knowledge", Journal of Computer and System Sciences,vol. 37, no. 2, October 1988, pp. 188-195.

[BCY] Brassard, G.,Crepeau, C.,Yung, M., "Everything in NP can be argued in a bounded number of rounds", Abstracts of Eurocrypt 89.

[Bl] Blum, M., "Three applications of the oblivious transfer: Part I: Coin flipping by telephone; Part II: How to exchange secrets; Part III: How to send certified electronic mail", Department of EECS, University of California, Berkeley, CA, 1981.

[Cr] Crepeau, C., "Verifiable all or nothing disclosure of secrets", Abstracts of Eurocrypt 89.

[GP] Graaf, J. de, Peralta, R., "A simple and secure way to show the validity of your public key", Proceedings Crypto 87, pp. 87-119.

[Ra] Rabin, M., "how to exchange secrets by oblivious transfer", Tech. Memo TR-81, Aiken Computation Laboratory, Harvard University, 1981.

[SP] Santis, A. de, Persiano, G., "Public-Randomness in Public-Key Cryptography", Abstracts of Eurocrypt 90.

Public-Randomness
in Public-Key Cryptography

E X T E N D E D   A B S T R A C T

Alfredo De Santis*

Dipartimento di Informatica ed Applicazioni

Università di Salerno

84081 Baronissi (Salerno), Italy

Giuseppe Persiano[†]

Aiken Comp. Lab.

Harvard University

Cambridge, MA 02138

Abstract

In this work we investigate the power of Public Randomness in the context of Public-key cryptosystems. We consider the Diffie-Hellman Public-key model in which an additional short random string is shared by all users. This, which we call Public-Key Public-Randomness (PKPR) model, is very powerful as we show that it supports simple non-interactive implementations of important cryptographic primitives.

We give the first *completely* non-interactive implementation of Oblivious Transfer. Our implementation is also secure against receivers with unlimited computational power.

We propose the *first* implementation of non-interactive nature for Perfect Zero-Knowledge in the dual model of Brassard, Crépeau, and Chaum for all NP-languages.

1   Introduction

The Public Key model, introduced by Diffie and Hellman [DH], suggests an elegant and efficient way to eliminate the need for preliminary secure interaction which is essential in Private-Key Cryptography. Each party A publishes in a Public File his encryption key P_A and keeps secret his decryption key S_A. Once the public file has been established, each user can receive any number of encrypted messages on a public channel from any other user that has access to the public file, without having to interact with them via exchanges of messages. Moreover, [DH] showed how to

*Part of this work was done while the author was visiting IBM Research Division, T. J. Watson Research Ctr, Yorktown Heights, NY 10598.

[†]Partially supported by ONR Grant #N00039-88-C-0163.

produce unforgeable signatures for messages and how two users could establish a private key in this (non-interactive) model.

The security of the protocols rests upon the very natural assumption that only limited computational resources are available to each user. The introduction of Complexity considerations in Cryptography caused much excitement and made possible applications never thought of before. Most notably, the *Oblivious Transfer* (OT) protocol by Rabin (see [HR]) and the *Zero-Knowledge Proof System* of [GMR] and [GMW1]. However, all proposed implementations of the OT require the ability to interact and therefore cannot be used in the Public-Key model. Different non-interactive or bounded-interaction scenarios in which Zero-Knowledge was possible have been proposed ([BFM], [DMP1], [DMP2], [DMP3], [BDMP], [K], [KMO]) but they all suffer of some practical drawbacks that limited their applicability.

In this paper, we consider the Diffie-Hellman model in which a short random string is shared beforehand by all users. We call this model Public-Key Public-Randomness Cryptosystem (PKPR Cryptosystem). The set up of the PKPR Cryptosystem does not require any preprocessing stage: Each user chooses and validates by himself his own public and private keys without any interaction. Moreover, no center or distributed fault-tolerant computation (in the sense of [GMW2, BGW, CCD, GHY, B, RB]) is ever invoked to protect against possible "cheating". Even though interaction is never allowed, the PKPR Cryptosystem is very powerful as we show that important cryptographic primitives have simple non-interactive implementations in this model.

Summary of the results. We give a *completely non-interactive* implementation of Oblivious Transfer in the PKPR model. This is the first non-interactive implementation of OT that does not require a trusted center or some distributed fault-tolerant computation.
Our implementation is essentially optimal. Indeed, a recent result of Ostrowsky and Yung [OY] shows that it is not possible to achieve a non-interactive OT from scratch. We prove that a PKPR setting is enough to achieve non-interactive OT.
Our implementation is also secure against receivers with unlimited computing power. We give the *first* implementation of non-interactive nature for Perfect Zero-Knowledge (in the dual model of [BC] and [Ch]) for all NP-languages. Unlike previous implementations of non-interactive Zero Knowledge with a common random string, our implementation is very simple and, most notably, allows any number of provers to be active (this solves the open problem of *many independent provers* (see [DMP1]), though in a slightly modified scenario).

Our results are based on the well known and widely used Quadratic Residuosity Assumption, and they demonstrate the added value of short Public Randomness in the context of Public key Cryptography.

2 Preliminaries

In this section we review some elementary facts from number theory about quadratic residues and the probabilistic encryption scheme based on the difficulty of deciding quadratic residuosity of [GM]. We follow the notation of [BDMP].

For each natural number x, the set of positive integers less than x and relatively prime to x form a group under multiplication modulo x denoted by Z_x^*. $y \in Z_x^*$ is a *quadratic residue* modulo x iff there is a $w \in Z_x^*$ such that $w^2 \equiv y \bmod x$. If this is not the case we call y a *quadratic non residue* modulo x. The *quadratic residuosity predicate* is defined as follows

$$Q_x(y) = \begin{cases} 0 & \text{if } y \text{ is a quadratic residue modulo } x \text{ and} \\ 1 & \text{otherwise.} \end{cases}$$

If $y_1, y_2 \in Z_x^*$, then

1. $Q_x(y_1) = Q_x(y_2) = 0 \implies Q_x(y_1 y_2) = 0$.

2. $Q_x(y_1) \neq Q_x(y_2) \implies Q_x(y_1 y_2) = 1$.

For any fixed $y \in Z_x^*$, the elements $\{yq \bmod x \mid q \text{ is a quadratic residue modulo } x\}$ constitute an equivalence class that has the same cardinality as the class of quadratic residues.

The set Z_x^{+1} is the subset of Z_x^* consisting of all elements with Jacobi symbol $+1$.

The problem of deciding quadratic residuosity consists of evaluating the predicate Q_x. This is easy when the modulus x is prime and appears to be hard when is composite. Indeed, no efficient algorithm is known for deciding quadratic residuosity modulo composite numbers whose factorization is not given. Actually, the fastest way known consists of first factoring x and then compute $Q_x(y)$. This fact has been first used in cryptography by Goldwasser and Micali [GoMi]. We use it in this paper with respect to the following special moduli.

Blum integers. Let $n \in \mathcal{N}$. The set of Blum integers of size n, $BL(n)$, is defined as follows: $x \in BL(n)$ if and only if $x = pq$, where p and q are primes of length n both $\equiv 3 \bmod 4$. These integers were introduced in Cryptography by M. Blum [B1]. Blum integers are easy to generate. There exists an efficient algorithm that, on input 1^n, outputs the factorization of a randomly selected $x \in BL(n)$. This class of integers constitutes the hardest input for any known efficient factoring algorithm. Since no efficient algorithm is known for deciding quadratic residuosity modulo Blum integers, this justifies the following

Quadratic Residuosity Assumption (QRA): For each efficient poly-size family of circuits $\{C_n\}_{n \in \mathcal{N}}$, all positive constants d, and all sufficiently large n,

$$Pr\left(x \leftarrow BL(n); \; y \leftarrow Z_x^{+1} : C_n(x,y) = Q_x(y)\right) < 1/2 + n^{-d}.$$

That is, no poly-size family of circuits can guess the value of the quadratic residuosity predicate substantially better than by random guessing. This assumption has been used for the first time by [GM] and is now widely used in Cryptography. For instance, the proof system of [BDMP] is based on it.

2.1 Encryption schemes

In a seminal paper, Goldwasser and Micali [GM] introduced a Public-key encryption scheme whose security is based on the quadratic residuosity assumption. The public key of user B contains a random integer x_B product of two primes of the same length and y_B, a random quadratic non residue modulo x_B with Jacobi symbol $+1$. B's secret key contains the prime factors of x_B. To secretly send B an l-bit message $M = m_1 \cdots m_l$, A just encrypts each bit m_i by computing $c_i = y_B^{m_i} r_i^2 \bmod x_B$, where r_i is a randomly chosen element of $Z_{x_B}^*$. It is easily seen that each c_i corresponding to a 0 bit is a quadratic residue and each c_i corresponding to a 1 bit is a quadratic non residue modulo x_B. This observation gives a very simple decryption algorithm: it is enough for B to compute the quadratic residuosity of the c_i's he has received.

What makes decryption possible in this scheme is the fact that y_B is a quadratic non residue. In fact, should y_B be a quadratic residue, then, independently of the value of m_i, each c_i is a quadratic residue, and therefore B has absolutely no *information* to recover the bits m_i's.

In what follows, we will denote by $E(x, y, m)$ the algorithm that returns a random encryption of m computed using the pair (x, y) and $D(p, q, c)$ the algorithms that returns the decryption of the ciphertext computed using x's prime factors p and q.

3 Public-Key Public-Randomness Cryptosystems

A *Public-Key Public-Randomness (PKPR) Cryptosystem* consists of:

(a) a random string σ, called the *reference* string;

(b) a set of transactions $\mathcal{T}$;

(c) a key-space K_σ, which is the set of keys associated with σ; (each key is a pair $< PK, SK >$, where PK is the *public key* and SK is the *private key*;)

(d) a probabilistic polynomial time algorithm *Key_Generator* which returns randomly chosen elements in K_σ;

(e) a poly-time algorithm *Verify_PublicKey* outputting VALID/NONVALID;

(f) for each $T \in \mathcal{T}$ a send/receive pair of probabilistic polynomial time algorithms (S_T, R_T).

Initialization of the PKPR cryptosystem.
Each user U chooses by himself his own public and private keys by following the procedure:

- Randomly select a key $< PK_U, SK_U >$ by running *Key-Generator* on input σ.

- PK_U is made public. The private key SK_U is kept secret.

Performing a transaction.
Let $T \in \mathcal{T}$ be a transaction (defined by a protocol program, e.g. oblivious transfer, secure message sending, Zero-Knowledge proof, etc.). User A, having x as input, performs T with another user B, by following the procedure:

- Verify B's public key PK_B by making sure that *Verify_PublicKey* (σ, PK_B) =VALID.

- Run the (possibly probabilistic) procedure $S_T(PK_B, x)$ and send B the output y.

B does not reply upon receiving y, he just privately computes $R_T(y, PK_A, SK_B)$.

Notice that each transaction is non-interactive. Also, for the set-up of the Cryptosystem no center or distributed fault-tolerant computation is required: each user chooses and validates by himself his own public key. Any other user can check the correctness of the construction of a public key, without any interaction. All is needed is the availability of a common random reference string, whose length does not depend on the number of users, but only on the desired level of security.

4 Oblivious Transfer

Oblivious Transfer has been introduced by Rabin (see [HR]), who first gave an implementation (for honest players) based on the difficulty of factoring. OT is a protocol for two parties: the *Sender* who has a string s, and the *Receiver*. Each of the following two events is equally likely to occur at the end of the protocol.

- The *Receiver* learns the string s.

- The *Receiver* does not get any information about s.

Moreover, at the end of the protocol, the *Sender* does not know whether the *Receiver* got s or not. The wide applicability of the OT was recognized since the early days of modern cryptography; the paper by Blum [B2] is a an example of how OT can be used to implement several other protocols.

A different flavor of OT, the *1-out-2 Oblivious Transfer* was later introduced by Even, Goldreich, and Lempel [EGL]. Here, the sender has two strings s_0 and s_1. Each of the following two events is equally likely to occur at the end of a 1-out-2 Oblivious Transfer:

- The *Receiver* learns the string s_0, and does not get any information about s_1.

- The *Receiver* learns the string s_1, and does not get any information about s_0.

The sender has no information on which string the receiver gets. It is clear that 1-out-2 Oblivious Transfer can be used to implement an Oblivious Transfer. Crépeau [Cr] showed how to achieve a 1-out-2 Oblivious Transfer by using Rabin's OT, thus establishing their equivalence. The rest of this section is organized as follows. In Section 4.1 we formally define what we mean by Non-Interactive Oblivious Transfer (NIOT) in the PKPR model and in Section 4.2 we give an implementation of NIOT in the PKPR model.

4.1 Non-Interactive Oblivious Transfer in the PKPR model

A Non-Interactive Oblivious Transfer is a quadruple of algorithms *(Key_Generator, Verify_PublicKey, Sender, Receiver)*. Suppose the sender A has two strings (s_0, s_1) that he wants to obliviously transfer to the receiver B. Informally, the mechanics of the transfer is the following. First, A verifies B's public file PK_B by making sure that $Verify\_PublicKey(\sigma, PK_B)$ =VALID. Then A computes and sends B the message $msg = Sender(PK_B, s_0, s_1)$. To retrieve one of the two strings, B computes $Receiver(PK_B, SK_B, \sigma, msg)$, where SK_B is its own private key.

Definition 1 A Non-Interactive Oblivious Transfer is a quadruple of algorithms (*Key_Generator, Verify_PublicKey, Sender, Receiver*), where *Key_Generator* and *Sender* are probabilistic polynomial time and *Verify_PublicKey* and *Receiver* are deterministic polynomial time, such that

1. *Meaningfullness: The receiver gets one of the two strings.*

2. *Verifiability: The validity of the construction can be efficiently verified.*

3. *1 out-of 2: The receiver gets only one string and not even a bit of the other.*

4. *Obliviousness: The sender cannot predict which string is going to be received.*

4.2 Implementing Oblivious Transfer in the PKPR Model

We show how to implement a (non-interactive) Oblivious Transfer in the PKPR setting. At this aim, we describe the *Key_Generator* algorithm and the *Verify_PublicKey* algorithm needed to initialize the public file and the *Sender* and *Receiver* algorithm to actually perform the OT.

Algorithm $Key\_Generator(\sigma)$

Input: An n^3-bit reference string $\sigma = \sigma_1 \circ \ldots \circ \sigma_{n^2}$, where $|\sigma_i| = n$ for $i = 1, 2, \ldots, n^2$.

1. *Select public and secret keys.*
   Randomly select two n-bit primes $p, q \equiv 3 \bmod 4$ and set $x = pq$.
   Randomly select $r \in Z_x^*$, $z \in Z_x^{+1}$ and compute $y = -r^2 \bmod x$.

2. *Validate public key.*
   Set $Val=$ empty string.
   For $i = 1, \ldots, n^2$
       if $\sigma_i \in Z_x^{+1}$ then
           if $\mathcal{Q}_x(\sigma_i) = 0$ then append $\sqrt{\sigma_i} \bmod x$ to Val.
           if $\mathcal{Q}_x(\sigma_i) = 1$ then append $\sqrt{y\sigma_i} \bmod x$ to Val.
       else append σ_i to Val.

3. Set $PK = (x, y, z, Val)$ and $SK = (p, q)$.

Output:(PK, SK).

Algorithm $Verify\_PublicKey(\sigma, PK)$

Input: An n^3-bit reference string $\sigma = \sigma_1 \circ \ldots \circ \sigma_{n^2}$, where $|\sigma_i| = n$ for $i = 1, 2, \ldots, n^2$. A Public Key $PK = (x, y, z, Val)$, where $Val = v_1 \circ \ldots \circ v_{n^2}$.

For $i = 1, \ldots, n^2$
    if $\sigma_i \in Z_x^{+1}$ then verify that either $\sigma_i = v_i^2 \bmod x$ or $y\sigma_i = v_i^2 \bmod x$.

Output: If all checks are successfully passed then output VALID else NONVALID.

The validation of the public key is just a non-interactive Zero-Knowledge proof of the NP statement "x is product of two primes and y is a quadratic non residue modulo x". This proof is obtained in a direct manner, that is without making use of reduction to $3SAT$ and, most importantly, the same string σ can be used by any number of users to certify their own public key entry.

To prove that the algorithm $Verify\_PublicKey$ satisfies the *Verifiability* requirement, we have to show that if either x is not product of two primes or y is not a quadratic non residue then with very high probability the algorithm $Verify\_PublicKey$ outputs NONVALID. As this is a non-interactive Zero-Knowledge proof, the proof of the verifiability can be obtained by using the techniques of [DMP1] and [BDMP].

Algorithm $Sender(PK, s_0, s_1)$
Input: A public key $PK = (x, y, z, Val)$. Two strings s_0, s_1.

Compute and send the pair $msg = (E(x, z, s_0), E(x, zy \bmod x, s_1))$.

Algorithm $Receiver(PK, SK, \sigma, msg)$
Input: A public key $PK = (x, y, z, Val)$ along with the corresponding secret key $SK = (p, q)$. An n^3-bit reference string $\sigma = \sigma_1 \circ \ldots \circ \sigma_{n^2}$, where $|\sigma_i| = n$ for $i = 1, 2, \ldots, n^2$. A pair $msg = (\alpha, \beta)$.

1. If z is a quadratic residue then
   If $D(\alpha, p, q) = 0$ then output $D(\beta, p, q)$ else STOP.

2. If z is a quadratic non residue then
   If $D(\beta, p, q) = 0$ then output $D(\alpha, p, q)$ else STOP.

Theorem 1 *Under the QRA, the above quadruple of algorithms (Key_Generator, Verify_PublicKey, Sender, Receiver) is a Non-Interactive Oblivious Transfer.*

Proof. We shall prove that the quadruple (*Key_Generator, Verify_PublicKey, Sender, Receiver*) meets the four requirements of Definition 1. We have already seen that the *Verifiability* requirement is met. Thus, all is left to prove is that *Meaningfullness, 1 out-of 2,* and *Obliviousness* are met too.

Meaningfullness and 1 out-of 2.
As y_B is a quadratic non residue and x_B is product of two primes, for each $z \in Z_{x_B}^{+1}$ exactly one between z_B and $y_B z_B \bmod x_B$ is a quadratic non residue. Therefore, B receives exactly one of s_0, s_1. If B receives s_0 (s_1), then s_1 (s_0) has been encrypted using a quadratic residue modulo x and all B gets is a sequence of random quadratic residues modulo x_B.

Obliviousness.
We prove that the existence of a pair of algorithms (ADV_0, ADV_1) such that for some $c > 0$, all sufficiently large n, and all (s_0, s_1)

$$Pr\,(\sigma \leftarrow \{0,1\}^n\,; \quad (PK, SK) \leftarrow Key\_Generator(\sigma);$$
$$msg \leftarrow ADV_0(\sigma, PK, s_0, s_1)$$
$$: Receiver(SK, PK, \sigma, msg) \neq \text{STOP}$$
$$\wedge\, ADV_1(PK, \sigma, s_0, s_1, msg) = Receiver(SK, PK, \sigma, msg)\,) \geq 1/2 + n^{-c}$$

contradicts the QRA. We shall in fact exhibit an algorithm $Q(\cdot, \cdot)$ that decides quadratic residuosity that uses ADV_0 and ADV_1 as subroutines.

Algorithm $Q(x, z)$.

Input: $x \in BL(n)$, $z \in Z_x^{+1}$

1. *Construct Public key*

   Set σ and *Val*=empty string.

   Randomly select $r \in Z_x^*$ and set $y = -r^2 \bmod x$.

   For $i = 1, \ldots, n^2$

       Randomly select an n-bit integer s_i.

       If $s_i \notin Z_x^{+1}$ append s_i to σ.

       else

           Toss a fair coin

           If HEAD then append $y s_i^2 \bmod x$ to σ and $y s_i$ to *Val*.

           If TAIL then append $s_i^2 \bmod x$ to σ and s_i to *Val*.

2. Set $PK = (x, y, z, Val)$ and select two strings s_0, s_1.

3. Set $(\alpha, \beta) = ADV_0(\sigma, PK, s_0, s_1)$

4. If $ADV_1(PK, \sigma, s_0, s_1, (\alpha, \beta)) = s_0$ then **Output(1)** else **Output(0)**.

Let us now compute

$$Pr(x \leftarrow BL(n); z \leftarrow Z_x^{+1} : Q(x,z) = Q_x(z)) = 1/2 \,(Pr(x \leftarrow BL(n); z \leftarrow QNR_x : Q(x,z) = 1) + Pr(x \leftarrow BL(n); z \leftarrow QR_x : Q(x,z) = 0))$$

where QR_x and QNR_x are the classes of quadratic residues and quadratic non residues modulo x, respectively.

Now, we observe that if $z \in QRN_x$ then certainly $Receiver(SK, PK, \sigma, msg) = s_0$, as s_0 has been encrypted using z. Therefore

$$Pr(x \leftarrow BL(n); z \leftarrow QNR_x^{+1} : Q(x,z) = 1) \geq 1/2 + n^{-c}$$

The same reasoning can be done for the case in which $z \in QR_x$, thus yielding

$$Pr(x \leftarrow BL(n); z \leftarrow Z_x^{+1} : Q(x,z) = Q_x(z)) \geq 1/2 + n^{-c}$$

which contradicts the QRA.

QED

Dependent Oblivious Transfers.

The same public key, $PK_B = (y_B, z_B)$, can be used to perform many OT's. However, these OT's will not be independent as the outcome of one of them determines the outcome of all of them. This is useful when we want that only one of two strings is received and not even one single bit of the other is leaked. If j independent OT's are desired, it is enough to have j different $z_B \in Z^{+1}_{x_B}$ in B's public key. The problem of obtaining j independent OT's using the same public key (whose size does not depend on j) has its own interest and is currently open. In the final paper we give a partial solution to it based on k-wise independence.

1-out-k Oblivious Transfer.
An immediate generalization of the 1-out-2 OT is the 1-out-k OT. This is a protocol where A transfer to B exactly one of k strings $s_0, s_1, ..., s_{k-1}$, in such a way that (1) B does not get any information on the other $k-1$ strings, and (2) A does not know which string B got. Our protocol can be easily modified to implement a 1-out-k OT. User B randomly chooses a Blum integer x_B, and $z^0_B, ..., z^{k-1}_B \in Z^{+1}_{x_B}$, such that exactly one z^i_B is a quadratic non residue. Then, B computes $PK_B = (x_B, z^0_B, ..., z^{k-1}_B, Val_B)$, $SK_B = $ (prime factors of x_B), where $Val_B = $ (Non-Interactive Zero-Knowledge proof, in the sense of [BFM], that PK_B has been correctly computed). To oblivious transfer one of $s_0, s_1, ..., s_{k-1}$, user A computes and sends B a random permutation π of them, computes and sends B the encryptions $E(x_B, z^i_B, s_{\pi(i)})$, $i = 0, ..., k-1$. Notice that if A does not permute the s_i this is a protocol for 1-out-k disclosure in which the receiver "secretly" chooses 1 out of k secrets (defined in [BCR]).

5 Non-Interactive Perfect Zero-Knowledge Arguments

Bit-Commitment in the Public-Key Public-Randomness Model.
A bit commitment protocol is a fundamental 2-party cryptographic protocol. It allows one party A to hide (commit) one bit b from the other party B and, later, to show (decommit) it to B. Even though B does not know which bit A has committed to, he is guaranteed that the bit decommitted is the bit A originally committed to. If B is poly-time, the bit commitment can be implemented as: A chooses a secure encryption scheme (in the sense of [GM]) and commits to a bit by encrypting it. B, being poly-time, cannot decrypt and compute b. To decommit b, A simply shows the random bit used for the encryption.
Suppose that A is polynomial time and B has infinite computing power. We use the unconditionally secure blobs of [BC] for the commitment of A to B. B publishes in his public file a random integer product of two primes x_B, s_B a random quadratic residue modulo x_B, and a NIZK proof of correctness of the publication. To commit to a bit b, A performs the procedure: (1) randomly select $r \in Z^*_{x_B}$; (2) compute and send B $w = s^b_B r^2 \bmod x_B$. If A wants to decommit the bit b committed by w, he reveals r, that must be the square of either w or $w/s_B \bmod x_B$. It is clear that B, by just looking at w, cannot figure out whether A knows a root of w or $w/s_B \bmod x_B$. On the other hand, if A knows a square root of both w and $w/s_B \bmod x_B$, then he can compute a square root of s_B as well. This implies that A can extract square roots of random squares and thus he is able to factor random composite numbers, in

contradiction with the Quadratic Residuosity Assumption. Notice that the same pair (x_B, s_B) can be used to perform any number of bit commitments to B. Later, we will use this bit commitment scheme to obtain Non-Interactive Perfect Zero-Knowledge Arguments for all NP languages.

Zero-Knowledge Proofs and Arguments.

Goldwasser, Micali, and Rackoff [GMR] introduced the notion of Interactive Zero-Knowledge Proof System. This allows an infinitely powerful (but not trusted) prover to convince a probabilistic polynomial time verifier that a certain theorem is true without revealing any other information. They also distinguished between *Computational Zero-Knowledge* and *Perfect Zero-Knowledge*. In a Computational Zero-Knowledge Proof all the information about the theorem and its proof is given, but it would take more than polynomial time for the verifier to extract it. On the other hand, a Perfect Zero-Knowledge Proof conveys no information at all and even an all powerful verifier could not extract any information from it. Perfect Zero-Knowledgeness is thus a desirable property in a proof as a prover can never be sure of the computational power of the verifier he is talking to. Unfortunately, although all NP languages have Computational Zero-Knowledge proofs [GMW1], Perfect Zero-Knowledge for an NP-complete language would cause the Polynomial Time Hierarchy to collapse [F, BHZ].

The concept of *Interactive Zero-Knowledge Argument* has been introduced by Brassard and Crépeau [BC] and Chaum [Ch]. Here, we have a probabilistic polynomial time prover that wants to convince a (possibly unlimited computing powerful) verifier that a certain theorem is true in Zero-Knowledge. (The term "Argument" is used instead of "Proof" as an all powerful prover could cheat the verifier.) As the verifier may have infinite computational power, in this setting all Zero-Knowledge Arguments must indeed be Perfect Zero-Knowledge. Surprisingly, Brassard and Crépeau [BC] were able to show that, in this scenario, all NP languages have Perfect Zero-Knowledge Arguments.

Although, different non-interactive scenarios in which Zero-Knowledge Proofs are possible have been proposed, all implementations of Perfect Zero-Knowledge Arguments require interaction. A protocol requiring only constant number of rounds is due to Brassard, Crépeau, and Yung [BCY].

We show that *Non-Interactive Zero-Knowledge Arguments* for all NP languages are possible in the PKPR setting, under the Quadratic Residuosity Assumption. Moreover, our implementation supports any polynomial (in the length of the reference string σ) number of distinct provers. Unlike [BFM], [DMP1], [BDMP], our proofs are directed to a single user, and thus are not publicly verifiable.

To prove that all NP languages have Perfect Zero-Knowledge Arguments, it is enough to prove it for an NP-complete language. We choose Hamiltonian graphs as NP-complete language. The ingredients for our construction are Blum's interactive protocol [B3], an efficient technique of [KMO], our implementations of OT and bit commitment schemes. More efficient protocols can be obtained by suitably using the interactive protocol of [IY] that simulates directly any given computation, or envelope-based protocols as in [KMO].

Non-Interactive Perfect Zero-Knowledge Arguments for Hamiltonian Graphs.
We show how to give a Perfect Zero-Knowledge Argument that a graph is Hamiltonian, in the Public-Key Public-Randomness Model. This in turn yields Perfect Zero-Knowledge Argument for all NP languages. We first describe the content of the public file and then show how the proof is actually performed.

The public file. V randomly chooses two primes p_V and q_V of the same length, computes $x_V = p_V q_V$, then V chooses s_V a random quadratic residue modulo x_V, y_V, a random quadratic non residue modulo x_V, and a random $z_V^1, ..., z_V^n \in Z_{x_V}^{+1}$. V publishes the public key $PK_V = (x_V, s_V, y_V, z_V^1, ..., z_V^n, Val_V)$, where $Val_V = $ (NIZK proof that PK_V has been correctly computed), and keeps secret the private key $SK_V = (p_V, q_V)$. The NIZK proof is computed on input the reference string and certifies that x_V is a Blum integer and s_V (y_V) is a quadratic residue (non residue) modulo x_V.

Proving that G is Hamiltonian. Suppose that the prover P wants to show that the graph G has an Hamiltonian path to a (possibly unlimited computing powerful) verifier V. He performs the following program.

For $i = 1, ..., n$ do

1. Randomly select a permutation π_i of the vertices of G and compute $\hat{G}_i = \pi_i(G)$, an isomorphic copy of G.

2. Using s_V, commit bit by bit the adjacency matrix of $\hat{G}_i$ and the permutation π_i.

3. Let α_i be the concatenation of the decommitment keys of the adjacency matrix of $\hat{G}_i$ and the permutation π_i and β_i the concatenation of the decommitment keys of the entry of the adjacency matrix of $\hat{G}_i$ that corresponds to edges in the Hamiltonian path.
   Oblivious transfer the pair (α_i, β_i) using y_V^i and z_V^i.

We now give an informal proof that the above protocol is a Perfect Zero-Knowledge Argument. A formal proof will be given in the final version. For each i, the verifier V gets either the adjacency matrix of $\hat{G}_i$ and the permutation π_i (and thus can check that $\hat{G}_i$ is indeed an isomorphic copy of G) or an hamiltonian path in $\hat{G}_i$ (and thus can check that $\hat{G}_i$ is Hamiltonian). By the properties of the Oblivious Transfer, P does not know which one V is going to receive. Therefore, if G is not Hamiltonian, either $\hat{G}$ is not isomorphic to G (in which case P is caught with probability $1/2$) or $\hat{G}$ is not Hamiltonian (again P is caught with probability $1/2$). Thus, under the QRA, there is only an exponentially low probability for a poly-time prover to convince a verifier that a non-hamiltonian graph is hamiltonian (soundness requirement of a proof system). The Perfect Zero-Knowledgeness of the protocol follows from the properties of the commitment scheme and of the NIZK proof contained in Val_V.

6 Related works

Non-Interactive Oblivious Transfer. Recently, Ostrowsky and Yung [OY] proved that it is not possible to achieve a non-interactive OT from scratch. That is, two parties that have never met beforehand need at least 2 rounds to perform an OT. It is then natural to ask: What additional resources are required for a "non-interactive OT" to be possible? We prove that a PKPR setting is enough to achieve non-interactive OT. Adopting a different perspective, we prove that 2 rounds are enough for an OT from A to B, if a random string is available. The rounds are (1) B selects and sends A his public key, and (2) A performs a non-interactive OT using B's public key. Thus, our implementation is essentially optimal in view of [OY].

Non-interactive Oblivious Transfer with a Center. Bellare and Micali [BM] introduced the problem of achieving non-interactive OT in a Public-key scenario where a trusted center is on duty and proposed an implementation based on a complexity assumption related to the Discrete Logarithm problem (the Diffie-Hellman assumption [DH]). In their model, the trusted center publishes a central public key and each user that wants to put his key on the public file has to validate it by interacting with the center. A center may be replaced by a collective distributed fault-tolerant computation if the majority of users is "honest". We show that a short random string can replace the trusted center (or a distributed fault-tolerant computation). Therefore, in our model, there is no need for interaction to validate the public file: each user can prove the correctness of the choice of his own public key by himself.

Non-interactive Zero-Knowledge. Blum, Feldman, and Micali [BFM] (with improvements in [DMP1] and [BDMP]) were the first to propose a non-interactive scenario in which Zero-Knowledge (in the sense of [GMR]) proofs of membership for all NP-languages were possible. A random short reference string is required for the proof-system. Their implementations can only support a limited number of provers. Recently, De Santis and Yung [DY] and Feige, Lapidot, and Shamir [FLS] proved that many provers can share the same random string. All these implementations are quite involved. Other models in which non-interactive (or bounded-interaction) Zero-Knowledge is achievable have been proposed. [DMP2] proved that after a preprocessing stage, it is possible to give a NIZK proof of any short NP-theorem. This proof-system is based on a very general assumption (existence of one-way functions) but has limited applicability. [K] showed how to achieve Zero-Knowledge proofs using only OT's. [KMO] gave protocols more efficient than [K], and furthermore showed that it is possible to move all the needed OT's in a short pre-processing stage. All these protocols are restricted to two parties.

Non-interactive Zero-Knowledge with a Center. Bellare and Micali [BM] showed how to implement NIZK proofs in a Public-key cryptosystem in which a center is available. In their implementation, the center has to compute and publish a central public key (which is common to all users). Then, each user chooses by himself its own public key, and must validate it by proving the correctness of its computation by shortly interacting with the center. Their implementation is based on a complexity assumption related to the Discrete Logarithm problem (the Diffie-Hellman assumption [DH]), whereas our implementation is based on the Quadratic Residuosity Assumption.

The Public-key Public-Randomness model shows how to dispose of an active center by having just a short random string known beforehand to all users. To validate his choice of the public key, each user computes by himself and publishes a proof of correctness in his public file.

<u>Perfect Zero-Knowledge Arguments.</u> Brassard and Crépeau [BC] and Chaum [Ch] consider a model where the prover is poly-bounded and the verifier has unlimited computational power. All NP-languages have Perfect Zero-Knowledge Arguments (that is Zero-Knowledge proofs in the BCC model). Actually Brassard, Crépeau, and Yung [BCY] proved that there are perfect Zero-Knowledge arguments that use a constant number of rounds (only 6) for any NP-language. In this paper, we prove that all NP-languages have Non-Interactive Perfect Zero-Knowledge Arguments in the PKPR setting. Our is the first implementation of non-interactive nature for Perfect Zero-Knowledge Arguments.

Note added in proof

Bert van Boer [Bo] has independently proved that it is possible to have a non-interactive Oblivious Transfer after an interactive preprocessing stage. His protocol relies on the Quadratic Residuosity Assumption. Using the techniques we discussed in this paper, his implementation can be adapted to work also in the PKPR setting.

7   Acknowledgments

We are very grateful to Moti Yung for encouragement and helpful discussions.

References

[B]   D. Beaver, *Secure Multiparty Protocols Tolerating Half Faulty Processors*, CRYPTO 1989.

[Bl1]   M. Blum, *Coin Flipping by Telephone*, IEEE COMPCON 1982, pp. 133–137.

[Bl2]   M. Blum, *Three Applications of the Oblivious Transfer*, Unpublished manuscript.

[Bl3]   M. Blum, *How to Prove a Theorem So No One Else Can Claim It*, Proceedings of the International Congress of Mathematicians, Berkeley, California, 1986, pp. 1444–1451.

[BC]   G. Brassard and C. Crépeau, *Non-transitive Transfer of Confidence: A Perfect Zero-Knowledge Interactive Protocol for SAT and Beyond*, Proceedings of the 27th IEEE Symp. on Foundation of Computer Science, 1986, pp. 188–195.

[BCC]   G. Brassard, C. Crépeau, and D. Chaum, *Minimum Disclosure Proofs of Knowledge*, Journal of Computer and System Sciences, vol. 37, no. 2, October 1988, pp. 156–189.

[BCR]   G. Brassard, C. Crépeau, and J.-M. Robert, *Information Theoretic Reductions among Disclosure Problems*, Proceedings of the 27th IEEE Symp. on Foundation of Computer Science, 1986, pp. 168–173.

[BCY]   G. Brassard, C. Crépeau, and M. Yung, *Everything in NP can be Proven in Perfect Zero-Knowledge in a Bounded Number of Rounds*, ICALP 89.

[BDMP] M. Blum, A. De Santis, S. Micali, and G. Persiano, *Non-Interactive Zero Knowledge*, MIT Research Report MIT/LCS/TM-430, May 1990.

[BFM]   M. Blum, P. Feldman, and S. Micali, *Non-Interactive Zero-Knowledge Proof Systems and Applications*, Proceedings of the 20th Annual ACM Symposium on Theory of Computing, Chicago, Illinois, 1988.

[BGW]   M. Ben-Or, S. Goldwasser, and A. Wigderson, *Completeness Theorems for Non-Cryptographic Fault-Tolerant Distributed Computations*, Proceedings of the 20th Annual ACM Symposium on Theory of Computing, 1988, pp. 1–10.

[BHZ]   R. Boppana, J. Hastad, and S. Zachos, *Does co-NP have Short Interactive Proofs?*, Information Processing Letters, vol. 25, May 1987, pp. 127–132.

[BM]   M. Bellare and S. Micali, *Non-interactive Oblivious Transfer and Applications*, CRYPTO 1989.

[Bo]   B. van Boer, *Oblivious Transfer Protecting Secrecy*, Eurocrypt 90.

[Ch]   D. Chaum, *Demonstrating that a Public Predicate can be Satisfied Without Revealing any Information About How*, in "Advances in Cryptology – CRYPTO 86", vol. 263 of "Lecture Notes in Computer Science", Springer Verlag, pp. 195–199.

[Cr]   C. Crépeau, *Equivalence Between Two Flavors of Oblivious Transfer*, in "Advances in Cryptology – CRYPTO 87", vol. 293 of "Lecture Notes in Computer Science", Springer Verlag, pp. 350–354.

[CCD]   D. Chaum, C. Crépeau, and I. Damgård, *Multiparty Unconditionally Secure Protocols*, Proceedings of the 20th Annual ACM Symposium on Theory of Computing, Chicago, Illinois, 1988, pp. 11–19.

[DH]   W. Diffie and M. E. Hellman, *New Directions in Cryptography*, IEEE Transactions on Information Theory, vol. IT-22, no. 6, Nov. 1976, pp. 644–654.

[DMP1] A. De Santis, S. Micali, and G. Persiano, *Non-Interactive Zero-Knowledge Proof Systems*, in "Advances in Cryptology – CRYPTO 87", vol. 293 of "Lecture Notes in Computer Science", Springer Verlag, pp. 52–72.

[DMP2] A. De Santis, S. Micali, and G. Persiano, *Non-Interactive Zero-Knowledge Proof-Systems with Preprocessing*, in "Advances in Cryptology - CRYPTO 88", Ed. S. Goldwasser, vol. 403 of "Lecture Notes in Computer Science", Springer-Verlag, pp. 269–282.

[DMP3] A. De Santis, S. Micali, and G. Persiano, *Removing Interaction from Zero-Knowledge Proofs,* in "Advanced International Workshop on Sequences", Positano, Italy, June 1988, Ed. R. M. Capocelli, Springer-Verlag, pp. 377–393.

[DY] A. De Santis and M. Yung, *Cryptographic Applications of Metaproofs,* CRYPTO 90.

[EGL] S. Even, O. Goldreich, and A. Lempel, *A Randomized Protocol for Signing Contracts,* CACM, vol. 28, 1985, pp. 637–647.

[F] L. Fortnow, *The Complexity of Perfect Zero-Knowledge,* Proceedings 19th Annual ACM Symposium on Theory of Computing, New York, 1987, pp. 204–209.

[FLS] U. Feige, D. Lapidot, and A. Shamir, *Multiple Non-Interactive Zero-Knowledge Proofs Based on a Single Random String,* FOCS 90.

[GHY] Z. Galil, S. Haber, and M. Yung, *Cryptographic Computation: Secure Fault-Tolerant Protocols and the Public-Key Model,* in "Advances in Cryptology – CRYPTO 87", vol. 293 of "Lecture Notes in Computer Science", Springer Verlag, pp. 135–155.

[GM] S. Goldwasser and S. Micali, *Probabilistic Encryption,* Journal of Computer and System Science, vol. 28, n. 2, 1984, pp. 270–299.

[GMR] S. Goldwasser, S. Micali, and C. Rackoff, *The Knowledge Complexity of Interactive Proof-Systems,* SIAM Journal on Computing, vol. 18, n. 1, February 1989.

[GMW1] O. Goldreich, S. Micali, and A. Wigderson, *Proofs that Yield Nothing but their Validity and a Methodology of Cryptographic Design,* Proceedings of 27th Annual Symposium on Foundations of Computer Science, 1986, pp. 174–187.

[GMW2] O. Goldreich, S. Micali, and A. Wigderson, *How to Play Any Mental Game,* Proceedings of the 19th Annual ACM Symposium on Theory of Computing, New York, 1987, pp. 218–229.

[HR] J. Halpern and M. O. Rabin, *A Logic to Reason about Likelihood,* Proceedings of the 15th Annual Symposium on the Theory of Computing, 1983, pp. 310–319.

[IY] R. Impagliazzo and M. Yung, *Direct Minimum Knowledge Computations,* in "Advances in Cryptology – CRYPTO 87", vol. 293 of "Lecture Notes in Computer Science", Springer Verlag pp. 40–51.

[K] J. Kilian, *Founding Cryptography on Oblivious Transfer,* Proceedings 20th Annual ACM Symposium on Theory of Computing, Chicago, Illinois, 1988, pp. 20–31.

[KMO] J. Kilian, S. Micali, and R. Ostrowsky, *Minimum-Resource Zero-Knowledge Proofs,* Proceedings of the 30th IEEE Symposium on Foundation of Computer Science, 1989.

[OY] R. Ostrowsky and M. Yung, *On Necessary Conditions for Secure Distributed Computation,* preprint 1989.

[RB]   T. Rabin and M. Ben-Or, *Verifiable Secret Sharing and Multiparty Protocols with Honest Majority*, Proceedings of the 21st Annual ACM Symposium on Theory of Computing, Seattle, Washington, 1989, pp. 73–85.

An Interactive Identification Scheme Based on Discrete Logarithms and Factoring*

(Extended Abstract)

Ernest F. Brickell   Kevin S. McCurley

Sandia National Laboratories
Albuquerque, NM 87185

Abstract. *We describe a modification of an interactive identification scheme of Schnorr intended for use by smart cards. Schnorr's original scheme had its security based on the difficulty of computing discrete logarithms. The modification that we present here will remain secure if either of two computational problems is infeasible, namely factoring a large integer and computing a discrete logarithm. For this enhanced security we require somewhat more communication and computational power, but the requirements remain quite modest, so that the scheme is well suited for use in smart cards.*

1 Introduction

In this note we describe an interactive identification scheme that is a variation of a scheme presented by Schnorr at Crypto '89 [9]. Schnorr's scheme has several features that make it advantageous for use in smart cards or other environments with limited computing power. Its security is based on the difficulty of the discrete logarithm problem in a subgroup of $\mathbb{Z}_p^*$. In this paper we shall describe a variation with the property that a successful attack on the scheme requires the ability to solve an instance of the discrete logarithm problem, and in addition to factor an integer that is divisible by two large primes.

Due to the current state of complexity theory, cryptographic schemes whose security is based on the difficulty of solving a specific computational problem are exposed to the danger that a fast algorithm may be found for the underlying computational problem. It therefore seems desirable to design systems with the property

---

*This work was performed under U. S. Department of Energy contract number DE-AC04-76DP00789

that breaking them requires the ability to solve two apparently dissimilar computational problems, both of which appear to be hard. An example of such a scheme was given in [7], where a key distribution scheme with this property was given. The key distribution scheme of [7] uses arithmetic modulo a number n that is a product of two primes. Breaking the system requires the factorization of n and the ability to solve the Diffie-Hellman problem modulo the prime factors of n. In the present paper we take a slightly different tack, by using arithmetic modulo a prime p. We choose p with the property that $p-1$ has at least two large prime factors, so that the factorization of $p-1$ is hard to recover. We then construct the system in such a way that breaking it requires both computing a discrete logarithm in a subgroup of $\mathbb{Z}_p^*$, and factoring $p-1$.

The extra security gained in this scheme extracts a penalty both in the computation time and the communication time, but the scheme still carries the advantage of allowing preprocessing of most of the computation, and should still be quite feasible for use in smart cards. The relative merits of the schemes will be discussed later, after we first present the schemes in detail.

2   Schnorr's Identification Scheme

We begin by describing the original Schnorr authentication scheme in terms a security parameter t. In this scheme, each person who wishes to use the scheme to prove his identity will visit a key authentication center (KAC) and register his or her public key. When the KAC is originally set up, it chooses

- primes p and q such that $q \mid p-1$, $q \geq 2^{140}$, and $p \geq 2^{512}$,

- α of order q in the group $\mathbb{Z}_p^*$,

- its own private and public keys.

The KAC publishes p, q, α, and its public key. When a user comes to the KAC for registration, the user chooses a secret $s \in \{1, \ldots, q\}$, computes $v \equiv \alpha^{-s} \pmod{p}$, and submits v to the KAC along with some form of identification. The KAC verifies the user's identity, generates an identification string I, and also generates a signature S of the pair (I, v). The KAC can use any secure digital signature scheme whatsoever for generating this signature.

We now describe the procedure by which party P (the prover) can prove its identity to V(the verifier). In a preprocessing phase, P should first have chosen a random number $r \in \{1, \ldots, q\}$ and computed $x \equiv \alpha^r \pmod{p}$. In the identification procedure, P first sends to V its identification string I, its public key v, the KAC's signature S of (I, v), and x. V then checks P's identification by verifying the signature S, chooses a random $e \in \{0, \ldots, 2^t - 1\}$, and transmits e to P. P sends to V the value $y := r + se \pmod{q}$. Finally, V checks that $x \equiv \alpha^y v^e \pmod{p}$ and accepts P's proof of identity if this holds.

Schnorr suggests using $t = 72$, although this can be reduced substantially for use in the identification scheme (Schnorr also proposed a companion signature scheme which requires the larger t). The parameter t is used to control the probability that an impostor will be able to guess a correct response to a challenge e. For use in an identification scheme, we need only choose t so large that the probability 2^{-t} of guessing the challenge e is negligible.

This scheme has a number of novel features. First of all, much of the arithmetic to be done by the prover can be done in a preprocessing phase, using idle time of the processor. This is well suited to the case of a smart card, where the processing power is relatively small. Second, the number of bits that must be communicated is considerably reduced over other schemes such as RSA or Fiat-Shamir. There is also a signature scheme based on the same choice of keys, but we shall not discuss it here.

Schnorr's scheme may be regarded as a practical refinement of the zero-knowledge protocols of Chaum et.al. [3], [2] for demonstrating possession of a discrete logarithm. In [3], the challenge e was either a zero or a one, and the basic protocol was repeated several times (requiring the prover to perform multiple exponentiations). Yet another interesting identification scheme based on discrete logarithms was proposed by Beth [1]. The security of the latter scheme is however more closely related to the ElGamal signature scheme.

3   The Modified Scheme

In this section we shall describe the modification of Schnorr's scheme. In the modified scheme, each user will have his own prime p and base element α, and these will need to be transmitted along with v during each identification session. Once again the KAC serves only to sign the public keys of each user, but now these include p and α. Rather than the single security parameter t, we describe the scheme in terms of the parameters k, t, and u.

When a user wishes to join the system, he chooses primes q and w with $q < w$, $2^{k-1} < q < 2^k$, and $qw > 2^{512}$. The user further chooses a prime $p \equiv 1 \pmod{qw}$, an element $\alpha \in \mathbb{Z}_p^*$ of order q, and a random number $s \in \{1, \ldots, q\}$. The user then computes $v \equiv \alpha^{-s} \pmod{p}$, and presents p, v, and α to the KAC along with some form of identification, but keeps q, w, and s secret. The KAC verifies the user's identity, generates an identification string I, and produces a signature S of the quadruple (I, v, p, α), which it provides to the user. Once again the KAC can use any digital signature scheme whatsoever.

In the identification procedure, P once again has a preprocessing phase, where P chooses a random number $r \in \{1, \ldots, q\}$ and computes $x \equiv \alpha^r \pmod{p}$. Then P sends to V the identification string I, its public keys v, p, and α, the KAC's signature S, and x. V checks P's identification by verifying the signature S of (I, v, p, α). If the keys are authentic, then V chooses a random $e \in \{0, \ldots, 2^t - 1\}$ and a random

$f \in \{0, \ldots, 2^u - 1\}$, and transmits the pair (e, f) to P. P then computes an integer y such that $y \equiv r + se \pmod{q}$ and $2^k f \leq y < 2^k(f+1)$, and sends y to V. V checks that $x \equiv \alpha^y v^e \pmod{p}$ and $2^k f \leq y < 2^k(f+1)$, and accepts P's proof of identity if these conditions are satisfied.

The parameters u and t can be adjusted to suit specific needs, but we suggest using $u = t = 20$. With this choice, there are 2^{40} possible challenges (e, f), and the probability of guessing the challenge ahead of time is therefore 2^{-40}. If an impostor somehow discovers the secret prime q, then a precomputed pair y, x that satisfies $\alpha^y v^e \equiv x \pmod{p}$ can always have the y adjusted to fit any challenge f, but the probability of guessing the e ahead of time is still only 2^{-20}. Similarly if an impostor knows a discrete logarithm of v to the base α, then the probability of success in guessing ahead of time is also 2^{-20}. We regard this as being acceptably low for use in an identification scheme.

Some care should be exercised in choosing the primes q and w, and in particular we should try to choose them in such a way as to thwart any known algorithms for factoring qw. The choice of $k > 140$ is probably marginal in avoiding a determined implementation of the elliptic curve method of H. W. Lenstra, Jr., but may suffice for applications of a commercial nature. At present the record for the largest factor found by the elliptic curve method has 38 decimal digits, or about 127 binary digits (this factor was found by Robert Silverman). On the other hand, choosing $k > 200$ will probably be safe against any conceivable implementation. The construction of p should be relatively easy, since heuristic evidence (see [10]) suggests that we should expect a prime $p \equiv 1 \pmod{qw}$ can be found with $p \leq qw \log^2(qw)$.

The recent results of Lenstra and Manasse [6] and Lenstra et. al. [5] have raised a question about how long a 512 bit modulus will remain safe from attack by current factorization methods. We suspect however that by the time anyone will have at their disposal enough computational power to factor a 512 bit modulus, the smart card technology will probably have advanced enough to allow easy use of a 1024 bit modulus. Moreover, the best known attack for breaking the scheme we present here requires in addition the computation of a discrete logarithm modulo a 512 bit prime, and current algorithms will probably have a much more difficult time with this problem.

4   Performance Analysis of the Modified Scheme

It is evident that the modified scheme suffers from a disadvantage in the number of bits that must be communicated. The following tables show the number of bits to be communicated in the two schemes, using the security parameters mentioned above. For the sake of comparison, we have assumed that 100 bits suffice for each of I and S. We have used a value of $k = 140$ in the original and $k = 200$ in the modified scheme.

	Original Scheme		Modified Scheme
		I	100
		v	512
I	100	p	512
v	512	α	512
S	100	S	100
x	512	x	512
e	40	(e, f)	40
y	140	y	220
total	1404	total	2508

The modified scheme therefore pays a penalty of an extra 1104 bits in communication, and possibly more if error correction is included. On the other hand, this is still well within the realm of possibility using present technology.

We now compare the computational requirements of the two schemes. In both the original Schnorr scheme and the modified scheme, numerous refinements can be devised to improve peformance. No matter what we do, however, the amount of arithmetic required in the new scheme appears to impose a slight penalty on speed. Part of the penalty comes from the fact that the prime q is larger for the modified scheme. Both schemes can use a 512 bit modular exponentiation with an exponent r of at most 140 bits in the preprocessing stage.

In the original Schnorr scheme, the prover is required to compute $y \equiv r + se$ (mod q), and the most obvious way to do this requires a multiplication, an addition, and a division by q. In the modified scheme, we require in addition a multiplication by q and an extra addition.

This does not however take into account any optimization. We now discuss a method for speeding the computations in both the original Schnorr scheme and the modified scheme. The idea here is to replace the divisions by q with multiplications (using shorter integers). This can be done by precomputing (only once, when the initial keys are selected) an approximation Q of s/q. If $0 < s/q - Q < 2^{-t-1}$, then

$$r - q < r + se - q[[Qe]] < r + q,$$

where $[[x]]$ denotes the nearest integer to x. Hence after computing $r + se - q[[Qe]]$, at most one subtraction or addition of q will be required to reduce $r + se$ modulo q. The overall improvement from performing the precomputation is to replace the division by q with a multiplication of Q and e (both of which are only t bits) followed by multiplication of q and a t bit integer, followed by at most two subtractions or additions involving k bit integers. Depending on the implementation, this may result in a significant speedup by eliminating the multiple precision division.

In the modified scheme, we can employ a similar approach. For the modified scheme we need to compute y so that $y \equiv r + se$ (mod q) and $2^k f \leq y < 2^k (f + 1)$. To do this, we precompute two sufficiently good approximations Q_1 and Q_2 of s/q

and $2^k/q$ respectively. We then compute $r + se + q[[Q_2 f - Q_1 e]]$, and if necessary adjust the result with at most one addition or subtraction of q.

Using these division-free algorithms for the computations, the only extra work required in the modified scheme is for an additional multiplication and subtraction, on numbers of approximately t bits. This should have a negligible effect on the overall computation speed. As we shall see in the next section, this slight degradation in performance brings in return the promise of an extra measure of security that cannot be achieved by simply increasing the key size.

We close this section with a final comment on the original Schnorr scheme. In that scheme, y is reduced modulo q before transmission. At first sight it may appear advantageous to remove the reduction of y modulo q in the original Schnorr scheme and thus gain a significant computational advantage in the on-line portion of the computation. In fact, this would be disastrous because if we know $r + se$ and e, then we can construct an interval of length approximately q/e containing s. An algorithm of Pollard [8] can then be used to compute s in only about $\sqrt{q/e}$ operations. For the parameters suggested by Schnorr, the expected value of this is only 2^{35}.

5 Security of the Modified Scheme

Like all cryptographic schemes, identification schemes can be attacked in a variety of ways. The purpose of introducing *interaction* to identification schemes is to protect against passive eavesdroppers recovering secret information that they can later use to impersonate the legitimate user. In this section, we will give evidence which indicates that our scheme does provide such protection. However, there are other kinds of attacks that might arise in applications that are not protected against by using an interactive identification scheme by itself.

In particular, Desmedt et.al. [4] have pointed out that an interactive identification scheme offers no protection against the situation in which the verifier cheats by passing on information provided to him by the prover to another cheating prover who (falsely) proves his identity at another location.

Furthermore, an interactive identification scheme does not offer any protection against a prover who gives away his secret information to another so that they may impersonate him, or against a prover who chooses weak secret keys that anyone can guess. A variant of this point was discussed by Burmester in the rump session at Eurocrypt '90.

Both of these attacks can be protected against if the system uses physical characteristic information to uniquely identify an individual. If the identification by physical characteristics offers perfect security, then there is no security gained by using an interactive identification scheme instead of simply using a digital signature (issued by the KAC) of the physical characteristics. However, if the identification by physical characteristics offers less that perfect security, then using an interactive identification scheme can in some cases result in increased total security of the system. For example, if two people share the same physical characteristics, then a digital signature of these characteristics could be transferred by a cheating verifier between these two people. With the use of interaction this will be impossible without the cooperation of the legitimate prover.

In the remainder of this section, we will consider only the security provided by the system against a passive eavesdropper. There are several basic attacks that can be mounted by a passive eavesdropper against identification schemes. For example, in the original Schnorr scheme, one kind of attack would be to try to construct a pair (I, α^{-s}) and a legitimate signature S of this pair for later use in identification. This would however require a successful attack on the signature scheme of the KAC. Another attack would involve observing a user identify himself several times, collecting a set of the tuples (x, e, f, y). It can be shown that a reasonable number of such tuples cannot provide any useful information, since the attacker could himself construct such tuples from a distribution that is very close to the legitimate user's distribution by first choosing y, then f, then e, and then x.

A more serious attack would involve observing a user going through the identification process, and for the pair (I, v) that is observed, try to later produce an x for which there is a reasonable chance of being able to answer the challenge by finding a suitable y. Schnorr proved that an attack of this kind for the original scheme would require the ability to compute the discrete logarithm of v. In the same spirit, we shall prove in Theorem 1 that an attack of this kind on the modified scheme would require the ability to factor $p - 1$ *and* the ability to find the discrete logarithm of v.

We should be careful to observe that an attack on the system has not been proved to be completely equivalent to the problem of simply factoring $p - 1$. While a successful attack requires the ability to factor $p - 1$, a cryptanalyst will be in possession of some side information. The most obvious information available is the knowledge of an element α whose order is the unknown factor q of $p - 1$. Whether this information can be used to discover the factor q is unknown.

Theorem 1. *Let p and α be as described in Section 3. Let $A = A_{p,\alpha,v,x}$ be an algorithm with running time bounded by T that receives an input (e, f), and attempts to compute an integer y such that $\alpha^y v^e \equiv x \pmod{p}$. If A will produce a correct output for at least $\epsilon 2^{u+t}$ of the possible challenge pairs (e, f) (where $\epsilon \geq \max(2^{1-t}, 2^{1-u})$), then there exists a probabilistic algorithm that with at least a constant probability, will compute the prime factor q of $p-1$ and a discrete logarithm of v in $O(\log^3 p + \frac{T}{\epsilon})$ bit operations.*

Proof. We first describe an algorithm for computing a discrete logarithm of v. The idea is to construct correct triples (e_1, f_1, y_1) and (e_2, f_2, y_2) with $e_1 \neq e_2$. We first choose random pairs (e_1, f_1) until one is found for which A gives a correct output y_1. We then choose random pairs (e_2, f_2) with $e_2 \neq e_1$ until we find one for which A gives a correct output y_2. We now have $\alpha^{y_1-y_2} \equiv v^{e_2-e_1} \pmod{p}$. We use the Euclidean algorithm to compute $d = \gcd(e_2 - e_1, p - 1)$. Assume first that

$d = 1$. Then the extended Euclidean algorithm gives an integer ℓ with $(e_2 - e_1)\ell \equiv 1$ (mod $p - 1$), so that $\alpha^{(y_1 - y_2)\ell} \equiv v$ (mod p). Hence $(y_1 - y_2)\ell$ is a discrete logarithm of v to the base α.

Suppose now that we found $d > 1$. In this case we let $d_1 = d$, $m_1 = p - 1$, and for $i = 2, \ldots$, we compute $m_i = m_{i-1}/d_{i-1}$ and $d_i = \gcd(e_2 - e_1, m_i)$. Since $|e_2 - e_1| < q < w$, we will eventually arrive at $d_i = 1$ and $q \mid m_i$. Applying the extended Euclidean algorithm, we then obtain an integer ℓ such that $\ell(e_2 - e_1) \equiv 1$ (mod m_i), and it follows that $(y_1 - y_2)\ell$ is a discrete logarithm of v.

Clearly, after examining $O(1/\epsilon)$ pairs (e_1, f_1) we have a probability of at least $1/2$ of getting an output from A. Even if all pairs (e_1, f) for $1 \leq f \leq 2^u$ are in the set of pairs on which A produces a correct output, the probability is still at least $\epsilon - 2^{-t}$ that a pair (e_2, f_2) with $e_1 \neq e_2$ will yield a correct output from A, so we have again a probability at least $1/2$ of success after we examine $O(1/(\epsilon - 2^{-t})) = O(1/\epsilon)$ pairs (e_2, f_2).

We now describe the algorithm for recovering the factor q. From the previous discussion, we may assume without loss of generality that we are already in possession of an integer L such that $\alpha^L \equiv v$ (mod p). We begin by choosing random (e_1, f_1) until a pair is found for which A produces a correct output y_1. After this we search for a second pair (e_2, f_2) for which A produces a correct output. Since $\alpha^{y_1 - y_2} \equiv v^{e_2 - e_1}$ (mod p), we have $y_1 - y_2 \equiv (e_2 - e_1)L$ (mod q). If it happens that $y_1 - y_2 \not\equiv (e_2 - e_1)L$ (mod w), then $\gcd(y_1 - y_2 - (e_2 - e_1)L, p - 1)$ will give a splitting of qw. On the other hand, for each e_2, the congruence $y_1 - y_2 \equiv (e_2 - e_1)L$ (mod w) has only one solution y_2 in the interval $[1, w]$, so there is at most one f_2 for each e_2 that can give such a solution y_2. Hence the number of pairs (e_2, f_2) that do not lead to a splitting of qw is at most 2^t, and therefore the probability of success in finding a pair (e_2, f_2) that will split qw is at least $\epsilon - 2^{-u}$. Hence we expect to split q and w after examining $O(1/(\epsilon - 2^{-u}) = O(1/\epsilon)$ pairs (e_2, f_2), and this completes the proof. $\square$

Acknowledgment. We would like to thank Jim Davis, John DeLaurentis, Peter Montgomery, Judy Moore, and C. P. Schnorr for helpful conversations during the course of this research.

References

[1] Thomas Beth, "Efficient zero-knowledge identification scheme for smart cards," *Advances in Cryptology (Proceedings of Eurocrypt '88)*, Lecture Notes in Computer Science **330** (1989), 77–84.

[2] David Chaum, Jan-Hendrik Evertse, Jeroen van de Graaf, and René Peralta, "Demonstrating possession of a discrete logarithm without revealing it," *Ad-*

vances in Cryptology (Proceedings of Eurocrypt '86) Lecture Notes in Computer Science **263** (1987), 200–212.

[3] David Chaum, Jan-Hendrik Evertse, and Jeroen van de Graaf, "An improved protocol for demonstrating possession of discrete logarithms and some generalizations," *Advances in Cryptology (Proceedings of Eurocrypt '87)* Lecture Notes in Computer Science **304** (1988), 127–141.

[4] Yvo Desmedt, Claude Goutier, and Samy Bengio, "Special uses and abuses of the Fiat-Shamir passport protocol," *Advances in Cryptology (Proceedings of Crypto '87)* Lecture Notes in Computer Science **293** (1988), 21–39.

[5] A. K. Lenstra, H. W. Lenstra, Jr., M. S. Manasse, and J. M. Pollard, "The Number Field Sieve", *Proceedings of the 22nd ACM Symposium on Theory of Computing*, Association for Computing Machinery, New York, 1990, 564–572.

[6] Arjen K. Lenstra and Mark S. Manasse, Factoring by Electronic Mail, *Proceedings of Eurocrypt '89, Lecture Notes in Computer Science*, to appear.

[7] Kevin S. McCurley, A Key Distribution System Equivalent to Factoring, *Journal of Cryptology* **1** (1988), 95–105.

[8] J. M. Pollard, "Monte Carlo Methods for Index Computation mod p," *Mathematics of Computation* **32** (1978), 918–924.

[9] C.P. Schnorr, Efficient Identification and Signatures for Smart Cards, *Proceedings of Crypto '89, Lecture Notes in Computer Science*, to appear.

[10] Samuel S. Wagstaff, Jr., *Greatest of the Least Primes in Arithmetic Progressions Having a Given Modulus*, Mathematics of Computation **33** (1979), 1073–1080.

Factoring with two large primes

(Extended Abstract)

Arjen K. Lenstra
Bell Communications Research, room 2Q334
435 South Street, Morristown, NJ 07960
email: lenstra@flash.bellcore.com

Mark S.Manasse
Digital Equipment Corporation, Systems Research Center
130 Lytton Avenue, Palo Alto, CA 94301
email: msm@src.dec.com

Factoring with two large primes

The study of integer factoring algorithms and the design of faster factoring algorithms is a subject of great importance in cryptology (cf. [1]), and a constant concern for cryptographers. In this paper we present a new technique that proved to be extremely useful, not only to achieve a considerable speed-up of an older and widely studied factoring algorithm, but also, and more importantly, to make practical application of a new factoring algorithm feasible. While this first application does not pose serious threats to factorization-based cryptosystems, the consequences of the second application could be very encouraging (from the cryptanalysts point of view).

The technique has led to various new factorization records. It took us 50 days to factor a 107 digit number using our new version of the multiple polynomial quadratic sieve, and 60 days to factor a 111 digit number. This is quite a bit faster than the 120 days we needed for our previous 106 digit record with the old version [8]. We factored a 138 digit number using a new special purpose factoring algorithm [7]. Combined with our new technique this took approximately 50 days; it would have been impossible without.

For the 107 and the 138 digit number reported above we used the network of approximately 300 CVAX processors at Digital Equipment Corporation's Systems Research Center (SRC). For the other numbers a substantial amount of the computation was carried out by the participants to our electronic mail factoring network, as reported in [8].

Let n be some large integer to be factored. In cryptanalytic applications it is usually known that n is the product of two unknown primes of approximately the same size. In other cases (cf. [2]) one first has to decide if n is prime or composite. It is well known that this is usually not hard to do: use a probabilistic compositeness test [6] to prove that n is composite, and *believe* that n is prime if several attempts to prove compositeness have failed. In the latter case it remains to *prove* the primality of n, a problem for which efficient algorithms have been designed and implemented [3, 5, 10]. In the former case we should remark that the compositeness proof for n does in general *not* provide information which makes it easier to find a non-trivial factorization of n.

Suppose that the compositeness of n has been established beyond doubt. How can we find a non-trivial factorization of n? Again, if n is the modulus in some cryptosystem, it will have two carefully constructed large factors of approximately the same size, but for the rest n, or its factors, will have no specific properties that could make factoring easier. In such cases one has to resort to a *general purpose* factoring algorithm, a factoring algorithm that works no matter how *difficult* the number might be and whose running time is solely determined by the size of n.

For other numbers one could try the various *special purpose* factoring algorithms. These come in essentially two flavors. In the first place there are the methods that make use of properties an unknown factor of n might have. Because the factors are unknown, success is uncertain. It explains however why care has to be taken when designing a difficult composite n: if for instance one of the primes p dividing n is such that $p \pm 1$ is built up from small primes only, then n can be factored quite easily. In case of failure of this type of special purpose algorithms, a general purpose method should be applied, unless the second type of special purpose algorithm can be applied. This concerns methods that make use of the special form that n might have. Their run time, however, does not depend on any properties of the factors of n. An example of such a method is the *number field sieve* [7], and will be discussed below.

For cryptanalytic applications the general purpose algorithms are the ones to be studied. Until the summer of 1989 the best practical general purpose factoring algorithm was one of the multiple polynomial variations of the *quadratic sieve* algorithm (mpqs, for short) [11, 12]. The heuristic expected run time of mpqs is given by

(1) $\exp((1 + o(1))(\log n)^{1/2}(\log\log n)^{1/2})$.

It is the only general purpose algorithm by which integers of more than 100 digits have been factored: a record factorization of a 106 digit integer in April 1989 took four months and used impressive computational resources [8].

Many general purpose factoring algorithms, and mpqs is no exception, work in two stages. In the first stage one collects so-called *relations*, in the second stage one uses the relations to find solutions $x, y \in \mathbf{Z}$ to $x^2 \equiv y^2 \bmod n$. Under reasonable assumptions each solution has a good chance to lead to a factorization of n by computing $\gcd(n, x \pm y)$. For mpqs a relation is an expression of the form

(2) $v^2 \equiv \prod p^{e_{vp}} \bmod n$,

where $e_{vp} \in \mathbf{Z}_{\geq 0}$ and the product ranges over the primes (including -1) in the factor base (i.e., the number -1 and the primes less than some bound B). In mpqs, relations are found by means of a process called *sieving* (cf. [11]). On a fixed number of processors the number of relations found after t units of time will behave as $c \cdot t$, for some positive constant c depending on n. For a typical 100 digit number the number of elements of the factor base would be set to 50,000.

If the number of relations (found in the first stage) is more than the number of elements of the factor base, then a dependency modulo 2 can be found among the exponent vectors (in the second stage). Each such dependency leads to a solution to $x^2 \equiv y^2 \bmod n$, and thus to a chance of factoring n.

For mpqs the two stages are in theory asymptotically equally hard (they both take time (1), see [6] for an analysis and further description of the algorithm). For numbers in our current range of interest however the run time is entirely dominated by the first stage: if relations could be found twice as fast, the algorithm would run twice as fast. Gaining a factor two in the run time means, roughly, that we can factor integers having three more digits (cf. [1]).

A considerable speed-up of the first stage of the basic mpqs method can indeed be achieved quite easily by using the *large prime variation*, an idea that is already quite old [11]. It appears that the sieving stage of the algorithm can easily be changed so that it not only finds relations of the form (2), but relations of the following form as well:

(3) $v^2 \equiv q_v \prod p^{e_{vp}} \bmod n$,

where q_v (the *large prime*) is a prime not in the factor base, and less than the square of the largest prime in the factor base. We will distinguish between relations (2) and (3) by calling them *small* (2) and *partial* (3) relations. So, a small relation is for instance given by a number v such that the least absolute residue $v^2 \bmod n$ completely factors over the factor base; for a partial relation there may still be *one* too large prime in the factorization. One would expect therefore that the sieving stage produces many more partials than smalls, and that is indeed what happens.

Unfortunately, however, the partials themselves are worthless for the rest of the factorization process, unlike the smalls. It is only by *combining* the partials that they can be made into something that is as useful as a small relation. Suppose that both u and w give rise to partial relations, and suppose that we are so lucky that $q_u = q_w$. Multiplication of the two relations then gives

(4) $(u \cdot w)^2 \equiv q_u^2 \prod p^{e_{up} + e_{wp}} \mod n$,

a so-called *big* relation. Since $u \cdot w$ can be divided by q_u (modulo n, unless n and q_u are not co-prime), such a big relation is just as useful as a small relation.

How much luck is involved in finding a big relation? The birthday paradox tells us that even a moderate number of partials will already lead to relations with matching large primes, and therefore big relations. To give an example (cf. [8]): of the approximately 320,000 relations gathered for a certain 100 digit number, 20,500 were small, and the remaining approximately 300,000 partials gave rise to 29,500 big relations. For this same 100 digit number the progress of the total number of relations as a function of the time is illustrated in Figure 1. For other number the graphs of the numbers of small and big relations behave similarly.

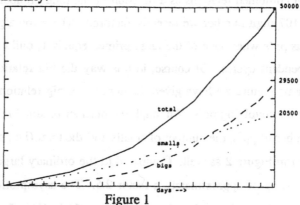

Figure 1

The large prime variation of mpqs affects the run time only in the $o(1)$ (cf. (1)). In practice it means a speed-up by a factor of approximately 2.5. Notice that it is straightforward to find the matching large primes: sort the partials according to their large prime, and match each pair of consecutive relations with the same large prime.

An obvious extension of the large prime variation is to allow *two* instead of only one large prime in a relation, i.e., a relation of the form

$$(5) \qquad v^2 \equiv q_{1v} q_{2v} \prod p^{e_{vp}} \bmod n,$$

where the q_{iv} are primes not in the factor base. These relations will be called *partial-partials* (or *pp*'s for short). Notice that there should be far more pp's than partials. Like the partials, the pp's can be combined into relations that are just as useful as small or big relations. This can for instance be seen as follows. Identify each large prime with a vertex in a graph, and put an edge between two vertices each time they occur in the same partial-partial relation. A cycle in the resulting graph corresponds to a combination of pp's where the large primes occur an even number of times, which makes that combination useful for factoring. Notice that we are only interested in independent cycles, because dependent cycles would give rise to trivial dependencies in the matrix of exponents.

If the pp's are only combined among themselves, then this process can be seen as drawing random edges in a (big) graph. It is well known that it takes many edges before a cycle can be expected, so that the yield will be quite low. But if the pp's are combined with the partials the picture changes dramatically as can be seen in Figure 2, where the total number of combined relations as a function of the total number of partials and pp's is given for a 107 digit number we recently factored. The same algorithm can be used: view partials as pp's where one of the large primes equals 1, build the same graph, and look for independent cycles. Of course, in this way the big relations will be found as well. In the same picture we have given the number of big relations, i.e., combinations of two partials not involving pp's. Although the number of small relations is not a function of the number of pp's, the number of smalls and the total (i.e., smalls plus combinations) are given in Figure 2 as well. Compared to the ordinary large prime variation we achieved a speed-up of approximately a factor 2.5. The asymptotic run time for mpqs remains the same, i.e., using pp's only affects the $o(1)$ in (1). The factorization of the 107 digit number took approximately 50 days, on many fewer machines than the previous 106 digit record. We used a factor base of 65,000 elements.

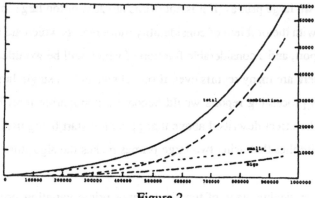

Figure 2

We got a similar picture for our current record general purpose factorization of a 111 digit composite factor of $2^{484}+1$:

$$2^{484}+1 = 17 * 353 * 209089 * 33186913 * 1251287137 * 2931542417 *$$
$$38608979869428210686559330362638245355335498797441 *$$
$$84694409197705740057696939084347325062258739942360856022665729.$$

The number we factored is the product of the last two factors. With a factor base consisting of 80,000 elements, we needed 14,300 small relations, and a total of 1,050,000 partials and pp's generating 66,100 cycles.

It can be seen in Figure 2 that combining partials and pp's only begins to pay off after finding an enormous number of relations. This could be one explanation for the fact that nobody used this method before: for smaller numbers fewer relations are needed, so that by the time the pp's have acquired enough weight, the other relations will already have done the job. For small numbers using pp's might even be counter-productive. This can be explained as follows. During the sieving stage the algorithm looks for so-called *reports*. For each report, the algorithm attempts to factor a certain number, using the primes in the factor base. If this attempt is successful, a small relation has been found. If the attempt fails, but the remaining cofactor is sufficiently small, then a partial relation has been found, because the cofactor is automatically prime. This implies that partial relations can be found at almost no extra cost; the only extra cost is caused by the fact that one has to allow for more reports to find more partials (and consequently more failures as well), but that will be made up for by the extra partials that will be found.

If one allows two large primes there should be even more reports, which makes the sieving stage slower, and there is the additional problem that the remaining cofactor must be factored into two large primes (if it is not prime already, or too large to be interesting). So, one is faced with the problem of considerably more reports which are more expensive to process per report, and a considerable fraction of which will be worthless.

For small n there are many reports even if one allows only a single large prime. With two large primes processing reports would become a dominating term in the run time. Combined with the effect described above that pp's only start being useful if there are a lot of them, this implies that using two large primes makes the algorithm slower for relatively small n.

For larger n the sieving stage of the single large prime variation produces very few reports. There one can easily afford more (and more expensive) reports, without noticeable effect on the total sieving time. Because many relations are needed, one gets the chance to build a huge database of pp's, and consequently the algorithm will run faster. In the full version of this paper we will describe how we proceeded to find the combinations (i.e., the cycles in the graph), and how we coped with the gigantic amounts of data.

Notice that using two large primes in mpqs can be very advantageous, but that it also works perfectly well, though slower, without it. We now discuss a more important application of the same idea, and of the same cycle finder, that could have far-reaching consequences. This other application existed before the one described above, and actually made us realize that it would be a useful idea for mpqs as well.

In [7] a new factoring algorithm, the *number field sieve* (nfs), is presented. This algorithm is an example of the second type of special purpose algorithms as described above, because it only applies to n of the form $r^e - s$, where r is a small positive integer, and s is a non-zero integer of small absolute value. This is precisely the type of number that can be found in [2]. To factor an n of this form, the number field sieve runs in heuristic expected time

(6) $\exp((c + o(1))(\log n)^{1/3}(\log\log n)^{2/3})$,

with $c = 2(2/3)^{2/3} \approx 1.526$. The algorithm has proved to be quite practical. Among others, we factored a 138 digit number that would have been absolutely impossible for mpqs. Using two large primes was of crucial importance to obtain this factorization.

The most exciting news about the nfs is that the algorithm can be generalized to integers of arbitrary form [4]. It is suspected [9] that the resulting general purpose factoring algorithm again runs in time (6), though with a slightly bigger value for c. The practical consequences remain to be seen. There is no doubt, however, that its chances of becoming practical are close to zero without the cycle finder.

For a description of the nfs we refer to [7]. Here we will only explain how two large primes can be used. Like mpqs, both the nfs and the generalized nfs consist of the two familiar stages, a relation collection stage, and an elimination stage. For nfs relations are expressions of the following form:

(7) $\prod \phi(g^{v_g}) \equiv \prod p^{e_p} \bmod n$.

The product at the left hand side ranges over elements g of some algebraic number field $K = Q(\alpha)$ of norm equal to 1 or of prime norm $\leq B$, for some bound B, and the product at the right hand side ranges over the primes $\leq B$. The exponents v_g and e_p are integral, and ϕ is some homomorphism from $Z[\alpha]$ to Z/nZ. We refer to [7] for the choice of K and ϕ. The form of n is important to be able to find a 'nice' number field. Relations are found by looking for coprime integers a and b such that the algebraic integer $a+b\alpha \in Z[\alpha]$ and the integer $a+bm \in Z$, with $m = \phi(\alpha)$, can be factored into small prime elements (and units) in $Z[\alpha]$ and primes in Z, respectively. This is done in the sieving stage of the nfs.

Suppose that the product on the left hand side ranges over B_1 elements, and the right hand side over B_2 elements. If we have more than B_1+B_2 relations, then we can, as in mpqs, use linear algebra to generate solutions to $x^2 \equiv y^2 \bmod n$, and thus factor n.

We have seen that for mpqs the number of small relations obtained is a linear function of the effort spent on finding relations. With the nfs the situation is different. There the yield becomes quite noticeably lower and lower, with the possibility of the unpleasant discovery, after spending years of CPU time, that the algorithm is not going to make it because the supply of solutions to (7) as generated by the nfs dries up. The theory simply tells us to start all over again with a bigger value for B. From a practical point of view this is less desirable: for numbers in our current range of interest B would have to be chosen so large that storing the exponent matrix, even in sparse form, becomes problematic, let alone finding a dependency among its rows.

So, to make the nfs practical, it is important to keep B as small as we can, while avoiding the problem of running out of solutions to (7). This can be achieved as follows. While sieving we not only collect relations as in (7), but we collect the following types of relations as well:

- As (7), but allow one large prime element at the left hand side, the *partial-fulls* or *pf*'s;

- As (7), but allow one large prime at the right hand side, the *full-partials* or *fp*'s;

- As (7), but allow both a large prime element at the left hand side and a large prime at the right hand side, the *partial-partials* or *pp*'s.

Relations as in (7) will be called *full-fulls* or *ff*'s.

The sieving stage can easily be changed so that it not only collects the ff's, but the pf's, fp's and pp's as well. For pf's and fp's this is trivial, as it was for partial relations in mpqs. For pp's this follows from the fact that the large prime element and the large prime involved come from different numbers (namely from $a+b\alpha$ and from $a+bm$, respectively). So, the problem of slower performance that mars finding pp's in mpqs when applied to comparatively small numbers does not occur here.

Clearly, the sieving stage should find many more pf's and fp's than ff's, and even more pp's. The pf's can be combined among themselves, just as the partials in mpqs, with the difference that we divide pf's with the same large prime element instead of multiplying them to avoid the problem of computing a generator for the large prime ideal. Similarly, fp's with the same large prime can be combined, either by multiplication or by division, to produce a useful relation. And the pp's, finally, can be used in almost the same way as the pp's in mpqs. The difference is that the pp's now give rise to a bipartite graph (with vertices identified with prime elements in $Z[\alpha]$ connected to vertices identified with primes in Z), plus one extra vertex (identified with 1) to put the pf's and the fp's in the same graph.

To give some examples, for a certain 122 digit number we needed a total of 49,000 relations. After two weeks sieving (on many machines simultaneously) we had gathered 10,688 ff's, 116,410 pf's, 103,692 fp's and 1,138,617 pp's. By that time it had become clear that our choice of B was too low to factor the number using only ff's, because the supply of ff's was drying up rapidly. The same was true for the pf's and fp's. Although the 116,410 pf's gave already 5,341 combinations, and the 103,692 fp's gave 5,058 relations, it was clear that they were coming in too slowly to make our choice of B feasible for this number, at least without using pp's. Using the cycle finder we found more than 28,000 independent cycles involving pp's, which was enough to factor the number. It took five days (on a single machine) to find a dependency in the resulting matrix. We are not sure what value for B we should have chosen to obtain this factorization without using pp's, but it is unlikely that we could have factored the number within a reasonable amount of time in that case.

For a 138 digit number, it took seven weeks to gather 17,625 ff's and a total of 1,741,365 pf's, fp's, and pp's, which gave 62,842 combinations. It took two weeks to process the resulting 80,000×80,000 matrix. Without pp's we would never have succeeded: B would have to be taken so large that the sieving would take almost forever, and we would not even be able to store the sparse representation of the resulting matrix.

Notice that relations that follow from combinations lead to denser rows in the matrix of exponents than the ff's. So, although the combinations are just as useful for factoring as the ff's, they lead to a denser matrix, and therefore to a slower second stage. The same holds for mpqs. However, this is a small price to pay if the only alternative leads to unsurmountable problems.

As remarked above, we gained our first experience with pp's because we had to while experimenting with the nfs. This naturally led us to the application of the same idea in mpqs.

We have seen that this relatively simple technique of finding cycles among partial and partial-partial relations is very useful for mpqs, and of great importance to make nfs practical. If the generalized nfs ever becomes practical, there can be little doubt that an important role will be played by the partial-partial relations. We therefore feel that it is an important technique that should be brought to the attention of everyone interested in factoring.

References

1   G. Brassard, *Modern Cryptology*, Lecture Notes in Computer Science, vol. 325, 1988, Springer-Verlag.

2   J.Brillhart, D.H. Lehmer, J.L. Selfridge, B. Tuckerman, S.S. Wagstaff, Jr., *Factorizations of $b^n \pm 1$, $b = 2, 3, 5, 6, 7, 10, 11, 12$ up to high powers, second edition*, Contemporary Mathematics, vol. 22, Amer. Math. Soc., Providence, Rhode Island 1988.

3   W. Bosma, M.-P. van der Hulst, A.K. Lenstra, "An improved version of the Jacobi sum primality test," in preparation.

4   J. Buhler, H.W. Lenstra, Jr., C. Pomerance, in preparation.

5 H. Cohen, A.K. Lenstra, "Implementation of a new primality test," *Math. Comp.*, v. 48, 1987, pp. 103-121.

6 A.K. Lenstra, H.W. Lenstra, Jr., "Algorithms in number theory," in: J. van Leeuwen, A. Meyer, M. Nivat, M. Paterson, D. Perrin (eds), *Handbook of theoretical computer science*, to appear.

7 A.K. Lenstra, H.W. Lenstra, Jr., M.S. Manasse, J.M. Pollard, "The number field sieve," to appear.

8 A.K. Lenstra, M.S. Manasse, "Factoring by electronic mail," Proceedings Eurocrypt '89.

9 H.W. Lenstra, Jr., personal communication.

10 F. Morain, "Primality testing: News from the front," Proceedings Eurocrypt '89.

11 C. Pomerance, "Analysis and comparison of some integer factoring algorithms," pp. 89-139 in: H.W. Lenstra, Jr., R. Tijdeman (eds), *Computational methods in number theory*, Mathematical Centre Tracts 154, 155, Mathematisch Centrum, Amsterdam, 1982.

12 R.D. Silverman, "The multiple polynomial quadratic sieve," *Math. Comp.*, v. 48, 1987, pp. 329-339.

Which new RSA signatures can be computed from some given RSA signatures?

(extended abstract)

Jan-Hendrik Evertse [‡]

Department of Mathematics and Computer Science, University of Leiden
P.O. Box 9512, 2300 RA Leiden, The Netherlands

Eugène van Heyst

CWI Centre for Mathematics and Computer Science
Kruislaan 413, 1098 SJ Amsterdam, The Netherlands

Abstract. We consider protocols in which a signature authority issues RSA-signatures to an individual. These signatures are in general products of rational powers of residue classes modulo the composite number of the underlying RSA-system. These residue classes are chosen at random by the signature authority. Assuming that it is infeasible for the individual to compute RSA-roots on randomly chosen residue classes by himself, we give, as a consequence of our main theorem, necessary and sufficient conditions describing whether it is feasible for the individual to compute RSA-signatures of a prescribed type from signatures of other types that he received before from the authority.

Key words. RSA scheme, RSA signature, cryptographic protocol.

1. Introduction

A cryptographic *protocol* can be taken to be a set of rules according to which messages are transmitted between parties. Generally the parties apply cryptographic operations (such as computation of digital signatures and encryption) to the messages sent and received, in order to protect their interests.

In this paper we consider *signature protocols* in which only one party, called the *signature authority*, can create signatures. The signature authority issues these signatures to an other party, called the *individual*. Such protocols are used, for instance, in credential systems (e.g. [CE86]) and payment systems (e.g. [CBHMS89]), in which a signature represents a credential or money.

[‡] This research has been made possible by a fellowship of the Royal Netherlands Academy of Arts and Sciences (K.N.A.W.)

Figure 1 shows a simple version based on the RSA-system with modulus N. Let e_1, e_2 be public exponents, known to both the signature authority Z and the individual A, and $1/e_2$ the secret exponent, known only to Z. Here $1/e_2$ is some integer such that $(x^{1/e_2})^{e_2} \equiv x \pmod{N}$, for all x coprime to N. (Note that this implies that only Z knows the factorization of the RSA modulus).

Individual A	**Signature authority Z**

chooses $x \pmod{N}$ randomly,   verifies if

computes $S \equiv x^{e_1/e_2} \pmod{N}$ $\xrightarrow{\quad x, S \quad}$ $S^{e_2} \equiv x^{e_1} \pmod{N}$

Fig. 1. A signature issuing protocol in which the individual has no influence on the choice of the integer.

The protocols we shall consider, are variations on or generalizations of the scheme in Figure 1. It will appear to be useful to consider variations in which Z does not send x to A, but only the signature (so then A can not verify the signature). In our most general protocols, the *RSA-signatures* are products of rational powers of residue classes modulo N, for instance $x_1^{2/5} \cdot x_2^{3/7} \pmod{N}$. It is reasonable to assume that an individual, not knowing the factorization of N, can not compute RSA-roots $x^{1/d} \pmod{N}$ on a randomly chosen x for $d > 1$ by himself. Yet it is possible that the individual learns some RSA-signatures computed by Z (e.g. by participating in some protocol or by eavesdropping) and can use these to compute some new signatures of a type not issued by Z. The purpose of this paper is to investigate which new types of RSA-signatures an individual can compute from the ones obtained from Z.

We give an example of the kind of problems we shall consider. Suppose A has received, by participating in some protocol (or by eavesdropping) two random integers x_1, x_2 and a signature $S \equiv x_1^{2/5} \cdot x_2^{3/7} \pmod{N}$. Then A can compute $x_1^{1/5}$, using that $x_1^{1/5} \equiv x_1^3 \cdot x_2^3 / S^7 \pmod{N}$. On the other hand we shall prove that for all positive integers d different from 1 and 5 (and relatively prime to $\varphi(N)$), it is infeasible for A to compute $x_1^{1/d}$ from (x_1, x_2, S). Another consequence of our results is a result of Shamir [Sh83] which states that it is feasible for A to compute $x^{1/m}$ from $(x, x^{1/a_1}, ..., x^{1/a_s})$ if and only if m divides the least common multiple of $(a_1, ..., a_s)$. In section 3 we give more detailed examples related to coin systems.

This paper is organized as follows. In section 2 the notation used in this paper is introduced. Section 3 contains descriptions of the RSA scheme and the four protocols that we want to investigate. We shall state four propositions related to the respective protocols and give some examples and applications to illustrate these propositions. With the lemmas of section 4, the four propositions will be proven in section 5.

The propositions of section 3 can not be considered as mathematical statements since they involve an intuitive notion of computational feasibility which we shall not formalize. Therefore in our main theorem in section 6, we will not use any assumption on the computational feasibility of RSA-roots by individuals. In this extended abstract we shall only state this theorem in words without using the formalism of Probabilistic Turing Machines, and we shall not prove this theorem here.

2. Notation

The following notation is used throughout this paper:

$\mathbb{N}, \mathbb{Z}, \mathbb{Q}$	the sets of positive integers, all integers and rational numbers respectively.
$(a_1, \ldots, a_t)$	the greatest common divisor of $a_1, \ldots, a_t$; also defined for rational numbers by $(a_1, \ldots, a_t) := \frac{(a_1 d, \ldots, a_t d)}{d}$, where $d \in \mathbb{N}$ such that $a_1 d, \ldots, a_t d \in \mathbb{Z}$; this definition is independent of the choice of d.
$\mathrm{lcm}(a_1, \ldots, a_t)$	the least common multiple of $a_1, \ldots, a_t \in \mathbb{Q}$ (this is defined for rational numbers analogously to the gcd).
$a \vert b$	there is an integer c such that $ac = b$; also defined for $a, b \in \mathbb{Q}$.
$a \equiv b \pmod{m}$	it holds that $m \vert (a-b)$, for $a, b \in \mathbb{Q}$, $m \in \mathbb{N}$; we shall omit the suffix $(\bmod\ m)$, if no confusion is likely to arise.
S^k	the set of k-dimensional column vectors with entries from the set S.
a	column vector $(a_1, \ldots, a_k)^T$; if $a \in S^k$, then $a_1, \ldots, a_k \in S$.
e_i	the ith unit vector $(0, \ldots, 0, 1, 0, \ldots, 0)^T$ which has a 1 on the ith place and zeros elsewhere (the dimension of these vectors will follow from the context).
$\langle a, b \rangle$	the scalar product of two column vectors $a = (a_1, \ldots, a_k)^T$ and $b = (b_1, \ldots, b_k)^T$, which is defined by $\langle a, b \rangle = a_1 b_1 + \ldots + a_k b_k$.
$[a_1 \ \ldots \ a_t]$	the matrix with columns $a_1, \ldots, a_t$.
$[C\ a]$	the matrix with column vector a concatenated at the right to matrix C.
$\mathrm{def}(a_1, \ldots, a_t; b)$	the defect of $a_1, \ldots, a_t; b \in \mathbb{Q}^k$; this is the smallest positive integer d such that $[a_1 \ \ldots \ a_t] y = db$ has a solution $y \in \mathbb{Z}^t$ (well defined if $[a_1 \ \ldots \ a_t] x = b$ has a solution $x \in \mathbb{Q}^t$). Examples: $\mathrm{def}(3;1) = 3$, $\mathrm{def}(5;1) = 5$, $\mathrm{def}(3,5;1) = 1$.
N	the RSA modulus used in all the protocols; N is a composite, odd number.
$\mathbb{Z}_N^*$	the set $\{a \vert\ a \in \mathbb{N},\ 1 \le a \le N,\ (a, N) = 1\}$.
$\varphi(N)$	Euler's Totient function; $\varphi(N) = \vert \mathbb{Z}_N^* \vert$.
$\mathbb{Q}_N$	the set $\{\frac{a}{b} \vert\ a, b \in \mathbb{Z},\ b > 0,\ (b, \varphi(N)) = 1\}$.

$$x^a \qquad x_1^{a_1} x_2^{a_2} \dots x_k^{a_k} \pmod{N}, \text{ for } x=(x_1,\dots,x_k)^{\mathrm{T}} \in (\mathbf{Z}_N^*)^k \text{ and}$$

$$a=(a_1,\dots,a_k)^{\mathrm{T}} \in (\mathbb{Q}_N)^k. \text{ Examples: } x^{e_i} \equiv x_i, \text{ and if}$$

$$x=(x^{b_1},\dots,x^{b_k}), \text{ then } x^a = x^{<a,b>}.$$

$$a/b \qquad (\frac{a_1}{b_1},\dots,\frac{a_k}{b_k})^{\mathrm{T}}, \text{ if } b_i \neq 0 \text{ for } i=1,\dots,k.$$

3. Protocols

In this paper we will consider 4 protocols, and each but the first is a generalization of the previous one. In each protocol, a signature authority Z issues one or more RSA-signatures of certain types to the individual A, who has no influence on the integers used. We deal with the problem to determine for which other types of RSA-signatures it is feasible for A to compute them from the types of signatures that he obtained from Z.

In order to avoid technical complications ,we shall not give a mathematically precise definition of the notion "*computational feasibility*", but only the following intuitive definition. If $a_1,\dots,a_t$ are binary strings chosen according to some prescribed probability distribution and b is a binary string with $b=f(a_1,\dots,a_t)$ for some function f, then we say that it is feasible to compute b from $a_1,\dots,a_t$ if there is an efficient probabilistic algorithm that outputs b with non-negligible probability, when it is given $a_1,\dots,a_t$ as input. In this section we shall freely use the notion of computational feasibility in statements of propositions, corollaries etc. We shall state four propositions, each related to a protocol.

First we briefly sketch the RSA scheme [RSA78]. The signature authority Z chooses two large "random" primes, each of 100 decimal digits say, and computes their product N, which will be used as RSA modulus.

Let $d \in \mathbf{Z}_{\varphi(N)}^*$. The equation $d\bar{d} \equiv 1 \pmod{\varphi(N)}$ [†] has a unique solution $\bar{d} \in \mathbf{Z}_{\varphi(N)}^*$ which can be computed by Z, because Z knows the factorization of N (and thus $\varphi(N)$). We define $x^{a/d} \pmod{N}$ to be the unique solution $y \in \mathbf{Z}_N^*$ to $y^d \equiv x^a \pmod{N}$, for $x \in \mathbf{Z}_N^*$ and $\frac{a}{d} \in \mathbb{Q}_N$. This solution y can be computed by $y \equiv x^{a\bar{d}} \pmod{N}$. We call $x^{1/d} \pmod{N}$ the dth *RSA-root* of $x \in \mathbf{Z}_N^*$.

Z makes N and d public, and keeps $\bar{d}$ and the factorization of N secret. The RSA-signatures issued by Z in the protocols are products of rational powers of residue classes. For all the signatures in this paper the same modulus is used. The case that an individual receives signatures with different moduli is partially solved in [Has85].

[†] The RSA-scheme can be made slightly more efficient by solving $\bar{d}$ from $d\bar{d} \equiv 1 \pmod{\lambda(N)}$, where $\lambda(N)$ is Carmichael's function. For instance, if $N=PQ$ for primes P,Q, then $\varphi(N)=(P-1)(Q-1)$ and $\lambda(N)=\varphi(N)/(P-1,Q-1)$.

We assume that it is computationally infeasible for an individual A to compute RSA-roots by himself: the only positive integer d with $(d,\varphi(N))=1$ for which A can feasibly compute $x^{1/d} \pmod{N}$ for uniformly chosen x from $\mathbf{Z}^*_N$, is $d=1$. In other words:

Assumption. *Let N be the used RSA-modulus. Then for every integer $d>1$ with $(d,\varphi(N))=1$ it is computational infeasible for A to compute $x^{1/d} \pmod{N}$ when given only N,d,x as input, where x is chosen uniformly from $\mathbf{Z}^*_N$*

We now describe the four protocols, the propositions and some examples (related to coin systems) to illustrate the propositions.

3.1. Protocol 1

Protocol 1. *Z makes public integers a,n with $a/n \in \mathbb{Q}_N$.*

(1) *Z chooses x uniformly from $\mathbf{Z}^*_N$ and computes the RSA-signature*

   $S \equiv x^{a/n} \pmod{N}$.
(2) *Z sends the pair (x,S) to A.*
(3) *A verifies the RSA-signature on x by checking if $S^n \equiv x^a \pmod{N}$.*

We consider the problem for which integers $m>0$ with $(m,\varphi(N))=1$, A is able to compute $x^{1/m} \pmod{N}$ from the pair (x,S) that he received from Z. Necessary and sufficient conditions are given in the next proposition.

Proposition 1. *Fix integers a,n,m with $n,m>0$ and $(n,\varphi(N))=(m,\varphi(N))=(a,n)=1$. Then the following three statements are equivalent:*

(i) *It is feasible for A to compute $x^{1/m}$ from $(x, x^{a/n})$, if Z chooses x uniformly from $\mathbf{Z}^*_N$.*
(ii) *There are integers v,w such that $1/m = v \cdot a/n + w$.*
(iii) *$m \mid n$.*

Proposition 1 can be applied to coin systems, such as in Figure 2. Here f is a fixed, public, "pseudo-random" function. In a coin system, different exponents s are used, each representing another coin value. Suppose that the exponents $s=3,5,7,9$ (assumed to be coprime with $\varphi(N)$) are used, and that they correspond to the coin values $8,4,2,1$ respectively. Now any user A can gain 7 money units simply by withdrawing a coin of value 1, which is of the form $C = f(y)^{1/9}$, and computing $C^3 = f(y)^{1/3}$, which is a coin of value 8. One can prevent users from gaining money by replacing $s=9$ for instance by $s=11$. Assume that A withdraws the coins $f(y)^{1/11}$, $f(y)^{1/7}$ and $f(y)^{1/5}$ of value 1,2 and 4 respectively. Then A can compute $f(y)^{1/(5 \cdot 7 \cdot 11)}$ by

$f(y)^{1/(5\cdot7\cdot11)}=\left(f(y)^{1/5}\right)^{3}\left(f(y)^{1/7}\right)^{13}\left(f(y)^{1/11}\right)^{-27}$. But by proposition 1, A cannot compute $f(y)^{1/3}$ from $f(y)^{1/(5\cdot7\cdot11)}$. So A cannot gain a money unit.

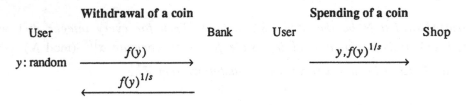

Fig. 2. A simple coin system

3.2. Protocol 2

In Protocol 2, Z issues to A *one* RSA-signature that is a product of powers of RSA-roots on integers (chosen by Z). Proposition 2 describes which new RSA-signatures are feasibly computable from the received ones.

Protocol 2. *Z makes public vectors* $a,n\in\mathbb{Z}^{k}$ *such that* $a/n\in\left(\mathbb{Q}_{N}\right)^{k}$. *Let* $n^{*}=\mathrm{lcm}(n_{1},\ldots,n_{k})$.

(1) *Z chooses x uniformly from* $(\mathbb{Z}^{*}_{N})^{k}$, *and computes the signature*

$$S\equiv x^{a/n}\,(\mathrm{mod}\ N).$$

(2) *Z sends* (x,S) *to A.*
(3) *A verifies the signature on x by checking whether*

$$S^{n^{*}}\equiv x_{1}^{a_{1}n^{*}/n_{1}}\cdot\ldots\cdot x_{k}^{a_{k}n^{*}/n_{k}}(\mathrm{mod}\ N).$$

Proposition 2. *Fix vectors* $a,n,b,m\in\mathbb{Z}^{k}$ *with* $(a_{i},n_{i})=(b_{i},m_{i})=(n_{i},\varphi(N))=(m_{i},\varphi(N))=1$ *for* $i=1,\ldots,k$. *Then the following three statements are equivalent:*

(i) *It is feasible for A to compute* $x^{b/m}$ *from* $(x,x^{a/n})$, *if Z chooses x uniformly from* $(\mathbb{Z}^{*}_{N})^{k}$.

(ii) *There are* $v\in\mathbb{Z}$ *and a vector* $w\in\mathbb{Z}^{k}$ *such that* $b/m=v(a/n)+w$.

(iii) $m_{i}|n_{i}$ *for* $i=1,\ldots,k$ *and*
$a_{i}b_{j}n_{j}/m_{j}\equiv a_{j}b_{i}n_{i}/m_{i}$ mod (n_{i},n_{j}) *for* $1\leq i,j\leq k$.

To illustrate this proposition, we consider the product $\prod_{i=1}^{10} x_{i}^{3^{i-1}}/17$. We are interested in the question whether it is feasible for an individual to change the order of the terms in the product, i.e. is it feasible for an individual to find a non-identical permutation τ such that

$\prod_{i=1}^{10} x_i^{3^{i-1}} / 17 = \prod_{i=1}^{10} x_i^{3^{\tau(i)-1}} / 17$? Using the next corollary (which can be derived from Proposition 2) we can prove that this is not feasible. So to each position in this product (i.e. to each exponent) we can assign a different coin value. This result is used in the offline check system of [CBHMS89].

Corollary 1. *Let p and q be different primes such that $(p,\varphi(N))=(q,\varphi(N))=1$ and let k,m be integers. Define the integral vectors $a=(q^m,...,q^{m+k-1})^T$ and $n=(p,...,p)^T$.*

The following statements are equivalent if x is chosen randomly from $(\mathbf{Z}_N^*)^k$.

(i) *There is a non-identical permutation τ of $(0,...,k-1)$, such that it is feasible for A to compute $x^{b/n}$ from $(x,x^{a/n})$ where $b=(q^{m+\tau(0)},...,q^{m+\tau(k-1)})^T$.*

(ii) *There is an i_0 with $1\le i_0\le k$ such that $q^{i_0} \equiv 1\,(\text{mod}\;\;p)$.*

3.3. Protocol 3

We now consider a general protocol, in which Z issues to A several signatures at once, together with the chosen vector x. Notice that sending x is exactly the same as sending $(x^{e_1},..., x^{e_k})$, where $e_1,...,e_k$ are the unit vectors of $(\mathbf{Q}_N)^k$.

Protocol 3. *Z makes public vectors $a_1,...,a_s\in (\mathbf{Q}_N)^k$.*

(1) *Z chooses x uniformly from $(\mathbf{Z}_N^*)^k$, and computes $S_i \equiv x^{a_i}(\text{mod}\;N)$ for $i=1,...,s$.*

(2) *Z sends $(x, S_1,...,S_s)$ to A.*

(3) *A verifies that $S_i^d \equiv x^{da_i}(\text{mod}\;N)$ for $i=1,...,s$, where d is a positive integer such that $da_1,...,da_s\in \mathbf{Z}^k$.*

We want to know for which vectors $b\in (\mathbf{Q}_N)^k$, it is feasible for A to compute $x^b\,(\text{mod}\;N)$ from $(x,x^{a_1},..., x^{a_s})$.

Proposition 3. *Fix vectors $a_1,...,a_s,b\in (\mathbf{Q}_N)^k$. Then the following four statements are equivalent:*

(i) *It is feasible for A to compute x^b from $(x,x^{a_1},..., x^{a_s})$, if Z chooses x uniformly from $(\mathbf{Z}_N^*)^k$.*

(ii) *There are $v_1,...,v_s\in \mathbf{Z}$ and a vector $w\in \mathbf{Z}^k$ such that $b=v_1a_1+...+v_sa_s+w$.*

(iii) *def$(a_1,...,a_s,e_1,...,e_k;b)=1$.*

(iv) Let $\lambda_1,...,\lambda_m$ be all the subdeterminants of $[a_1 ... a_s]$ of order between 1 and $\min(k,s)$, and $\lambda_{m+1},...,\lambda_n$ be all the subdeterminants of $[a_1 ... a_s\, b]$ of order between 1 and $\min(k,s+1)$, containing at least one entry from b. Then $(1,\lambda_1,...,\lambda_m)=(1,\lambda_1,...,\lambda_n)$ (i.e. $(1,\lambda_1,...,\lambda_m)\,|\,\lambda_i$, for $i=m+1,...,n$).

To illustrate how this proposition can be used, we consider the off-line coin system of [OO89]. In this system the bank uses a signature scheme which we do not specify here. The user makes RSA-signatures using his own modulus N whose factorization he keeps secret; so here the user plays the role of a signature authority. Let L be a fixed integer, and define $I \equiv (\text{account number user})^L \bmod N$. In Figure 3 the basic idea of the withdrawal (in which the user is able to blind and the bank to sign messages, cf. [OO89]) and spending protocol of a coin is given. Each shop sends the numbers it received to the bank and the bank verifies that these numbers have not been used before. Since the system is off-line, usually each shop first collects the numbers from several payments before sending them to the bank.

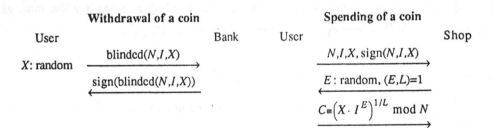

Fig. 3. The (simplified) off-line coin system of [OO89]

From Proposition 3 it follows that it is not feasible for the shop/bank to compute the identity of the user (i.e. $I^{1/L} \bmod N$) from N,I,X,E and $C = X^{1/L} \cdot I^{E/L}$. But if the user spends the same coin at two shops, then the bank receives the integers $N,I,X,\text{sign}(N,I,X)$, E_1,E_2 (coprime with L), $\left(X \cdot I^{E_1}\right)^{1/L} \bmod N$ and $\left(X \cdot I^{E_2}\right)^{1/L} \bmod N$. From Proposition 3 it follows that the bank can compute the users identity $I^{1/L} \bmod N$ from this if and only if $(E_1-E_2,L)=1$. Hence the probability that a double spender is caught by the bank is approximately $\varphi(L)/L$. This probability is close to 1 if L is a large prime, and close to 0 if L is the product of small primes. Therefore it is not wise to let the user choose L himself (which was the original suggestion of [OO89]), but to fix L as a large prime.

Suppose we modify protocol 3 in such a way that an individual A receives s signatures on *different vectors*, so suppose A has received $(x_1,...,x_s,$

$S_1 \equiv x_1^{a_1},..., S_s \equiv x_s^{a_s}$) and wants to compute $S_{s+1} \equiv x_{s+1}^{a_{s+1}}$, where $a_i \in (\mathbb{Q}_N)^{k_i}$ and $x_i \in (\mathbb{Z}_N^*)^{k_i}$,$(i=1,...,s+1)$. Define y to be the vector obtained by concatenating all the different entries of the vectors $x_1,...,x_{s+1}$. We can write S_i as y^{b_i}, where the exponent on the "new" y_j's (i.e. those y_j which were no entry of x_i) is zero. Now Proposition 3 can be used to determine for which a_{s+1} it is feasible for A to compute S_{s+1} from $(y,S_1,...,S_s)$.

3.4. Protocol 4

We now consider the most general protocol, in which Z issues to A several signatures at once, but without sending the used vector.

Protocol 4. Z *makes public vectors* $a_1,...,a_s \in (\mathbb{Q}_N)^k$.

(1) Z *chooses* x *uniformly from* $(\mathbb{Z}_N^*)^k$, *and computes* $S_i \equiv x^{a_i} (mod\ N)$ *for* $i=1,...,s$.
(2) Z *sends* $(S_1,...,S_s)$ *to* A.

If one does not accept this as a useful protocol (because A can in general not verify the signatures), then assume A has received $(x^{a_1},..., x^{a_s})$ during eavesdropping. We want to know for which vectors $b \in (\mathbb{Q}_N)^k$, it is feasible for A to compute $x^b (mod\ N)$ from $(x^{a_1},..., x^{a_s})$, and prove that the only b's for which x^b is computable from $(x^{a_1},..., x^{a_s})$, is the lattice generated by $a_1,...,a_s$.

Proposition 4. *Fix vectors* $a_1,...,a_s,b \in (\mathbb{Q}_N)^k$, *and assume that the equation* $[a_1 \ ... \ a_s]y=b$ *is solvable in* $y \in \mathbb{Q}^s$. *Then the following four statements are equivalent:*

(i) *It is feasible for A to compute* x^b *from* $(x^{a_1},..., x^{a_s})$, *if Z chooses* x *uniformly from* $(\mathbb{Z}_N^*)^k$.
(ii) *There are* $v_1,...,v_s \in \mathbb{Z}$ *such that* $b = v_1 a_1 + ... + v_s a_s$.
(iii) $def(a_1,...,a_s;b)=1$.
(iv) *Let* $\mu_1,...,\mu_m$ *be the subdeterminants of* $[a_1 \ ... \ a_s]$ *of order k and* $\mu_{m+1},...,\mu_n$ *be the subdeterminants of* $[a_1 \ ... \ a_s \ b]$ *of order k, containing at least one entry from* b.
Then $(1,\mu_1,...,\mu_m)=(1,\mu_1,...,\mu_n)$ *(i.e.* $(1,\mu_1,...,\mu_m) | \mu_i,$ *for* $i=m+1,...,n$).

This proposition implies that only the exponents must be investigated, and that the only reasonable computations an individual can do, in order to create a new signature from some received signatures, are the basic computations (add, subtract, multiply and divide). So applying the cosine or DES can not help an individual in creating more signatures.

The previous propositions can be used to prove the following results.

Corollary 2. *Let* $a, a_1, \ldots, a_s, b, d$ *be positive integers coprime with* $\varphi(N)$, c *be an integer, and* x, y *be chosen randomly from* $\mathbf{Z}^*_N$. *Then the following five results hold for A.*

(i) *It is feasible to compute* $x^{1/d}$ *from* $(x, x^{c/a})$ $\qquad \Leftrightarrow \quad d \mid \dfrac{a}{(a,\,c)}.$

(ii) *It is feasible to compute* $x^{1/d}$ *from* $(x, y, x^{1/a} \cdot y^{1/b})$ $\qquad \Leftrightarrow \quad d \mid \dfrac{a}{(a,\,b)}.$

(iii) *It is feasible to compute* $x^{1/d}$ *from* $(x, x^{1/a_1}, \ldots, x^{1/a_s})$ $\quad \Leftrightarrow \quad d \mid \mathrm{lcm}\,(a_1, \ldots, a_s)$
[Sh83].

(iv) *It is feasible to compute* $(xy)^{1/d}$ *from* $(x, y, x^{1/a}, y^{1/b})$ $\quad \Leftrightarrow \quad d \mid (a, b).$

(v) *It is feasible to compute* x^d *from* $(x^{a_1}, \ldots, x^{a_s})$ $\qquad\qquad \Leftrightarrow \quad \gcd(a_1, \ldots, a_s) \mid d.$

4. Auxiliary results

When we say that something is computable in polynomial time, we mean that it is computable by a polynomial time deterministic algorithm.

Lemma 1. *The following operations can be done in polynomial time:*
   (1) *computing* $\gcd(a,b)$ *from a and b,*
   (2) *computing the inverse of* $a \pmod{b}$ *from a and b, if* $(a,b)=1$,
   (3) *computing* $a^b \pmod{c}$ *from a,b and c, if* $(a,c)=1$,
   (4) *the Gaussian elimination method for a system of linear equations with rational coefficients,*
   (5) *determining the rank of a rational matrix,*
   (6) *determining the determinant of a given rational square matrix,*
   (7) *determining the inverse of a nonsingular rational square matrix,*
   (8) *testing rational vectors for linear independence,*
   (9) *computing the Hermite Normal Form of a matrix* [KaBa79],
   (10) *computing a unimodular matrix U, such that AU is the Hermite Normal Form of A, for a rational matrix A of full row rank,*
   (11) *deciding if a system of rational linear equations has an integral solution, and if so, finding one.*

References for the proofs can be found in Chapter 3 and 5 in [Schr86].

Lemma 2. ([Heg1858] page 111)
Let A be a rational matrix of full row rank, with k rows, and let b be a rational column k-vector. Then $Ax=b$ *has an integral solution* x, *if and only if the gcd of all subdeterminants of A of order k divides each subdeterminant of* [A b] *of order k.*

Lemma 3. *Let* $a_1,...,a_s,b \in (\mathbb{Q}_N)^k$, $A=[a_1 \ ... \ a_s]$ *of full row rank, and let* $d=\text{def}(a_1,...,a_s;b)$; *hence*

$$Av = db \qquad (4.1)$$

is solvable in $v \in \mathbb{Z}^s$.

Further, let $\mu_1,...,\mu_m$ *be the subdeterminants of A of order k and* $\mu_{m+1},...,\mu_n$ *be the subdeterminants of* [A b] *of order k, containing at least one entry from b.*

Then:

(i) $\quad d = \dfrac{(\mu_1,..., \mu_m)}{(\mu_1,..., \mu_m, \mu_{m+1},..., \mu_n)}$. $\qquad (4.2)$

(ii) *There is a polynomial time deterministic algorithm that computes d and a solution of* (4.1).

(iii) *There is a polynomial time deterministic algorithm that computes a* $z \in \mathbb{Q}^k$ *such that*

$$(d,<z,db>)=1 \ \text{and} \ A^T z \in \mathbb{Z}^s. \qquad (4.3)$$

Remark: Note that expression (4.2) does *not* yield a polynomial time algorithm to compute $\text{def}(a_1,...,a_s;b)$, because $m = \binom{s}{k}$, $n-m = \binom{s}{k-1}$, and $s \geq k$.

Proof. Matrix A has full row rank, so according to Lemmas 1.7 and 1.10, we can compute in polynomial time a matrix $[D0]$ (in Hermite Normal Form in which D is a nonsingular square matrix and 0 is a matrix consisting of zeros) and a unimodular matrix U such that $A=[D0]U$. The matrices $U,U^{-1},U^T,(U^T)^{-1}$ have integral entries and in this lemma matrix D is rational. Since $A^T z = U^T\begin{pmatrix} D \\ 0 \end{pmatrix}^T z = U^T\begin{pmatrix} D^T z \\ 0 \end{pmatrix}$, we have that $A^T z \in \mathbb{Z}^s$ if and only if $D^T z \in \mathbb{Z}^k$. Equation (4.1) has an integral solution if and only if

$$Dw=db \ \text{is solvable in} \ w \in \mathbb{Z}^k, \qquad (4.4)$$

because there is a 1-1 relationship between the solutions $v \in \mathbb{Z}^s$ of (4.1) and $w \in \mathbb{Z}^k$ of (4.4), defined by $Uv = \begin{pmatrix} w \\ 0 \end{pmatrix}$. Hence d is also the smallest positive integer such that $Dw=db$ has a solution in $w \in \mathbb{Z}^k$, in other words $\text{def}(A;b)=\text{def}(D;b)$. Combining the previous equations gives: $<z,db> = <z,Av> = <A^T z,v> = <U^T\begin{pmatrix} D^T z \\ 0 \end{pmatrix}, v> = $

$<\begin{pmatrix} D^T z \\ 0 \end{pmatrix}, Uv> = <\begin{pmatrix} D^T z \\ 0 \end{pmatrix},\begin{pmatrix} w \\ 0 \end{pmatrix}> = <D^T z,w>$ Hence (4.3) is equivalent to

$$D^T z \in \mathbb{Z}^k \ \text{and}$$

$$(d,<D^T z,w>)=1, \ \text{for every solution} \ w \ \text{of} \ (4.4). \qquad (4.5)$$

(i) For every integer δ, the subdeterminants of $[A \; \delta b]$ of order k are $\mu_1,\ldots,\mu_m, \delta\mu_{m+1},\ldots, \delta\mu_n$. Now Lemma 2 implies that the equation $Av = \delta b$ has a solution in $v \in \mathbf{Z}^s$ if and only if

$$(\mu_1,\ldots,\mu_m) \mid \delta\mu_i, \text{ for } i=m+1,\ldots,n.$$

This holds if and only if $(\mu_1,\ldots,\mu_m) \mid \delta \cdot (\mu_1,\ldots, \mu_m, \mu_{m+1},\ldots, \mu_n)$. Because d is the smallest positive integer for which (4.1) has an integral solution in v, we have

$$d = \frac{(\mu_1,\ldots,\mu_m)}{(\mu_1,\ldots,\mu_m,\mu_{m+1},\ldots,\mu_n)}.$$

(ii) Matrix D is a $k \times k$-matrix, so $\det(D)$ is the only subdeterminant of order k of D, and the matrix $[D \; b]$ has k subdeterminants $\eta_1,\ldots,\eta_k$ of order k containing an entry from b. If we apply Lemma 3.(i) on matrix D, we get $d = \dfrac{\det(D)}{(\det(D), \eta_1,\ldots,\eta_k)}$. The matrix D and the subdeterminants $\eta_1,\ldots,\eta_k$ can be computed in polynomial time from A. Hence d can be computed in polynomial time and with this d, a solution w of (4.4) can be computed by Gauss elimination. With this w a solution v of (4.1) can also be computed in polynomial time.

(iii) It is sufficient to prove that there is a polynomial time deterministic algorithm to compute a $z \in \mathbf{Q}^k$ such that (4.5) holds. Let w be a solution of (4.4) and define $d_1 = \gcd(w_1,\ldots,w_k)$. The equation $Dx = \dfrac{d}{(d,d_1)} b$ has $x = \dfrac{1}{(d,d_1)} w$ as an integral solution. But d was the smallest positive integer for which (4.4) is solvable, so we must have $(d,d_1)=1$. With (the extended) Euclid's algorithm we find in polynomial time an $y \in \mathbf{Z}^k$ such that $\langle y,w \rangle = d_1$. If we define $z := (D^{\mathrm{T}})^{-1}y$, then $z \in \mathbf{Q}^k$, $D^{\mathrm{T}}z \in \mathbf{Z}^k$ and $(d, \langle D^{\mathrm{T}}z,w \rangle) = (d, \langle y,w \rangle) = (d,d_1)=1$. Hence this z satisfies (4.5). $\square$

5. Proofs of the propositions

We derive the four propositions of section 3 from the previous lemmas (or from Theorem 1, using the assumption on the computability of RSA-roots by individuals).

Proof of Proposition 4.

(ii)⇔(iii)⇔(iv)

This follows from Lemma 3 and the definition of defect.

(i)⇒(iii)

Suppose that it is feasible for the individual to compute x^b from $(x^{a_1},\ldots, x^{a_s})$ for uniformly chosen x. Put $A=[a_1 \ldots a_s]$ and $d=\mathrm{def}(a_1, \ldots,a_s;b)$.

By Lemma 3 we can compute in polynomial time a vector $z=(z_1,\ldots,z_k) \in \mathbf{Q}^k$ such that $(d,\langle z,db \rangle)=1$ and $A^{\mathrm{T}}z \in \mathbf{Z}^s$. Hence $A^{\mathrm{T}}z=(c_1,\ldots,c_s)^{\mathrm{T}}$, where $c_i=\langle a_i,z \rangle \in \mathbf{Z}$ for

$i=1,\dots,s$. Using (the extended) Euclid's algorithm we can feasibly compute $\alpha,\beta\in\mathbb{Z}$ with $\alpha<z,db>+\beta d=1$, so $\alpha<z,b>+\beta=1/d$. Choose x uniformly from $\mathbb{Z}^*_N$ and put $x=(x^{z_1},\dots,x^{z_k})$. Hence $x^{a_i}\equiv x^{<z,a_i>}=x^{c_i}$ for $i=1,\dots,s$. So the individual can feasibly compute those x^{a_i}, and thus by assumption he can compute x^b. But then it is feasible to compute $(x^b)^\alpha x^\beta = x^{\alpha<z,b>+\beta} = x^{1/d}$. From the assumption on RSA-roots, it follows that $d=1$. This proves (iii).

(ii)$\Rightarrow$(i)

Suppose there are integers $v_1,\dots,v_s$ such that $v_1a_1+\dots+v_sa_s = db$. Lemma 3 states that it is feasible to compute such $v_1,\dots,v_s$. Now x^b can be computed in polynomial time from $x^b = (x^{a_1})^{v_1}\dots(x^{a_s})^{v_s}$. $\qquad\qquad$ ⬚

Proof of Proposition 3.

(i)$\Leftrightarrow$(ii)$\Leftrightarrow$(iii)

This follows from Proposition 4 with $A=[a_1 \dots a_s\ e_1 \dots e_k]$ (note that A has full row rank whence the equation $Ay=b$ is solvable in $y\in\mathbb{Q}^k$).

(iii)$\Leftrightarrow$(iv)

Define $\tilde{A}=[a_1 \dots a_s]$ and $I=[e_1 \dots e_k]$. Since each column of I has exactly *one* entry $\neq 0$, each subdeterminant of $[\tilde{A}\ I]$ containing q columns from I is a subdeterminant of $[\tilde{A}]$ of order $s-q$. Further $\det(I)=1$. Similarly, each subdeterminant of $[\tilde{A}\ I\ b]$ containing q columns from I and at least one entry from b is a subdeterminant of $[\tilde{A}\ b]$ of order $s-q$, containing at least one entry from b. $\quad$ ⬚

We leave the proofs of Propositions 1 and 2 to the reader.

6. Main theorem

In our propositions of section 3, we used the assumption on the computability of RSA-roots by individuals. These propositions can be generalized into a theorem, in which that assumption is not required anymore. To state this theorem we need the formalism of Probabilistic Turing Machines. But in this extended abstract we shall only state that theorem in words and therefore not prove it here.

Theorem 1. *Let $a_1,\dots,a_s,b\in(\mathbb{Q}_N)^k$, and assume that the equation $[a_1 \dots a_s]y=b$ is solvable in $y\in\mathbb{Q}^k$. Hence $d=\mathrm{def}(a_1,\dots,a_s;b)$ is defined.*

(1) *Suppose we have a "black box" which outputs x^b from the input $(x^{a_1},\dots,x^{a_s})$ with average probability (over x) $\geq\varepsilon_1>0$.*

With this black box we can build an algorithm which computes $u^{\frac{1}{d}}$ from input u, with probability $\geq\frac{1}{2}$, for every fixed $u\in\mathbb{Z}^*_N$. The running time of this algorithm is

$\frac{1}{\varepsilon_1}P_1$, where P_1 depends polynomially on the length of the input (which consists of N and the numerators and the denominators of the coordinates of $a_1,...,a_s,b$).

(2) Suppose we have a "black box" which outputs $u^{\frac{1}{d}}$ from the input with average probability (over u) $\geq \varepsilon_2 > 0$.

With this black box we can build an algorithm which computes x^b from input ($x^{a_1},..., x^{a_s}$), with probability $\geq \frac{1}{2}$, for every fixed $x \in (\mathbb{Z}^*_N)^k$. The running time of this algorithm is $\frac{1}{\varepsilon_3}P_2$, where P_2 depends polynomially on the same input as in (1).

More informally, this theorem states that computing x^b from ($x^{a_1},..., x^{a_s}$) is polynomial time reducible to computing $u^{\frac{1}{d}}$ from u, where $d=\mathrm{def}(a_1,...,a_s;b)$ and x,u random.

Under the assumption that it is infeasible for an individual to compute RSA-roots for random numbers, we can derive Proposition 4 (i)$\Leftrightarrow$(ii)$\Leftrightarrow$(iii) from Theorem 1.

References

[CBHMS89] David Chaum, Bert den Boer, Eugène van Heyst, Stig Mjølsnes and Adri Steenbeek, "Efficient Offline Electronic Checks", to appear in *Advances in Cryptology-EUROCRYPT '89*, Lecture Notes in Computer Science, Springer-Verlag.

[CE86] David Chaum and Jan-Hendrik Evertse, "A secure and privacy-protecting protocol for transmitting personal information between organizations", *Advances in Cryptology -CRYPTO '86*, A.M. Odlyzko ed., Lecture Notes in Computer Science 263, Springer-Verlag,.pp 118-167.

[Gill77] John Gill, "Computational Complexity of Probabilistic Turing Machines", *SIAM L. Comp.* **6** (1977) pp. 675-695.

[Has85] Johan Hastad, "On using RSA with low exponent in a public key network", *Advances in Cryptology -CRYPTO '85*, H.C. Williams ed., Lecture Notes in Computer Science 218, Springer-Verlag,.pp 403-408.

[Heg1858] I. Heger, "Über die Auflösung eines Systemes von mehreren unbestimmten Gleichungen des ersten Grades in ganzen Zahlen", *Denkschriften der Königlichen Akademie der Wissenschaften (Wien), Mathematisch-naturwissenschaftliche Klasse* **14** (2. Abth.) (1858) pp1-122.

[KaBa79] R. Kannan and A. Bachem, "Polynomial algorithms for computing the Smith and Hermite normal forms of an integer matrix", *SIAM Journal on Computing*, **8** (1979) pp 499-507.

[OO89] Tatsuaki Okamoto and Kazuo Ohta, "Disposable Zero-Knowledge Authentications and Their Applications to Untraceable Electronic Cash", to appear in *Advances in Cryptology -CRYPTO '89*, Lecture Notes in Computer Science, Springer-Verlag.

[RSA78]   R.L. Rivest, A. Shamir, and L. Adleman, "A Method for Obtaining Digital Signatures and Public Key Cryptosystems", *Comm. of the ACM* **21** (1978) pp 120-126.

[Schr86]   Alexander Schrijver, *Theory of Linear and Integer Programming*, John Wiley & Sons, 1986.

[Sh83]   Adi Shamir, "On the Generation of Cryptographically Strong Pseudorandom Sequences", *ACM Trans. on Computer Systems*, **1** (1983) pp 38-44.

Implementation of a Key Exchange Protocol Using Real Quadratic Fields

Extended Abstract

Renate Scheidler

Department of Computer Science

University of Manitoba

Winnipeg, Manitoba

Canada R3T 2N2

Johannes A. Buchmann

FB-10 Informatik

Universität des Saarlandes

6600 Saarbrücken

West Germany

Hugh C. Williams

Department of Computer Science

University of Manitoba

Winnipeg, Manitoba

Canada R3T 2N2

Implementation of a Key Exchange Protocol Using Real Quadratic Fields

Extended Abstract

1. Introduction

In [1] Buchmann and Williams introduced a key exchange protocol which is based on the Diffie-Hellman protocol (see [2]). However, instead of employing arithmetic in the multiplicative group F^* of a finite field F (or any finite Abelian group G), it uses a finite subset of an infinite Abelian group which itself is not a subgroup, namely the set of reduced principal ideals in a real quadratic field. As the authors presented the scheme and its security without analyzing its actual implementation, we will here discuss the algorithms required for implementing the protocol.

Let $D \in Z_+$ be a squarefree integer, $K = Q + Q\sqrt{D}$ the *real quadratic number field* generated by $\sqrt{D}$, and $O = Z + Z\dfrac{\sigma - 1 + \sqrt{D}}{\sigma}$ the *maximal real quadratic order* in K,

where $\sigma = \begin{cases} 1 & \text{if } D \equiv 2, 3 \pmod 4 \\ 2 & \text{if } D \equiv 1 \pmod 4 \end{cases}$.

A subset a of O is called an *ideal* in O if both $a + a$ and $O \cdot a$ are subsets of a. An ideal is said to be *primitive* if it has no rational prime divisors. Each primitive ideal a in O has a representation

$$a = \left[\frac{Q}{\sigma}, \frac{P + \sqrt{D}}{\sigma} \right] = Z\frac{Q}{\sigma} + Z\frac{P + \sqrt{D}}{\sigma},$$

where $P, Q \in \mathbf{Z}$, Q is a divisor of $D - P^2$ (see [5]). Let $\Delta = \dfrac{4}{\sigma^2} D$ denote the *discriminant* of K, set $d = \lfloor \sqrt{D} \rfloor$.

A *principal ideal* $\mathbf{a}$ of $\mathbf{O}$ is an ideal of the form $\mathbf{a} = \dfrac{1}{\alpha} \mathbf{O}$, $\alpha \in K\text{-}\{0\}$. Denote by $\mathbf{P}$ the set of primitive principal ideals in $\mathbf{O}$. An ideal $\mathbf{a} = \dfrac{1}{\alpha} \mathbf{O} \in \mathbf{P}$ is *reduced* if and only if α is a *minimum* in $\mathbf{O}$, i.e. if $\alpha > 0$ and there exists no $\beta \in \mathbf{O}\text{-}\{0\}$ such that $|\beta| < \alpha$ and $|\beta'| < \alpha$. Since the set $\{\log \alpha \mid \alpha \text{ is a minimum in } \mathbf{O}\}$ is discrete in the real numbers $\mathbf{R}$, the minima in $\mathbf{O}$ can be arranged in a sequence $(\alpha_j)_{j \in \mathbf{Z}}$ such that $\alpha_j < \alpha_{j+1}$ for all $j \in \mathbf{Z}$. If we define $\mathbf{a}_j = \dfrac{1}{\alpha_j} \mathbf{O}$ for all $j \in \mathbf{Z}$, then the set $\mathfrak{R}$ consisting of all reduced ideals in $\mathbf{P}$ is finite and can be written as $\mathfrak{R} = \{\mathbf{a}_1, ..., \mathbf{a}_l\}$ where $l \in \mathbf{Z}_+$.

Define an (exponential) *distance* between two ideals $\mathbf{a}, \mathbf{b} \in \mathfrak{R}$ as follows:

$\lambda(\mathbf{a}, \mathbf{b}) = \alpha$ where $\alpha \in K^{>0}$ is such that $\mathbf{b} = \dfrac{1}{\alpha} \mathbf{a}$ and $|\log \alpha|$ is minimal.

(The logarithm of this distance function is exactly the distance as defined in [1] and [4].) Similarly, let the distance between an ideal $\mathbf{a} \in \mathfrak{R}$ and a positive real number x be

$\lambda(\mathbf{a}, x) = \dfrac{e^x}{\alpha}$ where $\alpha \in K^{>0}$ is such that $\mathbf{a} = \dfrac{1}{\alpha} \mathbf{O}$ and $|x - \log \alpha|$ is minimal.

Throughout our protocol the inequalities $\eta^{-\frac{1}{4}} < \lambda(\mathbf{a}, \mathbf{b})$, $\lambda(\mathbf{a}, x) < \eta^{\frac{1}{4}}$ will be satisfied for all $\mathbf{a}, \mathbf{b} \in \mathfrak{R}$, $x \in \mathbf{R}_+$, where η is the *fundamental unit* of K.

Lemma 1: Let $\mathbf{b} \in \mathfrak{R}$ and write $\mathbf{b} = \mathbf{b}_j$, $\mathbf{b}_k = \left[\dfrac{Q_{k-1}}{\sigma}, \dfrac{P_{k-1} + \sqrt{D}}{\sigma} \right]$ for $k \geq j$. Then the following is true:

a) $\mathbf{b}_k \in \mathfrak{R}$ and $0 < P_k \leq d$, $0 < Q_k \leq 2d$ for $k \geq j$,

b) $\quad 1 + \dfrac{1}{\sqrt{\Delta}} < \lambda(\mathbf{b}_{j+1}, \mathbf{b}_j) < \sqrt{\Delta},$

c) $\quad \lambda(\mathbf{b}_{j+2}, \mathbf{b}_j) > 2,$

d) $\quad$ If $\mathbf{b} = \dfrac{1}{\beta}\mathbf{O},\ \beta \in K_{>0}$, then $\lambda(\mathbf{b}, x) = \dfrac{e^x}{\beta},$

e) $\quad \lambda(\mathbf{b}_k, \mathbf{b}_j) = \dfrac{\lambda(\mathbf{b}_k, x)}{\lambda(\mathbf{b}_j, x)}$ for any $x \in \mathbf{R}_+,\ k \geq j.$

Since principal ideal generators and distances are generally irrational numbers, we need to use approximations in our protocol. Denote by $\mathbf{a}(x)$ the reduced ideal *closest* to $x \in \mathbf{R}_+$, i.e. $|\log \lambda(\mathbf{a}(x), x)| < |\log \lambda(\mathbf{b}, x)|$ for any $\mathbf{b} \in \mathfrak{R},\ \mathbf{b} \neq \mathbf{a}$, and by $\mathring{\mathbf{a}}(x)$ the ideal actually computed by our algorithm. Define $\mathbf{a}_+(x)$ to be the reduced ideal such that its distance to x is maximal and < 1. Similarly, $\lambda(\mathbf{a}_-(x), x) > 1$ and minimal. Let $\lambda_1(x) = \lambda(\mathbf{a}(x), x)$, $\lambda_2(x) = \lambda(\mathring{\mathbf{a}}(x), x)$. Denote by $\hat{\lambda}(\mathbf{a}, x)$ the approximation of $\lambda(\mathbf{a}, x)$ computed by our algorithm; write $\hat{\lambda}(\mathbf{a}, x) = \dfrac{M(\mathbf{a}, x)}{2^p}$ where $M(\mathbf{a}, x) \in \mathbf{Z}_+$ and $p \in \mathbf{Z}_+$ is a *precision constant* to be determined later. $\hat{\lambda}_1(x)$, $M_1(x)$, $\hat{\lambda}_2(x)$, $M_2(x)$ are defined analogously to $\hat{\lambda}(x)$ and $M(x)$ with respect to $\lambda_1(x)$ and $\lambda_2(x)$. Set

$$G = 1 + \frac{1}{15(d+1)}, \qquad \gamma = \lceil G^{-1} 2^p \rceil, \qquad \chi = 1 + \frac{1}{2^{p-1}}.$$

The protocol can be outlined as follows: Two communication partners A and B agree publicly on a small number $c \in \mathbf{R}_+$ and an initial ideal $\mathring{\mathbf{a}}(c)$ with approximate distance $M_2(c)$ from c. A secretly chooses $a \in \{1, ..., d\}$, computes $\mathring{\mathbf{a}}(ac)$ and $M_2(ac)$ from $\mathring{\mathbf{a}}(c)$ and $M_2(c)$, and sends both to B. Similarly, B secretly chooses $b \in \{1, ..., d\}$, calculates $\mathring{\mathbf{a}}(bc)$ and $M_2(bc)$, and transmits both to A. Now both communication partners are able to determine an ideal $\mathring{\mathbf{a}}(abc)$. Although this ideal need not be the same for A and B (due to their different approximation errors in the computation), a little additional work will enable them to agree on a common ideal which is the secret key.

As pointed out in [1], we expect $l = |\Re| >> D^{\frac{1}{2} - \varepsilon}$ for arbitrary ε if D is chosen correctly and sufficiently large. This shows that an exhaustive search attack is infeasible. The authors conjecture that breaking the protocol enables one to factor. In [1] it is proved that solving the *discrete logarithm problem* for reduced principal ideals in real quadratic orders - given $a \in \Re$ find $\lambda(a, x)$ - in polynomial time implies being able to both break the scheme and factor D in polynomial time.

Throughout the protocol we will assume $M(a, x) \geq \gamma$ for all $a \in \Re$ and $x \in R_+$. Any number $\theta \in K$ is approximated by $\hat{\theta} \in Q$ such that $\chi^{-1}\theta \leq \hat{\theta} \leq \chi\theta$.

2. The Algorithms

For our protocol we need to perform arithmetic in both P and $\Re$. Our first algorithm enables us to compute any reduced ideal a_k from a given reduced ideal a_j by simply going through $\Re$ "step by step".

Algorithm 1 (*Neighbouring in* $\Re$): *Input:* $a_j \in \Re$.

Output: The neighbours $a_{j+1}, a_{j-1} \in \Re$ and ψ_+, ψ_- such that $a_{j\pm1} = \psi_\pm a_j$.

Algorithm: a_{j+1} is obtained by computing one iteration in the continued fraction expansion of the irrational number $\dfrac{P_{j-1} + \sqrt{D}}{Q_{j-1}}$. The algorithm for a_{j-1} is the inverse of the algorithm for a_{j+1}. In particular:

$$q_{j-1} = \left\lfloor \frac{P_{j-1} + d}{Q_{j-1}} \right\rfloor, \quad P_j = q_{j-1}Q_{j-1} - P_{j-1}, \quad Q_j = \frac{D - P_j^2}{Q_{j-1}}, \quad \psi_+ = \frac{\sqrt{D} - P_j}{Q_j},$$

$$Q_{j-2} = \frac{D - P_{j-1}^2}{Q_{j-1}}, \quad q_{j-2} = \left\lfloor \frac{P_{j-1} + d}{Q_{j-2}} \right\rfloor, \quad P_{j-2} = q_{j-2}Q_{j-2} - P_{j-1}, \quad \psi_- = \frac{\sqrt{D} + P_{j-1}}{Q_{j-2}}.$$

Algorithm 2 (*Multiplication in* **P**): *Input:* $\mathbf{a}, \mathbf{a}' \in \mathbf{P}$.

Output: $U \in \mathbf{Z}_{\geq 0}, \mathbf{c} \in \mathbf{P}$ such that $\mathbf{a}\mathbf{a}' = U\mathbf{c}$.

Algorithm: See [3], [4].

Lemma 2: If $\mathbf{a} = \mathbf{a}_s, \mathbf{a} = \mathbf{a}_t$ such that $\mathbf{a}_{s-1}, \mathbf{a}_{t-1} \in \mathfrak{R}$, then Algorithm 2 performs $O(\log D)$ arithmetic operations on numbers of input size $O(\log D)$.

Proof: By Lemma 1 all input numbers are polynomially bounded in D. The algorithm performs a fixed number of arithmetic operations plus two applications of the Extended Euclidean Algorithm which has complexity $O(\log D)$. ♦

Algorithm 3 (*Reduction in* **P**): *Input:* $\mathbf{c} = \left[\frac{Q}{\sigma}, \frac{P + \sqrt{D}}{\sigma} \right] \in \mathbf{P}$.

Output: $\mathbf{b} \in \mathfrak{R}, G, B \in \mathbf{Z}_{\geq 0}$ such that $\theta = \frac{G + B\sqrt{D}}{Q}$ and $\mathbf{b} = \theta\mathbf{c}$.

Algorithm: The algorithm is very similar to Algorithm 1 and uses again the continued fraction expansion of $\frac{P + \sqrt{D}}{Q}$ (see [3]).

Lemma 3: If $c = \frac{1}{U} a_s a_t$ where a_s, a_t are as in Lemma 2, then Algorithm 3 performs $O(\log D)$ arithmetic operations on numbers of input size $O(\log D)$.

Proof: By [5], Algorithm 2, and Lemma 1, the maximun number of iterations is $O(\log D)$. The bound on the input size follows from Lemma 1 and results in [4]. ◆

Algorithm 4: *Input:* $\hat{a}(x), \hat{a}(y) \in \mathfrak{R}$, $M_2(x), M_2(y)$ for $x, y \in \mathbf{R}_+$.

Output: $\hat{a}(x+y) \in \mathfrak{R}$, $M_2(x+y)$.

Algorithm: First use Algorithm 2 to compute $U \in \mathbf{Z}$, $c = \left[\dfrac{Q}{\sigma}, \dfrac{P + \sqrt{D}}{\sigma} \right] \in \mathbf{P}$ such that $(U)c = \hat{a}(x)\hat{a}(y)$. Then compute $b = \left[\dfrac{Q'}{\sigma}, \dfrac{P' + \sqrt{D}}{\sigma} \right] \in \mathfrak{R}$ and $G, B \in \mathbf{Z}_{\geq 0}$ such that $b = \theta c$, $\theta = \dfrac{G + B\sqrt{D}}{Q}$ using Algorithm 3. Finally apply Algorithm 1 to b a certain number of times to obtain $\hat{a}(x+y) = \zeta b = \dfrac{\zeta\theta}{U} \hat{a}(x)\hat{a}(y)$. Set

$$M_2(x+y) = \left\lceil \frac{\hat{\zeta}\hat{\theta}M_2(x)M_2(y)}{2^P U} \right\rceil,$$

where $\hat{\zeta}, \hat{\theta}$ are rational approximations to ζ, θ, respectively.

Lemma 4: If $\hat{a}(x) = a_s$, $\hat{a}(y) = a_t$ such that $a_{s-1}, a_{t-1} \in \mathfrak{R}$, then Algorithm 4 performs $O(\log D)$ arithmetic operations on inputs of size $O(\log D)$.

Proof:By Lemma 2, computing c takes $O(\log D)$ arithmetic operations on inputs of size $O(\log D)$. By Lemma 3, the same is true for the computation of b. From Lemma 1 it can be proved that, in obtaining $\hat{a}(x+y)$ from b, all numbers involved are polynomially bounded in D and $\hat{a}(x+y)$ can be obtained from b in $O(\log D)$ iterations. ◆

Both communication partners can determine the key by using the following algorithm which is based on the idea of a standard exponentiation method:

Algorithm 5: *Input:* $â(x) \in \Re$ for $x \in \mathbf{R}_+$, $M_2(x)$, $y \in \mathbf{Z}_+$.

Output: $â(xy)$, $M_2(xy)$.

Algorithm: 1) Determine the binary decomposition $y = \sum_{i=0}^{l} b_i\, 2^{l-i}$ of y, $b_i \in \{0,1\}$, $b_0 = 1$.

  2) Set $â(z_0) = â(x)$.

  3) for $i = 1$ to l do

    a) Compute $â(2z_{i-1})$, $M_2(2z_{i-1})$ using Algorithm 4.

        Set $â(z_i) := â(2z_{i-1})$, $M_2(z_i) := M_2(2z_{i-1})$.

    b) if $b_i = 1$ then compute $â(z_i+x)$, $M_2(z_i+x)$ using Algorithm 4.

    Set $â(z_i) := â(z_i+x)$, $M_2(z_i) := M_2(z_i+x)$.

  4) Set $â(xy) := â(z_l)$, $M_2(xy) = M_2(z_l)$.

Lemma 5: If $â(x) = a_s$ such that $a_{s-1} \in \Re$ and y is polynomially bounded in D, then Algorithm 5 performs $O((\log D)^2)$ arithmetic operations on inputs of size $O(\log D)$.

Proof: For each iteration, steps 3a and 3b each perform $O(\log D)$ operations on numbers of input size $O(\log D)$ by Lemma 4. So the number of operations needed for step 3 is $O(l \log D) = O((\log D)^2)$. ♦

3. The Protocol

Algorithm 6 (*Initial values*): *Input*: $r \in \{2,..., d\}$.

Output: $\mathbf{a} \in \mathfrak{R}$, $M \in \mathbf{Z}_+$, such that the ideal $\mathbf{a}$ and its distance M can be used as initial values for the protocol.

Algorithm: Set $\mathbf{a} = \hat{\mathbf{a}}(c) = \mathbf{O}$, $M = M_2(c) = \lceil 2^p r \rceil$, where $c = \log r$. Then $M \geq 2^{p+1} > \gamma$. Since $1 + \frac{1}{\sqrt{\Delta}} < r = \lambda_2(c) < \sqrt{\Delta}$, we have $\mathbf{a} = \mathbf{a}_-(c)$.

In order to find a unique key ideal, all approximation errors $\rho_2(x) = \dfrac{\hat{\lambda}_2(x)}{\lambda_2(x)}$ $(x \in \mathbf{R}_+)$ in Algorithms 4, 5, and 6 must be close to 1, i. e. p must be sufficiently large.

Theorem 1: Let $a, b \in \{1,...,d\}$, $\hat{\mathbf{a}}(c)$, $M_2(c)$ as in Algorithm 6. Let $\hat{\mathbf{a}}(abc)$ be computed by applying Algorithm 5 first to $\hat{\mathbf{a}}(c)$, $M_2(c)$, and b to obtain $\hat{\mathbf{a}}(bc)$ and $M_2(bc)$, then to $\hat{\mathbf{a}}(bc)$, $M_2(bc)$, and a to obtain $\hat{\mathbf{a}}(abc)$ and $M_2(abc)$. If $2^p \geq 1280d(d^2-1)$, then $\hat{\mathbf{a}}(abc) \in \{\mathbf{a}_-(abc), \mathbf{a}_+(abc)\}$ and $M_2(abc) \geq \gamma$.

The uniqueness of the key ideal is guaranteed by the following Lemma:

Lemma 6: Let $p, a, b, c, \hat{\mathbf{a}}(c), M_2(c)$ be as in Theorem 1. Set $x = abc$.

If $\lambda_1(x) > G^2$ or $\lambda_1(x) < G^{-2}$ then $\hat{\mathbf{a}}(x) = \mathbf{a}_-(x)$.

If $G^{-2} \leq \lambda_1(x) \leq G^2$ then $a(x)$ can be determined from $\hat{\mathbf{a}}(x)$.

Proof: Omit the argument x for brevity. If $\lambda_1 > G^2$ or $\lambda_1 < G^{-2}$ then $\hat{\lambda}_2 > G$ and hence $\lambda_2 = \dfrac{\hat{\lambda}_2}{\rho_2} > 1$, so $\hat{\mathbf{a}} = \mathbf{a}_-$.

If $G^{-2} \le \lambda_1 \le G^2$, then by Theorem 1 $\hat{a} \in \{a_+, a_-\}$, so a = $\hat{a}$ or a is one of the neighbours

of $\hat{a}$. From Theorem 1 it can be proved that $G^{-1} \le \rho_2 \le G$ and hence

$G^{-3} \le \hat{\lambda}_1 < \dfrac{1 + 2^{-p}}{1 - G^3 2^p} G^3$. So both communication partners can determine an ideal **b**

which is either $\hat{a}$ or a neighbour of $\hat{a}$ such that $G^{-3} \le \hat{\lambda}(\mathbf{b}, abc) < \dfrac{1 + 2^{-p}}{1 - G^3 2^p} G^3$. Then it

can be shown that $\dfrac{1}{1 + \dfrac{1}{\sqrt{\Delta}}} < \lambda(\hat{a}, \mathbf{b}) < 1 + \dfrac{1}{\sqrt{\Delta}}$, therefore by Lemma 1: $\hat{a}$ = a. ♦

We are now equipped to set up the protocol. We assume $2^p \ge 1280 d(d^2 - 1)$.

Protocol:

The two communication partners Alice and Bob perform the following steps:

1) Both Alice and Bob agree on D and a small positive integer r. They compute a = $\hat{a}(c)$, M = $M_2(c) \ge \gamma$ using Algorithm 6 where c = $\log r$. D, a, and M can be made public.

2) Alice secretly chooses $a \in \{1,..., d\}$ and from a, M computes $\hat{a}(ac)$, $M_2(ac) \ge \gamma$ using Algorithm 5. She sends both to Bob.

3) Bob secretly chooses $b \in \{1,..., d\}$ and from a, M computes $\hat{a}(bc)$, $M_2(bc) \ge \gamma$ using Algorithm 5. He sends both to Alice.

4) From $\hat{a}(ac)$, $M_2(ac)$, and b, Bob computes $\hat{a}(abc)$ and its two neighbours as well as their

approximate distances (i.e. M values) using Algorithms 5 and 1. If he finds among these an

ideal b such that $\dfrac{2^p}{G^3} \le M(\mathbf{b}, abc) < \dfrac{(1 + 2^p)G^3}{1 - 2^p G^3}$, then $\mathbf{b} = a(abc)$. In this case he sends

'0' back to Alice. If he cannot find such an ideal, then by Lemma 6 he can compute $a_-(abc)$. In this case he sends '1' to Alice.

5) From $\hat{a}(bc)$, $M_2(bc)$, and a, Alice computes $\hat{a}(abc)$, $M_2(abc)$ using Algorithm 5. If she received '0' from Bob, then she computes the neighbours of $\hat{a}(abc)$ and their M values and attempts to compute $a(abc)$. If successful, she sends '0' back to Bob. The common key is then $a(abc)$. Otherwise the ideal $\hat{a}(abc)$ she computed is $a_-(abc)$. In this case she sends '1' to Bob. If Alice received '1' from Bob, then he was unable to determine $a(abc)$, so we must have $\lambda_1(abc) < G^{-2}$ or $\lambda_1(abc) > G^2$ by Lemma 6, in which case the ideal $\hat{a}(abc)$ computed by Alice is $a_-(abc)$. This is then the key. In this case she sends '1' back to Bob.

6) If Bob receives the same bit he sent, then the ideal he computed in step 4 is the key. The only other possibility is that he sent '0' and received '1'. In this case Alice was unable to determine $a(abc)$. The key is then the ideal $\hat{a}(abc) = a_-(abc)$ initially computed by Bob.

References:

[1] J. A. Buchmann, H. C. Williams, *A key exchange system based on real quadratic fields*, extended abstract, to appear in: Proceedings of CRYPTO '89.

[2] W. Diffie, M. Hellman, *New directions in cryptography*, IEEE Trans. Inform. Theory, vol. 22, 1976.

[3] R. A. Mollin, H. C. Williams, *Computation of the class number of a real quadratic field*, to appear in: Advances in the Theory of Computation and Computational Mathematics (1987).

[4] A. J. Stephens, H. C. Williams, *Some computational results on a problem concerning powerful numbers*, Math. of Comp. vol. 50, no. 182, April 1988.

[5]   H. C. Williams, M. C. Wunderlich, *On the parallel generation of the residues for the continued fraction factoring algorithm*, Math. of Comp. vol. 48, no. 177, January 1987.

DISTRIBUTED PRIMALITY PROVING
AND
THE PRIMALITY OF $(2^{3539} + 1)/3$

François Morain [*†]

morain@inria.inria.fr

Abstract

We explain how the Elliptic Curve Primality Proving algorithm can be implemented in a distributed way. Applications are given to the certification of large primes (more than 500 digits). As a result, we describe the successful attempt at proving the primality of the 1065-digit $(2^{3539}+1)/3$, the first *ordinary* Titanic prime.

1 Introduction

For cryptographical purposes [7], it is desirable to generate large primes as fast as possible. This can be done via *ad hoc* techniques [30, 12, 14, 4] or by means of a general purpose primality testing algorithm such as that described in [1, 11, 10, 6] or the Elliptic Curve Primality Proving (ECPP) algorithm due to Atkin [2, 26, 24] (For a survey of primality testing, see [18]).

Another point is to certify large primes, such as the Cunningham numbers [8], which sometimes have more than 400 digits. The purpose of this paper is to explain how the ECPP algorithm has been implemented on a network of workstations and used to test some numbers with more than 500 digits for primality. In particular, it is now routine to test 800-digit numbers and it is not too hard to test 1000-digit numbers.

We first begin by a short introduction to ECPP and then, we explain the distributed process à *la* Lenstra-Manasse [19]. These ideas are exemplified by the certification of large primes and we also give the history of the primality of the record breaking $\mathcal{N}_{3539} = (2^{3539} + 1)/3$, which has 1065 digits.

*Institut National de Recherche en Informatique et en Automatique (INRIA), Domaine de Voluceau, B. P. 105 78153 LE CHESNAY CEDEX (France) / Département de Mathématiques, Université Claude Bernard, 69622 Villeurbanne CEDEX (France).

†On leave from the French Department of Defense, Délégation Générale pour l'Armement.

2 A brief description of ECPP

2.1 Elliptic curves

Let $\mathbf{K}$ be a field of characteristic prime to 6. An elliptic curve E over $\mathbf{K}$ is a non singular algebraic projective curve of genus 1. It can be shown [9, 34] that E is isomorphic to a curve with equation:

$$y^2 z = x^3 + axz^2 + bz^3, \tag{1}$$

with a and b in $\mathbf{K}$. The *discriminant* of E is $\Delta = -16(4a^3 + 27b^2)$ and the *invariant* is

$$j = 2^8 3^3 \frac{a^3}{4a^3 + 27b^2}.$$

We write $E(\mathbf{K})$ for the set of points with coordinates $(x : y : z)$ which satisfy (1) with $z = 1$, together with the point at infinity: $O_E = (0 : 1 : 0)$. We will use the well-known *tangent-and-chord* addition law on a cubic [16] over a finite field $\mathbf{Z}/p\mathbf{Z}$ as well as over a ring $\mathbf{Z}/N\mathbf{Z}$ with N composite (see [21] for a justification).

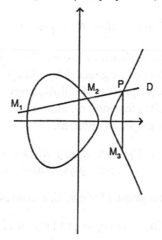

Figure 1: An elliptic curve over $\mathbf{R}$

In order to add two points $M_1 = (x_1, y_1)$ and $M_2 = (x_2, y_2)$ on E resulting in $M_3 = (x_3, y_3)$, the equations are

$$\begin{cases} x_3 &= \lambda^2 - x_1 - x_2 \\ y_3 &= \lambda(x_1 - x_3) - y_1 \end{cases}$$

where

$$\lambda = \begin{cases} (y_2 - y_1)(x_2 - x_1)^{-1} & \text{if } x_2 \neq x_1 \\ (3x_1^2 + a)(2y_1)^{-1} & \text{otherwise.} \end{cases}$$

We can compute kP using the binary method [17] (see also [10]) or addition-subtraction chains [29].

2.2   Primality proving

Let us recall one of the converses of Fermat's theorem.

Theorem 1 ([31]) *Let s be a divisor of $N-1$. Let a be an integer prime to N such that*

$$a^{N-1} \equiv 1 \bmod N \text{ and } \gcd(a^{(N-1)/q} - 1, N) = 1,$$

for each prime divisor q of s. Then each prime divisor p of N satisfies $p \equiv 1 \bmod s$.

Corollary 1 *If $s > \sqrt{N} - 1$ then N is prime.*

A similar theorem can be stated for elliptic curves.

Theorem 2 ([13, 20]) *Let N be an integer greater than 1 and prime to 6. Let E be an elliptic curve over Z/NZ, m and s two integers such that $s \mid m$. Suppose we have found a point P on E that satisfies $m P = O_E$, and that for each prime factor q of s, we have verified that $\frac{m}{q} P \neq O_E$. Then if p is a prime divisor of N, $\#E(Z/pZ) \equiv 0 \bmod s$.*

Corollary 2 *If $s > (\sqrt[4]{N} + 1)^2$, then N is prime.*

In order to use the preceding theorem, we need to compute the number of points m. This process is far from trivial in general (see [32]). From a practical point of view, it is desirable to use deep properties of elliptic curves over finite fields. This involves the theory of complex multiplication and class fields and requires a lot of theory [26]. We can summarize the principal properties.

Theorem 3 *Let p be an odd prime. Every elliptic curve E mod p has complex multiplication by an order of an imaginary quadratic field $K = Q(\sqrt{-D})$.*

From a very down-to-earth point of view, this comes down to saying

- p splits completely in K as $(p) = (\pi)(\pi')$ in K;

- $H_D(j(E)) \equiv 0 \bmod p$ for a fixed polynomial $H_D(X)$ in $Z[X]$;

- $m = \#E(Z/pZ) = (\pi - 1)(\pi' - 1) = p + 1 - t$, where $|t| \leq 2\sqrt{p}$ (Hasse's theorem).

The computation of the polynomials H_D is dealt with in [26] and [27].

2.3   Outline of ECPP

We now explain how the preceding theorems are used in a *factor and conquer* algorithm similar to the DOWNRUN process of [37]. The first phase of the algorithm consists in finding a sequence $N_0 = N > N_1 > \cdots > N_k$ of probable primes such that N_{i+1} prime $\Longrightarrow N_i$ prime. The second then proves that each number is prime, starting from N_k.

Procedure SearchN

1. $i := 0$; $N_0 := N$;

2. find a fundamental discriminant $-D$ such that (N_i) splits as the product of two principal ideals in $\mathbf{Q}(\sqrt{-D})$;

3. for each solution of $(N_i) = (\pi)(\pi')$, find all factors of $m_\pi = (\pi - 1)(\pi' - 1)$ less than a given bound B and let N_π be the corresponding cofactor;

4. if one of the N_π is a probable prime then set $N_{i+1} := N_\pi$, store $\{N_i, D, \pi, m_\pi\}$ set $i := i + 1$, and go to step 2 **else** go to step 3.

5. end.

The second phase consists in proving that the numbers N_i are indeed primes. This is done as follows.

Procedure Proof

for $i = k..0$

1. compute a root j of $H_{D_i}(X)$ mod N_i as described in [27, 28];

2. find an equation of the curve E_i whose invariant is j and cardinality m_i;

3. verify the condition of theorem (2).

end.

For more details, the reader is referred to [2].

3 Large primes

The author used ECPP to test about fifty numbers from the Cunningham tables [8] and some others, namely $S_p = ((1 + \sqrt{2})^p + (1 - \sqrt{2})^p)/2$ for $p \in \{1493, 1901\}$ with respectively 572 and 728 digits, in 30 and 40 days on a single SUN 3/60. Indeed, a simple extrapolation shows that testing a 1000-digit number would require about 6 months (at least). We must do something else to increase the bound on the largest number ECPP can test.

4 Distributed computations

From the preceding description, it is easy to see that this algorithm is very well suited for distributed computations. We can do the first phase in parallel and then the second one too. Let us see how I did this.

First of all, I implemented ECPP using the Le_Lisp language and the multiprecision described in [15]. Then the computations were done using a star network à la Caron-Silverman [33]. There are a *master* ($\mathfrak{M}$) and an indefinite number of workstations, called *slaves* ($\mathfrak{S}$).

The idea is that when dealing with very large numbers, the crucial part of ECPP is the first one, because it requires the factorization of very large numbers. There are basically two ways of doing that. The first one is to try to factor a single number using all the stations. The second is to let each station work on a different number. Actually, I use the latter scheme, because the first one would require more communications and also because it is not the *right* philosophy of the test: The less factoring power we use, the better.

We now describe the conditions required to do an optimal job.

4.1 Constraints

We want to use the idle time of a network of workstations. We do this in a way similar to that of [19]. We start a process on a machine in such a way that a legitimate user is not (too much) disturbed: If a user types on a console (in UNIX words, he changes the date of one of the tty's), then the program is stopped (by means of a kill -STOP) and restarted 10 minutes after the last action of the user (with a kill -CONT). The process is also stopped whenever the load climbs up some prescribed value (typically 1.5) and is subjected to the same restart conditions. All this is done with the shell scripts distributed by Mark Manasse for integer factorization. Another important feature of these programs is the ability to restart themselves after a small crash such as a Connection timed out from a server. Also, they do not depend on a particular machine (at least running UNIX or ULTRIX) or a particular language. It is possible to use a C program on a DEC station and a Le_Lisp program on a SUN.

4.2 The first phase

4.2.1 Role of the master

On $\mathfrak{M}$ (typically the author's own workstation), the program used does the following things for each N_i of the first phase

1. put in the file WHICHN the number to be tested;

2. find all fundamental discriminants D (from a finite subset $\mathcal{D}$) for which N_i is represented by a form of G_0 and put them in the file DSET;

3. initialize the rank of the next D to be examined to 1 in the file DRANK;

4. start finding a suitable D.

4.2.2 Role of the slaves

On $\mathfrak{S}$, the program looks like

1. read the number to be tested from WHICHN and call it N;

2. while N is equal to the content of WHICHN, select a new D in DSET, update DRANK and try to factor any of the m_π.

4.2.3 Tasks performed by every machine

Each machine does the following

1. find a D such that (N) splits completely in $\mathbf{Q}(\sqrt{-D})$;

2. try to factor each m_π using first trial division, then Pollard's ρ method, and finally the $p - 1$ method.

Inside each factoring algorithm, the program periodically tests whether something has happened. When this is so, it gives up on N_i and begins a new work on N_{i+1}. When using the ρ method [23], the test is done at each gcd (for our purposes, there are 10^4 iterations and a gcd each 1000 iterations). During $p - 1$, only once.

4.2.4 Communications between $\mathfrak{M}$ and $\mathfrak{S}$

The files DSET, DRANK and WHICHN have just been described. All this supposes the use of a distributed file system: Here it is NFS that does all the job. Special code has been written to handle the problems arising when one machine wants to read a file while another tries to write in it or to test whether the file can be accessed through NFS.

4.3 The second phase

For each N_i, it remains to check the primality conditions. Using a file containing the next number to be certified, each station takes the useful data and does its job. It should be noted that this phase can be started even if the first one is not complete.

4.4 Problems encountered

One of the major problem is the reliability of the NFS protocols, especially when using machines not depending from the same file server. The program is very well suited for testing the reliability of the network. Each time there is a connection problem, the process simply crashes.

Also, using a Le_Lisp executable requires a lot of memory and, sometimes, this resulted in a swap problem and also a crash.

5 Establishing a new frontier: the history of $\mathcal{N}_{3539}$

Last year, the 100-digit line was crossed for the first time for integer factorization [19]. In 1983, Yates [38] introduced the concept of *Titanic* primes, that is primes with at least 1000 decimal digits. This seemed to make a distinction between the real world of small primes and that of large primes. The frontier for primality testing was thus 1000 digits. The aim of this section is to describe how we went far beyond the line, thus making the testing of 1000-digit numbers a routine.

5.1 Entomology of a Record

The first thing to do was finding a good candidate. It had to be greater than the repunit R_{1031}, whose primality was proven by Williams and Dubner [36]. During their setting of the new Mersenne's conjecture [3], Bateman, Selfridge and Wagstaff tested some numbers of the form $\mathcal{N}_p = (2^p + 1)/3$ for primality. They found that $\mathcal{N}_p$ was a probable prime for $p \in \{1709, 2617, 3539\}$.

During EUROCRYPT '89 (April 10-13, 1989), it appeared that both ECPP and the Jacobi Sums test [11, 10, 6] were able to attack numbers as large as 1000 digits. This was the very start of a stimulating competition with W. Bosma and M.-P. van der Hulst.

Indeed, the first of these numbers ($p = 1709$, $\mathcal{N}_p$ with 514 digits) was the first number proven prime using ECPP in its distributed version. This was done on April 19, 1989 with three SUN's and four days of CPU.

Then, I decided to skip $p = 2617$ and try $\mathcal{N}_{3539}$. As shown by in the following figure, the factorization of $\mathcal{N}_{3539} - 1$ is not complete (up to now).

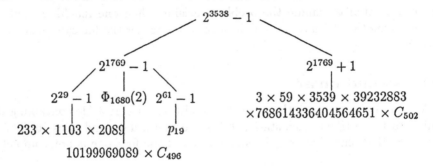

Using ECPP, I first launch the process on April 20, 1989. The set $\mathcal{D}$ mentioned above consisted of all D's with $h(-D) \leq 20$, sorted according to $(h/g, h, D)$. Following [2], the *difficulty* of testing N may be defined as

$$\Phi(N) = e^{-\gamma} \frac{\log N}{M(N)}$$

where γ is Euler's constant and $M(N)$ is defined as follows. Put

$$M(N) = \sum_{\substack{D \in \mathcal{D} \\ N \in G_0(-D)}} w(-D) \frac{g(-D)}{h(-D)},$$

where the summation is on all D for which N can be represented by a form of the principal genus of quadratic forms of discriminant $-D$, $g(-D) = 2^{t-1}$ with t the number of prime factors of D and $h(-D)$ the class number. As a matter of fact, $\Phi(N)$ yields the value $(\log B)$ of the upper bound on the largest factor of a number of points m we must factor in order to find a good candidate.

Coming back to $\mathcal{N}_{3539}$, I found that $M(\mathcal{N}_{3539}) = 55$ yielding $\log_{10} B = 11$. This implied in turn that the only way to achieve this was using ECM. At that time, I hadn't implemented this and so the program started using only Pollard ρ and the $p-1$ method *. This first attempt lasted till May 13, without any result: I couldn't even find a good N_1. There was something to be done. Moreover, some problems seemed to arise in the $p-1$ method, where the routine seemed to loop forever in some cases.

When looking at

$$\log B = \Phi(N), \tag{2}$$

there are two distinct ways of solving the problem. The first one is to use sophisticated factoring routines, the other one is to increase the value of $M(N)$. I used the second and decided to enlarge $\mathcal{D}$ with all D less than 2^{15} †. This increased $M(N)$ to the value of 174, yielding $\log_{10} B = 3.44$. This clearly said that ECM was no more necessary and that ρ was enough. After fixing some stupid bug in my ρ routine, I re-started the program on June 5 and it lasted till July 10, yet without any result.

Clearly, there was a problem. Using induction, it seemed clear that there was, somewhere, a deep bug that only appeared when dealing with large numbers, but not with small ones. So I decided to stop working on $\mathcal{N}_{3539}$, and began to reassure myself with a smaller one, namely $\mathcal{N}_{2617}$ (788 digits). Although this number could have been done by simply factoring $\mathcal{N}_{2617}-1$ (as remarked by Atkin), this attempt was designed to find this bug. So, the process started on July 21 and ended on August 19, proving the number to be prime, but without revealing any bug.

At this point, I decided to implement ECM, just to see if something would happen to change. I had problems with this, since it was only possible to use the first phase of the algorithm, all the second phases requiring too much memory (they all need about $(\log N)^2$ storage, making it infeasible for 1000-digit numbers). Moreover, I could only use 20 curves or so, again because the storage was making it prohibitive to use on workstations with not too much memory, such as a standard SUN 3/50 (4 Mo). This was quite a disappointment. The third attempt on $\mathcal{N}_{3539}$ was then started on September 12 and took two weeks. Nothing observable happened.

I decided then to replace all these D's less than 2^{15} with all D's with h/g small, irrespective of the size of D as soon as D fitted in a 32-bit word. More precisely, I computed and stored all D with $h(-D) \leq 50$ (plus some with $h = 64$) and ordered them according to $(h/g, h, D)$. This yielded $M(\mathcal{N}_{3539}) = 291$ and $\log_{10} B = 2.05$. What would appear as the last attempt then began on September 29.

A feeling of deep personal gratification came over me when, on October 5, 1989, I finally confirmed my initial impression that a part of the program was irretrievably

---

* As suggested by Atkin, $p-1$ is worth using when dealing with Cunningham numbers, because they have non-trivial arithmetic properties.

† This limitation comes from the language I used, Le_Lisp, which does not accept 32-bit integers.

bug ridden. This occurred when I thoroughly checked my factoring routines. I had simply forgotten to reduce the parameters of $\rho, p-1, \ldots$ after a factor was discovered ! The program was thus *asymptotically bugged*: When dealing with small numbers, I need maybe one large factor, but with large numbers, maybe two or more. This explained also the above mentioned problem with $p-1$ (because of the way the exponentiation routine was programmed, it wanted to find the first 1 in the binary expansion of a zero word).

And (not surprisingly ?), it began to work. The breakthrough occurred on October 6, when N_1 was reached, using $D = 97507$ (with $h = 36, g = 2$). After that, this was quite a quiet work, except that there happened to be a difficult client at one stage (namely N_{11}), requiring a D with $h = 56$ and $g = 2$. The building of the tower of primes was finally completed on November 8 at 830 pm (INRIA-Paris time).

Meantime, I had used one workstation for the second part of the process, proving the numbers to be primes. One week before the end of the first phase, I also used one of the SUN's on this $h/g = 28$ business, that is finding a root of a polynomial of degree 28 over a finite field with about 10^{991} elements. For that, I chose the most resistant SUN I could find. By this, I mean a station that was able to resist all network problems that could appear. Actually, this was a period of time where there was quite a lot of those. This computation took one week.

When the first phase ended, about 40 proving steps were done and I was able to launch the workstations on the remaining cases. On November 11, it was over, even the 28 degree stuff. It was it, I had sunk the Titanic, this time with an *ordinary* prime (as opposed to the Elliptic Mersenne Primes of [25]): The problem of testing 1000-digit numbers for primality was solved. Looking at the whole story, it took only one month and a half to do that. Moreover, it took only one week to come down from 700 digits to 10: This means that one can routinely test such numbers for primality. Some further experiments confirmed this [2].

The final result is a file of 500 kbytes consisting of the certificate of primality for N_{3539}. This file can be sent to anyone who wants to check it using the protocols described in [26].

5.2   Technical details

In order to prove the primality of N_{3539}, I used 12 SUN workstations, among which four 3/50, seven 3/60 and one 3/160 with a special chip designed for 512-bit multiplication [5]. Using the full power of the chip was done by using Montgomery's ideas on modular multiplications [22]: These ideas were only used for Pollard ρ, $p-1$ and pseudoprimality tests. The speedup for a modular exponentiation of 110 words of 32 bits is about 8.

The first phase took approximately 288 days of CPU (only one month and a half in real time). The second one 31 days of CPU. The total time is thus less than one year of CPU. The tower of primes consists of 162 numbers. In Figure 2, we print the number of digits of N_i versus the real time from the start of the job. The distribution of the *gains*, that is the number of digits we win in finding the following member is displayed in Figure 3: The mean value is 6.5, with minimum 0 and maximum 34. In

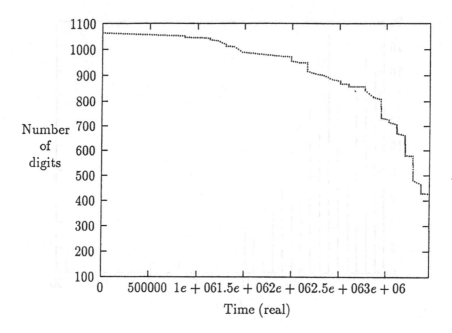

Figure 2: Number of digits reached vs. real time

Figure 4, we put the distribution of the values of h/g, the mean value being 3.49.

6 Conclusions

We see that ECPP in its distributed implementation is a very powerful tool to test arbitrary large numbers for primality. It should be able to deal with somewhat larger numbers (maybe with 1200 digits or so). The problem that is bound to arise is that there is a point where we need powerful factoring routines such as ECM. However, this would slow down the running time of the whole process. So it seems not possible to deal with 2000-digit numbers.

It should be noted that van der Hulst and Bosma finally succeeded in proving the same number to be prime (hopefully!). It took them [35] about three weeks and a half on a DEC 3100 (about five times faster than a SUN). They have made some improvements and now, it should just require one week and a half to do that size of number.

Acknowledgments. First of all, I thank the owners of the workstations that contributed to my record, namely L. Albert, L. Audoire, J.-J. Codani, V. Collette, P. Flajolet, P. Jacquet, P. Le Chenadec, M. Régnier, B. Serpette and P. Zimmermann. W. Bosma and M.-P. van der Hulst (and also A. K. Lenstra) must be thanked for their stimulating work to attack the supremacy of ECPP using the Jacobi Sums test.

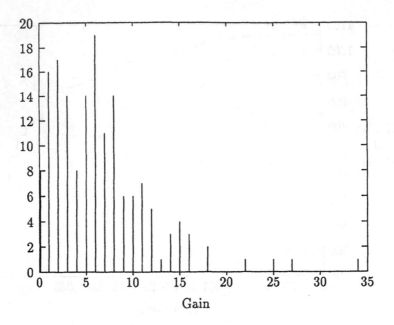

Figure 3: Number of digits gained at each step

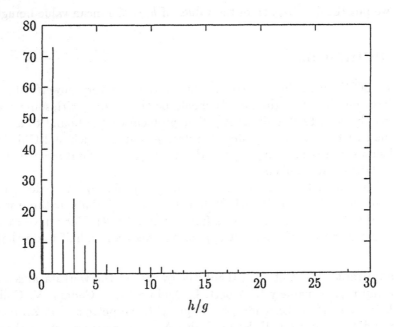

Figure 4: Distribution of h/g

Without the script-shells of M. Manasse, this job would have been less easy: special thanks to him, then. Thanks to R. Ehrlich who helped me modifying the above scripts and explained to me some of the magic properties of NFS. Thanks also to I. Vardi for (helpful or stylistic) comments about my manuscript.

Lastly, I'd like to thank the technical staff in Scanticon (where EUROCRYPT took place) for the wonderful job they did for me in doing my slides concerning some typical French comic character.

References

[1] L. M. ADLEMAN, C. POMERANCE, AND R. S. RUMELY. On distinguishing prime numbers from composite numbers. *Annals of Math. 117* (1983), 173–206.

[2] A. O. L. ATKIN AND F. MORAIN. Elliptic curves and primality proving. Rapport de Recherche 1256, INRIA, Juin 1990.

[3] P. T. BATEMAN, J. L. SELFRIDGE, AND S. S. WAGSTAFF, JR. The new Mersenne conjecture. *American Mathematical Monthly 96*, 2 (1989), 125–128.

[4] P. BEAUCHEMIN, G. BRASSARD, C. CRÉPEAU, C. GOUTIER, AND C. POMERANCE. The generation of random numbers that are probably prime. *J. Cryptology 1* (1988), 53–64.

[5] P. BERTIN, D. RONCIN, AND J. VUILLEMIN. Introduction to programmable active memories. In *Proc. of the Internat. Conf. on Systolic Arrays* (1989).

[6] W. BOSMA AND M.-P. VAN DER HULST. Faster primality testing. To appear in *Proc. Eurocrypt '89*.

[7] G. BRASSARD. *Modern Cryptology*, vol. 325 of *Lect. Notes in Computer Science*. Springer-Verlag, 1988.

[8] J. BRILLHART, D. H. LEHMER, J. L. SELFRIDGE, B. TUCKERMAN, AND S. S. WAGSTAFF, JR. *Factorizations of* $b^n \pm 1$, $b = 2, 3, 5, 6, 7, 10, 11, 12$ *up to high powers*, 2 ed. No. 22 in Contemporary Mathematics. AMS, 1988.

[9] J. W. S. CASSELS. Diophantine equations with special references to elliptic curves. *J. London Math. Soc. 41* (1966), 193–291.

[10] H. COHEN AND A. K. LENSTRA. Implementation of a new primality test. *Math. Comp. 48*, 177 (1987), 103–121.

[11] H. COHEN AND H. W. LENSTRA, JR. Primality testing and Jacobi sums. *Math. Comp. 42*, 165 (1984), 297–330.

[12] C. COUVREUR AND J. QUISQUATER. An introduction to fast generation of large prime numbers. *Philips J. Research 37* (1982), 231–264.

[13] S. GOLDWASSER AND J. KILIAN. Almost all primes can be quickly certified. In *Proc. 18th STOC* (Berkeley, 1986), pp. 316–329.

[14] D. GORDON. Strong primes are easy to find. In *Proc. Eurocrypt '84* (1984), Springer, pp. 216–223.

[15] J.-C. HERVÉ, F. MORAIN, D. SALESIN, B. SERPETTE, J. VUILLEMIN, AND P. ZIMMERMANN. Bignum: A portable and efficient package for arbitrary precision arithmetic. Rapport de Recherche 1016, INRIA, avril 1989.

[16] D. HUSEMÖLLER. *Elliptic curves*, vol. 111 of *Graduate Texts in Mathematics*. Springer, 1987.

[17] D. E. KNUTH. *The Art of Computer Programming: Seminumerical Algorithms*. Addison-Wesley, 1981.

[18] A. K. LENSTRA. *Cryptology and computational number theory*. AMS, 1989, ch. Primality testing. Lecture Notes, August 6–7, 1989, Boulder, Colorado.

[19] A. K. LENSTRA AND M. S. MANASSE. Factoring by electronic mail. To appear in Proc. Eurocrypt '89, 1989.

[20] H. W. LENSTRA, JR. Elliptic curves and number theoretic algorithms. Tech. Rep. Report 86-19, Math. Inst., Univ. Amsterdam, 1986.

[21] H. W. LENSTRA, JR. Factoring integers with elliptic curves. *Annals of Math. 126* (1987), 649–673.

[22] P. L. MONTGOMERY. Modular multiplication without trial division. *Math. Comp. 44*, 170 (April 1985), 519–521.

[23] P. L. MONTGOMERY. Speeding the Pollard and elliptic curve methods of factorization. *Math. Comp. 48*, 177 (January 1987), 243–264.

[24] F. MORAIN. Atkin's test: news from the front. To appear in Proc. Eurocrypt '89.

[25] F. MORAIN. Elliptic curves, primality proving and some Titanic primes. To appear in Actes des Journées Arithmétiques, Luminy 1989.

[26] F. MORAIN. Implementation of the Atkin-Goldwasser-Kilian primality testing algorithm. Rapport de Recherche 911, INRIA, Octobre 1988.

[27] F. MORAIN. Construction of Hilbert class fields of imaginary quadratic fields and dihedral equations modulo p. Rapport de Recherche 1087, INRIA, Septembre 1989.

[28] F. MORAIN. Résolution d'équations de petit degré modulo de grands nombres premiers. Rapport de Recherche 1085, INRIA, Septembre 1989.

[29] F. MORAIN AND J. OLIVOS. Speeding up the computations on an elliptic curve using addition-subtraction chains. Rapport de Recherche 983, INRIA, Mars 1989.

[30] D. A. PLAISTED. Fast verification, testing and generation of large primes. *Theoretical Computer Science 9* (1979), 1–16.

[31] H. C. POCKLINGTON. The determination of the prime and composite nature of large numbers by Fermat's theorem. *Proc. Cambridge Philos. Soc. 18* (1914-1916), 29–30.

[32] R. SCHOOF. Elliptic curves over finite fields and the computation of square roots mod p. *Math. Comp. 44* (1985), 483–494.

[33] R. D. SILVERMAN. Parallel implementation of the quadratic sieve. *The Journal of Supercomputing 1*, 3 (1987).

[34] J. T. TATE. The arithmetic of elliptic curves. *Inventiones Math. 23* (1974), 179–206.

[35] M.-P. VAN DER HULST. Timings for $(2^{3539} + 1)/3$. Email, February 1990.

[36] H. C. WILLIAMS AND H. DUBNER. The primality of $R1031$. *Math. Comp. 47*, 176 (1986), 703–711.

[37] M. C. WUNDERLICH. A performance analysis of a simple prime-testing algorithm. *Math. Comp. 40*, 162 (1983), 709–714.

[38] S. YATES. Titanic primes. *J. Recr. Math. 16* (1983/84), 250–260.

Properties of binary functions

Sheelagh Lloyd

Hewlett Packard Laboratories

1. INTRODUCTION

In this paper, we shall investigate the connections between three properties of a binary function : the Strict Avalanche Criterion, balance and correlation immunity. The strict avalanche criterion was introduced by Webster and Tavares [7] in order to combine the ideas of completeness and the avalanche effect. A cryptographic transformation is said to be complete if each output bit depends on each input bit, and it exhibits the avalanche effect if an average of one half of the output bits change whenever a single input bit is changed. Forré [1] extended this notion by defining higher order Strict Avalanche Criteria. A function is balanced if, when all input vectors are equally likely, then all output vectors are equally likely. This is an important property for many types of cryptographic functions. The idea of correlation immunity is also extremely important, especially in the field of stream ciphers, where combining functions which are not correlation immune are vulnerable to ciphertext only attacks (see, for example [4]). The concept of mth order correlation immunity was introduced by Siegenthaler [5] as a measure of resistance against such an attack.

In a previous paper [2], we found conditions under which a function satisfying the highest possible order Strict Avalanche Criterion was also balanced and/or correlation immune. Here we shall look at functions satisfying the next highest order Strict Avalanche Criterion. We shall also investigate higher orders of correlation immunity.

In Section 2, we establish some notation, define the properties to be examined and state characterisations of functions with the various properties. Section 3 is devoted to some preliminary calculations which will enable us to identify conditions the functions must satisfy. We present results on balance in Section 4, on correlation immunity in Section 5, and on simultaneous balance and correlation immunity in Section 6. In each of sections 4, 5 and 6, we shall produce necessary and sufficient conditions for a function to satisfy the criteria.

2. NOTATION AND DEFINITIONS

Although we are really dealing with functions of binary vectors of length n which take values in $\{-1, 1\}$, we shall find it convenient to identify a binary vector with its support, that is the set of positions in which it has a 1. We shall, therefore, deal instead with functions from subsets of $\{1, 2, .., n\}$ to $\{-1, 1\}$.

Let S be the set $\{1, 2, .., n\}$, and let $\mathcal{B}_S$ denote the set of functions which takes subsets of S to $\{-1, 1\}$. We formulate all the definitions and characterisations in terms of such functions.

2.1 Balance

This is the simplest of the three properties, and ensures that the number of 1's produced by f is the same as the number of -1's produced.

Definition 2.1.1 Let $f \in \mathcal{B}_S$. Then f is balanced if and only if

$$\sum_{V \subseteq S} f(V) = 0.$$

2.2 Correlation Immunity

Definition 2.2.1 $f \in \mathcal{B}_S$ is said to be first order correlation immune if, for any $i \in S$, the probability that $i \in V$, given that V satisfies $f(V) = 1$, is equal to $\frac{1}{2}$.

The definition is extended to higher orders as follows.

Definition 2.2.2 Let m be an integer with $1 \leq m \leq n$. Then $f \in \mathcal{B}_S$ is said to be mth order correlation immune if for any $J \subseteq S$ with $|J| = m$ and any $Y \subseteq J$, the probability that $V \cap J = Y$, given that $f(V) = 1$, is equal to $\frac{1}{2^m}$.

Note that, for any m with $2 \leq m \leq n$, mth order correlation immunity implies $(m-1)$th order correlation immunity.

In order to characterise correlation immune functions, we need to define the Hadamard-Walsh transform.

Definition 2.2.3 The Hadamard-Walsh transform of $f \in \mathcal{B}_S$ is defined by

$$H(U) = \sum_{V \subseteq S} f(V)(-1)^{|U \cap V|}.$$

There is a well known formula for inverting the Hadamard-Walsh transform, which we give below.

$$f(W) = \frac{1}{2^n} \sum_{U \subseteq S} H(U)(-1)^{|U \cap W|} \quad \text{for all } W \subseteq S.$$

Xiao and Massey [6] have proved the following theorem characterising correlation immune functions in terms of the values of their Hadamard-Walsh transforms.

Theorem 2.2.4 The function $f \in \mathcal{B}_S$ is mth order correlation immune if and only if $H(U) = 0$ for all $U \subseteq S$ with $1 \leq |U| \leq m$.

Let us define the integer valued function X by

$$X(W) = \sum_{V \subseteq W} f(V) \quad \text{for } W \subseteq S.$$

We will find it more convenient to express the characterisation of correlation immunity in terms of the function X. In order to do so, we need the following result.

Lemma 2.2.5 If X and H are defined as above, then

$$X(W) = \frac{1}{2^{n-|W|}} \sum_{\substack{U \subseteq \mathcal{S} \\ U \cap \bar{W} = \emptyset}} H(U) \quad \text{for all } W \subseteq \mathcal{S}.$$

Proof

Since H is the Hadamard-Walsh transform of f, we know that

$$f(W) = \frac{1}{2^n} \sum_{U \subseteq \mathcal{S}} H(U)(-1)^{|U \cap W|} \quad \text{for all } W \subseteq \mathcal{S}.$$

Substituting this into the definition of X, we obtain

$$X(W) = \sum_{V \subseteq W} \frac{1}{2^n} \sum_{U \subseteq \mathcal{S}} H(U)(-1)^{|U \cap V|}$$

$$= \frac{1}{2^n} \sum_{U \subseteq \mathcal{S}} H(U) \sum_{V \subseteq W} (-1)^{|U \cap V|}.$$

For any $V \subseteq W$, we can write $V = A \cup B$ with $A \subseteq (W \cap U)$ and $B \subseteq (W \setminus U)$. So

$$\sum_{V \subseteq W} (-1)^{|U \cap V|} = \sum_{B \subseteq (W \setminus U)} \sum_{A \subseteq (W \cap U)} (-1)^{|A|}$$

If $W \cap U \neq \emptyset$, there are as many subsets of odd size as of even size, so the sum is 0. If $W \cap U = \emptyset$, then the sum is just $2^{|W|}$. Hence

$$X(W) = \frac{1}{2^{n-|W|}} \sum_{\substack{U \subseteq \mathcal{S} \\ U \cap \bar{W} = \emptyset}} H(U) \quad \text{for all } W \subseteq \mathcal{S}.$$

Note that $X(\mathcal{S}) = H(\emptyset)$.

We shall now use this to produce a formulation of mth order correlation immunity in terms of X.

Lemma 2.2.6 If H and X are defined as above, then the following three conditions are equivalent:

(i) f is mth order correlation immune

(ii) $H(U) = 0$ for all $U \subseteq \mathcal{S}$ with $1 \leq |U| \leq m$.

(iii) $X(W) = 2^{|W|-n}X(\mathcal{S})$ for all $W \subseteq \mathcal{S}$ with $(n-m) \leq |W| \leq (n-1)$.

Proof

The equivalence of (i) and (ii) is given by Theorem 2.2.4. We shall now show the equivalence of (ii) and (iii), using Lemma 2.2.5.

Suppose that (ii) holds. Let $W \subseteq S$ be such that $(n - m) \leq |W| \leq (n - 1)$, and let $U \subseteq S$ be such that $W \cap U = \emptyset$. Then $0 \leq |U| \leq (n - |W|) \leq m$, so either $U = \emptyset$, or $H(U) = 0$. So

$$X(W) = \frac{1}{2^{n-|W|}} \sum_{\substack{U \subseteq S \\ U \cap \overline{W} = \emptyset}} H(U) = \frac{1}{2^{n-|W|}} H(\emptyset).$$

Now $X(S) = H(\emptyset)$, so we have (iii).

Now suppose that (iii) holds. We shall prove (ii) by induction on the size of U. Suppose first that $|U| = 1$. Let $W = S \setminus U$. Then $V \cap W = \emptyset$ if and only if either $V = \emptyset$ or $V = U$, so

$$X(W) = \frac{1}{2}(H(\emptyset) + H(U)).$$

But since $|W| = (n - 1)$, we know also that $X(W) = \frac{1}{2}H(\emptyset)$. Hence $H(U) = 0$.

Now suppose that $2 \leq |U| \leq m$ and that $H(V) = 0$ for all V with $1 \leq |V| < |U|$. Let $W = S \setminus U$, then $V \cap W = \emptyset$ if and only if $V \subseteq U$, so

$$X(W) = \frac{1}{2^{n-|W|}}(H(\emptyset) + H(U) + \sum_{V \subset U, V \neq \emptyset} H(V)).$$

Now, for any $V \subset U$, $V \neq \emptyset$, we see that $1 \leq |V| < |U|$, so $H(V) = 0$. Since $(n - m) \leq |W| \leq (n - 1)$, we know also that $X(W) = 2^{|W|-n}H(\emptyset)$. Thus we may conclude that $H(U) = 0$ as required.

Note that we may write the condition that f is balanced as $X(S) = 0$.

2.3 The Strict Avalanche Criterion

Definition 2.3.1 Let $f \in \mathcal{B}_S$. Then f satisfies the strict avalanche criterion (SAC) if and only if

$$\sum_{V \subseteq (S \setminus \{j\})} f(V)f(V \cup \{j\}) = 0 \qquad \text{for all } j, 1 \leq j \leq n.$$

We now define the higher order SAC. The SAC defined above is deemed to be the SAC of order 0, and the SAC of order m for $1 \leq m \leq n - 2$ is defined as follows.

Definition 2.3.2 [1] A function $f \in \mathcal{B}_S$ satisfies the SAC of order m, where $1 \leq m \leq (n - 2)$ if and only if given any subset T of S with $|T| = n - m$ and any subset P of $S \setminus T$, the function $g \in \mathcal{B}_T$ obtained from f by setting $g(V) = f(V \cup P)$ for each $V \subseteq T$ satisfies the SAC.

Let $\overline{f}$ denote the algebraic normal form of f (so $\overline{f}$ also takes subsets of S to $\{-1,1\}$). We shall sometimes find it convenient to write F for the function from S to $\{1, -1\}$ such that $F(x) = \overline{f}(\{x\})$. To reduce confusion between sets and elements of those sets, we shall use capital letters to denote subsets of S and small letters to denote elements of S.

In [3], we proved the following result characterising functions satisfying the SAC of order $(n-3)$. Note that, since we are dealing exclusively with functions satisfying the SAC of order $(n-3)$, we insist throughout that $n \geq 3$.

Theorem 2.3.3 [3] Suppose that $f \in \mathcal{B}_S$. Then f satisfies the SAC of order $(n-3)$ if and only if

$$f(V) = \prod_{U \subseteq V, |U| < 3} \overline{f}(U) \quad \text{for all } V \subseteq S$$

and for each $x \in S$, there is at most one $y \in S$ for which $\overline{f}(\{x, y\}) = 1$.

Suppose that f satisfies the SAC of order $(n-3)$. Then Theorem 2.3.3 tells us that for any $x \in S$, either $\overline{f}(\{x, y\}) = -1$ for all $y \in S$, or there is exactly one $y \in S$ for which $\overline{f}(\{x, y\}) = 1$. Given any $W \subseteq S$, and any $x \in W$, there can therefore be at most one $y \in W$ for which $\overline{f}(\{x, y\}) = 1$. Suppose there are exactly m pairs (x, y) in W with $\overline{f}(\{x, y\}) = 1$. Let us write these as $(x_1, y_1), .., (x_m, y_m)$ (where $0 \leq m \leq n/2$), and let us denote the remaining elements of W (if any) by $x_{2m+1}, .., x_{|W|}$. Then we have $W = \{x_1, x_2, .., x_m, y_1, y_2, .., y_m, x_{2m+1}, .., x_{|W|}\}$ where $\overline{f}(\{x_j, y_j\}) = +1$ for all $1 \leq j \leq m$ and $\overline{f}(\{a, b\}) = -1$ otherwise. Note that, although the elements may be numbered in various ways, the value of m is determined uniquely by W (and f). We shall find the following definition useful.

Definition 2.3.4 We shall write $A_W(n, m)$ $(W \subseteq S, 0 \leq 2m \leq |W|)$ for the set of functions $f \in \mathcal{B}_S$ satisfying the following conditions:

f satisfies the SAC of order $(n-3)$ and there exist $x_1, x_2, .., x_m, y_1, y_2, .., y_m, x_{2m+1}, .., x_{|W|}$ such that

$$W = \{x_1, x_2, .., x_m, y_1, y_2, .., y_m, x_{2m+1}, .., x_{|W|}\},$$

and

$$\overline{f}(\{x_j, y_j\}) = +1, \quad 1 \leq j \leq m$$
$$\overline{f}(\{a, b\}) = -1 \quad \text{otherwise.}$$

In what follows, we shall want to distinguish the cases where there exists a pair (x, y), such that $\overline{f}(\{x, y\}) = 1$ and $F(x) = -F(y)$, from those where no such pair exists. It turns out that the case where such pairs exist is much simpler than the other case. In order to be able to state some subsequent results concisely in the cases where no such pair exists, we introduce the following notation.

Definition 2.3.6 We shall write $C_W(n, m, r, t, q)$ $(W \subseteq S, 0 \leq 2m \leq |W|, 0 \leq r \leq m, 0 \leq t \leq |W| - 2m, q = 2r + 2t + m - |W|)$ for the set of functions $f \in \mathcal{B}_S$ satisfying the following conditions:

f belongs to $A_W(n, m)$

and

$$F(x_j) = F(y_j) = +1, \quad 1 \leq j \leq r$$
$$F(x_j) = F(y_j) = -1, \quad r+1 \leq j \leq m$$
$$F(x_j) = +1, \quad 2m+1 \leq j \leq 2m+t$$
$$F(x_j) = -1, \quad 2m+t+1 \leq j \leq |W|$$

For ease of notation, we shall write simply $C(n,m,r,t,q)$ for $C_S(n,m,r,t,q)$. So we see that, if $f \in \mathcal{B}_S$ satisfies the SAC of order $(n-3)$, then either there exists a pair (x,y), such that $\overline{f}(\{x,y\}) = 1$, and $F(x) = -F(y)$, or f belongs to $C(n,m,r,t,q)$ for some values of m, r, t and q.

3. PRELIMINARY CALCULATIONS

We want to express $f(V \cup \{x\})$ in terms of $f(V)$. We know that, given x and V, if f satisfies the SAC of order $(n-3)$, then there is at most one z in V with $\overline{f}(\{x,z\}) = 1$. We first deal with the case where no such z exists.

Proposition 3.1 Suppose that $f \in \mathcal{B}_S$ satisfies the SAC of order $(n-3)$. Suppose further that $x \notin V$, and that $\overline{f}(\{x,y\}) = -1$ for all $y \in V$. Then

$$f(V \cup \{x\}) = (-1)^{|V|} f(V) F(x).$$

Proof

Straightforward application of Theorem 2.3.3.

We turn now to the case where there is a unique element z in V with $\overline{f}(\{x,z\}) = 1$.

Proposition 3.2 Suppose that $f \in \mathcal{B}_S$ satisfies the SAC of order $(n-3)$. Suppose further that $x \notin V$, and that $\overline{f}(\{x,z\}) = 1$ (so $\overline{f}(\{x,y\}) = -1$ for all $y \in V$, $y \neq z$). Then

$$f(V \cup \{x\}) = (-1)^{|V|-1} f(V) F(x).$$

Proof

Straightforward application of Theorem 2.3.3.

We are now able to produce an expression for $X(W) = \sum_{V \subseteq W} f(V)$ in terms of the values of $\overline{f}$. In order to prove this, we need also to produce the corresponding expression for $X^-(W) = \sum_{V \subseteq W} (-1)^{|V|} f(V)$ as well.

Lemma 3.3 Suppose that f satisfies the SAC of order $(n-3)$. Let $W \subseteq S$, $x \in W$ and $U = W \setminus \{x\}$. Suppose that $\overline{f}(\{x,y\}) = -1$ for all $y \in U$. Then

$$X(W) = X(U) + F(x)X^-(U)$$

and

$$X^-(W) = X^-(U) - F(x)X(U)$$

Proof

$$X(W) = \sum_{V \subseteq W} f(V) = \sum_{V \subseteq U} f(V) + \sum_{V \subseteq U} f(V \cup \{x\}) = X(U) + \sum_{V \subseteq U} f(V \cup \{x\}).$$

To calculate the second sum, since $\overline{f}(\{x,y\}) = -1$ for all $y \in U$, we may use Proposition 3.1 to obtain

$$\sum_{V \subseteq U} f(V \cup \{x\}) = \sum_{V \subseteq U} (-1)^{|V|} f(V) F(x)$$

$$= F(x) \sum_{V \subseteq U} (-1)^{|V|} f(V)$$

$$= F(x) X^{-}(U)$$

as required. Similarly

$$X^{-}(W) = \sum_{V \subseteq U} (-1)^{|V|} f(V) - \sum_{V \subseteq U} (-1)^{|V|} f(V \cup \{x\}) = X^{-}(U) - \sum_{V \subseteq U} (-1)^{|V|} f(V \cup \{x\})$$

Using Proposition 3.1 again, we have

$$\sum_{V \subseteq U} (-1)^{|V|} f(V \cup \{x\}) = F(x) X(U).$$

Hence $X^{-}(W) = X^{-}(U) - F(x)X(U)$.

Lemma 3.4 Suppose that f satisfies the SAC of order $(n-3)$. Let $W \subseteq S$, and suppose that $x, y \in W$ are such that $\overline{f}(\{x,y\}) = 1$. Let $U = W \setminus \{x,y\}$, then

$$X(W) = X(U) + F(y)X^{-}(U) + F(x)X^{-}(U) + F(x)F(y)X(U)$$

and

$$X^{-}(W) = X^{-}(U) - F(y)X(U) - F(x)X(U) + F(x)F(y)X^{-}(U).$$

Proof

As before, we have

$$X(W) = \sum_{V \subseteq W} f(V) = \sum_{V \subseteq (U \cup \{y\})} f(V) + \sum_{V \subseteq (U \cup \{y\})} f(V \cup \{x\})$$

$$= X(U \cup \{y\}) + \sum_{V \subseteq (U \cup \{y\})} f(V \cup \{x\}).$$

By Propositions 3.1 and 3.2, we know that

$$f(V \cup \{x\}) = \begin{cases} (-1)^{|V|} f(V) F(x) & \text{if } y \notin V \\ (-1)^{|V|-1} f(V) F(x) & \text{if } y \in V \end{cases}.$$

So we have

$$\sum_{V \subseteq (U \cup \{y\})} f(V \cup \{x\}) = \sum_{\substack{V \subseteq (U \cup \{y\}) \\ y \notin V}} (-1)^{|V|} f(V) F(x) + \sum_{\substack{V \subseteq (U \cup \{y\}) \\ y \in V}} (-1)^{|V|-1} f(V) F(x)$$

$$= \sum_{V \subseteq U} (-1)^{|V|} f(V) F(x) + \sum_{V \subseteq U} (-1)^{|V|} f(V \cup \{y\}) F(x)$$

$$= F(x) X^{-}(U) + F(x) \sum_{V \subseteq U} (-1)^{|V|} f(V \cup \{y\})$$

Applying Proposition 3.1 again, since $\overline{f}(\{y, z\}) = -1$ for all $z \in U$, we have

$$\sum_{V \subseteq U} (-1)^{|V|} f(V \cup \{y\}) = \sum_{V \subseteq U} f(V) F(y) = F(y) X(U)$$

So

$$\sum_{V \subseteq (U \cup \{y\})} f(V \cup \{x\}) = F(x) X^{-}(U) + F(x) F(y) X(U).$$

Now by Lemma 3.3, $X(U \cup \{y\}) = X(U) + F(y) X^{-}(U)$, so putting these two together, we obtain the desired result.

We may now do exactly the same with $X^{-}(W)$ as follows.

$$X^{-}(W) = \sum_{V \subseteq (U \cup \{y\})} (-1)^{|V|} f(V) + \sum_{V \subseteq (U \cup \{y\})} (-1)^{|V|+1} f(V \cup \{x\})$$

$$= X^{-}(U \cup \{y\}) - \sum_{V \subseteq (U \cup \{y\})} (-1)^{|V|} f(V \cup \{x\}).$$

and

$$\sum_{V \subseteq (U \cup \{y\})} (-1)^{|V|} f(V \cup \{x\}) = \sum_{\substack{V \subseteq (U \cup \{y\}) \\ y \notin V}} f(V) F(x) - \sum_{\substack{V \subseteq (U \cup \{y\}) \\ y \in V}} f(V) F(x)$$

$$= F(x) X(U) - \sum_{V \subseteq U} f(V \cup \{y\}) F(x)$$

$$= F(x) X(U) - F(x) \sum_{V \subseteq U} f(V \cup \{y\})$$

Then

$$\sum_{V \subseteq U} f(V \cup \{y\}) = \sum_{V \subseteq U} (-1)^{|V|} f(V) F(y) = F(y) X^{-}(U)$$

so

$$\sum_{V \subseteq (U \cup \{y\})} (-1)^{|V|} f(V \cup \{x\}) = F(x) X(U) - F(x) F(y) X^{-}(U).$$

Now by Lemma 3.3, $X^{-}(U \cup \{y\}) = X^{-}(U) - F(y) X(U)$, so putting these two together, we obtain the desired result.

We are now able to prove our main result in this section.

Theorem 3.5 Suppose that $W \subseteq S$ and that f belongs to $A_W(n,m)$. Let i denote the square root of -1, and let

$$G_W = f(\emptyset) \prod_{j=1}^{m} (1 + F(x_j)F(y_j) + i(F(x_j) + F(y_j))) \prod_{j=2m+1}^{|W|} (1 + iF(x_j)),$$

then

$$X(W) = \Re(G_W) + \Im(G_W) \quad \text{and}$$
$$X^-(W) = \Re(G_W) - \Im(G_W)$$

where $\Re(x)$ and $\Im(x)$ denote the real and imaginary parts of x respectively.

Proof

The proof is by induction on the size of W. Firstly, we assume that $|W| = 0$, so that $W = \emptyset$. Then

$$G_W = f(\emptyset) \quad \text{and} \quad X(W) = f(\emptyset) \quad \text{and} \quad X^-(W) = f(\emptyset).$$

Now suppose the result true for all W with $|W| \leq K$, and let W be such that $|W| = K + 1$. Choose $x \in W$. We shall split the proof into two cases. Since f satisfies the SAC of order $(n-3)$, either $\overline{f}(\{x,y\}) = -1$ for all $y \in (W \setminus \{x\})$, or there exists a unique $y \in (W \setminus \{x\})$ for which $\overline{f}(\{x,y\}) = 1$.

Suppose that the first case holds, and let $U = W \setminus \{x\}$. Now, by Lemma 3.3,

$$X(W) = X(U) + F(x)X^-(U).$$

By the inductive hypothesis, since $|U| = K$, we have $X(U) = \Re(G_U) + \Im(G_U)$ and $X^-(U) = \Re(G_U) - \Im(G_U)$. We deduce, therefore, that

$$X(W) = \Re(G_U) + \Im(G_U) + F(x)(\Re(G_U) - \Im(G_U)) = \Re(G_W) + \Im(G_W).$$

since, in this case, $G_W = (1 + iF(x))G_U$.

We now turn to the second case, and let $U = W \setminus \{x,y\}$. By Lemma 3.4, we have

$$X(W) = X(U) + F(y)X^-(U) + F(x)X^-(U) + F(x)F(y)X(U).$$

By the inductive hypothesis, since $|U| = K$, we have

$$X(U) = \Re(G_U) + \Im(G_U) \quad \text{and} \quad X^-(U) = \Re(G_U) - \Im(G_U)$$

so we have

$$X(W) = \Re(G_U) + \Im(G_U) + F(y)(\Re(G_U) - \Im(G_U))$$
$$+ F(x)(\Re(G_U) - \Im(G_U)) + F(x)F(y)(\Re(G_U) + \Im(G_U))$$
$$= \Re(G_W) + \Im(G_W)$$

since, in this case, $G_W = (1 + F(x)F(y) + i(F(x) + F(y)))G_U$.

Corollary 3.6 Suppose that f belongs to $A_W(n,m)$. Suppose further that for some j, $1 \le j \le m$, we have $F(x_j) = -F(y_j)$. Then $\sum_{V \subseteq W} f(V) = 0$.

Proof

Suppose, without loss of generality, that $F(x_1) = -F(y_1)$. Then by Theorem 3.5, $\sum_{V \subseteq W} f(V) = \Re(G) + \Im(G)$, where

$$G = f(\emptyset) \prod_{j=1}^{m}(1 + F(x_j)F(y_j) + i(F(x_j) + F(y_j))) \prod_{j=2m+1}^{|W|}(1 + iF(x_j)).$$

But $1 + F(x_1)F(y_1) + i(F(x_1) + F(y_1)) = 0$, since $F(x_1) = -F(y_1)$, so $G = 0$. Hence $\sum_{V \subseteq W} f(V) = 0$.

Corollary 3.7 Suppose that f belongs to $C_W(n,m,r,t,q)$. Write k for $|W|$; then

$$X(W) = \begin{cases} f(\emptyset)(-1)^{\frac{1}{4}q}2^{\frac{1}{2}(m+k)} & q \equiv 0 \pmod 4 \\ f(\emptyset)(-1)^{\frac{1}{4}(q-1)}2^{\frac{1}{2}(m+k+1)} & q \equiv 1 \pmod 4 \\ f(\emptyset)(-1)^{\frac{1}{4}(q-2)}2^{\frac{1}{2}(m+k)} & q \equiv 2 \pmod 4 \\ 0 & q \equiv 3 \pmod 4 \end{cases}$$

Proof

Let G be defined as above; we shall examine each term in turn. Now if $1 \le j \le m$, then

$$(1 + F(x_j)F(y_j) + iF(x_j) + iF(y_j)) = \begin{cases} 2(1+i) & \text{if } F(x_j) = 1 \\ 2(1-i) & \text{if } F(x_j) = -1 \end{cases}$$

and if $2m+1 \le j \le n$, then

$$(1 + iF(x_j)) = \begin{cases} 1+i & \text{if } F(x_j) = 1 \\ 1-i & \text{if } F(x_j) = -1 \end{cases}$$

So

$$\begin{aligned} G &= f(\emptyset)2^r(1+i)^r 2^{m-r}(1-i)^{m-r}(1+i)^t(1-i)^{k-2m-t} \\ &= f(\emptyset)2^m(1+i)^{r+t}(1-i)^{k-m-r-t} \\ &= f(\emptyset)2^{k-r-t}(1+i)^{2r+2t+m-k} \end{aligned}$$

Now if $0 \le b \le 3$, then

$$\Re(1+i)^{4a+b} + \Im(1+i)^{4a+b} = \begin{cases} (-4)^a & \text{if } b = 0 \\ 2(-4)^a & \text{if } b = 1 \\ 2(-4)^a & \text{if } b = 2 \\ 0 & \text{if } b = 3 \end{cases}$$

Now $2r + 2t + m - k = q$, and so

$$\begin{aligned} X(W) &= \begin{cases} f(\emptyset)2^{k-r-t}(-4)^{\frac{1}{4}q} & \text{if } q \equiv 0 \pmod 4 \\ f(\emptyset)2^{k-r-t+1}(-4)^{\frac{1}{4}(q-1)} & \text{if } q \equiv 1 \pmod 4 \\ f(\emptyset)2^{k-r-t+1}(-4)^{\frac{1}{4}(q-2)} & \text{if } q \equiv 2 \pmod 4 \\ 0 & \text{if } q \equiv 3 \pmod 4 \end{cases} \\ &= \begin{cases} f(\emptyset)(-1)^{\frac{1}{4}q}2^{\frac{1}{2}(m+k)} & \text{if } q \equiv 0 \pmod 4 \\ f(\emptyset)(-1)^{\frac{1}{4}(q-1)}2^{\frac{1}{2}(m+k+1)} & \text{if } q \equiv 1 \pmod 4 \\ f(\emptyset)(-1)^{\frac{1}{4}(q-2)}2^{\frac{1}{2}(m+k)} & \text{if } q \equiv 2 \pmod 4 \\ 0 & \text{if } q \equiv 3 \pmod 4 \end{cases} \end{aligned}$$

4. BALANCE

We shall use the results of the preceding section to obtain necessary and sufficient conditions for a function satisfying the SAC of order $(n-3)$ to be balanced.

Theorem 4.1 Suppose that $f \in \mathcal{B}_S$ satisfies the SAC of order $(n-3)$. Then f is balanced if and only if either
(i) there exist x and y with $\overline{f}(\{x,y\}) = 1$ and $F(x) = -F(y)$ or
(ii) f belongs to $C(n,m,r,t,q)$ and $q \equiv 3 \pmod 4$.

Proof

Since f satisfies the SAC of order $(n-3)$, we know that either (i) holds, or there exist m, r, t and q such that f belongs to $C(n,m,r,t,q)$. We recall that f is balanced if and only if $X(S) = 0$.

If (i) holds, then by Corollary 3.6, we know that $X(S) = 0$.

If (ii) holds, then by Corollary 3.7, we have

$$
X(S) = \begin{cases}
f(\emptyset)(-1)^{\frac{1}{4}q}2^{\frac{1}{2}(m+n)} & \text{if } q \equiv 0 \pmod 4 \\
f(\emptyset)(-1)^{\frac{1}{4}(q-1)}2^{\frac{1}{2}(m+n+1)} & \text{if } q \equiv 1 \pmod 4 \\
f(\emptyset)(-1)^{\frac{1}{4}(q-2)}2^{\frac{1}{2}(m+n)} & \text{if } q \equiv 2 \pmod 4 \\
0 & \text{if } q \equiv 3 \pmod 4
\end{cases}
$$

So in this case, f is balanced if and only if $q \equiv 3 \pmod 4$, since $f(\emptyset) = +1$ or -1.

5. CORRELATION IMMUNITY

We shall now obtain necessary and sufficient conditions for a function satisfying the SAC of order $(n-3)$ to be correlation immune.

Proposition 5.1 Suppose $f \in \mathcal{B}_S$ satisfies the SAC of order $(n-3)$. Suppose there are exactly p pairs (x_j, y_j) such that $\overline{f}(\{x_j, y_j\}) = 1$ and $F(x_j) = -F(y_j)$. Then f is exactly $(p-1)$th order correlation immune.

Proof

Let us write $S = \{x_1, y_1, .., x_p, y_p, x_{2p+1}, .., x_n\}$ where, as usual, $\overline{f}(\{x_j, y_j\}) = 1$, and $\overline{f}(\{u, v\}) = -1$ otherwise. By Corollary 3.6, $X(W) = 0$ whenever there exists j, $1 \le j \le p$, with $x_j, y_j \in W$. Any W with $|W| > n-p$ must contain at least one such pair, so $X(W) = 0$ for any such W (including S). By Lemma 2.2.6, therefore, f is at least $(p-1)$th order correlation immune.

Let $U = S \setminus \{x_1, y_1, y_2, .., y_p\}$. Then U contains no pairs (x_j, y_j), and so $G_U \ne 0$. Let us write x for x_1 and y for y_1, and let $U_x = U \cup \{x\}$, and $U_y = U \cup \{y\}$. Since $F(x) = -F(y)$, we may assume without loss of generality that $F(x) = 1$ and $F(y) = -1$. Now

$$G_{U_x} = (1 + iF(x))G_U = (1+i)G_U,$$

so

$$X(U_x) = \Re(G_U) + \Im(G_U) + \Re(G_U) - \Im(G_U) = 2\Re(G_U).$$

On the other hand,

$$G_{U_y} = (1 + iF(y))G_U = (1 - i)G_U,$$

so

$$X(U_y) = \Re(G_U) + \Im(G_U) - \Re(G_U) + \Im(G_U) = 2\Im(G_U).$$

If f is pth order correlation immune, then $X(U_x) = X(U_y) = 0$. But this forces $G_U = 0$, which is not true. Hence f is not pth order correlation immune.

We shall now prove some results on the values of $X(W)$. In the four lemmas which follow, we assume that f belongs to $C_{W_j}(n, m_j, r_j, t_j, q_j)$ for $j = 1, 2$, and that $|W_j| = k_j$ for $j = 1, 2$. We shall calculate the relationship between $X(W_1)$ and $X(W_2)$ for various values of W_1 and W_2. The proofs of these results are straightforward applications of Corollary 3.7, and are omitted for brevity.

Lemma 5.2 Suppose that $W_1 \subseteq S$ and that $x, y \in W_1$ are such that $\overline{f}(\{x, y\}) = +1$. Let $W_2 = W_1 \setminus \{x\}$, then $X(W_2) = \frac{1}{2}X(W_1)$.

Lemma 5.3 Suppose that $W_1 \subseteq S$ and that $x \in W_1$ is such that $\overline{f}(\{x, y\}) = -1$ for all $y \in W_1$, and that $F(x) = +1$. Let $W_2 = W_1 \setminus \{x\}$. Then

$$X(W_1)/X(W_2) = \begin{cases} \infty & q_1 \equiv 0 \pmod 4 \\ 2 & q_1 \equiv 1 \pmod 4 \\ 1 & q_1 \equiv 2 \pmod 4 \\ 0 & q_1 \equiv 3 \pmod 4 \end{cases}$$

Lemma 5.4 Suppose that $W_1 \subseteq S$ and that $x \in W_1$ is such that $\overline{f}(\{x, y\}) = -1$ for all $y \in W_1$, and that $F(x) = -1$. Let $W_2 = W_1 \setminus \{x\}$. Then

$$X(W_1)/X(W_2) = \begin{cases} 1 & q_1 \equiv 0 \pmod 4 \\ 2 & q_1 \equiv 1 \pmod 4 \\ \infty & q_1 \equiv 2 \pmod 4 \\ 0 & q_1 \equiv 3 \pmod 4 \end{cases}$$

Corollary 5.5 Suppose that $W_1 \subseteq S$ and that $x \in W_1$ is such that $\overline{f}(\{x, y\}) = -1$ for all $y \in W_1$. Let $W_2 = W_1 \setminus \{x\}$. Then

$$X(W_2) = \frac{1}{2}X(W_1) \quad \text{if and only if} \quad q_1 \equiv 1 \pmod 4$$

Lemma 5.6 Suppose that $W_1 \subseteq S$ and that $x, y \in W_1$ are such that $\overline{f}(\{x, y\}) = +1$, and $F(x) = F(y) = +1$. Let $W_2 = W_1 \setminus \{x, y\}$, then

$$X(W_1)/X(W_2) = \begin{cases} \infty & q_1 \equiv 0 \pmod 4 \\ 2^2 & q_1 \equiv 1 \pmod 4 \\ 2 & q_1 \equiv 2 \pmod 4 \\ 0 & q_1 \equiv 3 \pmod 4 \end{cases}$$

Lemma 5.7 Suppose that $W_1 \subseteq S$ and that $x, y \in W_1$ are such that $\overline{f}(\{x,y\}) = +1$, and $F(x) = F(y) = -1$. Let $W_2 = W_1 \setminus \{x,y\}$, then

$$X(W_1)/X(W_2) = \begin{cases} 2 & q_1 \equiv 0 \pmod 4 \\ 2^2 & q_1 \equiv 1 \pmod 4 \\ \infty & q_1 \equiv 2 \pmod 4 \\ 0 & q_1 \equiv 3 \pmod 4 \end{cases}$$

Corollary 5.8 Suppose that $W_1 \subseteq S$ and that $x, y \in W_1$ are such that $\overline{f}(\{x,y\}) = +1$. Let $W_2 = W_1 \setminus \{x,y\}$, then

$$X(W_2) = \frac{1}{2^2} X(W_1) \quad \text{if and only if} \quad q_1 \equiv 1 \pmod 4$$

Proposition 5.9 Suppose that f belongs to $C(n,m,r,t,q)$. If $2m < n$, and $q \not\equiv 1$ (mod 4), then f is not correlation immune.

Proof

We must find W with $|W| = n - 1$, and $X(W) \neq \frac{1}{2}X(S)$. Let $W = S \setminus \{x_n\}$. Since $2m < n$, we may apply Corollary 5.5, with $W_1 = S$. Since $q \not\equiv 1$ (mod 4), we deduce that $X(W) \neq \frac{1}{2}X(S)$. So f is not correlation immune.

Proposition 5.10 Suppose that f belongs to $C(2m,m,r,0,q)$. If $q \not\equiv 1$ (mod 4), then f is exactly 1st order correlation immune.

Proof

Let us write $S = \{x_1, y_1, .., x_m, y_m\}$ where, as usual, $\overline{f}(\{x_j, y_j\}) = 1$, and $\overline{f}(\{u,v\}) = -1$ otherwise. We show first that if $|W| = n - 1$, then $X(W) = \frac{1}{2}X(S)$. Let W be such that $|W| = n - 1$. Then either $W = S \setminus \{x_j\}$ for some j or $W = S \setminus \{y_j\}$ for some j. By Lemma 5.2, therefore, with $W_1 = S$, $X(W) = \frac{1}{2}X(S)$.

So we have shown that f is at least 1st order correlation immune. We now need to find W with $|W| = n - 2$, and $X(W) \neq \frac{1}{2^2}X(S)$. We take $W = S \setminus \{x_1, y_1\}$. Then we may use Corollary 5.8, with $W_1 = S$. Since $q \not\equiv 1$ (mod 4), $X(W) \neq \frac{1}{2^2}X(S)$. So f is not 2nd order correlation immune.

We turn now to the case where $q \equiv 1$ (mod 4).

Lemma 5.11 Suppose that f belongs to $C(n,m,r,t,q)$ and that $q \equiv 1$ (mod 4). Then f is 1st order correlation immune.

Proof

Let us write $S = \{x_1, y_1, .., x_m, y_m, x_{2m+1}, .., x_n\}$ where, as usual, $\overline{f}(\{x_j, y_j\}) = 1$, and $\overline{f}(\{u,v\}) = -1$ otherwise. We must show that $X(W) = \frac{1}{2}X(S)$ for any W with $|W| = n - 1$. Choose any such W. Then we have the following possibilities for W :

$$W = S \setminus \{x_j\} \quad \text{for some } j, 1 \leq j \leq m \text{ or}$$
$$W = S \setminus \{y_j\} \quad \text{for some } j, 1 \leq j \leq m \text{ or}$$
$$W = S \setminus \{x_j\} \quad \text{for some } j, 2m + 1 \leq j \leq n$$

In either of the first two cases, we may apply Lemma 5.2, to obtain $X(W) = \frac{1}{2}X(S)$, while in the third case we may apply Corollary 5.5 to obtain $X(W) = \frac{1}{2}X(S)$. Hence f is 1st order correlation immune.

Lemma 5.12 Suppose that f belongs to $C(n, m, r, t, q)$ and that $q \equiv 1 \pmod 4$. Then f is 2nd order correlation immune if and only if $2m \geq n - 1$.

Proof

Let us write $S = \{x_1, y_1, .., x_m, y_m, x_{2m+1}, .., x_n\}$ where, as usual, $\overline{f}(\{x_j, y_j\}) = 1$, and $\overline{f}(\{u, v\}) = -1$ otherwise. We already know that f is 1st order correlation immune. We must show that $X(W) = \frac{1}{2^2} X(S)$ for any W with $|W| = n - 2$. Choose any such W. Then we have the following possibilities for W :

$$W = S \setminus \{x_j, y_k\}, j \neq k, 1 \leq j, k \leq m$$
$$W = S \setminus \{x_j, x_k\}, j \neq k, 1 \leq j, k \leq m$$
$$W = S \setminus \{y_j, y_k\}, j \neq k, 1 \leq j, k \leq m$$
$$W = S \setminus \{x_j, y_j\}, 1 \leq j \leq m$$
$$W = S \setminus \{x_j, x_k\}, 1 \leq j \leq m, 2m + 1 \leq k \leq n$$
$$W = S \setminus \{y_j, x_k\}, 1 \leq j \leq m, 2m + 1 \leq k \leq n$$
$$W = S \setminus \{x_j, x_k\}, j \neq k, 2m + 1 \leq j, k \leq n$$

In the first case, we may first apply Lemma 5.2 with $W_1 = S$ and $W_2 = S \setminus \{x_j\}$, and then apply Lemma 5.2 again with $W_1 = S \setminus \{x_j\}$ and $W_2 = W$ to obtain $X(W) = \frac{1}{2} X(S \setminus \{x_j\}) = \frac{1}{2^2} X(S)$ as required. This may also be done in the second and third cases. In the fourth case, we may apply Corollary 5.8, with $W_1 = S$ to obtain $X(W) = \frac{1}{2^2} X(S)$, as required. In the fifth and sixth cases, we may proceed in a similar manner as in the first case, applying Lemma 5.2, and then Corollary 5.5 to obtain the result (noting that q is unchanged after applying Lemma 5.2). When we come to the seventh case, however, we see that if we apply Corollary 5.5 with $W_1 = S$, and $W_2 = S \setminus \{x_j\}$, we obtain $X(W_2) = \frac{1}{2} X(W_1)$, but when we come to apply Corollary 5.5 again with $W_1 = S \setminus \{x_j\}$ and $W_2 = W$, we now have $q_1 \equiv 0 \pmod 4$, or $q_1 \equiv 2 \pmod 4$, according as $j \leq 2m_1 + t_1$ or $j > 2m_1 + t_1$, and so $X(W) \neq \frac{1}{2^2} X(S)$ in this case. This case can only occur when $2m + 1 < n$, so f is 2nd order correlation immune if and only if $2m \geq n - 1$.

Lemma 5.13 Suppose that f belongs to $C(n, m, r, t, q)$ and that $q \equiv 1 \pmod 4$. Then f is 3rd order correlation immune if and only if $2m = n$.

Proof

Let us write $S = \{x_1, y_1, .., x_m, y_m, x_{2m+1}, .., x_n\}$ where, as usual, $\overline{f}(\{x_j, y_j\}) = 1$, and $\overline{f}(\{u, v\}) = -1$ otherwise. Suppose first that $2m = n$. We therefore know that f is 2nd order correlation immune, since $2m \geq n - 1$. We must show that $X(W) = \frac{1}{2^3} X(S)$ for any W with $|W| = n - 3$. Let W be such that $|W| = n - 3$. If $W = S \setminus \{x_j, x_k, x_l\}$, with j, k and l all different, and $1 \leq j, k, l \leq m$, then we may apply Lemma 5.2 three times to obtain the result. The same method will also work in the cases $W = S \setminus \{x_j, x_k, y_l\}$, $W = S \setminus \{x_j, y_k, y_l\}$ and $W = S \setminus \{y_j, y_k, y_l\}$. The cases $W = S \setminus \{x_j, y_j, x_k\}$ and $W = S \setminus \{x_j, y_j, y_k\}$, where $1 \leq j, k \leq m$ may each be dealt with using first Corollary 5.8, and then Lemma 5.2. This means that when $2m = n$, f is 3rd order correlation immune.

When, however, $2m < n$, we must consider the case $W = \mathcal{S} \setminus \{x_1, y_1, x_n\}$. We apply Corollary 5.8 with $W_1 = \mathcal{S}$, and $W_2 = \mathcal{S} \setminus \{x_1, y_1\}$, and then apply Corollary 5.5 with $W_1 = \mathcal{S} \setminus \{x_1, y_1\}$, and $W_2 = W$. But this time either $q_1 \equiv 0 \pmod 4$ or $q_1 \equiv 2 \pmod 4$ according as $r > 0$ or $r = 0$. So in this case, f is not 3rd order correlation immune.

Lemma 5.14 Suppose that f belongs to $C(n, m, r, t, q)$ and that $q \equiv 1 \pmod 4$. Then f is not 4th order correlation immune.

Proof

We shall produce W with $|W| = n - 4$ but $X(W) \neq \frac{1}{2^4} X(\mathcal{S})$.

In order for f to be fourth order correlation immune, it must certainly be third order correlation immune. So, by Lemma 5.13, we must have $2m = n$. Let us write $\mathcal{S} = \{x_1, y_1, .., x_m, y_m\}$ where, as usual, $\overline{f}(\{x_j, y_j\}) = 1$, and $\overline{f}(\{u, v\}) = -1$ otherwise. We take $W = \mathcal{S} \setminus \{x_1, y_1, x_2, y_2\}$. (Note that this is possible since n is even and at least 3). Let us also denote $\mathcal{S} \setminus \{x_1, y_1\}$ by U. Then by Corollary 5.8, we see that $X(U) = \frac{1}{2^2} X(\mathcal{S})$, since $q_1 \equiv 1 \pmod 4$. We now apply Corollary 5.8 with $W_1 = U$. This time, however, we have $q_1 \equiv 0 \pmod 4$ or $q_1 \equiv 2 \pmod 4$, (according as $r \geq 1$ or not) so $X(W) \neq \frac{1}{2^2} X(U)$, and therefore $X(W) \neq \frac{1}{2^4} X(\mathcal{S})$. Hence f is not 4th order correlation immune.

We thus have, combining the preceding four lemmas.

Corollary 5.15 Suppose that f belongs to $C(n, m, r, t, q)$ and that $q \equiv 1 \pmod 4$. Then

(i) if $2m < n - 1$, then f is exactly 1st order correlation immune and

(ii) if $2m = n - 1$, then f is exactly 2nd order correlation immune and

(iii) if $2m = n$, then f is exactly 3rd order correlation immune.

Combining all the results of this section, we have the following theorems and corollaries.

Theorem 5.16 If $f \in \mathcal{B}_S$ satisfies the SAC of order $(n - 3)$, then f is not correlation immune if and only if either

(i) there is exactly one pair (x, y) with $\overline{f}(\{x, y\}) = 1$ and $F(x) = -F(y)$ or

(ii) f belongs to $C(n, m, r, t, q)$ and $2m < n$ and $q \not\equiv 1 \pmod 4$.

Theorem 5.17 If $f \in \mathcal{B}_S$ satisfies the SAC of order $(n - 3)$, then f is exactly 1st order correlation immune if and only if one of the following holds

(i) there are exactly two pairs (x, y) with $\overline{f}(\{x, y\}) = 1$ and $F(x) = -F(y)$ or

(ii) f belongs to $C(n, m, r, t, q)$ and $2m = n$ and $q \not\equiv 1 \pmod 4$ or

(iii) f belongs to $C(n, m, r, t, q)$ and $2m < n - 1$ and $q \equiv 1 \pmod 4$.

Theorem 5.18 If $f \in \mathcal{B}_S$ satisfies the SAC of order $(n-3)$, then f is exactly 2nd order correlation immune if and only if either

(i) there are exactly three pairs (x, y) with $\overline{f}(\{x, y\}) = 1$ and $F(x) = -F(y)$ or

(ii) f belongs to $C(n, m, r, t, q)$ and $2m = n - 1$ and $q \equiv 1 \pmod 4$.

Theorem 5.19 If $f \in \mathcal{B}_S$ satisfies the SAC of order $(n-3)$, then f is exactly 3rd order correlation immune if and only if either

(i) there are exactly four pairs (x, y) with $\overline{f}(\{x, y\}) = 1$ and $F(x) = -F(y)$ or

(ii) f belongs to $C(n, m, r, t, q)$ and $2m = n$ and $q \equiv 1 \pmod 4$.

Theorem 5.20 If $f \in \mathcal{B}_S$ satisfies the SAC of order $(n-3)$, then f is pth order correlation immune $(p > 3)$ if and only if there are exactly $(p+1)$ pairs (x, y) with $\overline{f}(\{x, y\}) = 1$ and $F(x) = -F(y)$.

6. BALANCE AND CORRELATION IMMUNITY

Combining the results in sections 4 and 5, we have the following result.

Theorem 6.1 If $f \in \mathcal{B}_S$ satisfies the SAC of order $(n-3)$, then f is both balanced and correlation immune if and only if either

(i) there exist at least two pairs (x, y) such that $\overline{f}(\{x, y\}) = 1$ and $F(x) = -F(y)$ or

(ii) f belongs to $C(n, m, r, t, q)$ and $n = 2m$ and $q \equiv 3 \pmod 4$.

7. REFERENCES

[1] Forré, R., "The Strict Avalanche Criterion : Spectral Properties of Boolean Functions and an Extended Definition", *Abstracts CRYPTO88*, 1988

[2] Lloyd, S.A, "Balance, uncorrelatedness and the Strict Avalanche Criterion", *Hewlett-Packard Research Laboratories, Bristol, Technical Memo no. HPL-ISC-TM-89-012*, 1989 (also submitted to Discrete Applied Mathematics)

[3] Lloyd, S.A, "Characterising and counting functions satisfying the Strict Avalanche Criterion of order $(n-3)$", to appear in *Proceedings of the Second IMA Conference on Cryptography and Coding*, 1989

[4] Siegenthaler, T., "Decrypting a class of stream ciphers using ciphertext only", *IEEE Transactions on Computers*, vol. C-34, (1985), pp.81-85

[5] Siegenthaler, T., "Correlation immunity of nonlinear combining functions for cryptographic applications", *IEEE Transactions on Information Theory*, vol. IT-30, (1984), pp776-780

[6] Xiao, G-Z, and Massey, J.L., "A Spectral Characterization of Correlation-Immune Combining Functions", *IEEE Transactions on Information Theory*, Vol. 34, No. 3 (1988), pp.569-571

[7] Webster, A.F. and Tavares, S.E., "On the design of S-boxes", *Advances in Cryptology, Proceedings CRYPTO85*, Springer Verlag, Heidelberg, 1986, pp. 523-534

How to Construct Pseudorandom Permutations from Single Pseudorandom Functions

Josef Pieprzyk *

Department of Computer Science
University College
University of New South Wales
Austarlian Defence Force Academy
Canberra, ACT 2600, AUSTRALIA

Abstract

The paper examines permutation generators which are designed using four rounds of the Data Encryption Standard and a single pseudorandom function. We have proved that such generators are pseudorandom only if the pseudorandom function is used internally at least five times. The proof is given using two different approaches: deterministic and probabilistic. Some cryptographic implications are also discussed.

1 Introduction

Random number generators are commonly used in many different areas of science. However truly random generators create some problems. The most evident is the unreproducibility of generated numbers, so that it is impossible to repeat the same experiment. Another problem is related to the assessment of randomness of generators.

Classical pseudorandom generators are deterministic algorithms that provide numbers which "look" like random ones. As they have well-defined mathematical structures, they can be analysed easily. The basic measurement of classical pseudorandom generators is the similarity of generated numbers to truly random ones. Yao [8] redefined the notion of pseudorandomness in terms of complexity theory. A generator is said to be pseudorandom if it is "indistinguishable" from the truly random one, assuming polynomially bounded computing resources.

Cryptographers have always been interested in how to extend truly random "seeds" (n-bit long) into n^k-bit output strings in such a way that the output is indistinguishable from a truly random string ($k = 2, 3, \cdots$). Blum and Micali [1] introduced the notion of cryptographically strong pseudorandom bit generators (CSB). Levin [3] proved that such generators exist if and only if one-way functions exist. There are several implementations of CSB generators (for details see for example [2]).

---

*Support for this project was provided in part by TELECOM Australia under the contract number 7027 and by the Australian Research Council under the reference number A48830241.

Goldreich, Goldwasser and Micali [2] showed that it is possible to construct pseudorandom functions using CSB generators. The next step in the theory of pseudorandomness is due to Luby and Rackoff [4]. They used pseudorandom functions to generate pseudorandom permutations using three rounds of the Data Encryption Standard (DES) and three different pseudorandom functions.

Ohnishi [9] proved that pseudorandom permutations can be obtained using three rounds of the DES and two different pseudorandom functions. During Eurocrypt'89 Zheng, Matsumoto and Imai [9] presented some new results about pseudorandom permutations. They gave a construction of distinguishing circuits for all permutation generators which use three rounds of DES and a single pseudorandom function. They also posed the following problem:

Prove or disprove that from one pseudorandom function, one can obtain in some way a pseudorandom (invertible) permutation applying four rounds of DES.

In this paper we solve the problem.

2 Notations and definitions

Let $I_n = \{0,1\}^n$ be the set of all 2^n binary strings of length n. For $a, b \in I_n$, $a \oplus b$ stands for bit-by-bit exclusive-or of a and b. The set of all functions from I_n to I_n is F_n, i.e.

$$F_n = \{f \mid f : I_n \rightarrow I_n\}$$

It consists of 2^{n2^n} elements. If we have two functions $f, g \in F_n$, their composition $f \circ g$ is defined as

$$f \circ g(x) = f(g(x))$$

for all $x \in I_n$. The set of all permutations from I_n to I_n is P_n, i.e.

$$P_n \subset F_n$$

Definition 2.1 *A permutation generator p with a key k of length $l(n)$ ($l(n)$ is a polynomial in n) is a (poly) collection of permutations*

$$p = \{p_n \mid n \in N, p_n \subseteq P_n\}$$

where for each key k of length $l(n)$ and any value of $\alpha \in I_n$, we can compute in polynomial time in n the element $p_{n,k}(\alpha)$ ($p_{n,k} \in p_n$) and N is the set of all positive integers.

Definition 2.2 *A distinguishing circuit C_n is an acyclic circuit which consists of Boolean gates (AND, OR, NOT), constant gates ("0" and "1") and oracle gates. The circuit has one bit output only. Oracle gates accept inputs of length n and generate outputs of the same length. Each oracle gate can be evaluated using some permutation from P_n.*

Definition 2.3 *A family of distinguishing circuits for a permutation generator p is an infinite sequence of circuits $C_{n_1}, C_{n_2}, \cdots$ ($n_1 < n_2 < \cdots$) such that for two constants c_1 and c_2 and for each parameter n, there exits a circuit C_n which has the following properties:*

- *The size of C_n is smaller than n^{c_1} (the size is defined as the number of all connections between gates).*

- Let $Pr[C_n(P_n)]$ be the probability that the output bit of C_n is one when a permutation is randomly selected from P_n and used to evaluate the oracle gates. Let $Pr[C_n(p_n)]$ be the probability that the output bit of C_n is one when a key k of length $l(n)$ is randomly selected and $p_{n,k} \in p_n$ is used to evaluate the oracle gates. The distinguishing probability for C_n:

$$| Pr[C_n(P_n)] - Pr[C_n(p_n)] | \geq \frac{1}{n^{c_2}}$$

Definition 2.4 *A permutation generator p is pseudorandom if there is no distinguishing circuit family for p.*

Definition 2.5 *For a function $f \in F_n$, we define the DES-like permutation associated with f as*

$$D_{2n,f}(L, R) = (R \oplus f(L), L)$$

where R and L are n-bit strings $(R, L \in I_n)$ and $D_{2n,f} \in P_{2n}$.

Having a sequence of functions $f_1, f_2, \cdots, f_i \in F_n$, we can determine the concatenation of their DES-like permutations ψ and

$$\psi(f_1, f_2, \cdots, f_i) = D_{2n,f_i} \circ D_{2n,f_{i-1}} \circ \cdots \circ D_{2n,f_1}$$

Of course, $\psi(f_1, f_2, \cdots, f_i) \in P_{2n}$.

We can now rephrase the previous results. Luby and Rackoff [4] proved that $\psi(g, h, f)$ is a pseudorandom permutation generator if g, h, f are different pseudorandom functions. Ohnishi [9] showed that both $\psi(g, f, f)$ and $\psi(f, f, g)$ are pseudorandom generators as well, where g, f are different pseudorandom functions. Note that $\psi(f, g, f)$ is not pseudorandom as it is symmetric. Zheng, Matsumto and Imai [9] considered permutations based on a single function $f \in F_n$ and they gave the description of distinguishing circuits for $\psi(f^i, f^j, f^k)$, where f^i is a composition $f \circ f^{i-1}$ and i, j, k are arbitrary positive integers.

Definition 2.6 *A function f is said to be (truly) random (denote this as $f \in_R F_n$) if for a fixed argument $x \in \{0, \cdots, 2^n - 1\}$, $f(x)$ is an independent and uniformly distributed random variable.*

Example 2.1 *Assume that $f \in_R F_2$. Then we have four independent uniformly distributed random variables $f(x_i)$ $(i = 0, 1, 2, 3)$ and*

$$Pr[f(x_i) = j] = \frac{1}{4} \text{ for } j = 0, 1, 2, 3$$

In this paper we are going to prove that $\psi(f, f, f, f^i)$ is pseudorandom permutation generator for $i = 2, 3, \cdots$. To shorten our considerations, we are going to use a reference permutation generator (instead of a truly random one). Its structure should be as close as possible to the structure of $\psi(f, f, f, f^i)$. As $\psi(f, f, g)$ is pseudorandom, we take $\psi(f, f, f, g)$ as the reference one. Obviously, it is pseudorandom (the proof is omitted). Intuitively, taking out the first f from $\psi(f, f, g)$, we get $\psi(f, g)$ which is no longer pseudorandom. Therefore, putting up an additional round for f to $\psi(f, f, g)$, we obtain $\psi(f, f, f, g)$ which must be pseudorandom (the additional round does not introduce symmetry).

The proof of pseudorandomness of $\psi(f, f, f, f^i)$ for $i = 2, 3, \cdots$ will be given in two different ways. In the next Section, it will be carried out using some deterministic arguments. Later on we show that the proof can also be based on some probabilistic ones.

3 Deterministic distinguishing circuits

To illustrate how deterministic distinguishing circuits (DDC) work, consider the following example.

Example 3.1 *Take a permutation generator $\psi(f, f, f^2)$. It is possible to show [9] that it has a DDC family for all n. The structure of such a distinguishing circuit is given in Figure 1.*

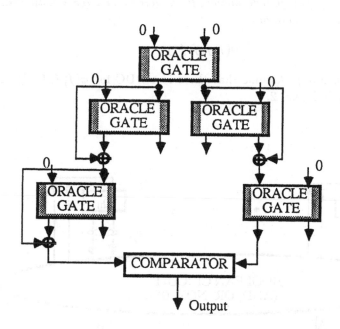

Figure 1. Deterministic distinguishing circuit for $\psi(f, f, f^2)$

The generator produces the following outputs:

$$(f^2(0), \; f(0) \oplus f^4(0)) \qquad \text{for input } (0,0)$$
$$(f(L \oplus f(0)), \; L \oplus f(0) \oplus f^3(L \oplus f(0))) \quad \text{for input } (L,0)$$
$$(R \oplus f^2(R), \; f(R) \oplus f^2(R \oplus f^2(R))) \qquad \text{for input } (0,R)$$

If $\psi(f, f, f^2)$ is used to evaluate the oracle gates, the comparator inputs can be described as $f^8(0)$. It is interesting what happens if we apply $\psi(f, f, g)$ for oracle gate evaluation. It can be verified that the left hand side input to the comparator does not change its description and is $f^8(0)$, while the right hand side one is represented as $fgfgf^2(0)$.

This example shows us some general properties of a DDC, namely:

P1 As it must work for all possible selections of f, it can be seen as a network which generates the same signal using two different paths. Each path can be expressed by equivalent algebraic formulae.

P2 If there is a DDC, it can always be designed using the oracle gate with input $(0,0)$ as the top one which initiates both paths.

P3 Modulo-2 operation $\oplus$ used outside oracle gates, removes input expression from the output.

Now we will prove the main lemma of the section. Without a loss of generality, we can consider generator $\psi(f,f,f,f^2)$ instead of $\psi(f,f,f,f^i)$ for $i = 2,3,\cdots$.

Lemma 3.1 *If* $\psi(f,f,f,g)$ *is pseudorandom, then there is no deterministic distinguishing circuit for a permutation generator*

$$\psi(f,f,f,f^2)$$

Proof (by contradiction): Assume that there is a DDC for $\psi(f,f,f,f^2)$. It may be represented as shown in Figure 2.

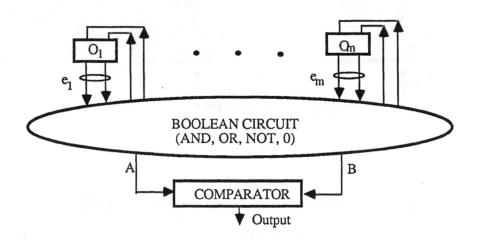

Figure 2. A general diagram of a deterministic distinguishing circuit (OG stands for oracle gate)

In general, for a given input (L, R), a oracle gate generates:

$$(L \oplus f(R) \oplus f(R \oplus f(L \oplus f(R))),$$
$$R \oplus f(L \oplus f(R)) \oplus f^2(f(R) \oplus f(R \oplus f(L \oplus f(R))))).$$

In particular, for input $(0,0)$, the oracle gate can be described as follows:

$$(f(0) \oplus f^3(0), \quad f^2(0) \oplus f^2(f(0) \oplus f^3(0))).$$

Take the last term of the second output of oracle gate for input (L, R) and denote it as $b(L, R)$, i.e.

$$b(L, R) = f^2(f(R) \oplus f(R \oplus f(L \oplus f(R)))).$$

Note that:

- f^2 in $b(L, R)$ does not commute with the function which is inside the brackets for all L and R (in particular, it does not commute for $L = 0$ and $R = 0$),

- it is impossible to separate a single expression $b(L, R)$ from the right hand side outputs of oracle gates so it always appears as a part of bigger expression which consists of at least two terms. In other words f^2 in $b(L, R)$ does not commute with the function f when $b(L, R)$ is a part of an expression placed inside the brackets of the function f.

It means that inputs to the comparator A and B are equal and can be presented as

$$A(e_1, ..., e_m) = A(e, b_1, ..., b_m)$$

where b_i is $b(L, R)$ for the i-th oracle gate $(i = 1, \cdots m)$ and e stands for the rest of oracle gate outputs. f^2 in all expressions $b_1, \cdots, b_m$ does not "mix" and stays in its original form.

Now if the DDC is used for $\psi(f, f, f, g)$, the Boolean circuit is fed with the same vector e and both A and B can be expressed as:

$$A = B = A(e, g_1, \cdots, g_m)$$

where g_i substitutes f^2 in b_i. It means that $\psi(f, f, f, g)$ has a DDC. This is the contradiction which proves the lemma.

□

Theorem 3.1 *If $\psi(f, f, f, g)$ is pseudorandom, then $\psi(f, f, f, f^2)$ is pseudorandom.*

Proof: Luby and Rackoff noted in [4] that any probabilistic distinguishing circuit can be converted to deterministic one. So if a generator has no DDC (see the lemma), it has no probabilistic distinguishing circuit and the generator is pseudorandom.

□

4 Probabilistic distinguishing circuits

In this section we are going to prove that $\psi(f, f, f, f^2)$ is pseudorandom using probabilistic arguments.

Lemma 4.1 *Let C_{2n} be an distinguishing circuit with m oracle gates. Then*

$$| Pr[C_{2n}(\psi(F))] - Pr[C_{2n}(\psi(F, G))] | \leq \frac{9m^3}{4^n}$$

where $Pr[C_{2n}(\psi(F))] = Pr[C_{2n}(\psi(f, f, f, f^2))]$ is the probability that the oracle circuit generates "1" when its oracle gates are evaluated using $\psi(f, f, f, f^2)$ (the function $f \in_R F_n$) and $Pr[C_{2n}(\psi(F, G))] = Pr[C_{2n}(\psi(f, f, f, g))]$ is the probability that the oracle circuit generates "1" when its oracle gates are evaluated using $\psi(f, f, f, g)$ $(f, g \in_R F_n)$.

Proof(sketch): Note that both f and g can be considered as two sequences of 2^n independent and uniformly distributed random variables. For two different arguments $a, b \in I_n$, $f(a)$ and $f(b)$ are independent random variables. When the input to an oracle gate is (L, R) $(L, R \in I_n)$, the gate produces the output:

$$f(R) \oplus f(R \oplus f(L \oplus f(R))), \quad f(L \oplus f(R)) \oplus f^2(L \oplus f(R) \oplus f(R \oplus f(L \oplus f(R))))$$

if it is evaluated using $\psi(f, f, f, f^2)$ or generates the output:

$$f(R) \oplus f(R \oplus f(L \oplus f(R))), \quad f(L \oplus f(R)) \oplus g(L \oplus f(R) \oplus f(R \oplus f(L \oplus f(R))))$$

when it is evaluated according to $\psi(f, f, f, g)$ (outputs of oracle gates are given after removing the input L and R, respectively). Of course the first components (random variables) of the outputs are the same.

Observe that if we have two oracle gates which are evaluated by $\psi(f, f, f, f^2)$ and whose inputs are different (an input (L, R) is different from (L', R') if either $L \neq L'$ or $R \neq R'$), their outputs yield four independent random variables.

A probabilistic distinguishing circuit generates 1 on its output with the same probability for both $\psi(f, f, f, f^2)$ and $\psi(f, f, f, g)$ if the random variables (the second part of the right hand side output of oracle gates) $f^2(\alpha_i)$ are independent from all internal random variables:

$$f(R_i);$$
$$f(L_i \oplus f(R_i));$$
$$\alpha_i = f(R_i \oplus f(L_i \oplus f(R_i)))$$

in all m oracle gates, where i is the number of the gate and (L_i, R_i) is the input to the i-th gate.

A probabilistic distinguishing circuit generates 1 on its output with different probabilities if there is at least one oracle gate in which $f^2(L_i \oplus f(R_i) \oplus \alpha_i)$ is expressable by a composition of two internal variables (there are $3m$ such random variables). Denote Pr_{diff} as the probability that there exists $f^2(L_i \oplus f(R_i) \oplus \alpha_i)$ that is expressable by two internal variables. Therefore:

$$| Pr[C_{2n}(\psi(F))] - Pr[C_{2n}(\psi(F, G))] | \leq Pr_{diff}$$

For fixed i $(1 \leq i \leq m)$, the probability that a single variable $f^2(L_i \oplus f(R_i) \oplus \alpha_i)$ can be constructed from two internal variables is:

$$\frac{9m^2}{4^n}$$

So:

$$Pr_{diff} = \frac{9m^3}{4^n}$$

and the final result follows. (A formal proof needs definitions of suitable probabilistic spaces to be introduced. Such spaces are the same as those in the paper by Luby and Rackoff [4]).

□

Denote that $p_F = Pr[C_{2n}(\psi(F))]$, $p_{F,G} = Pr[C_{2n}(\psi(F, G))]$ (those probabilities are defined above) and $p_R = Pr[C_{2n}(F_{2n})]$ is the probability that the oracle circuit generates "1" if the oracle gates are evaluated using a random function $f \in_R F_{2n}$.

Lemma 4.2 *If $\psi(f, f, f, g)$ is pseudorandom, then*

$$\mid p_F - p_R \mid \le \frac{\gamma(m)}{2^n}$$

where $\gamma(m)$ is a polynomial in m (m is the number of oracle gates in the distinguishing circuit).

Proof. The previous lemma stated that

$$\mid p_F - p_{F,G} \mid \le \frac{\gamma_1(m)}{2^n}$$

and the assumption about pseudorandomness of $\psi(f, f, f, g)$ is equivalent to

$$\mid p_{F,G} - p_R \mid \le \frac{\gamma_2(m)}{2^n}.$$

where $\gamma_2(m)$ is a polynomial in m. The following is true

$$\mid p_F - p_R \mid = \mid p_F - p_{F,G} + p_{F,G} - p_R \mid \le$$
$$\mid p_F - p_{F,G} \mid + \mid p_{F,G} - p_R \mid = \frac{\gamma(m)}{2^n}$$

where $\gamma(m) = \gamma_1(m) + \gamma_2(m)$.
□

Theorem 4.1 $\psi(f, f, f, f^2)$ *is pseudorandom permutation generator.*

Proof (by contradiction). Proof is similar to that given in [4]. Denote that $p_f = Pr[C_{2n}(\psi(f, f, f, f^2))]$, where $f \in F_n$ is a pseudorandom function. Assume that there is a distinguishing circuit family for $\psi(f, f, f, f^2)$. Let it be $C = \{C_{2n_1}, C_{2n_2}, \cdots\}$, where $n_1 < n_2 < \cdots$. It means that for large enough n, the following sequence is true:

$$\frac{1}{n^c} \le \mid p_f - p_R \mid = \mid p_f - p_F + p_F - p_R \mid \le \mid p_f - p_F \mid + \mid p_F - p_R \mid$$

where c is a constant. As $\mid p_F - p_R \mid$ is bounded by $\frac{\gamma(m)}{2^n}$, $\mid p_f - p_F \mid \ge \frac{1}{2n^c}$. In other words the family C defines a distinguishing circuit family for pseudorandom function f. This is the contradiction and the theorem is proved.
□

All above considerations are valid for $\psi(f, f, f, f^i)$ for fixed $i = 2, 3, \cdots$.

5 Towards provably secure block cryptosystems

The results obtained in the theory of pseudorandomness may be interpreted differently. As we know (see [4]), permutation generators $\psi(f, g, h)$ are pseudorandom if f, g, h are different pseudorandom functions. In general, if we use random functions instead pseudorandom ones then a permutation generator for which there is no distinguisher, is called a randomizer. Now we can rephrase the previous theorems.

Theorem 5.1 (Luby and Rackoff [4]) *Let $f, g, h \in_R F_n$ and a distinguisher C_{2n} have m oracle gates ($m < 2^n$), then:*

$$| Pr[C_{2n}(F)] - Pr[C_{2n}(\psi(f, g, h))] | \leq \frac{m^2}{2^n} \qquad (1)$$

and the randomizer ψ extends binary strings of length $3 \times n2^n$ into strings of length $2n2^{2n}$

To make the probability in (1) sufficiently small n should be larger than 64. For $n = 64$ the randomizer $\psi(f, g, h)$ extends 3.5×10^{21}-bit random string into a random string of length 4.4×10^{40} bits.

Theorem 5.2 (Ohnishi [9]) *Let $f, g \in_R F_n$ and a distinguisher C_{2n} have m oracle gates ($m < 2^n$), then:*

$$| Pr[C_{2n}(F)] - Pr[C_{2n}(\psi(f, f, g))] | \leq \frac{2(m+1)^2}{2^n} \qquad (2)$$

and the randomizer ψ extends binary strings of length $2 \times n2^n$ into strings of length $2n2^{2n}$

Theorem 5.3 *Let $f \in_R F_n$ and a distinguisher C_{2n} have m oracle gates ($m < 2^n$), then*

$$| Pr[C_{2n}(F)] - Pr[C_{2n}(\psi(f, f, f, f^i))] | \leq \frac{2(m+1)^2}{2^n} + \frac{3^i m^{i+1}}{2^{ni}} \qquad (3)$$

and the randomizer ψ extends binary strings of length $n2^n$ into strings of length $2n2^{2n}$

The last theorem gives the best known result and for $n = 32$, the randomizer $\psi(f, f, f, f^i)$ extends 1.4×10^{11} bits into binary string of length 3.5×10^{21}. Unfortunately the probability given by (3) can be too high even for relatively small m (for instance if $m = 10^5$ the probability is close to 1).

Zheng et al. [10] studied the practicability of provable secure cryptosystems (PSC) and they pointed out that there is still gap between the theory (which says that PSCs exist) and the practice (any implementation needs exponential size memory to memorize random functions).

Consider formulae (1,2,3) and note that both memory requirements and probabilities depend on the value of 2^n. Ideally, we would like to be able to select in some way memory requirements (the size of random function) and probabilities independently, possibly by introducing an additional parameter. In general, it is well known that increase of number of rounds in any cryptosystem improves its quality. Therefore, such a parameter may be the number of rounds.

There are three different approaches:

- direct concatenation of basic randomizers ψ, i.e. we create an randomizer $\Psi_j = \underbrace{\psi \circ \psi \circ \cdots}_{j}$;

- construction of complex randomizers using a basic randomizer ψ, i.e. $\theta = \psi(p, p, p, p^2)$ where $p = \psi(f, f, f, f^2)$ and f is a random function;

- concatenation of complex randomizers, i.e. $\Theta_j = \underbrace{\theta \circ \theta \circ \cdots}_{j}$, where θ is given above.

The basic randomizer uses a random function $f \in F_n$. The resulting randomizer Ψ_j generates a permutation from F_{2n}. Both randomizers θ and Θ_j are permutations in F_{4n}. The most important question is: *What are the corresponding probabilities bounds for the first two approaches ?* Ideally, we would expect that the probabilities are expressable as:

$$\frac{Q(n)}{2^{nj}}$$

where j is the number of iterations. In this case, it would be possible to select small enough n so the memory requirements could be met and at the same time, the probability (3) could be adjusted by selecting the number of concatenations.

Of course, all above randomizers are based on the DES structure. It is possible to generalize this structure as is shown in [10]. Maurer and Massey [5] considered codes as randomizers.

6 Conclusions

The generation of binary sequences that resemble truly random ones, is widely used in many different applications. Some of these applications impose especially strict requirements. One such application is cryptography. Most binary sequence generators turned to be useless from a cryptographic point of view.

Yao [8] defined a class of generators which are not distinguishable from truly random generators having polynomial size sample of output and polynomially-bounded computing resources. He called them pseudorandom.

The existance of PBG has been proved by Levin [3] providing existence of a way-one function. There several implementations of PBG using different one-way functions (see [2]). Goldreich et al. [2] showed that PBGs can be used to construct pseudorandom function generators. Later, Luby and Rackoff [4] described pseudorandom permutation generators using three DES rounds and three pseudorandom functions.

Ohnishi improved their result by proving that pseudorandom permutation generators (PPGs) can be made up from three DES rounds and two pseudorandom functions. This result is optimal in the sense that using a single pseudorandom function and three DES rounds, it is impossible to construct PPG (Zheng et al [9]).

Schnor [7] was the first to pose the problem of construction of pseudorandom permutation generators using single pseudorandom functions. Zheng et al. [9] and Rueppel [6] showed that Schnors generator $\psi(f, f, f)$ is not pseudorandom. In this paper we have proved that having a single pseudorandom function and four DES rounds it is possible to construct PPGs.

If we substitute pseudorandom functions by truly random ones in PPG, we obtain generators which are called randomizers. Their quality is expressable by their probability bound for distinguishing them from truly random permutations and does not rely on unproved assumptions (as the existence of one-way functions).

Randomizers of structure $\psi(f, f, f, f^2)$ (f is a random function from F_n) stretch $n \times 2^n$ input bits into $2n \times 2^{2n}$ output bits. Zheng et al [10] defined provable secure cryptosystems as randomizers. There is however, a gap between the practical implementation and the theory. To make this idea implementable, more research is necessary. Especially, it is interesting to examine the influence of the number of rounds on the distinguishing probability. Also a search for new "more" efficient randomizers could bring us closer to a practical implementation of PSC. Maurer and Massey [5] pointed out codes as one of possible randomizers.

ACKNOWLEDGMENT

I would like to thank Cathy Newberry, Dr Rei Safavi-Naini and Xian-Mo Zhang for their comments, discussions and assistance during the preparation of this work.

References

[1] M. Blum and S. Micali. How to generate cryptographically strong sequences of pseudo-random bits. *SIAM Journal on Computing*, 13:850–864, November 1984.

[2] O. Goldreich, S. Goldwasser, and S. Micali. How to construct random functions. *Journal of the ACM*, 33(4):792–807, October 1986.

[3] L. A. Levin. One-way function and pseudorandom generators. In *Proceedings of the 17th ACM Symposium on Theory of Computing*, pages 363–365, New York, 1985. ACM.

[4] M. Luby and Ch. Rackoff. How to construct pseudorandom permutations from pseudorandom functions. *SIAM Journal on Computing*, 17(2):373–386, April 1988.

[5] U.M. Maurer and J.L. Massey. Perfect local randomness in pseudorandom sequences. Astracts of CRYPTO'89, Santa Barbara, CA, August 1989.

[6] R.A. Rueppel. On the security of Schnorr's pseudo random generator. Astracts of EUROCRYPT'89, Houthalen, Belgium, April 1989.

[7] C.P. Schnorr. On the construction of random number generators and random function generators. In *Proc. of Eurocrypt 88, Lecture Notes in Computer Science*, New York, 1988. Springer Verlag.

[8] Andrew C. Yao. Theory and application of trapdoor functions. In *Proceedings of the 23rd IEEE Symposium on Fundation of Computer Science*, pages 80–91, New York, 1982. IEEE.

[9] Y. Zheng, T. Matsumoto, and H. Imai. Impossibility and optimality results on constructing pseudorandom permutations. Astracts of EUROCRYPT'89, Houthalen, Belgium, April 1989.

[10] Y. Zheng, T. Matsumoto, and H. Imai. On the construction of block ciphers provably secure and not relying on any unproved hypotheses. Astracts of CRYPTO'89, Santa Barbara, CA, July 1989.

Constructions of bent functions and difference sets

KAISA NYBERG

University of Helsinki
and
Finnish Defence Forces

1. Introduction. Based on the work of Rothaus [12], Olsen, Scholtz and Welch suggested the bent functions to be used as feed-forward functions to generate binary sequences which possess high linear complexity and very nearly optimum cross-correlation properties [10]. In [7] Meier and Staffelbach discovered, that binary bent functions give a solution to the correlation problem when used as combining functions of several binary linear shiftregister sequences. One of their results is that bent functions are at maximum distance to the set of affine functions. We refer to [7] for the cryptographic background and motivation. The general theory of the bent functions from Z_q^n to Z_q was developed by Kumar, Scholtz and Welch [2].

The main purpose of this paper is to consider the cryptographic properties of generalized bent functions. In §2 we give the basic definitions and properties of bent functions. For more details we refer to [2]. Our main results are concerned with the value distributions of p-ary bent functions, p prime, and their distances to the set of affine functions, and are given in §3. In §4 a method is given to produce all binary bent functions. We also consider the relation between difference sets and bent functions and review the previous construction methods and their properties.

2. Generalized bent functions. Let q be a positive integer and denote the set of integers modulo q by Z_q. Let

$$u = e^{i\frac{2\pi}{q}}$$

be the qth root of unity in $\mathbf{C}$, where $i = \sqrt{-1}$. Let f be a function from the set Z_q^n of n-tuples of integers modulo q to Z_q. Then the *Fourier transform* of u^f is defined as follows

$$F(\mathbf{w}) = \frac{1}{\sqrt{q^n}} \sum_{\mathbf{x} \in Z_q^n} u^{f(\mathbf{x}) - \mathbf{w} \cdot \mathbf{x}}, \quad \mathbf{w} \in Z_q^n.$$

DEFINITION 2.1. *A function $f : Z_q^n \to Z_q$ is bent if $|F(\mathbf{w})| = 1$ for all $\mathbf{w} \in Z_q^n$.*

Let f and g be two functions from Z_q^n to Z_q. Then their *convolution* is

$$(u^f * u^g)(\mathbf{w}) = \sum_{\mathbf{x} \in Z_q^n} u^{f(\mathbf{x}) + g(\mathbf{w} - \mathbf{x})},$$

and their *shifted cross-correlation*

$$c(f,g)(\mathbf{w}) = \frac{1}{q^n} \sum_{\mathbf{x} \in \mathbf{Z}_q^n} u^{f(\mathbf{x}+\mathbf{w})-g(\mathbf{x})} = \frac{1}{q^n}(u^f * u^{g_r})(\mathbf{w}),$$

where $g_r(\mathbf{x}) = -g(-\mathbf{x})$.

From these definitions we easily obtain the following

THEOREM 2.1. *A function $f : \mathbf{Z}_q^n \to \mathbf{Z}_q$ is bent if and only if*

$$|c(f,L)(\mathbf{w})| = \frac{1}{\sqrt{q^n}}$$

for all linear (or affine) functions $L : \mathbf{Z}_q^n \to \mathbf{Z}_q$ and $\mathbf{w} \in \mathbf{Z}_q^n$.

Analogously to the binary case it then follows that the q-ary bent functions have the minimum correlation to the set of all affine functions (see Theorem 3.5 in [7]).

In [2] also the following result can be found.

THEOREM 2.2. *A function $f : \mathbf{Z}_q^n \to \mathbf{Z}_q$ is bent if and only if*

$$c(f,f)(\mathbf{w}) = 0, \text{ for all } \mathbf{w} \neq \mathbf{0}.$$

This is in the binary case exactly the property of perfect nonlinearity used by Meier and Staffelbach to define bent functions. We make the following generalization.

DEFINITION 2.2. *A function $f : \mathbf{Z}_q^n \to \mathbf{Z}_q$ is perfect nonlinear if for all $\mathbf{w} \in \mathbf{Z}_q^n$, $\mathbf{w} \neq$ bf 0 and $k \in \mathbf{Z}_q$*

$$f(\mathbf{x}) = f(\mathbf{x}+\mathbf{w}) + k,$$

for exactly q^{n-1} values of $\mathbf{x} \in \mathbf{Z}_q^n$.

THEOREM 2.3. *A perfect nonlinear function from $\mathbf{Z}_q^n$ to $\mathbf{Z}_q$ is bent. The converse is true if q is a prime.*

PROOF: Let f be a function from $\mathbf{Z}_q^n$ to $\mathbf{Z}_q$. If f is perfect nonlinear, then

$$c(f,f)(\mathbf{w}) = \frac{1}{q} \sum_{k \in \mathbf{Z}_q} u^k = 0,$$

for all $\mathbf{w} \in \mathbf{Z}_p^n$, hence f is bent by Theorem 2.2.

Assume now that q is prime and f is bent. Then

$$0 = \frac{1}{q^n} \sum_{\mathbf{x} \in \mathbf{Z}_q^n} u^{f(\mathbf{x}+\mathbf{w})-f(\mathbf{x})} = \frac{1}{q^n} \sum_{k \in \mathbf{Z}_q} b_k u^k,$$

where $b_k = \#\{\mathbf{x} \in \mathbf{Z}_q^n \mid f(\mathbf{x} + \mathbf{w}) - f(\mathbf{x}) = k\}$. Since $\{u, u^2, \ldots, u^{q-1}\}$ is a basis for the qth cyclotomic field over the field of rational numbers (see, e.g., [4], Theorem 2.47 and Exercise 2.53), it follows that the numbers b_k are all equal, or what is the same, f is a perfect nonlinear function.

Example. Let q be an odd integer and $f : \mathbf{Z}_q \to \mathbf{Z}_q$ be the function $f(x) = x^2$ (mod q). Let $w \in \mathbf{Z}_q$ and take $\lambda \in \mathbf{Z}_q$ such that $w = 2\lambda$ (mod q). Then

$$F(w) = \frac{1}{\sqrt{q}} \sum_{x \in \mathbf{Z}_q} u^{x^2 - wx} = \frac{1}{\sqrt{q}} u^{-\lambda^2} \sum_{x \in \mathbf{Z}_q} u^{(x-\lambda)^2} = \frac{1}{\sqrt{q}} u^{-\lambda^2} \sum_{k \in \mathbf{Z}_q} u^{k^2}.$$

For q odd, the Gaussian quadratic sum takes the absolute value $\sqrt{q}$. Hence $|F(w)| = 1$ and f is bent. For $w \in \mathbf{Z}_q$, $w \neq 0$, the difference

$$f(x + w) - f(x) = 2wx + w^2 \pmod{q}$$

takes every value in $\mathbf{Z}_q$ equally many times if and only if w and q are relative primes. Consequently f is perfect nonlinear if and only if q is a prime.

3. Constructions and properties.

The values of a non-constant affine function from $\mathbf{Z}_p^n$ to $\mathbf{Z}_p$ are evenly distributed when p is a prime. Since for the functions

$$f(\mathbf{x}) = f(x_1, x_2, \ldots, x_{2m}) = x_1 x_{m+1} + x_2 x_{m+2} + \cdots + x_m x_{2m}$$

or

$$f(\mathbf{x}) = f(x_1, x_2, \ldots, x_n) = x_1^2 + x_2^2 + \cdots + x_n^2$$

their difference functions

$$\mathbf{x} \mapsto f(\mathbf{x} + \mathbf{w}) - f(\mathbf{x})$$

are non-zero affine functions for all $\mathbf{w} \neq 0$, it follows that these functions are perfect nonlinear. Hence bent functions from $\mathbf{Z}_p^n$ to $\mathbf{Z}_p$ exist for every prime p when n is even, and for every prime $p \geq 3$ when n is odd.

DEFINITION 3.1. A function $f : \mathbf{Z}_q^n \to \mathbf{Z}_q$ is a regular bent function if there is a function $g : \mathbf{Z}_q^n \to \mathbf{Z}_q$ such that

$$F(\mathbf{w}) = u^{g(\mathbf{w})}, \quad \text{for all } \mathbf{w} \in \mathbf{Z}_q^n.$$

The following theorem is due to Kumar, Scholtz and Welch [2]. For $q = 2$ it was first proved by Maiorana, see [1], generalizing the construction method of Rothaus [12].

THEOREM 3.1. *Let* $g : \mathbf{Z}_q^m \to \mathbf{Z}_q$ *be any function and* $\pi : \mathbf{Z}_q^m \to \mathbf{Z}_q^m$ *any bijective transformation. Then the function*

$$f : \mathbf{Z}_q^{2m} = \mathbf{Z}_q^m \times \mathbf{Z}_q^m \to \mathbf{Z}_q, \; f(\mathbf{x}_1,\mathbf{x}_2) = \pi(\mathbf{x}_1) \cdot \mathbf{x}_2 + g(\mathbf{x}_1)$$

is a regular bent function.

Clearly, different choices of π and g yield different bent functions. Hence we have a lower bound

$$q^{q^{\frac{n}{2}}}(q^{\frac{n}{2}}!)$$

for the number of bent functions in $\mathbf{Z}_q^n$.

Because of their good correlation properties with linear functions (see Theorem 2.1) bent functions could be used to combine several independently generated sequences. Then it would be important to know what is their distance to affine functions and how well balanced their value distributions are.

Let us make the notation

$$b_k = \#\{\mathbf{x} \in \mathbf{Z}_p^n \mid f(\mathbf{x}) = k\}, \; k \in \mathbf{Z}_p.$$

Then we say that the value distribution of $f : \mathbf{Z}_p^n \to \mathbf{Z}_p$ is the ordered p-tuple $(b_0, b_1, \ldots, b_{p-1})$.

THEOREM 3.2. *Let n be even and p a prime. Then the value distribution of a bent function $f : \mathbf{Z}_p^n \to \mathbf{Z}_p$ is $(b_0, b_1, \ldots, b_{p-1})$, where*

$$b_0 = p^{n-1} \pm (p-1)p^{\frac{n}{2}-1}$$
$$b_k = p^{n-1} \mp p^{\frac{n}{2}-1}, \; \text{for } k = 1, 2, \ldots, p-1,$$

or its cyclic shift. Here the $\pm$ signs are taken correspondingly. Moreover, a regular bent function has the upper signs.

PROOF: According to [2], Property 8, there is an integer s such that

$$F(0) = u^{\frac{s}{2}}.$$

Hence we have

$$\sum_{k=0}^{p-1} b_k u^k = p^{\frac{n}{2}} u^{\frac{s}{2}}.$$

If s is even this equation gets the form

$$\sum_{k=0}^{p-1} b_k u^{k-r} = p^{\frac{n}{2}}$$

for some integer r. This is always the case for a regular bent function. If s is odd, we choose $r = \frac{s-p}{2}$ to have

$$u^{\frac{s}{2}-r} = u^{\frac{p}{2}} = -1.$$

In this case the equation becomes

$$\sum_{k=0}^{p-1} b_k u^{k-r} = -p^{\frac{n}{2}}$$

for some integer r. Since p is a prime it then follows that

$$b_0 \mp p^{\frac{n}{2}} = b_1 = b_2 = \cdots = b_{p-1},$$

or a cyclic shift. On the other hand $\sum_{k=0}^{p-1} b_k = p^n$, from where we obtain the solution

$$b_0 \mp p^{\frac{n}{2}} = b_1 = b_2 = \cdots = b_{p-1} = p^{n-1} \mp p^{\frac{n}{2}-1}.$$

Let us give some examples of regular bent functions:

(1) the functions given in Theorem 3.1;
(2) all binary bent functions (which exist only for even dimensions);
(3) $f(\mathbf{x}) = x_1^2 + x_2^2 + \cdots + x_n^2$ for n even and $p \equiv 1 \pmod 4$;
(4) $f(\mathbf{x}) = x_1^2 + x_2^2 + \cdots + x_n^2$ for $n \equiv 0 \pmod 4$ and p odd.

An example of a function having the lower signs in Theorem 3.2. (and falling in the category (2) of Theorem 3.3 below) is $f(\mathbf{x}) = x_1^2 + x_2^2 + \cdots + x_n^2$ for $n \equiv 2 \pmod 4$ and $p \equiv 3 \pmod 4$ and so is, more generally, the sum $f + g : \mathbf{Z}_p^n \times \mathbf{Z}_p^m \to \mathbf{Z}_p$, where $g : \mathbf{Z}_p^m \to \mathbf{Z}_p$ is a regular bent function.

THEOREM 3.3. *Let n be even and p a prime. Then the Hamming distance of a bent function $f : \mathbf{Z}_p^n \to \mathbf{Z}_p$ to the nearest affine function is either*

(1) $(p-1)(p^{n-1} - p^{\frac{n}{2}-1})$; *or*
(2) $(p-1)p^{n-1} - p^{\frac{n}{2}-1}$.

Regular bent functions have the distance (1) to the nearest affine function.

PROOF: Let $A : \mathbf{Z}_p^n \to \mathbf{Z}_p$, $A(\mathbf{x}) = \mathbf{w} \cdot \mathbf{x} + r$, be an affine function and denote

$$a_k = \#\{\mathbf{x} \in \mathbf{Z}_p^n \mid f(\mathbf{x}) - \mathbf{w} \cdot \mathbf{x} - r = k\}.$$

We now make use of [2], Property 8, for $\mathbf{w} \neq 0$ and proceed exactly as in the proof of the preceding theorem to obtain

$$a_0 \mp p^{\frac{n}{2}} = a_1 = a_2 = \cdots = a_{p-1} = p^{n-1} \mp p^{\frac{n}{2}-1}.$$

The Hamming distance of f to the affine function A is then $\sum_{k=1}^{p-1} a_k$. This is minimized over the totality of all affine functions when we choose r such that a_0 obtains its maximal value which is either

(1) $p^{n-1} + (p-1)p^{\frac{n}{2}-1}$, or
(2) $p^{n-1} + p^{\frac{n}{2}-1}$.

From this theorem and the above example it follows that for $p = 3 \pmod 4$ there are two classes of bent functions and the first one, to which the regular bent functions belong, is closer to the set of affine functions than the second one, to which the square-sum function belongs.

To study the case where n is odd we need the following lemma from the theory of cyclotomic fields.

LEMMA. *For a prime p there is a unique integer solution $(a_1, a_2, \ldots, a_{p-1})$ to the equation*

$$a_1 u + a_2 u^2 + \cdots + a_{p-1} u^{p-1} = \begin{cases} \sqrt{p}, & \text{for } p = 1 \pmod 4 \\ i\sqrt{p}, & \text{for } p = 3 \pmod 4. \end{cases}$$

This solution is

$$a_k = \left(\frac{k}{p}\right), \quad k = 1, 2, \ldots, p-1.$$

PROOF: The proof is obtained by combining Gaussian quadratic sums, see, e.g., [2], formula (14), with the argument on the dimension of the pth cyclotomic field, [4], Theorem 2.47.

THEOREM 3.4. *Let n be odd and p an odd prime. The value distribution of a regular bent function from $\mathbf{Z}_p^n$ to $\mathbf{Z}_p$ is a cyclic permutation of $(b_0, b_1, \ldots, b_{p-1})$, where $b_0 = p^{n-1}$ and either*

(1)
$$b_k = p^{n-1} + \left(\frac{k}{p}\right) p^{\frac{n-1}{2}}, \quad \text{for all } k = 1, 2, \ldots, p-1, \text{ or}$$

(2)
$$b_k = p^{n-1} - \left(\frac{k}{p}\right) p^{\frac{n-1}{2}}, \quad \text{for all } k = 1, 2, \ldots, p-1.$$

PROOF: Let $f : \mathbf{Z}_p^n \to \mathbf{Z}_p$ be a bent function. We consider first the case $p = 1 \pmod 4$. By [2], Property 8, there is an integer s such that

$$F(0) = u^{\frac{s}{2}} = \frac{1}{\sqrt{p^n}} \sum_{k=0}^{p-1} b_k u^k.$$

Similarly as in the proof of Theorem 3.2. this equation becomes

$$b_0 + b_1 u + \cdots + b_{p-1} u^{p-1} = \pm\sqrt{p} \cdot p^{\frac{n-1}{2}}.$$

Now it follows from the preceding lemma that the solution is of the form

$$b_k = b_0 \pm \left(\frac{k}{p}\right) p^{\frac{n-1}{2}}, \quad k = 1, 2, \ldots, p-1.$$

From $\sum_{k=0}^{p-1} b_k = p^n$ we then get that $b_0 = p^{n-1}$.

Assume now that $p = 3 \pmod 4$. Then by [2], Property 8, there is an integer s such that

$$F(0) = u^{\frac{2s+1}{4}}.$$

Take $r = \frac{p^2-1}{4}(2s+1)$, which is an integer. Then we have

$$F(0) = u^{p(2s+1)\cdot\frac{p}{4}-r} = \pm i \cdot u^{-r},$$

since $p(2s+1)$ is odd. Now we proceed exactly as in the first case to obtain the solution.

By repeating this proof for $w \neq 0$ we get the following

THEOREM 3.5. *For n odd and p an odd prime the Hamming distance of a bent function from Z_p^n to Z_p to the nearest affine function is*

$$(p-1)p^{n-1} - p^{\frac{n-1}{2}}.$$

4. Difference sets and constructions. From Maiorana's construction we obtain a lower bound $2^{2^{\frac{n}{2}}}(2^{\frac{n}{2}}!)$ for the number of bent functions in Z_2^n. If f is a bent function $Z_2^n \to Z_2$ then so is $f \circ A + L$ for every affine bijective transformation A of Z_2^n and linear $L : Z_2^n \to Z_2$. We call two bent functions f and g equivalent if they are related to each other in this way, i.e., $g = f \circ A + L$. This equivalence relation devides the set of bent functions into disjoint equivalence classes each containing at most 2^{n^2+n} functions. The functions in a same class have the same nonlinear order which is bounded from above by $\frac{n}{2}$. It follows that for n large enough to satisfy

$$(2^{\frac{n}{2}}!)2^{2^{\frac{n}{2}}-n^2-n} > \frac{n}{2} - 1$$

i.e., for $n \geq 10$ Maiorana's construction gives non-equivalent bent functions of the same (highest) nonlinear order. For cryptographic purposes and unpredictability this is a very desirable property.

On the other hand, Rothaus made in [12] a complete list of bent functions in Z_2^6 and also verified using a computer program that all of them of nonlinear order 3 are obtainable from each other by an affine transformation of coordinates and the addition of a linear function. To clarify the situation for $n = 8$ other construction methods, especially those that give more bent functions for small n, might be useful.

The case $n = 8$ remains open also in[11]. There is also given another construction method which yields the same number of bent functions as Maiorana's method.

The following result can be used in the construction of the set of ones of a bent function.

THEOREM 4.1. *A function $f : Z_2^n \to Z_2$ with $2^{n-1} - 2^{\frac{n}{2}-1}$ ones is bent if and only if for every nonconstant linear function $L : Z_2^n \to Z_2$ the product Lf has either 2^{n-2} or $2^{n-2} - 2^{\frac{n}{2}-1}$ ones.*

Hence to construct the set of ones of any bent function in $\mathbb{Z}_2^n$ it is enough to find n vectors of length $2^{n-1} - 2^{\frac{n}{2}-1}$ each consisting of 2^{n-2} ones and $2^{n-2} - 2^{\frac{n}{2}-1}$ zeros such that every possible sum of these vectors has exactly 2^{n-2} or $2^{n-2} - 2^{\frac{n}{2}-1}$ ones. It is an open problem whether this method of construction can be made feasible for large n.

Example. The columns of the following matrix are constructed by means of the above principle.

x_4	x_3	x_2	x_1
0	1	0	1
0	1	1	0
1	0	1	0
1	0	1	1
1	1	0	1
1	1	1	1

Hence the row vectors form the set of the ones of a bent function. This bent function is $x_1 x_3 + x_2 x_3 + x_2 x_4$.

In the above matrix the sum of the first and the third row equals to the sixth row and adding up the second row with the fourth one gives the fifth. Hence this set of rows break up into two triples and its easily checked that these are the only existing triples. This implies that every row a can be expressed in two different ways as a difference of two other rows b and c, i.e., $a = b + c = c + b$

The rows of the above matrix form an example of a specific combinatorial structure called difference set.

DEFINITION 4.1. *Let G be an additive Abelian group of order v. A subset D of G is called a (v, k, λ)–difference set if it is of order k and if every nonzero element $a \in D$ can be expressed in λ different ways as a difference $a = b - c$, where $b \in D$ and $c \in D$.*

The following result was already known to Dillon [1].

THEOREM 4.2. *A function $f : \mathbb{Z}_2^n \to \mathbb{Z}_2$ is bent if and only if it is a characteristic function of a difference set.*

Let us mention that it was proved by Mann, [5] pp.72-73, that the parameters of a difference set in $\mathbb{Z}_2^n$ are $(2^n, 2^{n-1} \pm 2^{\frac{n}{2}-1}, 2^{n-2} \pm 2^{\frac{n}{2}-1})$.

There are some previous constructions of difference sets found in the literature. In [1] Dillon proved that the constructions of Menon [8] and [9] and that of Turyn [13] are special cases of Maiorana's binary bent functions. The main result of [1] is that for $m > 3$ there exist bent functions in $\mathbb{Z}_2^{2m}$ which are not equivalent to any Maiorana's functions.

McFarland gave in 1971 the following construction of difference set. It can be easily checked that a function f is a Maiorana bent function if and only if either its set of zeros or set of ones can be constructed by the method of McFarland.

Let $H_1, H_2, \ldots, H_r$ be the totality of different hyperplanes, i.e., $(m-1)$–dimensional subspaces in $\mathbf{Z}_2^m$; then $r = 2^m - 1$. Let $\mathbf{a}_j \in \mathbf{Z}_2^m$, $j = 1, \ldots, r$, be any elements and let $\mathbf{b}_j \in \mathbf{Z}_2^m$, $j = 1, \ldots, r$, be distinct elements.

THEOREM 4.3 (MCFARLAND [6]). *The union of the r subsets in $\mathbf{Z}_2^{2m}$ of the form*

$$\{\, (\mathbf{a}_j + \mathbf{x}, \mathbf{b}_j) \mid \mathbf{x} \in H_j\}, \quad j = 1, 2, \ldots, r,$$

is a difference set with parameters $(2^{2m}, 2^{2m-1} - 2^{m-1}, 2^{2m-2} - 2^{m-1})$.

In fact, when choosing the elements $\mathbf{a}_j$ the only thing that matters is whether $\mathbf{a}_j$ belongs to H_j or not. Moreover, we have the following property of McFarland's construction.

THEOREM 4.4. *The nonlinear order of a bent function constructed by McFarland's method is maximal if and only if the element $\mathbf{a}_j$ is chosen from H_j for an odd number of indices j.*

PROOF: Assume that f is a bent function whose set of ones is of the form of McFarland. Let π be a permutation which takes each element $\mathbf{b}_j$ to the nonzero element which is orthogonal to H_j. Let us define a function g such that

$$g(\mathbf{b}_j) = \begin{cases} 1, & \text{if } \mathbf{a}_j \in H_j \\ 0, & \text{if } \mathbf{a}_j \notin H_j, \end{cases}$$

and set $g(\mathbf{b}) = 0$ if $\pi(\mathbf{b}) = 0$. Then

$$f(\mathbf{x}_1, \mathbf{x}_2) = g(\mathbf{x}_1) + \pi(\mathbf{x}_1) \cdot \mathbf{x}_2$$

and its nonlinear order is maximal if and only if the number of ones of g is odd.

In the general p-ary case the connection between bent functions and difference sets is more complicated and remains to be studied. Let us only mention that the general constructions of McFarland do not produce difference sets in groups of order p^n. Under some conditions regular bent functions yield difference sets in $\mathbf{Z}_p^n$, n even, with parameters

$$(p^n, p^{n-1} + (p-1)p^{\frac{n}{2}-1}, p^{n-2} + (p-1)p^{\frac{n}{2}-1}).$$

also for $p > 2$.

Example. The zeros of the bent function $f : \mathbf{Z}_3^4 \to \mathbf{Z}_3$, $f(\mathbf{x}) = x_1 x_2 + x_3 x_4$, form a difference set with parameters $(81, 33, 15)$.

Acknowledgement. I wish to thank Rainer Rueppel for bringing [2] to my attention.

REFERENCES

1. J. F. Dillon, *Elementary Hadamard difference sets*, Proceedings of the Sixth Southeastern Conference on Combinatorics, Graph Theory and Computing, Boca Raton, Florida (1975), 237–249; Congressus Numerantium No. XIV, Utilitas Math., Winnipeg, Manitoba (1975).
2. P. V. Kumar, R. A. Scholtz and L. R. Welch, *Generalized bent functions and their properties*, J. Combinatorial Theory, Ser. A 40 (1985), 90–107.
3. A. Lempel and M. Cohn, *Maximal families of bent sequences*, IEEE Trans. Inform. Theory IT-28 (1982), 865–868.
4. R. Lidl and H. Niederreiter, "Introduction to finite fields and their applications," Cambridge University Press, Cambridge, 1986.
5. H. B. Mann, "Addition theorems," John Wiley & Sons, New York, 1965.
6. R. L. McFarland, *A family of difference sets in non-cyclic groups*, J. Combinatorial Theory, Ser. A 15 (1973), 1–10.
7. W. Meier and O. Staffelbach, *Nonlinearity criteria for cryptographic functions*, Advances in Cryptology, Proceedings of Eurocrypt '89 (to appear).
8. P. K. Menon, *Difference sets in Abelian groups*, Proc. Amer. Math. Soc. 11 (1960), 368–376.
9. _____, *On difference sets whose parameters satisfy a certain relation*, Proc. Amer. Math. Soc. 13 (1962), 739–745.
10. J. D. Olsen, R. A. Scholtz and L. R. Welch, *Bent function sequences*, IEEE Trans. Inform. Theory IT-28 (1982), 858–864.
11. B. Preneel et al., Propagation characteristics of Boolean bent functions, Proceedings of Eurocrypt '90 (to appear).
12. O. S. Rothaus, *On "bent" functions*, J. Combinatorial Theory, Ser. A 20 (1976), 300–305.
13. R. J. Turyn, *Character sums and difference sets*, Pacific J. Math. 15 (1965), 319–346.

Finnish Defense Forces, Signals Section, P.O. Box 919, SF–00101 Helsinki, Finland

Propagation Characteristics of Boolean Functions

Bart Preneel[1], Werner Van Leekwijck, Luc Van Linden,
René Govaerts and Joos Vandewalle

Katholieke Universiteit Leuven, Laboratorium ESAT
Kardinaal Mercierlaan, 94, B–3030 Heverlee, Belgium

Abstract

The relation between the Walsh-Hadamard transform and the auto-correlation function of Boolean functions is used to study propagation characteristics of these functions. The Strict Avalanche Criterion and the Perfect Nonlinearity Criterion are generalized in a Propagation Criterion of degree k. New properties and constructions for Boolean bent functions are given and also the extension of the definition to odd values of n is discussed. New properties of functions satisfying higher order SAC are derived. Finally a general framework is established to classify functions according to their propagation characteristics if a number of bits is kept constant.

1 Introduction

The design and evaluation of cryptographic functions requires the definition of design criteria. It is known that the security of schemes based on a combination of permutations and substitutions strongly depends on the characteristics of the substitution tables or S-boxes. The relevance of the criteria can be based on information theoretic grounds or on specific attacks that are possible if certain conditions are not fulfilled. It is possible to define characteristics for individual output bits of S-boxes as well as for the relation between the different functions. In this paper, only individual functions of n bits to 1 bit are studied, but it is certainly possible to extend the scope.

In the past following criteria have been proposed: the function must have a high nonlinear order (no affine functions are allowed), must be 0/1 balanced, complete, satisfy a strict avalanche criterion or be perfect non-linear (with respect to linear structures). These criteria can be extended by imposing the requirement that the functions created by *fixing a number of input bits* of the original function still satisfy certain criteria. A second extension is possible by not only specifying the average values but also the *extreme values*. It is clear that no function can satisfy all these criteria: a good function will be the golden mean.

---

[1]NFWO aspirant navorser, sponsored by the National Fund for Scientific Research (Belgium).

An example of the study of the design principles of the Data Encryption Standard can be found in [BMP86]. The analysis of key clustering for a limited number of rounds of the DES also used intensively properties of the S-boxes when two input bits are fixed [DQD84]. An attack on the same cipher uses linear structures [Eve87] and can be thwarted if the S-boxes are perfect non-linear [MS89]. The price paid is that perfect non-linear functions are not balanced.

In this paper, we will not attempt to study the link between design criteria and the cryptographic functions, but we will try to generalize certain criteria and to study the functions that satisfy them. In a first section, the importance of the autocorrelation function of a Boolean function will be stressed and a new propagation criterion will be defined. Next the bent functions will be studied, followed by the functions satisfying higher order SAC. These criteria are further generalized and functions satisfying these criteria are studied.

2 Definitions

2.1 Boolean functions

A Boolean function $f(\underline{x})$ is a function whose domain is the vector space $\mathbb{Z}_2^n$ of binary n-tuples $(x_1, x_2, \ldots, x_n)$ that takes the values 0 and 1. In some cases it will be more convenient to work with functions that take the values $\{-1, 1\}$. The functions $\hat{f}(\underline{x})$ is defined as $\hat{f}(\underline{x}) = 1 - 2 \cdot f(\underline{x})$.

The Hamming weight hwt of an element of $\mathbb{Z}_2^n$ is the number of components equal to 1. The Hamming distance $d(f, g)$ between two Boolean functions f and g is the number of function values in which they differ:

$$d(f,g) = hwt(f \oplus g) = 2^{n-1} - \frac{1}{2} \sum_{\underline{x}} \hat{f}(\underline{x}) \cdot \hat{g}(\underline{x}).$$

Here $\sum_{\underline{x}}$ denotes the summation over all $\underline{x} \in \mathbb{Z}_2^n$. The correlation $c(f,g)$ is closely related to the Hamming distance: $c(f,g) = 1 - d(f,g)/2^{n-1}$.

A Boolean function is said to be linear if there exists a $\underline{w} \in \mathbb{Z}_2^n$ such that it can be written as $L_{\underline{w}}(\underline{x}) = \underline{x} \cdot \underline{w}$ or $\hat{L}_{\underline{w}}(\underline{x}) = (-1)^{\underline{x} \cdot \underline{w}}$. Here $\underline{x} \cdot \underline{w}$ denotes the dot product of $\underline{x}$ and $\underline{w}$, defined as $\underline{x} \cdot \underline{w} = x_1 w_1 \oplus x_2 w_2 \oplus \ldots \oplus x_n w_n$. The set of affine functions $A_{\underline{w}, w_0}(\underline{x})$ is the union of the set of the linear functions and their complement: $A_{\underline{w}, w_0}(\underline{x}) = L_{\underline{w}}(\underline{x}) \oplus w_0, \ w_0 \in \mathbb{Z}_2$.

2.2 The Walsh-Hadamard transform

Definition 1 *Let $f(\underline{x})$ be any real-valued function with domain the vector space $\mathbb{Z}_2^n$. The **Walsh-Hadamard transform** of $f(\underline{x})$ is the real-valued function over the vector space $\mathbb{Z}_2^n$ defined as*

$$F(\underline{w}) = \sum_{\underline{x}} f(\underline{x}) \cdot (-1)^{\underline{x} \cdot \underline{w}}.$$

The function $f(\underline{x})$ can be recovered by the **inverse Walsh–Hadamard transform:**

$$f(\underline{x}) = \frac{1}{2^n} \sum_{\underline{w}} F(\underline{w}) \cdot (-1)^{\underline{x} \cdot \underline{w}}.$$

The relationship between the Walsh-Hadamard transform of $f(\underline{x})$ and $\hat{f}(\underline{x})$ is given by [For88] $\hat{F}(\underline{w}) = -2F(\underline{w}) + 2^n\, \delta(\underline{w})$ and $F(\underline{w}) = -\frac{1}{2}\hat{F}(\underline{w}) + 2^{n-1}\, \delta(\underline{w})$, where $\delta(\underline{w})$ denotes the Kronecker delta ($\delta(\underline{0}) = 1, \delta(\underline{k}) = 0\ \forall \underline{k} \neq \underline{0}$).

As the Walsh-Hadamard transform is linear, an alternative definition based on a matrix product is possible. The function values of $f(\underline{x})$ and $F(\underline{x})$ are written in the column matrices $[f]$ and $[F]$ respectively

$$[F] = H_n \cdot [f],$$

where H_n is the Walsh-Hadamard matrix of order n that can be recursively defined as

$$H_n = \begin{bmatrix} 1 & 1 \\ 1 & -1 \end{bmatrix} \otimes H_{n-1}, \qquad H_0 = [1].$$

Here $\otimes$ denotes the Kronecker product between matrices. It is easily seen that $H_n^2 = 2^n \cdot I_n$.

In case of the Fourier transform, the energy spectrum (the square modulus of the Fourier transform) is used in a wide range of applications. A very important relation, known as the *Wiener-Khintchine theorem*, is that the inverse transform of the energy spectrum results in the autocorrelation function. A similar relation can be established in case of the Walsh-Hadamard transform. The autocorrelation function $\hat{r}(\underline{a})$ is defined as

$$\hat{r}(\underline{a}) = \sum_{\underline{x}} \hat{f}(\underline{x}) \cdot \hat{f}(\underline{x} \oplus \underline{a}).$$

Note that $\hat{r}(\underline{0})$ equals 2^n. The Walsh-Hadamard energy spectrum of any real-valued function is defined as $\hat{F}^2(\underline{w})$. The Walsh-Hadamard transform of $\hat{r}(\underline{a})$ will be denoted with $\hat{R}(\underline{w})$.

Theorem 1 *The Walsh-Hadamard transform of the autocorrelation function of any real-valued function is equal to the Walsh-Hadamard energy spectrum of that function:*
$\hat{R}(\underline{w}) = \hat{F}^2(\underline{w}), \quad \forall \underline{w} \in \mathbb{Z}_2^n.$

In case $\hat{f}(\underline{x})$ is Boolean, the importance of the autocorrelation function can be seen as follows:

$$\Pr\left(\hat{f}(\underline{x}) \neq \hat{f}(\underline{x} \oplus \underline{a})\right) = \frac{1}{2} - \frac{\hat{r}(\underline{a})}{2^{n+1}}.$$

For large values of n this theorem allows for an efficient computation of these probabilities, requiring $O(n2^n)$ operations (in stead of $O(2^{2n})$ for a straightforward computation). Summing these probabilities for all $\underline{a} \neq \underline{0}$ yields:

$$\sum_{\underline{a} \neq \underline{0}} \Pr\left(\hat{f}(\underline{x}) \neq \hat{f}(\underline{x} \oplus \underline{a})\right) = 2^{n-1} - \frac{\hat{F}^2(0)}{2^{n+1}}$$

If all probabilities are equal to $\frac{1}{2}$, then $|\hat{F}(0)| = 2^{\frac{n}{2}}$. If $f(\underline{x})$ is 0/1 balanced then $\hat{F}(0) = 0$ and the average of the probabilities is $(2^{n-1})/(2^n - 1) > \frac{1}{2}$.

2.3 The algebraic normal transform

The Walsh-Hadamard transform writes a Boolean function $\hat{f}(\underline{x})$ as a sum of linear functions of the form $(-1)^{\underline{x} \cdot \underline{w}}$. It can also be of interest to write a Boolean function as the sum of all products of the variables:

$$f(\underline{x}) = a_0 \oplus \sum_{1 \leq i \leq n} a_i x_i \oplus \sum_{1 \leq i < j \leq n} a_{ij} x_i x_j \oplus \ldots \oplus a_{12\ldots n} x_1 x_2 x_n.$$

This form is called the algebraic normal form of a Boolean function f, denoted with $\mathcal{F}$ and the corresponding transformation is called the algebraic normal transform [Rue86]. When the natural order of the variables is used, namely $1, x_1, x_2, x_1 x_2, x_3, x_1 x_3, \ldots, x_1 x_2 \ldots x_n$ following definition can be stated [Jan89]:

Definition 2 *The* **algebraic normal transform** *of a Boolean function f is a linear transformation (with respect to addition modulo 2) defined as*

$$[\mathcal{F}] = A_n \cdot [f]$$

$$\textit{with} \quad A_n = \begin{bmatrix} 1 & 0 \\ 1 & 1 \end{bmatrix} \otimes A_{n-1}, \qquad A_0 = [1].$$

The transform is an involution: $A_n^2 = I_n$.

Definition 3 *The* **nonlinear order** *of a Boolean function (notation: $\mathrm{ord}(f)$) is defined as the degree of the highest order term in the algebraic normal form.*

The affine Boolean functions are thus the functions with nonlinear order < 2.

2.4 Definition of propagation properties

The definitions of strict avalanche criterion and of perfect non-linearity can be formulated with the help of the autocorrelation function.

A Boolean function $f(\underline{x})$ satisfies the *strict avalanche criterion (SAC)* if and only if $f(\underline{x})$ changes with a probability of one half whenever a single input bit of $\underline{x}$ is complemented [WT85], i.e.

$$\hat{r}(\underline{a}) = 0 \quad \text{for} \quad hwt(\underline{a}) = 1.$$

As a consequence, theorem 1 in [For88], stating that $\hat{f}(\underline{x})$ fulfills the SAC if and only if

$$\sum_{\underline{w}} \hat{F}^2(\underline{w}) \cdot (-1)^{w_i} = 0 \quad 1 \le i \le n$$

is a special case of our theorem 1.

A Boolean function $f(\underline{x})$ is *perfect non-linear (with respect to linear structures)* if $f(\underline{x})$ changes with a probability of one half whenever i ($1 \le i \le n$) bits of $\underline{x}$ are complemented [MS89]:

$$\hat{r}(\underline{a}) = 0 \text{ for } 1 \le hwt(\underline{a}) \le n.$$

The perfect non-linear functions or bent functions will be studied in section 3. Note that if functions from $\mathbb{Z}_q^n$ to $\mathbb{Z}_q$ are considered, the definition of perfect non-linear and bent can be extended. Only if q is prime both concepts coincide [Nyb90].

These definitions can be generalized in a natural way as follows. A Boolean function $f(\underline{x})$ satisfies the *propagation criterion of degree k (PC of degree k)* if $f(\underline{x})$ changes with a probability of one half whenever i ($1 \le i \le k$) bits of $\underline{x}$ are complemented:

$$\hat{r}(\underline{a}) = 0 \text{ for } 1 \le hwt(\underline{a}) \le k.$$

Note that SAC is equivalent to PC of degree 1 and perfect non-linear is PC of degree n.

When we are interested in propagation properties, it is important to be able to construct different functions that satisfy the same property starting from one function. A trivial example of such a construction is a complementation. The following theorem shows that other constructions are possible for functions satisfying PC of degree k by means of a dyadic shift in the Walsh-Hadamard domain.

Theorem 2 *Let $f(\underline{x})$ be a Boolean function. Then the function $g(\underline{x})$ defined as $\hat{G}(\underline{w}) = \hat{F}(\underline{w} \oplus \underline{s})$ is also Boolean for all $\underline{s}$. Moreover, the autocorrelation function of $\hat{g}(\underline{x})$ has the same absolute values (and thus the same zeroes) as the autocorrelation function of $\hat{f}(\underline{x})$ and for $\underline{s} \ne \underline{0}$ the distance $d(f,g)$ equals 2^{n-1}.*

Corollary 1 *A dyadic shift and a complementation in the Walsh-Hadamard domain generate all 2^{n+1} possible changes of terms in the algebraic normal form with degree smaller than 2.*

Note that a dyadic shift in the original domain generates a Boolean function with the same autocorrelation function, inducing only changes of terms of degree $< ord(f)$. However, the maximal distance condition is only fulfilled if the original function is bent.

3 Properties and constructions of bent functions

In [Rot76] bent functions are first defined and the importance for cryptographic applications was explained in [MS89]. In this section we give a brief overview of the known properties of bent functions. Some new properties will be stated without proof. A new construction method for bent functions is derived, and upper and lower bounds on the number of bent functions are computed accordingly. Finally, an extension of the definition for odd n is considered.

3.1 Properties of bent functions

In the following, unless mentioned explicitly, we will consider only Boolean bent functions. Bent functions only exist for even values of n. They have a flat energy spectrum with value 2^n. As a consequence bent functions are never balanced and the difference between the number of ones and the number of zeroes equals $\pm 2^{\frac{n}{2}}$. This also implies that their autocorrelation function is an impulse function. In [MS89] it is shown that bent functions have maximum distance 2^{n-2} to all linear structures and maximum distance $2^{n-1} - 2^{\frac{n}{2}-1}$ to all affine functions and thus they have minimum correlation $\pm 2^{-\frac{n}{2}}$ to all affine functions. The nonlinear order of bent functions is bounded from above by $\frac{n}{2}$ for $n > 2$.

A first important property is based on the similarity of the Walsh-Hadamard transform and its inverse. A bent function has a constant energy spectrum, but not all functions with a constant energy spectrum are Boolean (i.e. have an absolute value of 1). The conditions for a function to be Boolean or bent are dual, as shown by the following theorem.

Theorem 3 *Any real-valued function over $\mathbb{Z}_2^n$ is* **Boolean** *if and only if*

$$\mid \hat{f}(\underline{x}) \mid = 1 \quad \Longleftrightarrow \quad \sum_{\underline{w}} \hat{F}(\underline{w})\hat{F}(\underline{w} \oplus \underline{a}) = 2^{2n}\delta(\underline{a}).$$

Any real-valued function over $\mathbb{Z}_2^n$ is **bent** *if and only if*

$$\sum_{\underline{x}} \hat{f}(\underline{x})\hat{f}(\underline{x} \oplus \underline{a}) = 2^n \delta(\underline{a}) \quad \Longleftrightarrow \quad \mid \hat{F}(\underline{w}) \mid = 2^{\frac{n}{2}}.$$

It is easily seen that the number of all (non necessarily Boolean) bent functions over $\mathbb{Z}_2^n$ equals the number of Boolean functions over $\mathbb{Z}_2^n$, namely 2^{2^n}. On the other hand, the number of Boolean bent functions for n even remains an open problem. This duality leads to following theorem [Rot76] and definition:

Theorem 4 *Let $\hat{f}(\underline{x})$ be a Boolean bent function. Then $\hat{g}(\underline{x})$ and $\hat{h}(\underline{x})$, defined as $\hat{g}(\underline{x}) = 2^{-\frac{n}{2}}\hat{F}(\underline{x})$ and $\hat{h}(\underline{x}) = -2^{-\frac{n}{2}}\hat{F}(\underline{x})$ are also Boolean bent functions.*

Definition 4 *The* **dual** *and* **anti-dual** *bent functions are the Boolean eigenvectors of the Walsh-Hadamard matrix of order n with eigenvalues $2^{\frac{n}{2}}$ and $-2^{\frac{n}{2}}$ respectively.*

Hence these bent functions are the Boolean solutions of the equation $H_n[\hat{f}] = \pm 2^{n/2}[\hat{f}]$, that is studied in more detail in [YH82].

The alternative formula for the distance $d(f,g) = 2^{n-1} - 2^{-n-1} \sum_{\underline{w}} \hat{F}(\underline{w})\hat{G}(\underline{w})$ shows that the minimum distance to affine functions equals $2^{n-1} - 1/2 \max_{\underline{w}} \mid \hat{F}(\underline{w})\mid$. The maximal value of the spectrum is thus a measure for the minimum distance to affine functions. The degree of nonlinearity with respect to linear structures is defined to be 0 for affine functions and 1 for bent functions.

Definition 5 *The* **degree of nonlinearity with respect to linear structures DNL** *is defined as*

$$DNL = \frac{2^{n-1} - 1/2 \max_{\underline{w}} |\hat{F}(\underline{w})|}{2^{n-1} - 2^{\lfloor n/2 \rfloor - 1}}$$

For $n = 4$, this results in following distribution, where abstraction is made of constant and linear terms. Every function in the table corresponds to 32 different functions.

number of functions	1	16	120	560	875	448	28
DNL	0	1/6	2/6	3/6	4/6	5/6	1

The sum of all the values of the Walsh-Hadamard transform of a Boolean function is restricted to two distinct values. This observation can be extended in case l bits of $\underline{w}$ are fixed through considering the sum over the 2^{n-l} remaining possibilities (notation: $\sum_{\underline{w}}[l]$). Following theorem gives the values that these sums can take.

Theorem 5 *Let* $\hat{f}(\underline{x})$ *be any Boolean function. Then*

$$\sum_{\underline{w}}[l]\,\hat{F}(\underline{w}) = -2^n + k \cdot 2^{n+1-l} \quad with\ 0 \le k \le 2^l.$$

Let $\hat{f}(\underline{x})$ *be any bent function (non necessarily Boolean). Then*

$$\sum_{\underline{w}}[l]\,\hat{F}(\underline{w}) = -2^{3n/2-l} + k \cdot 2^{n/2+1} \quad with\ 0 \le k \le 2^{n-l}.$$

Let $\hat{f}(\underline{x})$ *be any Boolean bent function. Then* $\sum_{\underline{w}}[l]\,\hat{F}(\underline{w}) = 2^{n/2}\sum_{\underline{x}}[l]\hat{f}(\underline{x}) =$

$$\begin{cases} -2^n + k \cdot 2^{n+1-l} & with\ 0 \le k \le 2^l, & if\ 0 \le l \le \frac{n}{2} \\ -2^{3n/2-l} + k \cdot 2^{n/2+1} & with\ 0 \le k \le 2^{n-l}, & if\ \frac{n}{2} \le l \le n. \end{cases}$$

The previous theorem makes it theoretically possible to construct *all* bent functions of n variables without performing an exhaustive search. The Walsh spectrum of a (Boolean) bent function has $2^{n-1} \pm 2^{n/2-1}$ positive values. Starting from the previous theorem, this can be extended for the case that l bits are kept constant.

Lemma 1 *The number of positive entries of the Walsh-Hadamard transform of a Boolean bent function when l bits of $\underline{w}$ are fixed equals*

$$P_+(l) = 2^{n-l-1} + 2^{-n/2-1} \cdot \sum_{\underline{w}}[l]\,\hat{F}(\underline{w}).$$

For $n = 4$, this results in following values for $\{P_+(0)\} = \{6, 10\}$, $\{P_+(1)\} = \{2, 4, 6\}$. For $l \ge n/2$ no additional restrictions are imposed. With a recursive construction starting from $l = 0$ the number of positive entries at level l can be composed of the combination of two entries of level $l + 1$. Additional restrictions have to be imposed, and for the time being it is only feasible for $n = 4$ to construct all bent functions in this way.

3.2 Construction of bent functions

Most known constructions are enumerated in [YH89]: the two Rothaus constructions [Rot76], the eigenvectors of Walsh-Hadamard matrices [YH82], constructions based on Kronecker algebra, concatenation, dyadic shifts and linear transformations of variables. The generalization of the Rothaus construction by Maiorana and McFarland is discussed in [Nyb90]. In this section new constructions will be given based on Walsh-Hadamard matrices and concatenation. Conditions are stated for a function of nonlinear order 2 to be bent.

Theorem 6 *For $n = 2m$, consider the rows of the Walsh-Hadamard matrix H_m. The concatenation of the 2^m rows or their complement in arbitrary order results in $(2^m)! \, 2^{2^m}$ different Boolean bent functions of n variables.*

This construction results in the same lower bound as the Maiorana construction. An upper bound for $n > 2$ can be computed based on the restriction on the nonlinear order:

$$2^{2^{n-1}+d(n)} \quad \text{with} \quad d(n) = \frac{1}{2}\binom{n}{n/2}.$$

Following table gives the lower bound, the number of bent functions (if known), the upper bound and the number of Boolean functions. For $n = 6$, the number of bent functions equals $5,425,430,528$.

n	lower bound	# bent	upper bound	# Boolean
2	8	8	8	16
4	384	896	2048	65536
6	$2^{23.3}$	$2^{32.3}$	2^{42}	2^{64}
8	$2^{60.3}$	?	2^{163}	2^{256}

The construction of a Boolean function $\hat{f}(\underline{x})$ through concatenation implies that the vector $[\hat{f}]$ is obtained as a concatenation of different vectors $[\hat{g}_i]$.

Theorem 7 *The concatenation $\hat{f}$ of dimension $n + 2$ of 4 bent functions $\hat{g}_i$ of dimension n is bent if and only if*

$$\hat{G}_1(\underline{w}) \cdot \hat{G}_2(\underline{w}) \cdot \hat{G}_3(\underline{w}) \cdot \hat{G}_4(\underline{w}) = -2^{2n}, \quad \forall \underline{w} \in \mathbb{Z}_2^n.$$

Corollary 2 *If $\hat{f}$, $\hat{g}_1$, $\hat{g}_2$ and $\hat{g}_3$ are bent then $\hat{g}_4$ is also bent.*

- The order of the $\hat{g}_i$ has no importance.
- In case $\hat{g}_1 = \hat{g}_2$, the theorem reduces to $\hat{g}_4 = -\hat{g}_3$, and if $\hat{g}_1 = \hat{g}_2 = \hat{g}_3$, then $\hat{g}_4 = -\hat{g}_1$. These special cases are considered in [YH89].

Theorem 8 *If the concatenation of 4 arbitrary vectors of dimension n is bent, then the concatenation of all 4! permutations of these vectors is bent.*

Note that through dyadic shifting of f only 4 of the 24 permutations can be obtained. This theorem is best possible in the sense that if the function is split up in 8 vectors, all permutations will in general not result in a bent function.

When the nonlinear order is restricted to 2, counting and construction of all functions is possible. In that case the autocorrelation function can take only the values $\{0, \pm 2^n\}$. Following theorem was stated in chapter 15 of [MWS78].

Theorem 9 *Let $\hat{f}(\underline{x})$ be a Boolean function with $\text{ord}(f) = 2$ and second order coefficients of the algebraic normal form denoted by a_{ij}. Then $\hat{f}(\underline{x})$ is bent if and only if the matrix V, defined as $v_{ij} = a_{ij}$ for $i \neq j$ and $v_{ii} = 0$, has full rank. For n even, the number of second order bent functions equals $\prod_{i=0}^{(n/2)-1}(2^{2i+1} - 1)2^{2i}$. For odd n, no bent functions exist.*

3.3 Extension of bent functions for odd n

As there exist no bent functions for odd n, it is interesting to extend the concept. In [MS89], following proposal for extension was made:

$$f(\underline{x}) = \bar{x}_n f_0(x_1, \ldots, x_{n-1}) \oplus x_n f_1(x_1, \ldots, x_{n-1}), \quad \text{with} \quad f_0, f_1 \text{ bent}.$$

It is clear that $ord(f) \leq (n+1)/2$. It is also shown that for half of the values of $\underline{w}$, $\hat{F}(\underline{w}) = 0$ and for the other half $|\hat{F}(\underline{w})| = 2^{(n+1)/2}$. The maximum distance to affine functions equals $2^{n-1} - 2^{(n+1)/2-1}$. However, these functions do not necessarily satisfy PC of degree 1: e.g. if $f_0 = f_1$, then $\hat{r}(0\ldots01) \neq 0$, but for all other values of $\underline{a} \neq \underline{0}$, $\hat{r}(\underline{a}) = \underline{0}$. It is clear that $\hat{r}(\underline{a}) = 0$ if the most significant bit of $\underline{a}$ equals zero. If the right choice for f_0 and f_1 is made, $\hat{r}(\underline{a})$ will also be zero for $\underline{a} = 10\ldots0$, and only in these cases f will satisfy PC of degree 1. Other constructions that will be discussed in a forthcoming paper result in functions that satisfy PC of degree $n - 1$. A second observation is that f will be balanced if and only if f_0 and f_1 have a different number of zeroes and ones. An example of a function satisfying both properties will be given in section 5.

4 Higher order SAC

Higher order SAC is defined in [For88]. In [Llo89] the definition is simplified and it is shown that exactly 2^{n+1} functions satisfy the SAC of maximal order $n - 2$.

Definition 6 *A Boolean function $f(\underline{x})$ satisfies the **strict avalanche criterion of order m** (SAC of order m) if any function obtained from $f(\underline{x})$ by keeping m of its input bits constant satisfies the SAC.*

A first theorem gives an upper bound on the non-linear order of functions satisfying SAC of order m.

Theorem 10 *Let f be a Boolean function of n variables, $n > 2$.*

1. *if f satisfies SAC of order $n - 2$, then $\text{ord}(f) = 2$.*
2. *if f satisfies SAC of order m $(0 \leq m < n - 2)$, then $\text{ord}(f) \leq n - m - 1$.*

This results in following table:

m (order of SAC)	0	1	...	$n-4$	$n-3$	$n-2$
non-linear order $\leq$	$n-1$	$n-2$	...	3	2	2

The next theorem gives a characterization of all second order functions satisfying SAC.

Theorem 11 *Let f be a Boolean function of n variables, with $n > 2$ and $\mathrm{ord}(f) = 2$. f satisfies SAC of order m ($0 \leq m \leq n - 2$), if and only if every variable x_i occurs in at least $m + 1$ second order terms of the algebraic normal form.*

Definition 7 *The function $s_n(\underline{x})$ is defined as the function with the algebraic normal form containing all second order terms:*

$$s_n(\underline{x}) = \sum_{1 \leq i < j \leq n} x_i x_j.$$

It is clear from theorem 10 and 11 that if abstraction is made of affine terms, this is the only function satisfying SAC of order $n - 2$. Hence the sum modulo 2 of 2 functions satisfying SAC of order $n - 2$ is always affine. The observations in this paragraph were independently made by Bert den Boer [dBo90].

Theorem 10 and 11 characterize all functions satisfying SAC of order $n - 3$ and enable to count them. A different characterization and counting method is explained in [Llo90].

5 Propagation characteristics

It seems plausible to study what happens if m bits are kept constant in functions that satisfy PC of degree k, as was done for the strict avalanche criterion. This allows for a more general classification of propagation characteristics of Boolean functions. We impose the restriction $k + m \leq n$: if m bits are kept constant at most $n - m$ bits can be changed.

Definition 8 *A Boolean function $f(\underline{x})$ of n variables satisfies the **propagation criterion of degree k and order m** (PC of degree k and order m) if any function obtained from $f(\underline{x})$ by keeping m input bits constant satisfies PC of degree k.*

Definition 9 *The **propagation matrix** N_n for all Boolean functions of n variables is the $n \times n$ matrix: $N_n(k, m) = \# \{f \mid f \text{ satisfies PC of degree } k \text{ and order } m\} / 2^{n+1}$. with f a Boolean function and $k + m \leq n$.*

The division by 2^{n+1} implies that abstraction is made of linear and constant terms, that have no influence on propagation properties. It is clear that $N_n(k, m) \leq N_n(k, l)$ for $m \geq l$, $N_n(k, m) \leq N_n(l, m)$ for $k \geq l$, and $N_n(1, n - 1) = 0$. The entry $N_n(n, 0)$ contains the number of bent functions divided by 2^{n+1}. Note that theorem 10 also imposes restrictions on the nonlinear order of functions for values of $k > 1$.

The functions satisfying PC of degree 1 (= higher order SAC) were studied in section 4. Theorems 9 and 11 can be generalized resulting in a classification of the functions $s_n(\underline{x})$.

Theorem 12 *The functions $s_n(\underline{x})$ satisfy PC of degree k and order m if $k+m \leq n-1$ and also if $k + m = n$ and k is even. The functions $s_n(\underline{x})$ are the only functions satisfying PC of degree 2 and order $n - 2$.*
For n even the functions $s_n(\underline{x})$ are bent (theorem 9), while for n odd these functions are an example of balanced functions that satisfy PC of degree $n - 1$.

The matrices $N_3(k,m)$, $N_4(k,m)$ and $N_5(k,m)$ are given in Table 1. The entries $N_4(2\text{-}4,0)$ contain only the bent functions. The entry $N_4(1,0)$ contains 88 third order functions. The entries $N_4(1,2)$, $N_4(2,1)$ and $N_4(2,2)$ consist of the function s_4.

$k \backslash m$	0	1	2
1	4	1	0
2	1	1	–
3	0	–	–

$k \backslash m$	0	1	2	3
1	129	10	1	0
2	28	1	1	–
3	28	0	–	–
4	28	–	–	–

$k \backslash m$	0	1	2	3	4
1	430040	813	26	1	0
2	3568	28	1	1	–
3	168	28	0	–	–
4	28	28	–	–	–
5	0	–	–	–	–

Table 1: The matrices $N_3(k,m)$, $N_4(k,m)$ and $N_5(k,m)$

The concept of PC can be extended if the restriction $k + m \leq n$ is removed. This means that a certain value is given to m bits and subsequently k bits are changed. However, the set of bits that are given a certain value and the set of those that are changed can have common elements. This leads to a definition based on information theoretic grounds.

Definition 10 *A Boolean function $f(\underline{x})$ of n variables satisfies the **extended propagation criterion of degree k and order m (EPC of degree k and order m)** if knowledge of m bits of $\underline{x}$ gives no information on $f(\underline{x}) \oplus f(\underline{x} \oplus \underline{a})$, $\forall$ $\underline{a}$ with $1 \leq \text{hwt}(\underline{a}) \leq k$.*

Definition 11 *The **extended propagation matrix** N_n^* for all Boolean functions of n variables is the $n \times n$ matrix:*
$$N_n^*(k,m) = \# \{f \mid f \text{ satisfies EPC of degree } k \text{ and order } m\} / 2^{n+1}.$$

The definition can be restated in terms of balancedness and correlation immunity [Sie84] of the directional derivative of a Boolean function.

Definition 12 *The **directional derivative** of a Boolean function f in direction $\underline{a}$ is defined as:*
$$d_{f,\underline{a}}(\underline{x}) = f(\underline{x}) \oplus f(\underline{x} \oplus \underline{a}).$$

The relation between the directional derivative and the autocorrelation function is given by:
$$\hat{r}(\underline{a}) = \sum_{\underline{x}} \hat{d}_{f,\underline{a}}(\underline{x}).$$

This results in an equivalent formulation of the EPC.

Theorem 13 *Let f be a Boolean function of n variables.*

1. *f satisfies EPC of degree k and order 0 if and only if the directional derivative $\hat{d}_{f,\underline{a}}(\underline{x})$ of f is balanced $\forall \, \underline{a} : 1 \leq \mathrm{hwt}(\underline{a}) \leq k$.*

2. *f satisfies EPC of degree k and order $m > 0$ if and only if the directional derivative $\hat{d}_{f,\underline{a}}(\underline{x})$ of f is balanced and mth order correlation immune $\forall \, \underline{a} : 1 \leq \mathrm{hwt}(\underline{a}) \leq k$.*

The relation between PC and EPC is given in following theorem:

Theorem 14 *Let f be a Boolean function of n variables.*

1. *If f satisfies EPC of degree k and order m then f satisfies PC of degree k and order m.*

2. *If f satisfies PC of degree k and order 0 or 1 then f satisfies EPC of degree k and order 0 or 1.*

3. *If f satisfies PC of degree 1 and order m then f satisfies EPC of degree 1 and order m.*

Following example shows that this theorem is tight. Consider the function $s_n(\underline{x})$. The directional derivative of this function for a vector $\underline{a}$ with a 1 in position i and j is equal to $x_i \oplus x_j \oplus 1$. This function is balanced and correlation immune of order 1, but not of order 2. Hence $s_n(\underline{x})$ does not satisfy EPC of degree 2 and order 2. In this particular example, it is clear that if $x_i = x_j$, then $f(\underline{x} \oplus \underline{a}) \neq f(\underline{x}), \forall \underline{x}$ and if $x_i \neq x_j$, then $f(\underline{x} \oplus \underline{a}) = f(\underline{x}), \forall \underline{x}$. For the ordinary PC the average of both cases is considered: $f(\underline{x})$ changes on average with a probability of 0.5. For $n \leq 5$, the functions $s_n(\underline{x})$ are the only functions for which PC and EPC are not equivalent.

6 Summary

The importance of the relation between the autocorrelation function and the Walsh-Hadamard energy spectrum has been shown. This concept can be extended in the case of S-boxes to crosscorrelation properties. Existing propagation criteria were generalized to the propagation criterion PC of degree k and order m and an extended propagation criterion EPC. Functions satisfying these criteria were studied, especially the bent functions, the functions satisfying higher order SAC and the functions s_n consisting of all second order terms. The results show that functions satisfying PC of highest order and degree are only second order functions that clearly have some other weaknesses. Further research is necessary to characterize and count functions that satisfy the propagation criteria, to construct functions that are a compromise between different criteria and to extend the concepts to the design of S-boxes.

Acknowledgement

We would like to thank the people that showed interest in our work and especially Antoon Bosselaers, Kaisa Nyberg and Herman J. Tiersma for their helpful comments.

References

[BMP86] E.F. Brickell, J.H. Moore and M.R. Purtill, "Structures in the S-boxes of the DES", *Advances in Cryptology, Proc. Crypto 86*, Springer Verlag, 1987, p. 3–8.

[dBo90] B. den Boer, personal communication.

[DQD84] Y. Desmedt, J.-J. Quisquater and M. Davio, "Dependence of output on input in DES: small avalanche characteristics", *Advances in Cryptology, Proc. Crypto 84*, Springer Verlag, 1985, p. 359–376.

[Eve87] J.-H. Evertse, "Linear Structures in block ciphers", *Advances in Cryptology, Proc. Eurocrypt 87*, Springer Verlag, 1988, p. 249–266.

[For88] R. Forré, "The strict avalanche criterion: spectral properties of Boolean functions and an extended definition", *Advances in Cryptology, Proc. Crypto 88*, Springer Verlag, 1990, p. 450–468.

[Jan89] C.J.A. Jansen, *"Investigations on nonlinear streamcipher systems: construction and evaluation methods"*, PhD. Thesis, Technical University Delft, 1989.

[Llo89] S. Lloyd, "Counting functions satisfying a higher order strict avalanche criterion", *Advances in Cryptology, Proc. Eurocrypt 89*, Springer Verlag, to appear.

[Llo90] S. Lloyd, "Characterising and counting functions satisfying the strict avalanche criterion of order $(n-3)$".

[MWS78] F.J. MacWilliams and N.J.A. Sloane, *"The theory of error-correcting codes"*, North-Holland Publishing Company, Amsterdam, 1978.

[MS89] W. Meier and O. Staffelbach, "Nonlinearity criteria for cryptographic functions", *Advances in Cryptology, Proc. Eurocrypt 89*, Springer Verlag, to appear.

[Nyb90] K. Nyberg, "Constructions of bent functions and difference sets", *These Proceedings*.

[Rot76] O.S. Rothaus, "On bent functions", *Journal of Combinatorial Theory (A)*, Vol. 20, p. 300–305, 1976.

[Rue86] R.A. Rueppel, *"Analysis and design of stream ciphers"*, Springer Verlag, 1986.

[Sie84] T. Siegenthaler, "Correlation immunity of non-linear combining functions for cryptographic applications", *IEEE Trans. Inform. Theory*, Vol. IT-30, p. 776–780, Oct. 1984.

[WT85] A.F. Webster and S.E. Tavares, "On the design of S-boxes", *Advances in Cryptology, Proc. Crypto 85*, Springer Verlag, 1986, p. 523–534.

[GM88] X. Guo-Zhen and J.L. Massey, "A spectral characterization of correlation-immune combining functions", *IEEE Trans. Inform. Theory*, Vol. IT-34, p. 569–571, May 1988.

[YH82] R. Yarlagadda and J.E. Hershey, "A note on the eigenvectors of Hadamard matrices of order 2^n", *Linear Algebra & Appl.*, Vol. 45, p. 43–53, 1982.

[YH89] R. Yarlagadda and J.E. Hershey, "Analysis and synthesis of bent sequences", *Proc. IEE*, Vol. 136, Pt. E, p. 112–123, March 1989.

The Linear Complexity Profile and the Jump Complexity of Keystream Sequences

HARALD NIEDERREITER

Institute for Information Processing

Austrian Academy of Sciences

Sonnenfelsgasse 19, A–1010 Vienna, Austria

Abstract

We study the linear complexity profile and the jump complexity of keystream sequences in arbitrary finite fields. We solve counting problems connected with the jump complexity, establish formulas for the expected value and the variance of the jump complexity, and prove probabilistic theorems on the jump complexity profile of random sequences. We also extend earlier work on frequency distributions in the linear complexity profile to joint frequency distributions.

1. Introduction

An important tool for the assessment of keystream sequences in the context of stream ciphers is the linear complexity profile introduced by Rueppel [9], [10, Ch. 4]. Let F_q be the finite field with q elements, where q is an arbitrary prime power, let n be a positive integer, and let S be a finite or infinite sequence $s_1, s_2, \ldots$ of elements of F_q which contains at least n terms. Then the nth *linear complexity* $L_n(S)$ is defined to be the least integer k such that $s_1, s_2, \ldots, s_n$ form the first n terms of a kth-order linear feedback shift register (LFSR) sequence, where we observe the convention that the zero sequence is viewed as an LFSR sequence of order 0. The sequence $L_1(S), L_2(S), \ldots,$ extended as long as $L_n(S)$ is defined, is called the *linear complexity profile* of S. We note that we have $0 \leq L_n(S) \leq n$ and $L_n(S) \leq L_{n+1}(S)$ as long as $L_n(S)$ and $L_{n+1}(S)$ are defined. Detailed studies of the linear complexity profile were carried out in [1], [5], [6], [7], [8], [9], [10, Ch. 4], [11], [12].

A new way of looking at the linear complexity profile was recently introduced by Wang [11], [12]. If S and n are as above, then the nth *jump complexity* $P_n(S)$ is defined as the number of positive integers among $L_1(S), L_2(S) - L_1(S), \ldots, L_n(S) - L_{n-1}(S)$. Thus $P_n(S)$ is the number of "jumps" in the first n terms of the linear complexity profile of S. Wang [11], [12] proved some combinatorial and elementary statistical results on the jump complexity in the special case $q = 2$, and related results were also shown by Carter [1]. In a sense, the quantity $P_n(S)$ already appeared in Niederreiter [6], since it is clear in view of [6, Lemma 1] that the quantity $j(n, S)$ defined in the proof of [6, Theorem 11] is identical with $P_n(S)$. In analogy with the terminology introduced in the previous paragraph, we call the sequence $P_1(S), P_2(S), \ldots$, extended as long as $P_n(S)$ is defined, the *jump complexity profile* of S.

The main aims of the present paper are to study the jump complexity in greater detail, to solve counting problems connected with the jump complexity, to prove probabilistic theorems on the jump complexity profile of random sequences, and to derive all these results for arbitrary q. We also extend the work in [6, Sect. 6] on frequency distributions in the linear complexity profile to joint frequency distributions.

2. Enumeration Formulas for the Jump Complexity

It is clear that we have $0 \le P_n(S) \le L_n(S) \le n$ for any S containing at least n terms. We obtain a stronger upper bound from the following well-known lemma (see [4], [10, p. 34]) which holds if S contains at least $n + 1$ terms.

Lemma 1. If $L_n(S) > n/2$, then $L_{n+1}(S) = L_n(S)$. If $L_n(S) \le n/2$, then $L_{n+1}(S) = L_n(S)$ for exactly one choice of $s_{n+1} \in F_q$ and $L_{n+1}(S) = n + 1 - L_n(S)$ for exactly $q - 1$ choices of $s_{n+1} \in F_q$.

Lemma 2. $P_n(S) \le \min(L_n(S), n - L_n(S) + 1)$.

Proof. We already noted that $P_n(S) \le L_n(S)$, so it suffices to show $P_n(S) \le n - L_n(S) + 1$. Consider the last jump in the first n terms of the linear complexity profile of S, say $L_m(S) < L_{m+1}(S) = \ldots = L_n(S)$. By Lemma 1 we must have $L_{m+1}(S) = m + 1 - L_m(S)$, therefore

$$P_n(S) - 1 = P_m(S) \le L_m(S) = m + 1 - L_n(S) \le n - L_n(S). \qquad \square$$

We now establish an explicit formula for the number $N_n(L, r)$ of sequences S of elements of F_q with fixed length n and with prescribed values $L_n(S) = L$ and $P_n(S) = r$ for the nth linear complexity and the nth jump complexity, respectively. Such a formula was earlier shown by Carter [1] for $q = 2$ and Wang [11], [12] for $q = 2$ and $L = \lceil n/2 \rceil$.

Theorem 1. Let n be a positive integer and let L and r be integers with $0 \le L, r \le n$. Then:

(i) $N_n(L, r) = 1$ if $L = r = 0$;

(ii) $N_n(L, r) = 0$ if $L \ge 1$ and $r = 0$;

(iii) $N_n(L, r) = 0$ if $r > \min(L, n - L + 1)$;

(iv) $N_n(L, r) = \binom{\min(L-1, n-L)}{r-1} (q - 1)^r q^{\min(L, n-L)}$ if $1 \le r \le \min(L, n - L + 1)$.

Proof. (i) is valid since $L_n(S) = P_n(S) = 0$ holds exactly for the sequence S of n zeros. (ii) holds since $L_n(S) \geq 1$ implies $P_n(S) \geq 1$. (iii) follows from Lemma 2. To prove (iv), we proceed by induction on n. The case $n = 1$ is checked immediately. Suppose the formula is shown for length n, and now consider length $n + 1$. We take $1 \leq r \leq \min(L, n-L+2)$ with $1 \leq L \leq n+1$, where L and r are the prescribed values of $L_{n+1}(S)$ and $P_{n+1}(S)$, respectively. If $L \leq n/2$, then Lemma 1 implies $N_{n+1}(L, r) = N_n(L, r)$, and the induction hypothesis yields the desired formula. If $L = (n + 1)/2$, so that n is odd, then we must have $L_n(S) = (n+1)/2$ by Lemma 1, and so by induction hypothesis

$$N_{n+1}(L, r) = qN_n\left(\frac{n+1}{2}, r\right) = \binom{(n-1)/2}{r-1}(q-1)^r q^{1+(n-1)/2}$$

$$= \binom{\min(L-1, n+1-L)}{r-1}(q-1)^r q^{\min(L, n+1-L)}.$$

If $L \geq (n+2)/2$, then by Lemma 1 we either have $L_n(S) = L, P_n(S) = r$, or $L_n(S) = n+1-L, P_n(S) = r-1$. Together with the induction hypothesis we get

$$N_{n+1}(L, r) = qN_n(L, r) + (q-1)N_n(n+1-L, r-1)$$

$$= \binom{n-L}{r-1}(q-1)^r q^{n-L+1} + \binom{n-L}{r-2}(q-1)^r q^{n+1-L} = \binom{n+1-L}{r-1}(q-1)^r q^{n+1-L}$$

$$= \binom{\min(L-1, n+1-L)}{r-1}(q-1)^r q^{\min(L, n+1-L)}. \quad \Box$$

From Theorem 1 we get the following alternative proof of the formula of Gustavson [2] for the number $M_n(L)$ of sequences S of elements of F_q with fixed length n and $L_n(S) = L$.

Corollary 1. $M_n(0) = 1$ and $M_n(L) = (q-1)q^{\min(2L-1, 2n-2L)}$ for $1 \leq L \leq n$.

Proof. We have $M_n(0) = N_n(0, 0) = 1$. For $1 \leq L \leq n$ we get by Theorem 1,

$$M_n(L) = \sum_{r=1}^{\min(L, n-L+1)} \binom{\min(L-1, n-L)}{r-1}(q-1)^r q^{\min(L, n-L)}$$

$$= (q-1)q^{\min(L, n-L)} \sum_{r=0}^{\min(L-1, n-L)} \binom{\min(L-1, n-L)}{r}(q-1)^r$$

$$= (q-1)q^{\min(L, n-L)}(1 + (q-1))^{\min(L-1, n-L)}$$

$$= (q-1)q^{\min(L, n-L)+\min(L-1, n-L)} = (q-1)q^{\min(2L-1, 2n-2L)}. \quad \Box$$

Theorem 1 also yields a formula for the number $N_n(r)$ of sequences S of elements of F_q with fixed length n and $P_n(S) = r$. The special case $q = 2$ was treated by Carter [1].

Theorem 2. We have $N_n(0) = 1$ and $N_n(r) = 0$ for $r > \lceil n/2 \rceil$. For $1 \leq r \leq \lceil n/2 \rceil$ we have

$$N_n(r) = (q+1)(q-1)^r \sum_{L=r-1}^{(n-2)/2} \binom{L}{r-1}q^L \quad \text{for even} \quad n,$$

$$N_n(r) = (q+1)(q-1)^r \sum_{L=r-1}^{(n-1)/2} \binom{L}{r-1} q^L - \binom{(n-1)/2}{r-1}(q-1)^r q^{(n+1)/2} \quad \text{for odd} \quad n.$$

Proof. $N_n(0) = 1$ follows from Theorem 1 (i), (ii). If $r > \lceil n/2 \rceil$, then $r > \min(L, n - L + 1)$ for $0 \leq L \leq n$, hence $N_n(r) = 0$ for $r > \lceil n/2 \rceil$ by Theorem 1 (iii). If $1 \leq r \leq \lceil n/2 \rceil$, then by Theorem 1 we have $N_n(L, r) > 0$ if and only if $r \leq \min(L, n - L + 1)$, i.e. if and only if $r \leq L \leq n + 1 - r$. Therefore

$$N_n(r) = \sum_{L=r}^{n+1-r} N_n(L, r).$$

Now let n be even. Then by Theorem 1 (iv),

$$N_n(r) = \sum_{L=r}^{n/2} \binom{L-1}{r-1}(q-1)^r q^L + \sum_{L=(n+2)/2}^{n+1-r} \binom{n-L}{r-1}(q-1)^r q^{n-L}$$

$$= q(q-1)^r \sum_{L=r-1}^{(n-2)/2} \binom{L}{r-1} q^L + (q-1)^r \sum_{L=r-1}^{(n-2)/2} \binom{L}{r-1} q^L$$

$$= (q+1)(q-1)^r \sum_{L=r-1}^{(n-2)/2} \binom{L}{r-1} q^L.$$

For odd n we obtain by Theorem 1 (iv),

$$N_n(r) = \sum_{L=r}^{(n-1)/2} \binom{L-1}{r-1}(q-1)^r q^L + \binom{(n-1)/2}{r-1}(q-1)^r q^{(n-1)/2} +$$

$$+ \sum_{L=(n+3)/2}^{n+1-r} \binom{n-L}{r-1}(q-1)^r q^{n-L}$$

$$= q(q-1)^r \sum_{L=r-1}^{(n-3)/2} \binom{L}{r-1} q^L + \binom{(n-1)/2}{r-1}(q-1)^r q^{(n-1)/2} + (q-1)^r \sum_{L=r-1}^{(n-3)/2} \binom{L}{r-1} q^L$$

$$= (q+1)(q-1)^r \sum_{L=r-1}^{(n-1)/2} \binom{L}{r-1} q^L - \binom{(n-1)/2}{r-1}(q-1)^r q^{(n+1)/2}. \quad \square$$

3. Expected Value and Variance of the Jump Complexity

We can view $P_n = P_n(S)$ as a random variable on the space of all sequences of elements of F_q with fixed length n, where each such sequence is equiprobable (i.e., has probability q^{-n}). In the following two theorems we extend the formulas for the expected value and the variance of P_n given by Carter [1] and Wang [11], [12] for $q = 2$ to the general case of arbitrary q.

Theorem 3. The expected value $E(P_n)$ of P_n is given by

$$E(P_n) = \frac{(q-1)n}{2q} + \frac{(q+1)^2 - (-1)^n(q-1)^2}{4(q^2+q)} - \frac{1}{(q+1)q^n}.$$

Proof. With the notation in Theorem 2 we have

$$E(P_n) = q^{-n} \sum_{r=0}^{n} r N_n(r) = q^{-n} \sum_{r=1}^{\lceil n/2 \rceil} r N_n(r).$$

If n is even, then by Theorem 2 we get

$$E(P_n) = (q+1)q^{-n} \sum_{r=1}^{n/2} r(q-1)^r \sum_{L=r-1}^{(n-2)/2} \binom{L}{r-1} q^L$$

$$= (q+1)q^{-n} \sum_{L=0}^{(n-2)/2} q^L \sum_{r=1}^{L+1} \binom{L}{r-1} r(q-1)^r$$

$$= (q^2-1)q^{-n} \sum_{L=0}^{(n-2)/2} q^L \sum_{r=0}^{L} \binom{L}{r} (r+1)(q-1)^r.$$

To treat the inner sum, we differentiate the identity $\sum_{r=0}^{L} \binom{L}{r} z^{r+1} = z(z+1)^L$ with respect to z and then put $z = q-1$ to obtain

$$\sum_{r=0}^{L} \binom{L}{r} (r+1)(q-1)^r = q^L + (q-1)Lq^{L-1}. \tag{1}$$

This yields

$$E(P_n) = (q^2-1)q^{-n} \sum_{L=0}^{(n-2)/2} \left(q^{2L} + \frac{q-1}{q} L q^{2L} \right)$$

$$= q^{-n}(q^n - 1) + (q^2-1)(q-1)q^{-n-1} \sum_{L=0}^{(n-2)/2} L q^{2L}.$$

For any integer $k \geq 1$, differentiation of $\sum_{L=0}^{k-1} z^L = (z^k - 1)/(z-1)$ with respect to z and then multiplication by z yields

$$\sum_{L=0}^{k-1} L z^L = \frac{(k-1)z^{k+1} - kz^k + z}{(z-1)^2} \quad \text{for} \quad z \neq 1. \tag{2}$$

Putting $k = n/2, z = q^2$ in (2) we get

$$E(P_n) = 1 - q^{-n} + \frac{q}{q+1} q^{-n} \left(\frac{n-2}{2} q^n - \frac{n}{2} q^{n-2} + 1 \right)$$

$$= \frac{(q-1)n}{2q} + \frac{1}{q+1} \left(1 - q^{-n} \right)$$

by simple algebraic manipulations. For odd n we use Theorem 2 to obtain

$$E(P_n) = (q+1)q^{-n} \sum_{L=0}^{(n-1)/2} q^L \sum_{r=1}^{L+1} \binom{L}{r-1} r(q-1)^r - q^{(1-n)/2} \sum_{r=1}^{(n+1)/2} \binom{(n-1)/2}{r-1} r(q-1)^r$$

$$= (q^2 - 1)q^{-n} \sum_{L=0}^{(n-1)/2} q^L \sum_{r=0}^{L} \binom{L}{r} (r+1)(q-1)^r$$

$$- (q-1)q^{(1-n)/2} \sum_{r=0}^{(n-1)/2} \binom{(n-1)/2}{r} (r+1)(q-1)^r.$$

We apply (1) and get

$$E(P_n) = (q^2 - 1)q^{-n} \sum_{L=0}^{(n-1)/2} \left(q^{2L} + \frac{q-1}{q} L q^{2L} \right)$$

$$- (q-1)q^{(1-n)/2} \left(q^{(n-1)/2} + (q-1) \frac{n-1}{2} q^{(n-3)/2} \right)$$

$$= q^{-n} \left(q^{n+1} - 1 \right) + (q^2 - 1)(q-1)q^{-n-1} \sum_{L=0}^{(n-1)/2} L q^{2L} - q + 1 - \frac{(q-1)^2(n-1)}{2q}.$$

Putting $k = (n+1)/2, z = q^2$ in (2) we obtain

$$E(P_n) = 1 - q^{-n} + \frac{q}{q+1} q^{-n} \left(\frac{n-1}{2} q^{n+1} - \frac{n+1}{2} q^{n-1} + 1 \right) - \frac{(q-1)^2(n-1)}{2q}$$

$$= \frac{(q-1)n}{2q} + \frac{q^2+1}{2(q^2+q)} - \frac{1}{(q+1)q^n}$$

by simple algebraic manipulations. $\square$

Theorem 4. The variance $\mathrm{Var}(P_n)$ of P_n is given by

$$\mathrm{Var}(P_n) = \frac{(q-1)n}{2q^2} - \frac{q}{(q+1)^2} + \frac{(q-1)n}{(q+1)q^{n+1}} + \frac{1}{(q+1)q^n} - \frac{1}{(q+1)^2 q^{2n}} \quad \text{for even} \quad n,$$

$$\mathrm{Var}(P_n) = \frac{(q-1)n}{2q^2} + \frac{q^3 - 5q^2 + q + 1}{2q^2(q+1)^2} + \frac{(q-1)n}{(q+1)q^{n+1}} + \frac{2q^2 - q + 1}{(q+1)^2 q^{n+1}} - \frac{1}{(q+1)^2 q^{2n}} \quad \text{for}$$

odd n.

Proof. With the notation in Theorem 2 we have

$$E(P_n^2) = q^{-n} \sum_{r=0}^{n} r^2 N_n(r) = q^{-n} \sum_{r=1}^{\lceil n/2 \rceil} r^2 N_n(r).$$

Using Theorem 2 and the same techniques as in the proof of Theorem 3 we get

$$E(P_n^2) = \frac{(q-1)^2 n^2}{4q^2} + \frac{(3q^2 - 2q - 1)n}{2q^2(q+1)} - \frac{q-1}{(q+1)^2} + \frac{q-1}{(q+1)^2 q^n} \quad \text{for even} \quad n,$$

$$E(P_n^2) = \frac{(q-1)^2 n^2}{4q^2} + \frac{(q^3 + q - 2)n}{2q^2(q+1)} + \frac{q^4 + 2q^3 - 8q^2 + 2q + 3}{4q^2(q+1)^2} + \frac{q-1}{(q+1)^2 q^n} \quad \text{for odd} \quad n.$$

Now $\mathrm{Var}(P_n) = E(P_n^2) - E(P_n)^2$, and so an application of Theorem 3 yields the desired formulas by straightforward algebraic manipulations. $\square$

4. The Jump Complexity Profile of Random Sequences

In this section we study the behavior of the jump complexity profile for random infinite sequences. Note that by Theorem 3 we expect $P_n(S)$ to be close to $(q-1)n/(2q)$ and that by Theorem 4 there should be deviations that are at least of the order of magnitude $n^{1/2}$. We set up a suitable probabilistic model as in earlier probabilistic studies (see [6], [7], [8]). Let F_q^∞ be the set of all infinite sequences of elements of F_q. A probability measure h on F_q^∞ is defined by first considering the uniform probability measure μ on F_q which assigns the measure $1/q$ to each element of F_q and then letting h be the complete product measure on F_q^∞ induced by μ. We say that a property π of sequences $S \in F_q^\infty$ holds *with probability 1* if the set of all $S \in F_q^\infty$ which have the property π has h-measure 1.

Theorem 5. If f is a nonnegative function on the positive integers such that $\sum_{n=1}^{\infty} n e^{-qf(n)^2/n} < \infty$, then with probability 1 we have

$$|P_n(S) - \frac{(q-1)n}{2q}| \le f(n) \quad \text{for all sufficiently large} \quad n.$$

Proof. Choose $c \in (0, \frac{1}{2})$ such that

$$\frac{2c}{3(1-2c)^3} - c + \frac{c}{(q-1)^2} = \frac{1}{2(q-1)}. \tag{3}$$

This is possible since the left-hand side of (3) tends to 0 as $c \to 0$ and it tends to ∞ as $c \to \frac{1}{2}$ from the left. Put

$$F(n) = \min\left(f(n), \frac{cn}{q} - 1\right),$$
(4)

where we assume that n is so large that $F(n) \geq \frac{1}{2}$ (note that the condition on f implies $\lim_{n \to \infty} f(n) = \infty$). For fixed n consider

$$D_n = \left\{ S \in F_q^\infty : P_n(S) > \frac{(q-1)n}{2q} + F(n) \right\}.$$

Let $k(n)$ be the least integer satisfying

$$k(n) > \frac{(q-1)n}{2q} + F(n).$$
(5)

For even n we get by Theorem 2,

$$h(D_n) = q^{-n} \sum_{r=k(n)}^{n/2} N_n(r) = (q+1)q^{-n} \sum_{r=k(n)}^{n/2} (q-1)^r \sum_{L=r-1}^{(n-2)/2} \binom{L}{r-1} q^L.$$

For the inner sum we have

$$\sum_{L=r-1}^{(n-2)/2} \binom{L}{r-1} q^L \leq q^{(n-2)/2} \sum_{L=r-1}^{(n-2)/2} \binom{L}{r-1} = q^{(n-2)/2} \binom{n/2}{r},$$

thus

$$h(D_n) < 2q^{-n/2} \sum_{r=k(n)}^{n/2} \binom{n/2}{r} (q-1)^r.$$

Since $r \geq k(n) > (q-1)n/(2q)$, the terms of the last sum form a decreasing function of r, hence

$$h(D_n) < 2q^{-n/2} \left(\frac{n}{2} - k(n) + 1\right) \binom{n/2}{k(n)} (q-1)^{k(n)} \leq nw$$
(6)

with

$$w = \binom{n/2}{k(n)} (q-1)^{k(n)} q^{-n/2}.$$

We use Stirling's formula in the form

$$\log(n!) = \left(n + \frac{1}{2}\right) \log n - n + O(1).$$

Then

$$\log w = \frac{n+1}{2}\log\frac{n}{2} - \left(k(n)+\frac{1}{2}\right)\log k(n) - \left(\frac{n+1}{2}-k(n)\right)\log\left(\frac{n}{2}-k(n)\right)$$

$$+ k(n)\log(q-1) - \frac{n}{2}\log q + O(1)$$

$$\leq \frac{n}{2}\log\frac{n}{2} - k(n)\log k(n) - \left(\frac{n}{2}-k(n)\right)\log\left(\frac{n}{2}-k(n)\right) + k(n)\log(q-1) - \frac{n}{2}\log q + O(1)$$

$$=: u + O(1).$$

Put

$$d(n) = k(n) - \frac{(q-1)n}{2q}.$$

Then by straightforward manipulations

$$u = -\left(\frac{n}{2q} - d(n)\right)\log\left(1 - \frac{2qd(n)}{n}\right) - \left(\frac{(q-1)n}{2q} + d(n)\right)\log\left(1 + \frac{2qd(n)}{(q-1)n}\right).$$

Note that by the definitions of $F(n)$ and $k(n)$ in (4) and (5) we have $0 \leq d(n) \leq cn/q$, hence $0 \leq 2qd(n)/n \leq 2c$. For $0 \leq z \leq 2c$ we have

$$\log(1-z) \geq -z - \frac{z^2}{2} - \frac{z^3}{3(1-2c)^3}$$

by Taylor's theorem. Using also $\log(1+z) \geq z - \frac{1}{2}z^2$ for $z \geq 0$, we get

$$u \leq \left(\frac{n}{2q} - d(n)\right)\left(\frac{2qd(n)}{n} + \frac{2q^2d(n)^2}{n^2} + \frac{8q^3d(n)^3}{3(1-2c)^3n^3}\right)$$

$$- \left(\frac{(q-1)n}{2q} + d(n)\right)\left(\frac{2qd(n)}{(q-1)n} - \frac{2q^2d(n)^2}{(q-1)^2n^2}\right)$$

$$< - \frac{q^2d(n)^2}{(q-1)n} + \frac{2q^2d(n)^3}{n^2}\left(\frac{2}{3(1-2c)^3} - 1 + \frac{1}{(q-1)^2}\right)$$

$$= - \frac{q^2d(n)^2}{(q-1)n} + \frac{q^2d(n)^3}{c(q-1)n^2} = \frac{q^2d(n)^2}{(q-1)n}\left(-1 + \frac{d(n)}{cn}\right) \leq -\frac{qd(n)^2}{n}$$

by (3) and $d(n) \leq cn/q$. It follows that

$$\log w < -\frac{qd(n)^2}{n} + O(1),$$

and so (6) and $d(n) > F(n)$ yield

$$h(D_n) = O\left(ne^{-qF(n)^2/n}\right). \tag{7}$$

For odd n we get by Theorem 2,

$$h(D_n) = q^{-n}\sum_{r=k(n)}^{(n+1)/2} N_n(r) \leq (q+1)q^{-n}\sum_{r=k(n)}^{(n+1)/2}(q-1)^r\sum_{L=r-1}^{(n-1)/2}\binom{L}{r-1}q^L.$$

Proceeding as in the case of even n, we obtain

$$h(D_n) < \left(q^{1/2} + 1\right) q^{-n/2} \sum_{r=k(n)}^{(n+1)/2} \binom{(n+1)/2}{r}(q-1)^r.$$

Since $r \geq k(n) > (q-1)n/(2q) + 1/2$, the terms of the last sum form a decreasing function of r, hence

$$h(D_n) < \left(q^{1/2} + 1\right) q^{-n/2} \left(\frac{n+1}{2} - k(n) + 1\right) \binom{(n+1)/2}{k(n)}(q-1)^{k(n)} < C(q)nw_1 \quad (8)$$

with $C(q)$ depending only on q and

$$w_1 = \binom{(n+1)/2}{k(n)}(q-1)^{k(n)}q^{-n/2}.$$

By Stirling's formula,

$$\log w_1 = \frac{n+2}{2}\log\frac{n+1}{2} - \left(k(n) + \frac{1}{2}\right)\log k(n) - \left(\frac{n+2}{2} - k(n)\right)\log\left(\frac{n+1}{2} - k(n)\right)$$

$$+ k(n)\log(q-1) - \frac{n}{2}\log q + O(1)$$

$$\leq u + O(1),$$

where u is as before. Using (8) and the bound $u < -qd(n)^2/n$ shown above, we see that (7) also holds for odd n. From (4) we deduce

$$n\, e^{-qF(n)^2/n} < n\, e^{-qf(n)^2/n} + n\, e^{1-(c^2n/q)},$$

and so the hypothesis on f implies $\sum_{n=1}^{\infty} n\, e^{-qF(n)^2/n} < \infty$. In view of (7) this shows that $\sum_{n=1}^{\infty} h(D_n) < \infty$. An application of the Borel–Cantelli lemma [3, p. 228] yields that the set of all S for which $S \in D_n$ for infinitely many n has h-measure 0. In other words, with probability 1 we have $S \in D_n$ for at most finitely many n. From the definition of D_n it follows then that with probability 1 we have

$$P_n(S) \leq \frac{(q-1)n}{2q} + F(n) \leq \frac{(q-1)n}{2q} + f(n) \quad \text{for all sufficiently large } n, \quad (9)$$

where we applied (4) in the second inequality.

To obtain a corresponding lower bound, we proceed by similar arguments. Choose $c_1 \in (0, \frac{1}{2})$ such that

$$\frac{2c_1}{3(1 - 2c_1)^3(q-1)} - \frac{c_1}{q-1} + (q-1)c_1 = \frac{1}{2(q-1)}. \quad (10)$$

Put

$$G(n) = \min\left(f(n), \frac{(q-1)c_1n}{q} - 1\right), \quad (11)$$

where we assume that n is so large that $G(n) \geq 0$. For fixed n consider

$$E_n = \left\{ S \in F_q^\infty : P_n(S) < \frac{(q-1)n}{2q} - G(n) \right\}.$$

Let $m(n)$ be the largest integer satisfying

$$m(n) < \frac{(q-1)n}{2q} - G(n). \tag{12}$$

For even n we get by Theorem 2,

$$h(E_n) = q^{-n} \sum_{r=0}^{m(n)} N_n(r) = q^{-n} + (q+1)q^{-n} \sum_{r=1}^{m(n)} (q-1)^r \sum_{L=r-1}^{(n-2)/2} \binom{L}{r-1} q^L.$$

By treating the inner sum as before we obtain

$$h(E_n) < q^{-n} + 2q^{-n/2} \sum_{r=1}^{m(n)} \binom{n/2}{r} (q-1)^r.$$

Since $r \leq m(n) < (q-1)n/(2q)$, the terms of the last sum form an increasing function of r, hence

$$h(E_n) < 3q^{-n/2} m(n) \binom{n/2}{m(n)} (q-1)^{m(n)} < \frac{3n}{2} w_2 \tag{13}$$

with

$$w_2 = \binom{n/2}{m(n)} (q-1)^{m(n)} q^{-n/2}.$$

Put

$$e(n) = \frac{(q-1)n}{2q} - m(n).$$

Then as before

$$\log w_2 \leq u_1 + O(1)$$

with

$$u_1 = -\left(\frac{n}{2q} + e(n) \right) \log \left(1 + \frac{2qe(n)}{n} \right) - \left(\frac{(q-1)n}{2q} - e(n) \right) \log \left(1 - \frac{2qe(n)}{(q-1)n} \right).$$

By the definitions of $G(n)$ and $m(n)$ in (11) and (12) we have $0 \leq e(n) \leq (q-1)c_1 n/q$, hence $0 \leq 2qe(n)/(q-1)n \leq 2c_1$. Using the same lower bound for $\log(1-z)$ as before and also $\log(1+z) \geq z - \frac{1}{2}z^2$ for $z \geq 0$, we get

$$u_1 \leq -\left(\frac{n}{2q} + e(n) \right) \left(\frac{2qe(n)}{n} - \frac{2q^2e(n)^2}{n^2} \right) +$$

$$+ \left(\frac{(q-1)n}{2q} - e(n) \right) \left(\frac{2qe(n)}{(q-1)n} + \frac{2q^2e(n)^2}{(q-1)^2n^2} + \frac{8q^3e(n)^3}{3(1-2c_1)^3(q-1)^3n^3} \right)$$

$$< -\frac{q^2e(n)^2}{(q-1)n} + \frac{2q^2e(n)^3}{n^2} \left(\frac{2}{3(1-2c_1)^3(q-1)^2} - \frac{1}{(q-1)^2} + 1 \right)$$

$$= -\frac{q^2e(n)^2}{(q-1)n} + \frac{q^2e(n)^3}{(q-1)^2c_1n^2} = \frac{q^2e(n)^2}{(q-1)n} \left(-1 + \frac{e(n)}{(q-1)c_1n} \right) \leq -\frac{qe(n)^2}{n}$$

by (10) and $e(n) \le (q-1)c_1 n/q$. Therefore

$$\log w_2 < -\frac{qe(n)^2}{n} + O(1),$$

and so (13) and $e(n) > G(n)$ yield

$$h(E_n) = O\left(n\, e^{-qG(n)^2/n}\right).$$

It can again be proved that this bound also holds for odd n. Together with the hypothesis on f this shows that $\sum_{n=1}^{\infty} h(E_n) < \infty$. By applying the Borel–Cantelli lemma as before we deduce that with probability 1 we have

$$P_n(S) \ge \frac{(q-1)n}{2q} - G(n) \ge \frac{(q-1)n}{2q} - f(n) \quad \text{for all sufficiently large}\quad n.$$

Together with (9) this yields the result of the theorem. $\square$

Theorem 6. With probability 1 we have

$$\overline{\lim_{n\to\infty}}\,(n\log n)^{-1/2}\left|P_n(S) - \frac{(q-1)n}{2q}\right| \le \left(\frac{2}{q}\right)^{1/2}.$$

Proof. For a positive integer m consider the function

$$f(n) = \left(\frac{2+m^{-1}}{q}\right)^{1/2} (n\log n)^{1/2}.$$

Then

$$n\, e^{-q\, f(n)^2/n} = n^{-1-m^{-1}},$$

and so $\sum_{n=1}^{\infty} n\, e^{-qf(n)^2/n} < \infty$. Thus Theorem 5 shows that with probability 1 we have

$$\left|P_n(S) - \frac{(q-1)n}{2q}\right| \le \left(\frac{2+m^{-1}}{q}\right)^{1/2} (n\log n)^{1/2} \quad \text{for all sufficiently large}\quad n.$$

This property holds simultaneously for all m with probability 1 since the countable intersection of sets of h-measure 1 has again h-measure 1. The desired conclusion follows. $\square$

Corollary 2. With probability 1 we have

$$\lim_{n\to\infty} \frac{P_n(S)}{n} = \frac{q-1}{2q}.$$

We note that the result of Corollary 2 was already shown in [6, eq.(6)] by using the deeper methods of that paper. It suffices to observe, as we already did in Section 1 of the present paper, that the quantity $j(n, S)$ in [6] is identical with $P_n(S)$.

5. Joint Frequency Distributions in the Linear Complexity Profile

In [6, Sect. 6] we studied frequency distributions in the linear complexity profile by considering the quantity $Z(N; c; S)$ defined as follows. For any integers c and N with $N \geq 1$ and any $S \in F_q^\infty$, $Z(N; c; S)$ is given as the number of integers n with $1 \leq n \leq N$ for which $L_n(S) = (n + c)/2$. It was shown in [6, Theorem 11] that with probability 1 we have

$$\lim_{N \to \infty} \frac{Z(N; c; S)}{N} = \frac{1}{2}(q - 1)q^{-|c-(1/2)|-(1/2)} \quad \text{for all integers} \quad c. \tag{14}$$

We now extend this work to joint frequency distributions. For $S \in F_q^\infty$ and integers c_0, c_1, N with $N \geq 1$ let $Z(N; c_0, c_1; S)$ be the number of integers n with $1 \leq n \leq N$ for which $L_n(S) = (n + c_0)/2$ and $L_{n+1}(S) = (n + c_1)/2$. Then we are interested in the existence of

$$\lim_{N \to \infty} \frac{Z(N; c_0, c_1; S)}{N}.$$

Since $L_n(S) \leq L_{n+1}(S)$, we can assume that $c_0 \leq c_1$, for otherwise we have the trivial case where $Z(N; c_0, c_1; S) = 0$ for all N and S. We have to distinguish the cases $c_0 = c_1$ and $c_0 < c_1$.

Theorem 7. With probability 1 we have

$$\lim_{N \to \infty} \frac{Z(N; c, c; S)}{N} = \frac{1}{2}(q - 1)q^{-|c-1|-1} \quad \text{for all integers} \quad c.$$

Proof. First let $c \geq 1$. If $L_n(S) = (n + c)/2$, then $L_n(S) > n/2$, and so by Lemma 1 we have $L_{n+1}(S) = (n + c)/2$. Therefore $Z(N; c, c; S) = Z(N; c; S)$, and the desired result follows from (14). Now let $c \leq 0$. Suppose $n, 1 \leq n \leq N$, is such that $L_n(S) = L_{n+1}(S) = (n + c)/2$. We go to the next jump, say

$$\frac{n + c}{2} = L_n(S) = L_{n+1}(S) = \ldots = L_{n+k}(S) < L_{n+k+1}(S),$$

where $k \geq 1$. Then by Lemma 1,

$$L_{n+k+1}(S) = n + k + 1 - L_{n+k}(S) = \frac{n + 2k + 2 - c}{2}.$$

It follows that for $1 \leq i \leq k$ we have

$$L_{n+k+i}(S) \geq L_{n+k+1}(S) = \frac{n + 2k + 2 - c}{2} \geq \frac{n + 2k + 2}{2} > \frac{n + k + i}{2},$$

and so Lemma 1 yields $L_{n+2k}(S) = L_{n+k+1}(S) = (n+2k+2-c)/2$. If $n+2k \leq N$, this shows that $n+2k$ counts towards $Z(N; 2-c; S)$. If $n+2k > N$, then n is the largest value in $[1, N]$ which counts towards $Z(N; c, c; S)$. Therefore

$$Z(N; c, c; S) \leq Z(N; 2-c; S) + 1. \tag{15}$$

Now suppose $n, 1 \leq n \leq N$, is such that $L_n(S) = (n+2-c)/2$, where again $c \leq 0$. We go to the previous jump, say

$$L_{n-k}(S) < L_{n-k+1}(S) = \ldots = L_n(S) = \frac{n+2-c}{2},$$

where $k \geq 1$. Then by Lemma 1,

$$L_{n-k}(S) = n - k + 1 - L_{n-k+1}(S) = \frac{n-2k+c}{2}.$$

We claim that

$$L_{n-2k+i}(S) = L_{n-k}(S) \quad \text{for} \quad 0 \leq i \leq k. \tag{16}$$

For suppose that for some $i, 0 \leq i < k$, we had $L_{n-2k+i}(S) < L_{n-2k+i+1}(S) = L_{n-k}(S)$. Then Lemma 1 yields

$$L_{n-2k+i}(S) = n - 2k + i + 1 - L_{n-2k+i+1}(S) = n - 2k + i + 1 - \frac{n-2k+c}{2}$$
$$= \frac{n-2k-c}{2} + i + 1 > \frac{n-2k+i}{2}.$$

Another application of Lemma 1 implies $L_{n-2k+i+1}(S) = L_{n-2k+i}(S)$, a contradiction. Thus (16) is shown. Using (16) with $i = 0, 1$, we get $L_{n-2k}(S) = L_{n-2k+1}(S) = (n-2k+c)/2$. If $n-2k \geq 1$, this shows that $n-2k$ counts towards $Z(N; c, c; S)$. If $n-2k < 1$, then n is the smallest value in $[1, N]$ which counts towards $Z(N; 2-c; S)$. Thus

$$Z(N; 2-c; S) \leq Z(N; c, c; S) + 1. \tag{17}$$

It follows from (14), (15), and (17) that if $c \leq 0$, then with probability 1 we have

$$\lim_{N \to \infty} \frac{Z(N; c, c; S)}{N} = \lim_{N \to \infty} \frac{Z(N; 2-c; S)}{N} = \frac{1}{2}(q-1)q^{c-2} = \frac{1}{2}(q-1)q^{-|c-1|-1}. \ \square$$

Now let $c_0 < c_1$. If $L_n(S) = (n+c_0)/2$ and $L_{n+1}(S) = (n+c_1)/2$, then it follows from Lemma 1 that $L_{n+1}(S) = n+1 - L_n(S)$, and so we must have $c_1 = 2 - c_0$.

Theorem 8. If $c_0 < c_1 = 2 - c_0$, then with probability 1 we have

$$\lim_{N \to \infty} \frac{Z(N; c_0, c_1; S)}{N} = \frac{1}{2}(q-1)^2 q^{-c_1}.$$

If $c_0 < c_1 \neq 2 - c_0$, then $Z(N; c_0, c_1; S) = 0$ for all N and S.

Proof. The second part was shown above. Now let $c_0 < c_1 = 2 - c_0$ and note that this implies $c_1 \geq 2$. From Lemma 1 we see that $L_n(S) = (n + c_0)/2$ and $L_{n+1}(S) = (n + c_1)/2$ hold simultaneously if and only if $L_{n+1}(S) - L_n(S) = c_1 - 1$. Therefore $Z(N; c_0, c_1; S) = J(N; c_1 - 1; S)$, where the latter denotes the number of integers n with $1 \leq n \leq N$ and $L_{n+1}(S) - L_n(S) = c_1 - 1$. Thus with probability 1 we have

$$\lim_{N \to \infty} \frac{Z(N; c_0, c_1; S)}{N} = \lim_{N \to \infty} \frac{J(N; c_1 - 1; S)}{P_N(S)} \cdot \frac{P_N(S)}{N}$$

$$= (q - 1)q^{1 - c_1} \frac{q - 1}{2q} = \frac{1}{2}(q - 1)^2 q^{-c_1},$$

where we applied a result in [6, p.208] and Corollary 2. $\square$

Since there are just countably many pairs (c_0, c_1) of integers, it follows that with probability 1 we have the asymptotic frequency distributions in Theorems 7 and 8 simultaneously for all choices of c_0 and c_1.

References

[1] G.D. Carter: *Aspects of Local Linear Complexity*, Ph.D. thesis, Univ. of London, 1989.

[2] F.G. Gustavson: Analysis of the Berlekamp-Massey linear feedback shift-register synthesis algorithm, *IBM J. Res. Develop.* **20**, 204-212 (1976).

[3] M. Loève: *Probability Theory*, 3rd ed., Van Nostrand, New York, 1963.

[4] J.L. Massey: Shift-register synthesis and BCH decoding, *IEEE Trans. Information Theory* **15**, 122-127 (1969).

[5] H. Niederreiter: Sequences with almost perfect linear complexity profile, *Advances in Cryptology - EUROCRYPT '87*, Lecture Notes in Computer Science, Vol. **304**, pp. 37-51, Springer, Berlin, 1988.

[6] H. Niederreiter: The probabilistic theory of linear complexity, *Advances in Cryptology -EUROCRYPT '88*, Lecture Notes in Computer Science, Vol. **330**, pp. 191-209, Springer, Berlin, 1988.

[7] H. Niederreiter: A combinatorial approach to probabilistic results on the linear-complexity profile of random sequences, *J. of Cryptology*, to appear.

[8] H. Niederreiter: Keystream sequences with a good linear complexity profile for every starting point, *Advances in Cryptology - EUROCRYPT '89*, Lecture Notes in Computer Science, Springer, Berlin, to appear.

[9] R.A. Rueppel: Linear complexity and random sequences, *Advances in Cryptology - EUROCRYPT '85*, Lecture Notes in Computer Science, Vol. **219**, pp. 167-188, Springer, Berlin, 1986.

[10] R.A. Rueppel: *Analysis and Design of Stream Ciphers*, Springer, Berlin, 1986.

[11] M.Z. Wang: *Cryptographic Aspects of Sequence Complexity Measures*, Ph.D. dissertation, ETH Zürich, 1988.

[12] M.Z. Wang: Linear complexity profiles and jump complexity, submitted to *J. of Cryptology*.

Lower Bounds for the Linear Complexity of Sequences over Residue Rings

Zong-duo Dai[1]) Thomas Beth[2]) Dieter Gollmann[2])

[1]) University of Linköping, Sweden
on leave from Academia Sinica, Beijing, China

[2]) E.I.S.S., University of Karlsruhe, Germany

Abstract

Linear feedback shift registers over the ring Z_{2^e} can be implemented efficiently on standard microprocessors. The most significant bits of the elements of a sequence in $Z_{2^e}^\infty$ constitute a binary pseudo-random sequence. We derive lower bounds for the linear complexity over F_2 of these binary sequences.

1   Sequences over Residue Rings

For a positive integer e let Z_{2^e} denote the residue ring $Z/(2^e)$, that is the set of integers $\{0, 1, \ldots, 2^e - 1\}$ with arithmetic operations carried out modulo 2^e. Computation in Z_{2^e} differs from computation in F_{2^e} in the way overflows are handled. In a polynomial basis representation of F_{2^e} we choose an irreducible polynomial from $F_2[x]$ and thus define how overflows, i.e. terms of degree larger or equal e, are fed back, i.e. reduced to polynomials of degree less than e. In Z_{2^e} overflows modulo 2^e are simply discarded.

Sequences in $Z_{2^e}^\infty$ are of particular interest from an application point of view as they can be generated very efficiently on microprocessors when e is the word length of the processor. A sequence $\alpha = (a_t)$ generated by a linear feedback shift register over Z_{2^e} obeys a linear recursion of the form

$$a_{t+n} = \sum_{j=0}^{n-1} c_j a_{t+j} \pmod{2^e} \quad \text{for } t \geq 0 \, ,$$

where n is the length of the shift register and $a_t, c_j \in Z_{2^e}$ (see Fig.1). Due to a result by Ward [7] the upper bound for the period of linear recursive sequences of degree n over Z_{2^e} is $2^{e-1}(2^n - 1)$.

Definition 1.1   Linear recursive sequences of degree n over Z_{2^e} with period $2^{e-1}(2^n - 1)$ are called maximal length linear sequences, in short MLL-sequences ([1,4]).

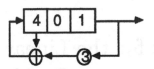

Figure 1: A linear feedback shift register over $\mathbf{Z}_8$ with feedback polynomial $f(x) = x^3 - x^2 - 3$

The period of a polynomial $f(x) \in \mathbf{Z}_{2^e}[x]$ is defined as follows.

Definition 1.2 Let $(f(x))$ denote the ideal in $\mathbf{Z}_{2^e}[x]$ generated by some polynomial $f(x) \in \mathbf{Z}_{2^e}[x]$. The period of $f(x)$ is given by

$$\mathrm{per}(f(x)) := \min\{T \geq 1 | \exists d \geq 0 : x^{T+d} - x^d \in (f(x))\} \ .$$

A polynomial $f(x) \in \mathbf{Z}_{2^e}[x]$ is called primitive if it has period $2^{e-1}(2^n - 1)$.

2 Binary Sequences Generated from Sequences over Residue Rings

Binary sequences α_i, $0 \leq i < e$, can be derived from sequences α over $\mathbf{Z}_{2^e}$ by

$$a_t = \sum_{i=0}^{e-1} a_{ti} 2^i \ , \quad a_{ti} \in \{0, 1\} \ ,$$

$$\alpha_i = (a_{0i}, a_{1i}, \ldots, a_{ti}, \ldots) \ .$$

As an example, the sequence $(1, 0, 4, 7, \ldots) \in \mathbf{Z}_{2^e}^\infty$ is represented by

$$\alpha_0 = (1, 0, 0, 1, \ldots)$$
$$\alpha_1 = (0, 0, 0, 1, \ldots)$$
$$\alpha_2 = (0, 0, 1, 1, \ldots) \ .$$

Let L be the left shift operator on the sequences $\alpha \in \mathbf{Z}_{2^e}^\infty$. For any polynomial

$$f(x) = \sum_{i=0}^{n} c_i x^i \in \mathbf{Z}_{2^e}[x]$$

define

$$f(L)\alpha := \sum_{i=0}^{n} c_i L^i \alpha \ .$$

Definition 2.1 The minimal polynomial of α is the monic polynomial $f(x) \in \mathbf{Z}_{2^e}[x]$ of lowest degree so that $f(L)\alpha = 0$.

The polynomial $f(x)$ can be decomposed into polynomials $f_i(x) \in \mathbf{F}_2[x]$ by

$$f(x) = \sum_{i=0}^{e-1} f_i(x) 2^i \ .$$

If $f(x)$ is a primitive polynomial then $f_0(x)$ is a primitive polynomial in $\mathbf{F}_2[x]$ and the roots of $f_0(x)$ have period $2^n - 1$.

We quote the following results relating to the period of binary sequences derived from MLL-sequences over $\mathbf{Z}_{2^e}$ ([1,4,7]).

Theorem 2.1 Let α be a MLL-sequence over $\mathbf{Z}_{2^e}$ with minimal polynomial $f(x)$ of degree n. We have

$$\text{per}(\alpha_{e-1}) = \text{per}(\alpha) = 2^{e-1}(2^n - 1) \ .$$

Let $f(x) \in \mathbf{Z}_{2^e}[x]$ be a primitive polynomial of degree n. There exist

$$2^{(e-1)n}(2^n - 1)$$

different MLL-sequences with minimal polynomial $f(x)$.

Let α and β be two MLL-sequences with minimal polynomial $f(x)$. The probability that α is a cyclic shift of β is $2^{-(n-1)(e-1)}$.

Dai [1] and Huang [4] also give upper bounds for the linear complexity of the sequence α_{e-1}. In this paper we will bound the linear complexity from below.

3 Main Lemma

Let $\alpha \in \mathbf{Z}_{2^e}^\infty$ be a MLL-sequence generated by some LFSR over $\mathbf{Z}_{2^e}$. A new element of the sequence α_r will be computed from the previous elements of this sequence, from the lower bit sequences, and from the carries generated by the lower bit sequences. Let β_{ij} denote the sequence of carries propagated from sequence α_i to sequence α_j and let

$$\Phi_k(x_0, x_1, \ldots, x_n) := \sum_{0 \leq i_1 < i_2 < \cdots < i_k \leq n} x_{i_1} x_{i_2} \cdots x_{i_k}$$

be the symmetric function of order k applied to n arguments ([6], p. 182 ff). Fig. 2 describes the decomposed computation of the sequences α_r and β_{ij}. Note that the coefficients of $f_0(x)$ act as a filter for the inputs to the Φ_k, i.e. for

$$f_0(x) = x^n + \sum_{i=0}^{n-1} c_{0,i} x^i$$

we have

$$\beta_{ij} = \Phi_{2^j}(c_{00}\alpha_i, c_{01}L\alpha_i, \ldots, c_{0,n-1}L^{n-1}\alpha_i) \ .$$

Let α be a MLL-sequence with minimal polynomial $f(x)$ of degree n and let θ be a root of $f_0(x)$. For any

$$e = \sum_{i=0}^{n-1} e_i 2^i \ , \ \ e_i \in \{0, 1\} \ ,$$

define the weight of $\rho = \theta^e$ as

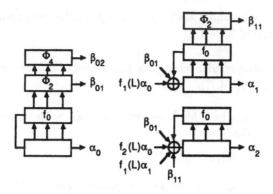

Figure 2: The carries in the generation of a sequence in $\mathbf{Z}_8^\infty$

$$w(\rho) = \sum_{i=0}^{n-1} e_i \in \mathbf{Z} .$$

Let $\mathbf{K}$ be the algebraic closure of $\mathbf{F}_2$. Define

$$g_m^+(x) = \prod_{\rho, w(\rho)=m} (x - \rho) ,$$

$$g_m^-(x) = \prod_{\rho, w(\rho)<m} (x - \rho) ,$$

and

$$v_\rho = (1, \rho, \rho^2, \ldots, \rho^t, \ldots) \in \mathbf{K}^\infty .$$

We will sketch the main step in deriving lower bounds for the linear complexity of α_{e-1}. A detailed account can be found in [2]. The sequences α_r will be decomposed into

$$\alpha_r = \alpha_r^+ + \alpha_r^- ,$$

where α_r^+ and α_r^- are chosen so that the minimal polynomial of α_r^+, is a divisor of $g_{2r}^+(x)$ and the minimal polynomial of α_r^- and is a divisor of $g_{2r}^-(x)$. Hence, we have for the linear complexity $\mathcal{L}(\alpha_r)$ of α_r

$$\mathcal{L}(\alpha_r) = \mathcal{L}(\alpha_r^+) + \mathcal{L}(\alpha_r^-) .$$

We will bound the linear complexity of α_r from below by determining the sequences α_r^+, $0 \le r \le e - 1$. Writing $\beta_{ij} = \beta_{ij}^+ + \beta_{ij}^-$ as above, the sequence α_r^+ can be related to the carry sequences $\beta_{i,r-i}^+$ and we obtain finally

Lemma 3.1 Let $\alpha \in \mathbf{Z}_{2^e}^\infty$ be a MLL-sequence with minimal polynomial $f(x) \in \mathbf{Z}_{2^e}[x]$ of degree n, let θ be a root of $f_0(x)$. We have for any r, $2^r \le n$,

$$\alpha_r^+ = L^\tau \sum_{\rho, w(\rho)=2^r} v_\rho .$$

for some $\tau \ge 0$.

4 Lower bounds for the linear complexity

Let $m(x)$ be the minimal polynomial of α_{e-1}. The degree of $m(x)$ will be examined by means of the discriminant of $f(x)$ and Lemma 3.1:

Definition 4.1 The discriminant of $f(x)$, $\Delta_f(x)$, is given by

$$\Delta_f(x) \equiv \begin{cases} h_f(x) & (\mathrm{mod}\ (f_0(x), 2)) \quad \text{if } e = 2 \\ h_f(x)(h_f(x) + 1) & (\mathrm{mod}\ (f_0(x), 2)) \quad \text{if } e \geq 3 \end{cases}$$

with $h_f(x)$ determined by

$$x^{2^n - 1} \equiv 1 + 2h_f(x) \quad (\mathrm{mod}\ (f(x), 2^2)).$$

Lemma 4.1 ([4]) For any e, r, $2 \leq r < e - 1$, we have

$$\left(L^{2^n - 1} - 1\right)^{2^{e-2} - 2^{r-1}} \alpha_{e-1}^+ = (\Delta_f(L)\alpha_0)\alpha_r^+ .$$

With these preparations we embark on the analysis of the sequences α_r^+. We may assume

$$\alpha_0 = \sum_{\rho, w(\rho) = 1} v_\rho = \sum_{i=0}^{n-1} v_{\theta^{2^i}} ,$$

hence

$$\Delta_f(L)\alpha_0 = \sum_{i=0}^{n-1} \Delta_f(\theta^{2^i}) v_{\theta^{2^i}} ,$$

and by Lemma 3.1

$$(\Delta_f(L)\alpha_0)\alpha_r^+ = \sum_{\rho, w(\rho) = 2^r + 1} \sigma_\rho v_\rho \tag{1}$$

with

$$\sigma_\rho = \sum_{j=1}^{2^r + 1} \Delta_f(\theta^{2^{i_j}}) \quad \text{for } \rho = \theta^{2^{i_1} + 2^{i_2} + \cdots + 2^{i_{2^r}} + 1} .$$

Write

$$\sigma_r(x) = \prod_{\substack{\rho, w(\rho) = 2^r + 1 \\ \sigma_\rho \neq 0}} (x - \rho) .$$

The root θ of $f_0(x)$ has period $2^n - 1$, therefore

$$\sigma_r(x) \mid \left(x^{2^n - 1} - 1\right) .$$

We define $M := \min(e - 1, \log_2(n - 1))$ and derive from Lemma 4.1 and from (1)

$$\sigma_r(x)^{2^{e-2} - 2^{r-1} + 1} \mid m(x), \ 2 \leq r \leq M .$$

This gives

$$\mathcal{L}(\alpha_{e-1}^+) \geq \sum_{k=2}^{M} (2^{e-2} - 2^{k-1} + 1) \cdot \deg \sigma_k(x) . \tag{2}$$

With $\Delta = \Delta_f(\theta)$ and

$$\mathcal{N}_k(\Delta) = \left\{ (i_1, \ldots, i_{2^k + 1}) \mid 0 \leq i_1 < \cdots < i_{2^k + 1} < n , \sum_{j=1}^{2^k + 1} \Delta^{2^{i_j}} \neq 0 \right\}$$

we may rewrite (2) as follows,

Theorem 4.2 Let $\alpha \in \mathbf{Z}_{2^e}^{\infty}$ be a ML-sequence and let M be defined as above. We have

$$\mathcal{L}(\alpha_{e-1}) \geq \sum_{k=2}^{M}(2^{e-2} - 2^{k-1} + 1)|\mathcal{N}_k(\Delta)| \ .$$

It can be shown that

$$|\mathcal{N}_k(\Delta)| \leq \binom{n}{2^k+1}$$

for $2^k < n$. In fact, $|\mathcal{N}_k(\Delta)|$ is very close to this upper bound for any Δ. We consider a special case where $|\mathcal{N}_k(\Delta)|$ obtains its maximal value.

Theorem 4.3 If $\mathbf{F}_2(\Delta) = \mathbf{F}_{2^d}$ and $\left\{\Delta, \Delta^2, \ldots, \Delta^{2^{d-1}}\right\}$ is a normal basis of $\mathbf{F}_2(\Delta)$ over $\mathbf{F}_2$, then we have

$$\mathcal{L}(\alpha_{e-1}) \geq \begin{cases} \displaystyle\sum_{k=2}^{\lfloor \log_2 n-1\rfloor}(2^{e-2} - 2^{k-1} + 1)\binom{n}{2^k+1} & \text{if } n < 2^{e-1} \\[4mm] \displaystyle\binom{n}{2^{e-1}} + \sum_{k=2}^{e-2}(2^{e-2} - 2^{k-1} + 1)\binom{n}{2^k+1} & \text{if } n \geq 2^{e-1} \ . \end{cases}$$

Proof. It suffices to prove for $k < e - 1$

$$\Delta(\underline{i}) = \sum_{j=1}^{2^k+1}\Delta^{2^i j} \neq 0$$

for any $\underline{i} = (i_1, i_2, \ldots, i_{2^k+1})$, $0 \leq i_1 < i_2 < \cdots < i_{2^k+1} < n$. In fact, if $\mathbf{F}_2(\Delta) = \mathbf{F}_{2^d}$, $\Delta(\underline{i})$ can be reduced to

$$\Delta(\underline{i}) = \Delta^{2^{k_1}} + \Delta^{2^{k_2}} + \cdots + \Delta^{2^{k_t}} \ ,$$

where $0 \leq k_1 < k_2 < \cdots < k_t < d$, $t \equiv 1 \pmod 2$. Since $\left\{\Delta, \Delta^2, \ldots, \Delta^{2^{d-1}}\right\}$ is a normal basis we get $\Delta(\underline{i}) \neq 0$. ∎

The lower bound given in Theorem 4.3 is quite large. For example, if $n = 2(2^k+1) < 2^{e-1}$ for some k, just one term in the right hand sum will be, ref. [5],

$$(2^{e-2} - 2^{k-1} + 1)\binom{n}{2^k+1} \geq 2^{e-3}\binom{n}{\frac{n}{2}} \geq 2^{e-3}\frac{\sqrt{\pi}}{2} \cdot \frac{2^{n+1}}{\sqrt{2\pi n}} = \frac{2^{n+e-3}}{\sqrt{2n}} \ .$$

MLL-sequences have period $2^{e-1}(2^n - 1)$ so we get

$$L(\alpha_{e-1}) \geq \frac{2^{n+e-1}}{4\sqrt{2n}} \geq \frac{\text{per}(\alpha_{e-1})}{4\sqrt{2(\log(\text{per}(\alpha_{e-1}))-e+1)}} \ .$$

5 Conclusions

Binary sequences generated from MLL-sequences over Z_{2^e} are of particular interest from an application point of view as they can be implemented very efficiently on microprocessors when e is the word length. We have shown that the lower bounds for the linear complexity of these sequences are reasonably high. However, the reader should be aware that there exist algorithms to reconstruct sequences generated by linear congruences from truncated outputs [3] if the congruence is known. Thus, the results on sequences over residue rings rather should help to evaluate their contribution to more sophisticated designs than be taken as an argument for their security.

References

[1] Z.D.Dai, *The Binary Sequences Derived from the Sequences over the Integral Residue Ring $Z/(2^e)$. I*, Internal Report, Academia Sinica, Beijing, China

[2] Z.D.Dai, D.Gollmann *Lower Bounds for the Linear Complexity of Sequences over Residue Rings*, E.I.S.S. Report, No. 90/7, 1990

[3] A.M.Frieze, J.Hastad, R.Kannan, J.C.Lagarias, A.Shamir, *Reconstructing Truncated Integer Variables Satisfying Linear Congruences*, SIAM J. Comput., Vol.17, No.2, pp.262-280, April 1988

[4] M.Q.Huang, *The Binary Sequences Derived from the Sequences over the Integral Residue Ring $Z/(2^e)$. II*, Internal Report, Academia Sinica, Beijing, China

[5] W.Peterson, E.Weldon, *Error Correcting Codes*, 2nd ed., M.I.T.Press, Cambridge, Berlin, 1986

[6] R.A.Rueppel, *Analysis and Design of Stream Ciphers*, Springer, Cambridge, Mass., 1972

[7] M.Ward, *The arithmetical theory of linear recurring sequences*, Transactions of the American Mathematical Society, Vol.35, pp.600-628, July 1933

On the Construction of Run Permuted Sequences

CEES J.A. JANSEN

Philips Crypto B.V.

P.O. Box 218, 5600 MD Eindhoven

The Netherlands

Abstract

This paper describes the construction of classes of binary sequences, which are obtained by permuting the runs of zeroes and ones of some given periodic binary sequence $\underline{s} = (s_0, s_1, \ldots, s_{p-1})^{\infty}$, $s_i \in GF(2)$. A large class of sequences is constructed by permuting the runs of zeroes and ones of a DeBruijn sequence of given order. The properties of the sequences in this class are discussed. As is known, in this way all DeBruijn sequences of given order are obtained, but also many more sequences with higher complexities, all satisfying Golomb's first and second randomness postulates. It is shown how to generate the sequences in this class with the use of enumerative coding techniques. A more efficient sequence generator, employing shift registers is also introduced. The binary sequence generator obtained in this way can be useful for cryptographic purposes, e.g. in streamcipher systems.

1 Introduction

Run permuted sequences were introduced in [4] as sequences obtained through independently permuting the runs of ones and zeroes of a DeBruijn sequence of given order. DeBruijn sequences are well-known for their properties, i.e. for order n their period is 2^n, every n-tuple occurs exactly once in the sequence and by a result of DeBruijn [2] it is known that there are exactly $2^{2^{n-1}-n}$ such sequences. We will recall the properties of this sequence class, i.e. the maximum order complexities as introduced in [4] and the number of sequences obtained in this way. We will show how to generate the sequences in this class with the use of enumerative coding techniques. Also, a more efficient sequence generator, using shift registers, is given.

2 The Sequence Class $\mathcal{C}_n$

In [4, Ch. 6] a class of sequences is constructed through permuting the runs of ones and zeroes of a DeBruijn sequence of given order. The procedure described there is the following: the given sequence is written in its run-length representation and then the integers representing the runs of ones and the integers representing the runs of zeroes are permuted independently. The new sequence of integers obtained in this way is then transformed back into a binary sequence. The reason for doing this is twofold:

- The described procedure preserves the R1 and R2 properties of the original sequence ([3]), i.e. the number of ones and zeroes as well as the distribution of the runs remain unaltered.

- The sequences obtained in this way have interesting properties such as a good maximum order complexity.

Definition 1 *The class $\mathcal{C}_n$ of binary periodic sequences is defined as the class of cyclicly inequivalent sequences obtained through independently permuting the runs of ones and zeroes of a DeBruijn sequence of order n.*

It can be shown that the number of sequences in this class is given by the following equation:

$$|\mathcal{C}_n| = 2^{-n+2} \prod_{l=1}^{n-2} \binom{2^l}{2^{l-1}}^2. \tag{1}$$

If the binomial coefficients in (1) are approximated using Stirling's approximation formula, the result becomes:

$$|\mathcal{C}_n| = \rho_n \left(\frac{4}{\pi}\right)^{n-2} \prod_{k=1}^{n} G_k, \tag{2}$$

where $G_k = 2^{2^{k-1}-k}$ the number of binary DeBruijn sequences of order k and ρ_n is a correction factor which is less than 1 for all n and which converges to $0.61\cdots$ for large n.

Equation (2) clearly shows that the fraction of DeBruijn sequences contained in $\mathcal{C}_n$ goes to zero for large n. However, all DeBruijn sequences of order n are contained in $\mathcal{C}_n$, as DeBruijn sequences are by definition all those sequences in which all subsequences of length n occur exactly once. Hence, it is demonstrated that Golomb's statement [3, pg. 113] that the number of sequences in this class "is slightly larger" than the number of DeBruijn sequences of order n, is not very careful.

3 Complexity Properties of Sequences in $\mathcal{C}_n$

As all DeBruijn sequences of order n (i.e. maximum order complexity n) constitute only a small fraction of the entire class, the other sequences in $\mathcal{C}_n$ must necessarily have maximum order complexities higher than n. Concerning this complexity the following results are given in [4].

Proposition 1 *For all sequences $\underline{s} \in \mathcal{C}_n$, $n > 2$, the maximum order complexity $c(\underline{s})$ satisfies the inequality: $n \leq c(\underline{s}) \leq 2^{n-1} - 1$.*

Clearly, the lowerbound is attained by the DeBruijn sequences. For $n = 3$ the lowerbound coincides with the upperbound. For $n > 3$ the upperbound is obtained by sequences constructed as follows. All runs are divided into two sets; this is possible for all runs except for the two longest runs of both ones and zeroes, which are unique. With these two sets two identical sequences are constructed. Then the longest runs of ones and zeroes are placed in front of the first sequence and the longest but one runs in front of the second sequence. The two sequences are now concatenated to one sequence. The longest subsequence which occurs twice in this sequence has length $2^{n-1} - 2$, as can be seen from the next example.

Example 1 $\underline{s} \in C_5$, $c(\underline{s}) = 15$

Two constructions:

$$11111 \quad 00000 \quad 11001010 \qquad 00000 \quad 11001010 \quad 11111$$
$$111 \quad 000 \quad 11001010 \qquad 000 \quad 11001010 \quad 111$$

Corollary 2 *For all sequences $\underline{s} \in C_n$, $n > 2$, the maximum order complexity $c(\underline{s})$ cannot be equal to $2^{n-1} - 2$.*

The number of sequences with maximum order complexity of $2^{n-1} - 1$ (which is the maximum value) is given by the next proposition.

Proposition 3 *Let $\mathcal{M}_n \subset C_n$ denote the subset of sequences in C_n having maximum order complexity of $2^{n-1} - 1$. The number of sequences in this set satisfies $|\mathcal{M}_n| = 2^{-n+5}|C_{n-1}|$, for all $n \geq 4$.*

It appears not to be straightforward to find a general expression for the number of sequences in C_n with given complexity, other than for the three values shown above. However, some small numerical examples can give an impression of the distribution of complexity.

Example 2

$$
\begin{array}{lll}
G_3 = 2 & |C_3| = 2 & |\mathcal{M}_3| = 2 \\
G_4 = 16 & |C_4| = 36 & |\mathcal{M}_4| = 4 \\
G_5 = 2048 & |C_5| = 88200 & |\mathcal{M}_5| = 36 \\
G_6 = 67108864 & |C_6| = 7304587290000 & |\mathcal{M}_6| = 44100
\end{array}
$$

$n = 4$:

c	4	5	6	7
$\#(\underline{s})$	16	16	0	4

Average complexity $\bar{c} = 4.7778$

$n = 5$:

c	5	6	7	8	9	10	11	12	13	14	15
$\#(\underline{s})$	2048	17376	37824	19824	8048	1840	860	256	88	0	36

Average complexity $\bar{c} = 7.2892$

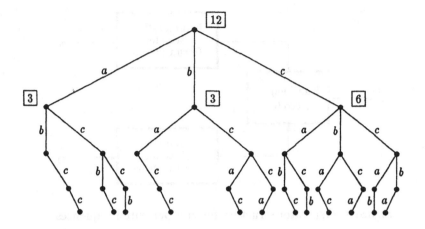

Figure 1: Ternary code tree for all permutations of *abcc*

4 Generation of Run Permuted Sequences

The sequences we have discussed so far in this section can also be generated efficiently. In order to be able to generate *all* sequences in C_n enumerative coding techniques (as in [1]) can be applied to generate the permutations.

An example demonstrates the enumerative encoding and decoding of permutations. Consider all sequences which comprise one a, one b and two c's; clearly there are $4!/2! = 12$ of these sequences. Let us assume a lexicographic ordering $a < b < c$ and define the leftmost character in the sequence to be the most significant character. All 12 sequences are represented by the ternary tree of Figure 1. The numbers written with the nodes of the tree, denote the number of leaves that can be reached from that node, i.e. the number of sequences starting with a, b, c, aa, ab, ... These numbers can easily be determined. For example, if N_x denotes the number of sequences starting with an x, where x is a sequence of length less than 4 from $\{a, b, c\}$, we have $N_a = 3!/2!$, $N_b = 3!/2!$, $N_c = 3!/1!$, $N_{ab} = 2!/2!$, and so on. From the code tree it can be seen that *cabc* is coded into $(3 + 3) + 0 + 0 = 6$. The codewords are often called the indices, written as $i(cabc) = 6$. The general expression for the indices is:

$$i(\xi_0\xi_1\xi_2\xi_3) = \sum_{n=0}^{2} \sum_{\nu_n < \xi_n} N_{\xi_0\cdots\xi_{n-1}\nu_n},$$

where $\nu_n, \xi_n \in \{a, b, c\}$, for $n = 0, 1, 2, 3$. Using the above expression we see that $i(abcc) = 0$ and $i(ccba) = 11$.

The decoding process uses again the node numbers of the code tree. As an example, consider the decoding of an index with value 5. Clearly, the sequence cannot start with an a, as all such sequences have indices less than 3. Also, the sequence cannot start with a c, as all of these sequences have indices $\geq 3 + 3$. Hence, the sequence starts with a b. Next, the index is decreased by $N_a = 3$, yielding a new index with value 2. Applying the same procedure, c is obtained as the second character, yielding

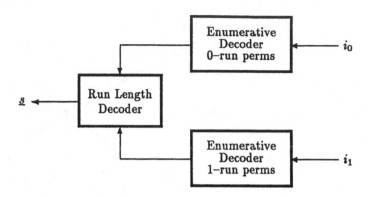

Figure 2: Generator structure for run permuted sequences

a new index with value 1. Continuing in this way we find that the sequence with index 5 is *bcca*.

It is obvious that the enumerative decoding process can be applied twice, once for the runs of ones and once for the runs of zeroes. In this way, all different run permuted sequences can be generated in their run-length representations. To generate the corresponding binary sequences it suffices to apply the run-length decoding algorithm. The structure of such a sequence generator is depicted in Figure 2

5 A Shift Register Construction

Although the enumerative decoding techniques are well understood, their implementational complexity is still quite substantial. Therefore, in this section we present a run permuted sequence generator, based on feedback shift registers; binary counting devices that are easy to implement.

The proposed method works because of the fact that the generator needs not be able to generate the entire class C_n, but rather a smaller subclass of sequences, with sequences possibly occuring more than once, i.e. the class may contain phase shifted versions of one and the same sequence. Under these relaxed conditions the enumerative decoders, as shown in Figure 2, can be replaced by integer-sequence generators, being capable of generating a number of different sequences of integers.

These sequences of integers have to satisfy a number of conditions, viz.:

1. the period must be 2^{n-2}, as there are exactly that many runs of both ones and zeroes in a run permuted sequence of order n

2. because of the perfect run distribution, there must be 2^{n-3} integers with value 1, 2^{n-4} with value 2, etc.

Sequences satisfying these conditions can be constructed very efficiently by means of a nonlinear feedback shift register, generating a DeBruijn sequence of order $n-2$. An example is shown in Figure 3, which shows the complete binary sequence generator of

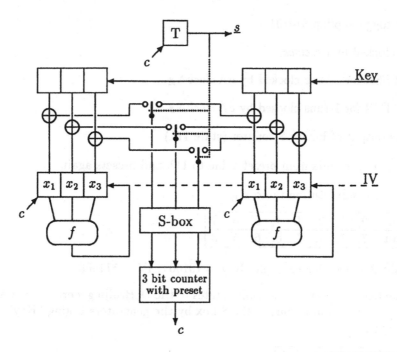

Figure 3: Generator example for run permuted sequences of order 5

order 5, being capable of generating 512 different run permuted sequences of period 32. The complexities of these sequences are listed below:

Total $\#(\underline{s}) = 512$		(88200)
$c = 6$	128	(17376)
$c = 7$	168	(37824)
$c = 8$	168	(19824)
$c = 9$	48	(8048)
$\bar{c} = 7.2656$		(7.2892)

The numbers between brackets are for the entire class C_5.

From Figure 3 the two identical DeBruijn sequence generators of order 3 can be identified, i.e. one for the runs of ones and one for the runs of zeroes. The statesequences of each of these generators are permuted by the contents of the Key registers, thus giving rise to 8 different statesequences each. The initial vector IV causes the two statesequences to combine in 8 different ways. The statesequences are converted into appropriate sequences of integers by means of the S-box. Alternatively this S-box may be implemented separately for the two statesequences, thus allowing additional permutations. The following details apply to the generator of Figure 3:

- $f(x_1, x_2, x_3) = x_1 + x_3 + \overline{x_2 \vee x_3}$

- T: toggle flipflop $010101\cdots$

- T clocked by $c \wedge clock$

- NLFSR for 0-runs clocked by $c \wedge clock \wedge \underline{s}$

- NLFSR for 1-runs clocked by $c \wedge clock \wedge \bar{\underline{s}}$

- c is output of binary 8-counter (state 111)

- 8-counter counts from preset value to 111 and presets again

- S-box is substitution table, e.g.:

in:	0,	1,	2,	3,	4,	5,	6,	7
out:	7,	7,	7,	7,	6,	6,	5,	3

Example 3 Operating example: Key = 101001, IV = 111000

The table below shows the successive states of the DeBruijn generators and the corresponding integer values input to the S-box by the generators taking "Key" and "IV" into account.

DB–gen	G-r	G-l
000	0	2
001	1	3
011	3	0
111	7	7
110	6	1
101	5	5
010	2	4
100	4	6

Gr: 10.2.6.......7.....4...3..5....1.

Gl: 2.3.0...7.........1..5..4...6...2

ct: 7....5673456734567.6767.6767567..

$\underline{s}$: 0101011100000111101100100100110010

From the foregoing it should be clear that higher order sequences can be obtained analogously. In particular, for n^{th} order sequences, having period 2^n, the DeBruijn sequence generators must be of order $n - 2$ and the presetable counter must comprise $\lceil \log_2 n \rceil$ bits. In this way 2^{3n-6} different sequences can be generated efficiently.

6 Conclusions

In this paper a construction method for binary sequences was described and its properties investigated. This construction method is based on source coding techniques. Starting with a DeBruijn sequence of given order we construct an entire class of sequences by permuting the runs of ones and zeroes. The class of sequences contains all DeBruijn sequences of the given order, but many other sequences as well, all satisfying Golomb's first and second randomness postulates.

It was demonstrated that all the sequences in the class can be generated by enumerative decoding techniques. Also, an efficient sequence generator based on shift registers was shown, which seems very well suited for high speed applications. In the way described in this paper, many more classes of sequences can be constructed, e.g. if the starting sequence is a *Maximum Length Linear FSR sequence* or a *Legendre sequence* ([3, pg. 47]), which may be useful for cryptographic purposes. To this end, it suffices for the the shift register generator to simply use other S-boxes.

References

[1] T. Cover. "Enumerative Source Coding", *IEEE Trans. on Info. Theory*, vol. IT–19, pp. 73–76, January 1973.

[2] N. G. de Bruijn. "A Combinatorial Problem", *Nederl. Akad. Wetensch. Proc.*, vol. 49, pp. 758–754, 1946.

[3] S. W. Golomb. *Shift Register Sequences*, Holden–Day Inc., San Francisco, 1967.

[4] C. J. A. Jansen. *Investigations On Nonlinear Streamcipher Systems: Construction and Evaluation Methods*, PhD. Thesis, Technical University of Delft, Delft, April 1989.

Correlation Properties of Combiners with Memory in Stream Ciphers

(Extended Abstract) [1]

Willi Meier [2]   Othmar Staffelbach [3]

[2] HTL Brugg-Windisch
CH-5200 Windisch, Switzerland

[3] GRETAG, Althardstrasse 70
CH-8105 Regensdorf, Switzerland

Abstract

In stream cipher design pseudo random generators have been proposed which combine the output of one or several LFSRs in order to produce the key stream. For memoryless combiners it is known that the produced sequence has correlation to sums of certain LFSR–sequences whose correlation coefficients c_i satisfy the equation $\sum_i c_i^2 = 1$. It is proved that a corresponding result also holds for combiners with memory.

If correlation probabilities are conditioned on side information, e.g. on known output digits, it is shown that new or stronger correlations may occur. This is exemplified for the summation cipher with two LFSRs where such correlations can be exploited in a known plaintext attack. A cryptanalytic algorithm is given which is shown to be successful for LFSRs of considerable length and with arbitrary feedback connection.

1 Introduction

Cryptographic transformations are usually designed by appropriate composition of nonlinear functions. In stream cipher design such functions have been applied to combine the output of linear feedback shift registers (LFSRs) in order to produce the key stream. In this design the combining functions should not leak information about the individual LFSR-sequences into the key stream. For this purpose the concept of correlation immunity has been introduced in [7,5] in order to prevent divide and conquer correlation attacks. For nonlinear combiners there is a tradeoff between the nonlinear order of the Boolean function and its order of correlation immunity. As has been pointed out in [5], this tradeoff can be avoided if the function is allowed to have memory.

---

[1] Full paper to appear in the *Journal of Cryptology*.

For a memoryless combiner the output always has correlation to certain linear functions of the inputs, and the "total" correlation is independent of the combining function. In fact it has been shown in [3] that the sum of the squares of the correlation coefficients is always 1. If such a combiner is applied to the output of LFSRs, there result correlations to sums of certain LFSR–sequences whose correlation coefficients c_i satisfy

$$\sum_i c_i^2 = 1 \qquad (1)$$

Choosing the combiner to be correlation immune of some order means that certain of these c_i's vanish. In particular, to prevent divide and conquer, one can achieve that there is no correlation to sums of outputs of only few LFSRs. However by (1) there must be stronger correlation to certain other sums of LFSR–sequences, which have to be considered with regard to the cryptanalytic algorithms described in [2]. A first goal of the present paper is to show that a result similar to (1) remains valid for combiners with memory. This implies that memory does not affect the total correlation. In fact the total correlation is independent of the combining functions as for memoryless combiners.

The correlation coefficients in (1) are derived from unconditioned probabilities. However it is often the case that the cryptanalyst has access to side information, e.g. he may know portions of the output sequence. In fact, if correlation is conditioned on the output, new or much stronger correlations may occur. This is exemplified for the basic summation combiner with two inputs, where the resulting correlations can be cryptanalytically exploited in a known plaintext attack on the basic summation cipher with two LFSRs. A cryptanalytic algorithm is given which is shown to be successful for LFSRs of considerable length and with arbitrary feedback connection. As a consequence for the design of summation ciphers, it is recommended to take several LFSRs of moderate length rather than just few long LFSRs.

2 Basic Summation Combiner

The summation combiner has been considered in [5] as a main example of a combiner with memory, in order to generate cryptographically strong binary sequences out of given (cryptographically weak) sequences. It is based on integer addition which, when viewed over $GF(2)$, defines a nonlinear function with memory whose correlation immunity is maximum.

To describe this combiner, consider two binary sequences $\mathcal{A} = (a_0, a_1, \ldots)$ and $\mathcal{B} = (b_0, b_1, \ldots)$. For every n the first n digits are viewed as the binary representation of an integer, i.e. $a = a_{n-1}2^{n-1} + \cdots + a_1 2 + a_0$ and $b = b_{n-1}2^{n-1} + \cdots + b_1 2 + b_0$. Then the integer sum $z = a + b$ defines the first n digits of the resulting sequence $\mathcal{Z} = (z_0, z_1, \ldots)$. If $\mathcal{A}$ and $\mathcal{B}$ are semi-infinite then $\mathcal{Z}$ is also defined as a semi-infinite sequence. The digit z_j is recursively computed by

$$z_j = f_0(a_j, b_j, \sigma_{j-1}) = a_j + b_j + \sigma_{j-1} \qquad (2)$$

$$\sigma_j = f_1(a_j, b_j, \sigma_{j-1}) = a_j b_j + a_j \sigma_{j-1} + b_j \sigma_{j-1} \qquad (3)$$

where in (2) σ_{j-1} denotes the carry bit, and $\sigma_{-1} = 0$. For this basic summation combiner it is possible to give an explicit description of all correlations that exist between its output and linear functions of the inputs.

If $\mathcal{A}$ and $\mathcal{B}$ are considered to be independent and uniformly distributed sequences of random variables, the output sequence $\mathcal{Z} = (z_0, z_1, z_2, \ldots)$ is also uniformly distributed. Moreover z_j is independent of a_j, b_j and the sum $a_j + b_j$. However it can be shown (cf. [4]) that z_j is correlated to $a_j + b_j + a_{j-1}$ and $a_j + b_j + b_{j-1}$, with probability

$$p = P(z_j = a_j + b_j + a_{j-1}) = P(z_j = a_j + b_j + b_{j-1}) = 0.75 \tag{4}$$

This means that the corresponding correlation coefficient (cf. [3]) amounts to $c = 2p - 1 = 0.5$. More generally, it is proved in [4] that for every i, $1 \le i \le j$, there are correlations to $N = 2^{i+1} - 2$ linear functions of the form

$$s = \sum_{k=j-i}^{j} \alpha_k a_k + \beta_k b_k, \tag{5}$$

and the corresponding correlation coefficients c_h satisfy

$$\sum_{h=1}^{N} c_h{}^2 = 1 - \frac{1}{2^i} \tag{6}$$

Note that the right hand side of (6) tends to 1 as i tends to ∞. This means that the total correlation for the basic summation combiner approaches 1, similar to the case of memoryless combiners. This remarkable fact can be extended to completely general combiners with 1 bit memory.

3 General Combiner with One Bit Memory

A general combiner with 1 bit memory is described by two balanced functions f_0 and f_1 as follows:

$$z_j = f_0(x_{1j}, \ldots, x_{nj}, \sigma_{j-1}) \tag{7}$$

$$\sigma_j = f_1(x_{1j}, \ldots, x_{nj}, \sigma_{j-1}) \tag{8}$$

Hereby σ_j denotes the state of the memory, and the inputs $\mathcal{X}_m = (x_{m0}, x_{m1}, x_{m2}, \ldots)$, $1 \le m \le n$, are assumed to be independent and uniformly distributed sequences of random variables.

In order to study the correlation properties of this combiner we investigate correlations of the combining functions $f_0, f_1 : GF(2)^{n+1} \to GF(2)$ to linear functions. The correlation of an arbitrary function $f : GF(2)^{n+1} \to GF(2)$ to the linear function $L_{\mathbf{w}}(\mathbf{x}) = \mathbf{w} \cdot \mathbf{x}$ ($\mathbf{w}, \mathbf{x} \in GF(2)^{n+1}$) is computed by the Walsh transform

$$F(\mathbf{w}) = \sum_{\mathbf{x} \in GF(2)^{n+1}} f(\mathbf{x})(-1)^{\mathbf{w} \cdot \mathbf{x}} \tag{9}$$

Here, in connection with Walsh transforms, all Boolean functions are considered with values $+1$ and -1 (i.e. $f(\mathbf{x})$ is replaced by $(-1)^{f(\mathbf{x})}$). Then the correlation between f and $L_{\mathbf{w}}$ (cf. [3]) is computed as

$$c(f, L_{\mathbf{w}}) = \frac{F(\mathbf{w})}{2^{n+1}} \tag{10}$$

For the combining functions $f_0(\mathbf{x}, \sigma)$ and $f_1(\mathbf{x}, \sigma)$, $\mathbf{x} \in GF(2)^n$, we distinguish between correlation to linear functions of the form

$$L(\mathbf{x}, \sigma) = \mathbf{w} \cdot \mathbf{x} \tag{11}$$

and

$$L(\mathbf{x}, \sigma) = \mathbf{w} \cdot \mathbf{x} + \sigma \tag{12}$$

For the function f_0 the corresponding correlation coefficients are given by $c_0(\mathbf{w}) = F_0(\mathbf{w}, 0)/2^{n+1}$ and $c_1(\mathbf{w}) = F_0(\mathbf{w}, 1)/2^{n+1}$, where F_0 denotes the Walsh transform of f_0. In order to distinguish between linear functions of the form (11) and (12) we introduce

$$C_0{}^2 = \sum_{\mathbf{w} \in GF(2)^n} c_0(\mathbf{w})^2, \quad C_1{}^2 = \sum_{\mathbf{w} \in GF(2)^n} c_1(\mathbf{w})^2 \tag{13}$$

In a similar way, for the function f_1, we introduce $d_0(\mathbf{w}) = F_1(\mathbf{w}, 0)/2^{n+1}$, $d_1(\mathbf{w}) = F_1(\mathbf{w}, 1)/2^{n+1}$ and

$$D_0{}^2 = \sum_{\mathbf{w} \in GF(2)^n} d_0(\mathbf{w})^2, \quad D_1{}^2 = \sum_{\mathbf{w} \in GF(2)^n} d_1(\mathbf{w})^2 \tag{14}$$

Then by Parseval's theorem

$$C_0{}^2 + C_1{}^2 = 1 \quad \text{and} \quad D_0{}^2 + D_1{}^2 = 1 \tag{15}$$

In this framework we can determine all correlations of the output z_j of the general combiner (7,8) to linear functions of the form

$$s = \sum_{k=j-i}^{j} \sum_{m=1}^{n} w_{mk} x_{mk}. \tag{16}$$

There are $N = 2^{(i+1)n}$ such functions. As a generalization of equation (6) we obtain the following theorem, proved in [4].

Theorem 1 *Let $1 \leq i \leq j$. Then the output digit z_j of the general combiner with one bit memory is correlated to linear functions $s_1, s_2, \ldots, s_N$ of the form (16), and the corresponding correlation coefficients c_h satisfy*

$$\sum_{h=1}^{N} c_h{}^2 = C_0{}^2 + C_1{}^2(1 - (D_1{}^2)^i). \tag{17}$$

Theorem 1 is our main result for general combiners with 1 bit memory and has several implications on the design of stream ciphers using combiners with memory. First observe that the sum of the squares of the correlation coefficients (17) converges to 1 as i tends to ∞, except in the (singular) case $D_0 = 0$, where the limit is $C_0{}^2$.

If the input sequences $\mathcal{X}_m = (x_{m0}, x_{m1}, x_{m2}, \ldots)$, $1 \leq m \leq n$, to a combiner with memory are generated by LFSRs, the correlation of z_j to linear functions of the form (16) leads to correlation to sums of LFSR–sequences. By (16) these sums are given by $s = \sum_{m=1}^{n} (\sum_{k=j-i}^{j} w_{mk} x_{mk})$. Note that for each m the inner sum $s_m = \sum_{k=j-i}^{j} w_{mk} x_{mk}$ is in fact a phase of the m–th LFSR. If certain of these s_m's vanish, a divide and conquer correlation attack is possible. To prevent divide and conquer *maximum order correlation immunity* has been postulated in [6,3]. According to Theorem 1 the combiner is maximum order correlation immune if for every linear function of the form (16) with nonvanishing correlation coefficient, and for every m, $1 \leq m \leq n$, there is at least one index k with $w_{mk} \neq 0$. Note that this coincides with condition MCI as introduced in [3] for memoryless combiners.

Theorem 1 extends Rueppels's treatment of maximum order correlation immunity in [5], as it covers every kind of correlation to LFSR–sequences. Such correlations exist even if the combiner is chosen to be maximum order correlation immune. In fact in the case $D_0 \neq 0$ the "total" correlation is independent of the combiners f_0 and f_1 as expression (17) converges to 1. This generalizes a corresponding result in [3] for memoryless combiners.

Motivated by these results one might be tempted to choose a maximum order correlation immune combiner (7,8) satisfying $D_0 = 0$. However in this case it can be shown that the sequence $z'_j = z_j + z_{j-1}$ is generated even by a memoryless combiner. Hence by a result in [3] z'_j is correlated to LFSR–sequences with correlation coefficients c_i with the property that

$$\sum_i c_i{}^2 = 1$$

Thus one can well achieve that all correlation coefficients in Theorem 1 vanish. However the sequence $\mathcal{Z}'$, which is easily obtained from $\mathcal{Z}$, has correlation to LFSR–sequences, as in the case where $D_0 \neq 0$.

4   Correlation Conditioned on Known Output Sequence

So far correlation was not conditioned on the output of the combiner. A completely different situation results if correlation is conditioned on the events $z_j = 0$ or $z_j = 1$. This is exemplified for the basic summation combiner with two inputs, where knowledge of portions of the output sequence can considerably reduce the uncertainty about the carry bit. This effects correlation of z_j to the input sum $a_j + b_j$, although z_j and $a_j + b_j$ are uncorrelated in the average.

It can be shown that in a run of s consecutive output digits 0 the carries tend to be 1. For example assume that $z_{j+1} = z_{j+2} = \cdots = z_{j+s} = 0$. Then at the end of the run the carry bit σ_{j+s} is 1 with probability at least $1 - 2^{-s}$. More generally for every t with $1 \le t \le s$, the conditional probability satisfies $P(\sigma_{j+s} = \sigma_{j+s-1} = \cdots = \sigma_{j+t} = 1) \ge 1 - 2^{-t}$. Similarly, in a run of s consecutive output digits 1 the carries tend to be 0. As a consequence the following result can be derived (cf. [4]).

Theorem 2 *(1) Suppose that the output of the basic summation combiner satisfies $z_{j+1} = z_{j+2} = \cdots = z_{j+s} = 0$, and $z_{j+s+1} = 1$. Then for every t with $1 \le t \le s$ the following $s - t + 2$ equations*

$$
\begin{aligned}
z_{j+t+1} &= a_{j+t+1} + b_{j+t+1} + 1 = 0 \\
z_{j+t+2} &= a_{j+t+2} + b_{j+t+2} + 1 = 0 \\
&\vdots \\
z_{j+s+1} &= a_{j+s+1} + b_{j+s+1} + 1 = 1 \\
z_{j+s+2} &= a_{j+s+2} + b_{j+s+2} + a_{j+s+1}
\end{aligned}
\tag{18}
$$

are simultaneously satisfied with probability at least $1 - 2^{-t}$.

(2) Suppose that the output of the basic summation combiner satisfies $z_{j+1} = z_{j+2} = \cdots = z_{j+s} = 1$, and $z_{j+s+1} = 0$. Then for every t with $1 \le t \le s$ the following $s - t + 2$ equations

$$
\begin{aligned}
z_{j+t+1} &= a_{j+t+1} + b_{j+t+1} = 1 \\
z_{j+t+2} &= a_{j+t+2} + b_{j+t+2} = 1 \\
&\vdots \\
z_{j+s+1} &= a_{j+s+1} + b_{j+s+1} = 0 \\
z_{j+s+2} &= a_{j+s+2} + b_{j+s+2} + a_{j+s+1}
\end{aligned}
\tag{19}
$$

are simultaneously satisfied with probability at least $1 - 2^{-t}$.

Observe that Theorem 2, stating that the equations (18, 19) are simultaneously satisfied with a certain probability, is much stronger than the statement that these equations are individually satisfied with the same probability. This fact can be cryptanalytically exploited as described in the next section.

5 Cryptanalysis of the Summation Cipher with Two LFSRs

In the basic summation cipher the two input sequences to the adder are produced by LFSRs. Then the systems of equations in Theorem 2 can be cryptanalytically exploited in a known plaintext attack.

Suppose that a run $z_{j+1}, \ldots, z_{j+s}$ of s consecutive 0's or 1's has been observed in the key stream sequence. By considering the digits $z_{j+t+1}, \ldots, z_{j+s+2}$ one obtains $s - t + 2$ equations of the form (18) or (19) which are simultaneously satisfied with probability at least $1 - 2^{-t}$. The actual value of t, which is a parameter for the reliability of the equations, may be chosen depending on the length of known portion of the key stream. Since the digits a_j and b_j in (18, 19) are linearly expressed in terms of the initial state of the two LFSRs, one obtains a system of $s-t+2$ linear equations for the initial digits of the LFSRs. Our aim is to find sufficiently many such systems with highest reliability, which are suitably combined to a system of linear equations for the initial digits.

Let N be the length of the known key stream sequence and k be the key size (which is the sum of the LFSR–lengths). The key stream is scanned for runs of at least s consecutive 0's or 1's. Suppose that a total number of n such runs have been found. According to the desired reliability we choose the parameter t, and we obtain, as described in Theorem 2, a "block" of $d = s-t+2$ equations for each run. Thus we get at least nd equations for the initial digits. We assume that $nd > k$, i.e. $nd = \alpha k$ where $\alpha > 1$. To solve for the key we only need $m = \lceil k/d \rceil \approx \alpha^{-1} n$ "correct" blocks of equations. In order to find m correct blocks proceed as follows.

1. Randomly choose m out of the n available blocks, and solve the resulting system of linear equations for the k unknowns.

2. Test all possible solutions obtained in step 1 whether they produce the correct key stream. If there is a correct solution terminate, else go to step 1.

The complexity of this cryptanalytic algorithm is dominated by the total number of trials. To get an estimate of this number, observe that each block has probability ρ of being incorrect, where $\rho \leq 2^{-t}$. Then the expected number of trials needed is the reciprocal value of the probability q that, by sampling without replacement, m randomly chosen blocks are correct. We estimate this probability in a typical situation where ρn blocks are incorrect. (Assume here for simplicity that ρn is an integer.) Then q is estimated as

$$q = \left(1 - \frac{\rho n}{n}\right)\left(1 - \frac{\rho n}{n-1}\right) \cdots \left(1 - \frac{\rho n}{n-(m-1)}\right) \tag{20}$$

$$> \left(1 - \frac{\rho n}{n-m}\right)^m = \left(1 - \frac{\alpha \rho}{\alpha - 1}\right)^m \tag{21}$$

As an illustration we consider the following example.

Example. Consider a basic summation cipher with two LFSRs of length approximately 200, i.e. $k = 400$. Suppose that we have $N = 50,000$ digits of the key stream sequence. If this sequence is scanned for runs, then in the average

$$n \approx \frac{N}{2^s} \tag{22}$$

runs of length at least s are to be expected (see [1], p. 322 ff.). If we choose $s = 7$ we obtain $n = 390$ runs of length at least 7. Take $t = 4$. Then $d = s - t + 2 = 5$ is the length of a block, and $\rho = 2^{-4} = 1/16$ is the probability of a block being incorrect.

Moreover $m = k/d = 80$ blocks of equations are needed to solve for the key. The value of α is obtained as $\alpha = n/m = 390/80 = 4.88$. Thus

$$q > \left(1 - \frac{4.88}{3.88} \cdot \frac{1}{16}\right)^{80} = 0.0014 \quad \text{and} \quad q^{-1} < 699$$

and therefore less than 700 trials are already sufficient in a typical situation.

This example shows that the summation cipher with two LFSRs can be successfully cryptanalyzed for LFSRs of considerable length with arbitrary feedback connection. Note that our algorithm also works if the known portion of the key stream has some (but not too many) gaps.

6 Comments on the Cryptanalytic Algorithm

The success of our algorithm rests on the weakness of the basic summation combiner as observed in Theorem 2. It is shown in [8] that a similar cryptanalysis is no longer possible for a summation cipher with more than 2 LFSRs. From this point of view it is recommended to take several LFSRs of moderate length rather than just few long LFSRs.

The method of our algorithm can be described in more general terms. Basically the cryptanalytic problem consists in finding a k–bit key. Observing that this key is determined by a number $m = k/d$ of (correct) blocks of equations, we search for $m = k/d$ such blocks instead of the k unknown bits. Since a block is correct with probability at least $1 - 2^{-t} \geq 0.5$, this procedure may be compared with an exhaustive search over only k/d bits instead of k bits. This is similar to the effect of a reduction of the key size by the factor d. However if a block is incorrect it cannot be corrected by complementing like a single bit. Therefore a set S of more than m blocks is required in order to find m correct blocks.

Let n denote the total number of available blocks and ρ the probability of a block being incorrect. Then $n - \rho n$ blocks are expected to be correct. Hence it is necessary that $n - \rho n$ is larger than m. Therefore it is favourable to have ρ small and $\alpha = n/m$ large. In fact already for $\alpha \approx 5$ and ρ only slightly smaller than 0.5, our cryptanalytic algorithm is much faster than an exhaustive search, even if the blocks consist of single bits, i.e. if $d = 1$ and $k = m$.

In the case $d = 1$ our method leads to a procedure to find k correct bits out of a set of n bits, where each bit in the set is assumed to be incorrect with probability ρ. This is exactly the situation one is faced with in the general correlation problem in cryptanalysis. In this direction our method applies to increase the efficiency of a cryptanalytic algorithm described in [2].

Algorithm A in [2] addresses the problem of determining the initial digits of a k–bit LFSR (with few feedback taps) from a disturbed output sequence of the LFSR. The algorithm describes a method to find k digits with highest probability of being undisturbed. These digits are taken as an estimate of the LFSR–sequence at the corresponding positions. Then the correct sequence is found by testing modifications of this estimate. Again denote by ρ the probability of a selected bit to be incorrect. It is shown in [2] that in the average

$$W_0 = 2^{h(\rho)k} \tag{23}$$

trials are necessary, where $h(\rho)$ denotes the binary entropy function.

We can improve algorithm A by applying the method as introduced in Section 5. According to this method we start with a set S of more than k digits having high probability of being undisturbed. Then we randomly choose k digits from this set and test these whether they are correct, i.e. whether they determine the correct LFSR–sequence. This process is repeated until k correct digits have been found.

We express the cardinality of the set S as a multiple of k, i.e. $|S| = \alpha k$ where $\alpha > 1$. According to (21) it is favourable to choose S (or α) as large as possible. On the other hand for increasing cardinality of S the reliability of the selected digits will decrease (cf. [2]). However for moderate α (e.g. $\alpha = 4$ or 5) the error probability ρ turns out to be roughly the same as for $\alpha = 1$. Therefore according to (21), in a typical situation the average number of trials is less than

$$W_1 = \left(1 - \frac{\alpha}{\alpha-1}\rho\right)^{-k} = 2^{-\log_2(1-\frac{\alpha}{\alpha-1}\rho)\,k} \tag{24}$$

For a comparison of the two work factors W_0 and W_1 we may assume that the fraction $\alpha/(\alpha - 1)$ in (24) is close to 1. Thus (24) can be replaced by

$$W_1 = 2^{\ell(\rho)k} \tag{25}$$

where $\ell(\rho) = -\log_2(1 - \rho)$. Formulas (23) and (25) show that both methods have exponential complexity. However the exponent in (25) is smaller than that in (23). In particular, for small ρ the value $\ell(\rho)$ is a small fraction of $h(\rho)$. In fact

$$\lim_{\rho \to 0} \frac{h(\rho)}{\ell(\rho)} = \infty \tag{26}$$

Thus for small ρ the method of Section 5 leads to a substantial improvement of algorithm A, as is also illustrated in the following example.

Example. Consider a LFSR of length k $= 200$ with few feedback taps. Then with the method of algorithm A it is feasible to find e.g. a set S of 1000 digits with error probability lower than 0.1, i.e. with $\alpha = 5$ and $\rho \leq 0.1$. Then, in order to find the LFSR–sequence with a search as in the original algorithm A, formula (23) shows that 2^{94} trials would be necessary in the average. However if the improved algorithm A is applied, the number of trials according to (20) can be estimated as 2^{34}.

In order to find sufficiently many digits with small ρ, it has to be assumed (as in [2]) that the number of feedback taps is small. In fact for LFSRs with more than 10 feedback taps the feasibility of the improved algorithm is roughly limited to the same conditions as the original algorithm A.

References

[1] W. Feller, *An Introduction to Probability Theory and its Applications*, Vol 1, John Wiley & Sons, Inc., 1968.

[2] W. Meier, O. Staffelbach, *Fast Correlation Attacks on Certain Stream Ciphers*, Journal of Cryptology, Vol 1, No. 3, pp. 159–176, 1989.

[3] W. Meier, O. Staffelbach, *Nonlinearity Criteria for Cryptographic Functions*, Proceedings of Eurocrypt'89, Springer-Verlag, to appear.

[4] W. Meier, O. Staffelbach, *Correlation Properties of Combiners with Memory in Stream Ciphers*, full paper to appear in the Journal of Cryptology.

[5] R.A. Rueppel, *Correlation Immunity and the Summation Generator*, Advances in Cryptology—Crypto'85, Proceedings, pp. 260–272, Springer-Verlag, 1986.

[6] R.A. Rueppel, *Analysis and Design of Stream Ciphers*, Springer-Verlag, 1986.

[7] T. Siegenthaler, *Correlation–Immunity of Nonlinear Combining Functions for Cryptographic Applications*, IEEE Trans. Inform. Theory, Vol IT-30, pp. 776–780, 1984.

[8] O. Staffelbach, W. Meier, *Cryptographic Significance of the Carry for Ciphers Based on Integer Addition*, Proceedings of Crypto'90, Springer-Verlag, to appear.

Correlation Functions of Geometric Sequences

AGNES HUI CHAN, MARK GORESKY AND ANDREW KLAPPER

Northeastern University
Boston, Massachusetts 02115

ABSTRACT

This paper considers the cross-correlation function values of a family of binary sequences obtained from finite geometries. These values are shown to depend on the intersection of hyperplanes in a projective space and the cross-correlation function values of the nonlinear feedforward functions used in the construction of the geometric sequences.

1. Introduction

Maximum period linear feedback shift register sequences with nonlinear feedforward functions have been used in modern communication systems. Many of these sequences are required to have high linear complexities, good autocorrelation and/or cross-correlation function values. Recently, Chan and Games [1] introduced a class of binary sequences obtained from finite geometries using nonlinear feedforward function $\rho : GF(q) \rightarrow GF(2)$, with q odd. They showed that these sequences have high linear complexities. Brynielsson [2] had studied similar problem with q even and established the linear complexities of these sequences in terms of the polynomial expression of the function ρ. In this paper, we consider the autocorrelation and cross-correlation functions of these sequences, and establish their values in terms of the autocorrelation and cross-correlation values of the sequence obtained from $(\rho(\beta^0), \rho(\beta), \ldots, \rho(\beta^{q-2}))$, where β is a primitive element of $GF(q)$. In the case where q is even, we show that the autocorrelation and cross-correlation function values are vastly different, making these geometric sequences viable candidates for applications in spread spectrum communications.

2. Geometric Sequences

Let q be a prime power and $GF(q^n)$ be the field of q^n elements. A q-ary m-sequence $\mathbf{R}$ of span n and period $q^n - 1$ can be generated by choosing a primitive polynomial $f(x)$ over $GF(q)$. A binary sequence $\mathbf{S}$ can be obtained from the m-sequence $\mathbf{R}$ for any choice of mapping $\rho : GF(q) \rightarrow GF(2)$ (sometimes called a "nonlinear feedforward function") by defining $S_i = \rho(R_i)$ for all $i \geq 0$. Such a sequence $\mathbf{S}$ is closely related to finite geometry and is called a *binary geometric sequence*.

It is well known that the sequence **R** can be represented as $(Tr(\alpha^0),$
$Tr(\alpha), Tr(\alpha^2), \ldots)$, where α is a root of the primitive polynomial $f(x)$ and $Tr :$
$GF(q^n) \to GF(q)$ is the trace function. Thus

$$S_i = \rho(Tr(\alpha^i))$$

Let $v = (q^n - 1)/(q - 1)$ and $\beta = \alpha^v$. Then β is a primitive element in the base field
$GF(q)$ and $Tr(\alpha^{v+i}) = \beta Tr(\alpha^i)$, so $R_{i+v} = \beta R_i$ for all $i \geq 0$. In [1], Chan and Games
studied the linear complexities of these binary sequences with q odd, and proved the
following result.

THEOREM. *Let* **S** *be a binary sequence obtained from finite geometries with q odd.*
Then

$$linear\ complexity\ (\mathbf{S}) = v * linear\ complexity\ (\mathbf{s}).$$

where $\mathbf{s} = (\rho(\beta^0), \rho(\beta), \ldots, \rho(\beta^{q-1})).$

By choosing ρ appropriately, the linear complexity of $\mathbf{s}$ can be made as high as $q-1$,
and so the linear complexity of **S** can reach $q^n - 1$. In [2] Brynielsson considered the
linear complexities of binary finite geometric sequences with q even, and proved a
similar result:

THEOREM. *Let* $\rho : GF(2^e) \to GF(2)$ *be represented as a polynomial* $\sum_{i=0}^{e-1} A_i x^i$ *over*
$GF(2^e)$. *Then*

$$linear\ complexity\ (\mathbf{S}) = \sum_{A_i \neq 0} n^{|i|}$$

where $|i|$ *denotes the dyadic weight of the integer i (i.e. the weight of the binary*
vector representation of i).

In the latter case, the linear complexity of **S** is maximal if the polynomial representation of ρ has nonzero terms of every degree. Thus, for q even, the linear complexity
of **S** has $\sum_{d=0}^e \binom{e}{d} n^d$ as an upper bound.

In this paper we consider the crosscorrelation of binary geometric sequences **S** and
Z, each of period $q^n - 1$, where $S_i = \rho(Tr(A\alpha^i))$ and $Z_i = \gamma(Tr(B\alpha^i))$, and where
A, B are fixed elements in $GF(q^n)$. Note that **S** and **Z** are geometric sequences with
same linear feedback functions but different nonlinear feedforward functions ρ and γ.
(In a later paper we will consider the crosscorrelation of geometric sequences with
different feedback functions.)

3. Hyperplanes in $GF(q^n)$

The geometric sequences are based on the geometry of hyperplanes in the finite
field $GF(q^n)$. The crosscorrelation of these geometric sequences is calculated by
counting the number of elements in the intersections of two hyperplanes. The use of
intersecting hyperplanes for evaluating crosscorrelation of pseudorandom sequences
was considered by Games in [3] and our method is similar to his. In this section we
review some of the basic facts concerning hyperplanes and their intersections.

Let Tr: $GF(q^n) \to GF(q)$ denote the trace function. For any $U \in GF(q)$ we define

$$H_U = \{x \in GF(q^n) | Tr(x) = U\}$$

Then H_U is an (affine) hyperplane, i.e. it is an $n-1$ dimensional vector subspace of $GF(q^n)$ which does not necessarily pass through the origin. If $V \in GF(q)$ then the hyperplanes H_U and H_V are parallel, i.e. they have no points of intersection unless $U = V$, in which case they are equal. Now let $b \in GF(q^n)$, $V \in GF(q)$, and consider the hyperplane

$$\begin{aligned} b^{-1}H_V &= \{b^{-1}y | y \in H_V\} \\ &= \{b^{-1}y | Tr(y) = V\} \\ &= \{x | Tr(bx) = V\}. \end{aligned}$$

LEMMA 1. *The hyperplanes H_U and $b^{-1}H_V$ are parallel if and only if $b \in GF(q)$.*

PROOF: If $b \in GF(q)$ then

$$\begin{aligned} b^{-1}H_V &= \{x | Tr(bx) = V\} \\ &= \{x | bTr(x) = V\} \\ &= H_{b^{-1}V}. \end{aligned}$$

Since both b and V are in $GF(q)$, $b^{-1}V \in GF(q)$, so $H_{b^{-1}V}$ is parallel to H_U.

On the other hand, if H_U and $b^{-1}H_V$ are parallel, then we must show that $b \in GF(q)$. Let us first consider the special case when $U = 0$ and the two parallel hyperplanes $H_U = H_0$ and $b^{-1}H_V$ actually coincide. Thus,

$$x \in H_0 \quad \text{iff} \quad Tr(x) = 0 \quad \text{iff} \quad Tr(bx) = V.$$

By taking $x = 0$ we see immediately that $V = 0$. Now choose $z \in GF(q^n) - H_0$. Since H_0 is a hyperplane, the addition of this one more linearly independent element will span all of $GF(q^n)$. Therefore bz may be written as a linear combination involving z and H_0,

$$bz = az + h$$

for some $a \in GF(q)$ and $h \in H_0$. We will show that $b = a \in GF(q)$. If this were false, we would have $z = h/(b-a)$. But multiplication by a preserves H_0, and multiplication by b also preserves H_0, so multiplication by $(b-a)$ preserves H_0, and so multiplication by $(b-a)^{-1}$ preserves H_0. Therefore $z \in H_0$, and this is a contradiction.

Next we consider the general case of U arbitrary and H_U not necessarily equal to $b^{-1}H_V$. As above, let $H_0 = \{x | Tr(x) = 0\}$. Then H_U, $b^{-1}H_V$, and H_0 are parallel. Thus there are translations x_1, $x_2 \in GF(q^n)$ such that

$$H_0 = H_U - x_1 = b^{-1}H_V - x_2$$

Define $V' = V - Tr(bx_2)$. Then

$$b^{-1}H_V - x_2 = \{b^{-1}x - x_2 | Tr(x) = V\}$$
$$= \{y | Tr(by + bx_2) = V\}$$
$$= \{y | Tr(by) = V'\}$$
$$= b^{-1}H_{V'}$$

Thus $b^{-1}H_{V'} = H_0$ and the preceding special case applies to this situation, from which we conclude that $b \in GF(q)$.

LEMMA 2. If $b \in GF(q^n) - GF(q)$ then for any $U, V \in GF(q)$, the number of elements in the intersection $H_U \cap b^{-1}H_V$ is precisely q^{n-2}.

PROOF: By lemma 1, the hyperplaces H_U and $b^{-1}H_V$ are not parallel. If two hyperplanes are not parallel, then their intersection is a hyperplane inside each, i.e. it is an $n - 2$ dimensional (affine) subspace of $GF(q^n)$. Therefore it contains q^{n-2} points.

4. Cross-Correlation Functions

In the notation of section 2, we consider a primitive element $\alpha \in GF(q^n)$ and two geometric sequences based on this element,

$$S_i = \rho(Tr(A\alpha^i)), \quad Z_i = \gamma(Tr(B\alpha^i))$$

Recall that the cross-correlation function associated with the sequences $\mathbf{S}$ and $\mathbf{Z}$ is given by:

$$C_{\mathbf{S},\mathbf{Z}}(\tau) = \sum_{t=0}^{q^n-2} (-1)^{S_t}(-1)^{Z_{t+\tau}},$$

where $0 \le \tau \le q^n - 2$. Using the notation $\Phi(\mu) = (-1)^{\rho(\mu)}$ and $\Gamma(\mu) = (-1)^{\gamma(\mu)}$ for $\mu \in GF(q)$, and denoting by $\beta = \alpha^v$ the corresponding primitive element of $GF(q)$, we have the following definitions.

DEFINITION. The *short cross-correlation* function is defined as

$$c_{\rho,\gamma}(m) = \sum_{\mu \in GF(q)} \Phi(\mu)\Gamma(\mu\beta^m).$$

DEFINITION. The *imbalance* of ρ, denoted by $I(\rho)$, is defined by

$$I(\rho) = \sum_{\mu \in GF(q)} (-1)^{\rho(\mu)}.$$

The imbalance of a nonlinear function ρ measures the difference in the number of 0-images and the number of 1-images under the mapping ρ. Let d represent the phase displacement of the two binary sequences $\mathbf{S}$ and $\mathbf{Z}$, that is, $\alpha^d = B/A \in GF(q^n)$, then we prove

THEOREM. *Let* S *and* Z *be two binary geometric sequences of span* n *with period* $q^n - 1$ *as above. Let* d *denote their phase shift and let* $v = (q^n - 1)/(q - 1)$. *Then the cross-correlation function* $C_{S,Z}(\tau)$ *is given by:*

$$C_{S,Z}(\tau) = q^{n-1}c_{\rho,\gamma}(m) - \Phi(0)\Gamma(0) \qquad \text{if } d + \tau = mv$$

and

$$C_{S,Z}(\tau) = q^{n-2}I(\rho)I(\gamma) - \Phi(0)\Gamma(0) \qquad \text{otherwise.}$$

Observe that if q is even then it is possible to choose $\rho : GF(2^e) \to GF(2)$ such that exactly half of the elements in $GF(2^e)$ are mapped to 0 and the other half to 1. Then $C_{S,Z}(\tau) = \Phi(0)\Gamma(0) = \pm 1$ for $d + \tau \neq iv$. However if q is odd then the imbalance is always at least 1, so the crosscorrelation is always greater than or equal to $q^{n-2} - 1$.

PROOF OF THEOREM:

The cross correlation is

$$C_{S,Z}(\tau) = \sum_{t=1}^{p^{n}-1} \Phi(Tr(A\alpha^t))\Gamma(Tr(B\alpha^{t+\tau}))$$

Substituting $x = A\alpha^t$, $B/A = \alpha^d$, and $b = \alpha^{d+\tau}$ we obtain,

$$C_{S,Z}(\tau) = \sum_{x \in GF(q^n)} \Phi(Tr(x))\Gamma(Tr(bx)) - \Phi(Tr(0))\Gamma(Tr(0))$$

To each $x \in GF(q^n)$ there corresponds unique elements $U = Tr(x)$ and $V = Tr(bx)$ in $GF(q)$. Thus the elements of $GF(q^n)$ are divided into disjoint subsets of the form $H_U \cap b^{-1}H_V$, so the above sum may be rewritten as,

$$C_{S,Z}(\tau) = \sum_{U \in GF(q)} \sum_{V \in GF(q)} |H_U \cap b^{-1}H_V|\Phi(U)\Gamma(V) - \Phi(0)\Gamma(0)$$

According to lemma 2, the number of points $|H_U \cap b^{-1}H_V|$ in this intersection is q^{n-2} unless $b \in GF(q)$, i.e. unless $d + \tau$ is a multiple of $v = (q^n - 1)/(q - 1)$. So in the first case we obtain

$$C_{S,Z}(\tau) = q^{n-2}\sum_{U} \Phi(U)\sum_{V} \Gamma(V) - \Phi(0)\Gamma(0)$$
$$= q^{n-2}I(\rho)I(\gamma) - \Phi(0)\Gamma(0)$$

as claimed. In the second case, if $b \in GF(q)$, then $d + \tau$ is some multiple, say m, of $v = (q^n - 1)/(q - 1)$. Thus

$$b = \alpha^{d+\tau} = \beta^m$$

where $\beta = \alpha^v$ is the primitive element of $GF(q)$. As observed above,

$$\beta^{-m} H_V = H_{\beta^{-m}V}$$

which has no points in common with H_U unless $U = \beta^{-m}V$. Therefore the only nonzero terms in the above double sum give

$$C_{S,Z}(\tau) = q^{n-1} \sum_{U \in GF(q)} \Phi(U)\Gamma(\beta^m U) - \Phi(0)\Gamma(0)$$

$$= q^{n-1} c_{\rho,\gamma}(m) - \Phi(0)\Gamma(0)$$

as claimed. ∎

Recall that the autocorrelation function of a sequence S is given by

$$A_S(\tau) = \sum_{t=0}^{q^n-2} (-1)^{S_t}(-1)^{S_{t+\tau}}.$$

To compute the values of $A_S(\tau)$, we simply substitute S with Z in $C_{S,Z}(\tau)$ and obtain the following result.

COROLLARY. *The autocorrelation function of the sequence S is given by:*

$$A_S(\tau) = q^{n-1} c_\rho(m) - 1 \qquad if \ \tau = mv$$

and

$$A_S(\tau) = q^{n-2} I(\rho)^2 - 1 \qquad otherwise.$$

where $c_\rho(m)$ corresponds to the short autocorrelation function, defined as

$$c_\rho(m) = \sum_{mu \in GF(q)} \Phi(\mu)\Phi(\mu\beta^m).$$

5. Absolute Correlation Functions

The notion of "absolute" cross correlation between two pseudorandom sequences with period $q^n - 1$ has also been studied in the literature [3]. The absolute cross correlation counts only the coincident *ones* in the sequences.

DEFINITION. *The absolute cross correlation function between two sequences S and Z is defined as*

$$B_{S,Z}(\tau) = \sum_{t=0}^{q^n-2} S_t Z_{t+\tau}.$$

To consider the absolute cross correlation functions of geometric sequences, the same argument as above works, but we must replace the "short" cross correlation with the "absolute short" cross correlation,

$$a_{\rho,\gamma}(m) = \sum_{\mu \in GF(q)} \rho(\mu)\gamma(\mu\beta^m)$$

and we replace the imbalance $I(\rho)$ by the weight, $W(\rho)$, defined by

$$W(\rho) = \sum_{\mu \in GF(q)} \rho(\mu).$$

Then theorem 1 becomes

THEOREM 1'. *With the same hypotheses as theorem 1, the absolute crosscorrelation function of* S *and* Z *is*

$$B_{S,Z}(\tau) = \begin{cases} q^{n-1}a_{\rho,\gamma}(m) - \rho(0)\gamma(0) & if \quad d+\tau = mv \\ q^{n-2}W(\rho)W(\gamma) - \rho(0)\gamma(0) & otherwise \end{cases}$$

6. Applications

G.M.W. Sequences. In [3], R. Games calculated the crosscorrelation of an m-sequence and a GMW sequence having the same primitive polynomial. His method involved intersecting hyperplanes, and our theorem 1 is similar to his. In this paragraph, we show how to recover his result.

Suppose a, b, and r are integers, with a dividing b, and with r relatively prime to $2^a - 1$. Fix a primitive element $\alpha \in GF(2^b)$. The sequence GMW$(b,a;r)$ is the sequence given by

$$S_i = Tr_1^a(Tr_a^b(\alpha^i)^r).$$

The GMW sequence is a geometric sequence in the sense of §2: take $q = 2^a$, $n = b/a$, and $\rho(\mu) = Tr_1^a(\mu^r)$ for any $\mu \in GF(2^a)$. In the notation of §2 we have

$$S_i = \rho(Tr(\alpha^i)).$$

Similarly the m-sequence

$$Z_i = Tr_1^a(Tr_a^b(\alpha^i)) = Tr_1^b(\alpha^i)$$

is the geometric sequence corresponding to $\gamma(\mu) = Tr_1^a(\mu)$. If we apply theorem 1' to find the absolute crosscorrelation between these two sequences, we obtain

COROLLARY 2 [3]. *Given integers* a, b, *and* r, *with* a *dividing* b *and with* $(r, 2^a - 1) = 1$, *let* S_i *be the sequence* GMW$(b,a;r)$ *and let* Z_i *be the m-sequence based on the same primitive polynomial. Then*

$$B_{S,Z}(\tau) = \begin{cases} (2^a)^{\frac{b}{a}-1}a_{\rho,\gamma}(k) & = 2^{b-a}B_{w,u}(k) \quad if \tau = kv \\ (2^a)^{\frac{b}{a}-2}2^{a-1}2^{a-1} & = 2^{b-2} \quad otherwise \end{cases}$$

where $v = (2^b - 1)/(2^a - 1)$, *where* u *and* w *are the m-sequences of span* a *given by*

$$u_i = Tr_1^a(\beta^i) \qquad w_i = Tr_1^a(\beta^{ir}).$$

with $\beta = \alpha^v$ *a primitive element of* $GF(2^a)$.

We remark that for many values of r, these "short" crosscorrelation values are known, or can be estimated [6] [7] .

Bent Sequences. The method in this paper may be used to calculate crosscorrelation values of Bent Sequences [5], the computation is fairly straightforward and will not be carried out here.

ACKNOWLEDGEMENT

We would like to thank R. Games for reading a first draft of this paper and for making several valuable suggestions.

REFERENCES

1. A. H. Chan and R. A. Games, *On the Linear Span of Binary Sequences from Finite Geometries, q Odd*, Proceedings of Crypto86, page 405–417.
2. L. Brynielsson, *On the Linear Complexity of Combined Shift Register*, Proceedings of Eurocrypt84, page 156–160.
3. R. A. Games, *Crosscoreelation of m-Sequences and GMW- Sequences With the Same Primitive Polynomial*, Discrete Applied Mathematics 12 (1985), pages 139–146.
4. R. A. Games, *The Geometry of m-Sequences: Three-Valued Cross-correlations and Quadrics in Finite Projective Geometry*, SIAM J. Alg. Disc. Mathematics, vol 7 (1986), pages 43–52.
5. J. Olson, R. A. Scholtz and L. R. Welch, *Bent Function Sequences*, IEEE Trans. on Information Theory, vol. IT–28 (1982), pages 858–864.
6. T. Helleseth, *Some Results About the Cross-Correlation Function Between Two Maximal Linear Sequences*, Discrete Math 16 (1976), pages 209–232.
7. D. Sarwate and M. Pursley, *Crosscorrelation Properties of Pseudorandom and Related Sequences*, IEEE Proceedings, vol. 68 (1980), pages 593–619.

EXPONENTIATING FASTER WITH ADDITION CHAINS

Y. Yacobi
Bellcore, 445 South St.
Morristown, NJ 07962
yacov@bellcore.com

The similarity between fast computations with huge numbers and data compression is investigated. For example, in data-compression, frequent messages are assigned short codes, while in fast computations we store and reuse the results of frequent computations. In compression we sometimes send the difference of consecutive messages, if its usually small (Δ modulation), while in fast computation we compute $x^{n+\Delta}=x^n \cdot x^\Delta$, if x^n is already known, and $\Delta \ll n$.

We demonstrate the similarity by applying a modification of the Lempel-Ziv data compression algorithm to fast exponentiation, to result runtime which is comparable to the best known methods, in the practical range (500 –1000 bits), for random exponents. For compressible exponents we gain on the average the compression ratio. This may be applicable to Diffie-Hellman like cryptographic systems. It is an open question whether compressible exponents are safe.

1. Introduction

The similarity between fast computations with huge numbers and data compression is investigated. For example, in data-compression, frequent messages are assigned short codes, while in fast computations we store and reuse the results of frequent computations. In compression we sometimes send the difference of consecutive messages, if its usually small (Δ modulation), while in fast computation we compute $x^{n+\Delta}=x^n \cdot x^\Delta$, if x^n is already known, and $\Delta \ll n$.

We demonstrate the similarity by applying a modification of the Lempel-Ziv data compression algorithm to fast exponentiation, to result runtime which is comparable to the best known methods, in the practical range (500 –1000 bits), for random exponents [Q]. For compressible exponents we gain on the average the compression ratio. This may be applicable to Diffie-Hellman like cryptographic systems. It is an open question whether compressible exponents are safe.

Suppose we are required to compute x^n. We view n as a binary string $<n_L,...n_1,n_0>$, i.e. $n = \sum\limits_{i=0}^{L} n_i \cdot 2^i$. When exponentiating using the binary method [K] there are two kinds of operations, squaring and multiplications. For example, if we want to raise x^{1101} (the exponent is written here in binary) we can compute $x^{1\cdot 2^3} \cdot x^{1\cdot 2^2} \cdot x^{0\cdot 2^1} \cdot x^{1\cdot 2^0} = (((((x)^2 \cdot x)^2) \cdot 1)^2) \cdot x$. In this example we had three squarings and two multiplications. This is a special case of *addition chain*. In general, an addition chain for n ([K] page 444) is a sequence of integers $1=a_0, a_1, \cdots, a_r=n$ with the property that $a_i=a_j+a_k$ for some k and j, $k \leq j < i$, for all $i=1,2,..r$. Binary addition chains as seen in the above example, determine an upper bound on the number of operations (squarings+multiplications) needed for the exponentiation, namely, $l(n) \leq L + v(n) - 1$. Here $l(n)$ is the length of the addition chain, which equals the number of operations, and $v(n)$ is the Hamming weight of n, i.e. the number of 1's in the binary representation of n. Erdös [E] proved that for minimum length addition chains,

$$l(n) = L + L/\log(L) + o(L/\log(L))$$

There are some heuristics for finding addition chains [BC], but generally finding the minimum length addition chain is an NP-hard problem [D], and therefore all practical methods which work for arbitrary exponents, or very long exponents, avoid it, and use a longer addition chain. For our method, we have on the average

$$l(n) = L + (\log(L) - \log\log(L))/2 + 1.5 \cdot L/\log(L)$$

The first two expressions are the number of squarings, and the last expression is the number multiplications. The m-ary addition chain method ([K] page 444, [CL] pp. 110-111) runs in time $l(n) = L + m/2 + L/m + 2^{m-1}$ (the first two terms are squarings, the last two terms are multiplications), where m is an optimization parameter. Asymptotically it is optimal (choose $m=c\log(L)$, such that $c^{-1}=1+\epsilon$, for ϵ as small as you wish). However, in the practical range the average running times of this method and of the new method are comparable.

Both methods use about L squarings, but squarings are usually cheaper than multiplications. This is true not only for the naive schoolboy multiplication method, where squaring is twice as fast as multiplying, but also for all FFT based multiplication methods. To begin with, when squaring one needs only one FFT transformation (and one inverse), while for multiplication two transformations are needed (and one inverse). Also, in the transformed domain multiplication is pairwise, which means that squaring is broken into many small squarings. The size of squaring table is the square root of the size of multiplication table. This means that for the same space complexity for some machines we can gain another factor of 2 for squaring. Altogether squaring may be 3 times faster than multiplying for FFT algorithms, in some machines. In some other applications the situation may be reversed. For example, for elliptic curve arithmetic it seems that squaring is more expensive than multiplication.

We can use a very similar method to multiply large numbers.

2. The Method

We first explain the m-ary method, and introduce few of our computational improvements. This may help following the description of the new algorithm.

2.1 The m-ary Method

Let m be an optimization parameter. Given base X, compute the set S of exponents $\{X^{e_i}\}$, for all odd e_i of length m bits. To compute X^n, write the binary representation of n as follows. $n = e_k 0^{Z_k}, e_{k-1} 0^{Z_{k-1}}, \ldots, e_1 0^{Z_1}$, Here 0^{Z_i} means a run of Z_i zeroes. For all i $Z_i \geq 0$. The most significant segment, e_k, may be shorter than m bits. The computation of X^n is accomplished by

$$X^n = (\ldots((X^{e_k})^{2^{Z_k + |e_{k-1}|}} \cdot X^{e_{k-1}})^{2^{Z_{k-1} + |e_{k-2}|}} \cdot \ldots \cdot X^{e_1})^{2^{Z_1}}.$$

To compute the set S we can use the following ideas, which we later use in our method. Create a complete half binary tree, such that the root has only one son, with arc labeled "1", and all the rest of the nodes have two sons each. The tree is of depth m, and each node is associated with the path which leads from the root to that node in the natural way. Hence each leaf corresponds to one of the e_i's. Our aim is to store in each leaf the value of "its" X^{e_i}. To do so we start from the root and store in each node the value of its $X^{e'_i}$ ($e'_i < m$ for non leaf nodes). Suppose that we completed this task for level (depth) i. The computation of level $i+1$ is done as follows. Rights sons (arcs labeled 0) are copied from their fathers (a most significant new bit, which equals 0 does not change the result). A left son is computed by $X^{2^i} \cdot father$. So each left son requires one multiplication, and the whole level requires one squaring, since $X^{2^i} = (X^{2^{i-1}})^2$.

Altogether we get m squarings, and 2^{m-1} multiplications, to create the tree. Using the tree requires L/m multiplications and $L - m/2$ squarings (on the average $|e_k| = m/2$).

The new method is very similar to the m-ary method. The difference is that we compute in the above tree only those e_i's which actually appear in the exponent, and we don't truncate the tree to a prespecified depth.

2.2 The New Method

The intuition behind the method is that if some patterns in the binary representation of the exponent reappear, then we don't have to recompute their contribution, if we already stored it. The method is similar to the m-ary method, with two differences. First, in the m-ary method we may do some precomputations, which are not used later, while in the new method we precompute exactly those subexponents that we are about to use. Second, we do not bound our precomputations to a fixed predetermined length. Rather our algorithm is adaptive to the particular exponent at hand. A similar intuition led to the Lempel-Ziv [LZ] compression algorithm.

We parse n right to left (i.e., from low order to high order bits) the way Lempel and Ziv [LZ] do in their compression method, i.e. we create a binary "compression" tree, where the path e_i from the root to node i is a segment of the exponent, and node i contains the partial result x^{e_i}. The structure of the tree makes it easy to add a new leaf, given its parent. We describe first a subroutine that creates the binary tree, and then the main program, which uses the tree to compute x^n.

The exponent is called here seq, and is treated sometimes as an integer, and sometimes as a sequence of bits. Each segment e_i of the exponent is an odd number of length $L_i \geq 1$, and is preceded (to the right) by $Z_i \geq 0$ zeroes.

subroutine **build-tree**
   begin
- **Init:** Store $e_0 = 1$; $Z_0 \leftarrow L_{-1} \leftarrow 0$ in node #0, and set $L_0 \leftarrow 0$. put parse # 0 right of seq; $i \leftarrow 1$;
- **While** there are more symbols in seq **do**
- **begin**
  - $L_i \leftarrow 0$; $Z_i \leftarrow 0$.
  - Scan seq from the $i-1_{th}$ parse to the left.
  - While the new symbol (ns) $= 0$ $Z_i \leftarrow Z_i + 1$.
  - (ns=1) $L_i \leftarrow L_i + 1$; start following the path of the tree defined by seq , incrementing L_i, until a leaf is reached.
  - Scan one more symbol (ns), add a new arc from the last visited leaf, label it ns, add a new leaf and a new parse, and label them i. e_i is the integer represented by the segment which starts at parse $i-1$ and ends at parse i, not including leading zeroes to the right.
  - Compute x^{e_i}, and store it together with $L_{i-1} = |e_{i-1}|, e_{i-1}$, and Z_i in leaf i.
  - $i \leftarrow i + 1$;
- **end;**
end.

Let $k = max(i)$. As in the m-ary method, the exponent looks as follows.
$$n = e_k 0^{z_k} e_{k-1} 0^{z_{k-1}} \dots e_1 0^{z_1}$$

Algorithm 1:

begin

— Call **build-tree** subroutine.

— $x^n \leftarrow ((\dots((x^{e_k})^{2^{z_k+L_{k-1}}} \cdot x^{e_{k-1}})^{2^{z_{k-1}+L_{k-2}}} \cdot x^{e_{k-2}}\dots)^{2^{z_2+L_1}} \cdot x^{e_1})^{2^{z_1}}$

end.

Remark: e_{i-1} is stored in "node e_i" as a back pointer, later used by algorithm 1 to determine the right sequencing of the partial results.

The method is general, and applies to any multiplicative group.
A numerical example appears in the appendix.

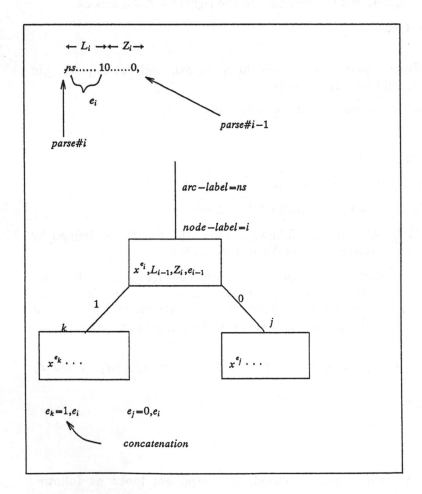

fig 1. The general structure of the tree

3. Complexity

3.1 Time complexity

We first estimate the tree size for a random exponent. Then we use it to count the number of operations needed in the entire computation. On the average we expect to get balanced trees (applying the Lempel-Ziv algorithm to random sequences result balanced trees).

Let h be the depth of the tree, k the number of nodes in the tree (maybe one less than the k of line 2 algorithm 1, if $e_k = e_j$ for some $j < k$), and, as before, $L = log(n)$ the length of the exponent.

Lemma 1: on the average $h = log(k)$, and $L = k \cdot log(k)$.

Proof: Our tree is half a binary tree (and we assume that on the average that half is complete. This assumption is true for random exponent). Hence at depth i we have 2^{i-1} nodes. The path to each node at depth i covers on the average (including leading zeroes) a segment of length $i+1$ of the exponent (because the probability of a run of exactly j zeroes in a random string is $2^{-(j+1)}$, hence the expected length of runs of zeroes is $\sum_{j=1}^{\infty} j \cdot 2^{-(j+1)} = 1$). So,

$$L = \sum_{i=1}^{h} (i+1) \cdot 2^{i-1}, \text{ and by simple induction on } h \text{ we get } L = h \cdot 2^h.$$

Since a full binary tree of depth h has $2^{h+1} - 1$ nodes, our tree has $k = 2^h$ nodes. So, $h = log_2 k$, and $L = k \cdot log_2(k)$.

Q.E.D.

Corollary: $k = L / log(L) + o(L / log(L))$. (i.e. $\lim\limits_{L \to \infty} \dfrac{k}{L / log(L)} = 1$.)

Lemma 2: The average time complexity of Algorithm 1 is

$$l(n) = L - (log(L) - loglog(L))/2 + 1.5 \cdot (L / log(L) + o(L / log(L)).$$

Proof: There are four types of operations involved in the above computation. Squarings and multiplications during the construction of the tree, denoted TS and TM respectively, and exponentiations and multiplications which appear in line (2) of algorithm 1., denoted S and M, respectively. #S=$L - h/2$, this follows from the fact that the last segment e_k is of expected length $h/2$. $h = log(k)$, hence by the corollary of Lemma 1, $h \sim log(L) - loglog(L)$.

#M=#nodes in the tree= k.

As for the cost of building the tree, note that for each depth i, we have to compute once x^{2^i}, which is done using one squaring, then we apply it on the average to half the nodes using $x^{(1,z)} = x^{2^{|z|}} \cdot x^z$. Here $(1,z)$ denotes 1 concatenated to the left of z. The other half of the nodes we get for free, since $x^{(0,z)} = x^z$. So, #TM=$\frac{1}{2} \cdot k$ on the average $(=v(n))$, and #TS=h. We conclude that the cost of our method is,

$$l(n) = \#S + \#M + \#TM + \#TS = (L - h/2) + k + k/2 + h = l + h/2 + 1.5 \cdot k =$$
$$L + (log(L) - loglog(L))/2 + 1.5 \cdot (L/log(L) + o(L/log(L))).$$

Q.E.D.

Asymptotically the m-ary method is superior. It achieves the lower bound. However, for a practical exponent size like 512, or 1024 the new method is comparable for random exponents, and superior for compressible exponents. On the average the gain in the number of multiplications is the compression ratio. However, for individual compressible sequences it may differ from that ratio. For compressible exponents the tree is always skewed. If it is skewed to the left the gain is smaller than the compression ratio, while if it is skewed to the right the gain is greater than the compression ratio.

Open problem: What is the complexity of the discrete log problem, given that the exponent is compressible (with a given compression ratio)?

If the fact that the exponent is compressible does not affect the complexity of discrete-log then more efficient Diffie-Hellman like key distribution systems can be designed.

3.2 Space complexity

The tree has $k = L/log(L)$ nodes. If the exponentiation is modular, as is the case in modern cryptography, then each node stores $L = log(n)$ bits. All together we have space complexity $L^2/log(L)$. (We assumed here that the modulus and the exponent are of the same size.)

Acknowledgments

I would like to thank Shimon Even, Rich Graveman, Stuart Haber, and Arjen Lenstra for many helpful discussions.

4. References

[BC] J. Bos, M. Coster: "Addition Chain Heuristics", *Crypto'89*.

[CL] H. Cohen, and A.K. Lenstra: "Implementation of a New Primality Test", *Math. of Computation* , Vol. 48, No. 177, Jan. 1987, pp.103-121.

[D] Downey, Leony, and Sethi: "Computing Sequences With Addition Chains", *SIAM Jour, Comput.* 3, (1981), pp.638-696.

[E] P. Erdös: "Remarks on Number Theory III, on Addition Chains", *Acta Arith. VI* (1960), pp.77-81.

[GKP] Graham, Knuth and Patashnik: " Concrete Mathematics, a Foundation for Computer Science", *Addison Wesley*, 1988.

[LZ] Lempel and Ziv: "Compression of Individual Sequences via Variable Rate Coding" *IEEE Trans of Inf. Theory* , IT-24 No. 5 (September 1978) pp.530-536.

[K] Knuth: "The Art of Computer Programming, Vol. 2, Seminumerical Algorithms", *Addison-Wesley*, 1980, pp.441-462.

[Q] J.J. Quisquater: (A talk on m-ary exponentiation at the Rump session of Eurocrypt'90)

[S] Schönhage: "A Lower Bound on The Length of Addition Chains", *Theoretical Computer Science* Vol. 1 (1975), pp.229-242.

5. Appendix

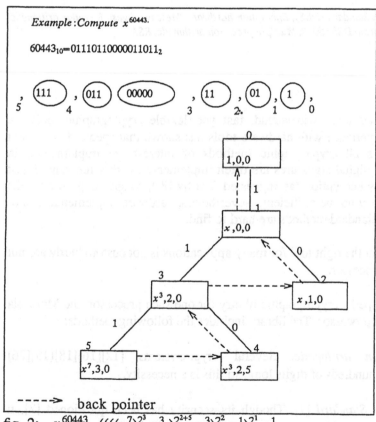

fig. 2: $x^{60443} = ((((x^7)^{2^3} \cdot x^{3\cdot})^{2^{2+5}} \cdot x^3)^{2^2} \cdot x^1)^{2^1} \cdot x^1$

A Cryptographic Library for the Motorola DSP56000

Stephen R. Dussé
Burton S. Kaliski Jr.
RSA Data Security Inc.
Redwood City, CA

Abstract. *We describe a cryptographic library for the Motorola DSP56000 that provides hardware speed yet software flexibility. The library includes modular arithmetic, DES, message digest and other methods. Of particular interest is an algorithm for modular multiplication that interleaves multiplication with Montgomery modular reduction to give a very fast implementation of RSA.*

Key words. *Data Encryption Standard (DES), Encryption hardware, Message digest, Modular arithmetic, Montgomery reduction, Motorola DSP56000, Multiple-precision arithmetic, RSA.*

1. Introduction

As cryptography becomes more widespread, fast yet flexible cryptographic tools are becoming important. Experience with hardware tools has shown that speed often cannot fully be realized unless all cryptographic methods of interest are implemented in hardware. For example, digital signatures are often implemented with a message digest followed by a public key encryption (as suggested first by [8]), so speeding up only the public key encryption may not be sufficient. Nevertheless, hardware implementations of many important yet nonstandard methods are hard to find.

We therefore propose that the right tool for many applications is not custom hardware but a fast general-purpose processor.

We have recently developed a cryptographic library for one such processor, the Motorola DSP56000 digital signal processor. The library includes the following methods:

Multiple-precision arithmetic. Several cryptosystems [12][16][18][19][26] involve integers hundreds of digits long, so this is a necessity.

Data Encryption Standard [7]. Though its security has been questioned [2], it remains an important tool.

Message digest. This operation is essential to almost every signature scheme. Flexibility is important as there is no widely accepted, secure, standard message digest; our choices include FIPS 113 MAC [6] and RSA-MD2, both of which were proposed for Internet electronic mail [22]. We are also considering RSA-MD4 [25].

In evaluating various general-purpose processors we found that the DSP56000 is especially well-suited because it can multiply two 24-bit integers and add the product to a 56-bit integer in 100 ns [14]. Such an operation is important not only in digital signal processing but also in multiple-precision arithmetic. The 24-bit word size also matches the 48-bit round keys of DES nicely. However, we expect that most of our results can be applied on other general-purpose processors.

This paper is organized as follows. We begin by describing our algorithms for RSA and DES. Then we present the design of a "crypto-accelerator card" for the IBM PC. Finally, we summarize the performance of the cryptographic library.

2. Related work

Work that motivated ours is Barrett's, Wiener's and Davio *et al*'s. Barrett observed the effectiveness of digital signal processors for cryptography and presented an implementation of RSA on Texas Instruments' TMS32010 [1]. Wiener developed a general software implementation of RSA on the DSP56000 that achieves 10.2K bits/s for 512-bit modular exponentiation with the Chinese remainder theorem [4]. An implementation specific to 512-bit moduli is even faster [30]. Davio *et al* made considerable progress in efficient techniques for DES [9], some of which we apply in our implementation.

Among other recent work on fast cryptography are Buell and Ward's implementation of multiple-precision arithmetic on a a Cray computer [5] and Laurichesse's fast implementations of RSA on conventional processors [21]. A number of fast hardware implementations can be found in Brickell's 1989 survey [4].

Currently the record for the fastest implementation of RSA is held by Shand, Bertin and Vuillemin of Digital Equipment Corporation's Paris Research Laboratories, who have achieved 226K bits/s for 508-bit modular exponentiation with the Chinese remainder theorem [29].

3. RSA algorithm

We now describe our implementation of RSA on the Motorola DSP56000. This section addresses the algorithms; performance is dealt with in Sec. 6.

In the RSA cryptosystem [26] one performs *modular exponentations*: computations of the form $C = M^E \bmod N$ where C, M, E and N are multiple-precision integers. This computation is central to several other cryptosystems [12][16][18][19] so our results apply to those as well.

Speeding modular exponentiation has been of interest for some time, and there are a number of speedups [3][20][24][27]. We focus on one particular aspect, the integration of multiple-precision multiplication with modular reduction according to Montgomery's method [23]. Our speedup is complementary to others that focus on reducing the number of multiplications and reductions so ours and the others can be applied concurrently.

Our algorithm is most effective on a processor on which multiplication is fast relative to shifting, for then the convolution-sum approach described below outperforms the conventional shift-and-add method. We believe our algorithm will result in some speed improvement on every processor, but given that it is a little more complicated than conventional methods, the algorithm may not be justified on all processors.

3.1 Montgomery's method

We now outline Montgomery's method for modular arithmetic. Readers familiar with the topic may skip to Sec. 3.2.

In Montgomery's method we represent residue classes in an unusual way and redefine modular arithmetic within this representation. Specifically, let N be an integer (the modulus) and let R be an integer relatively prime to N. We represent the residue class A mod N as AR mod N and redefine modular multiplication as

$$\text{MONTGOMERY-PRODUCT}(A,B,N,R) = ABR^{-1} \bmod N$$

It is not hard to verify that Montgomery multiplication in the new representation is isomorphic to modular multiplication in the ordinary one:

$$\text{MONTGOMERY-PRODUCT}(AR \bmod N, BR \bmod N, N, R) = (AB)R \bmod N$$

We can similarly redefine modular exponentiation as repeated Montgomery multiplication. This "Montgomery exponentiation" can be computed with all the usual modular exponentiation speedups. To compute ordinary modular exponentiation $C = M^E$ mod N, we compute $M' = MR$ mod N (ordinary modular reduction), $C' = (M')^E R^{1-E}$ mod N (Montgomery exponentiation), and $C = C'R^{-1}$ mod N (Montgomery reduction).

The practicality of Montgomery's method rests on the following nice theorem, which leads directly to an algorithm for Montgomery multiplication.

Theorem 1 (Montgomery, 1985)

Let N and R be relatively-prime integers, and let $N' = -N^{-1} \bmod R$. Then for all integers T, $(T+MN)/R$ is an integer satisfying

$$(T+MN)/R \equiv TR^{-1} \bmod N \tag{1}$$

where $M = TN' \bmod R$.

Proof Equation 1 is straightforward. The fact that $(T+MN)/R$ is an integer can be shown by substituting M. ■

If we choose the right R—say, a power of the base in which we represent multiple-precision integers—then division by R and reduction modulo R are trivial. With such an R Montgomery reduction is no more expensive than two multiple-precision products, and we can make it even easier.

3.2 Computing the Montgomery product

We now describe our algorithm for the Montgomery product. For the discussion we will let b be the base in which multiple-precision integers are represented. That is, we will represent an integer A as a sequence of digits $\langle a_0,\dots,a_{n-1}\rangle$ where

$$A = a_{n-1}b^{n-1} + a_{n-2}b^{n-2} + \cdots + a_1 b + a_0 \tag{2}$$

We will further require that all inputs to our algorithms can be represented in n base b digits, and that $R = b^n$. In Sec. 3.3 we determine limitations on the individual digits a_0, ..., a_{n-1}.

We derive our algorithm by successive improvements, beginning with the following algorithm taken directly from Theorem 1. (We note that our algorithm does not "normalize" its output to the range $[0,N-1]$. Sec. 3.3 shows why.)

MONTGOMERY-PRODUCT(A,B,N,R)
1     $N' \leftarrow -N^{-1} \bmod R$
2     $T \leftarrow AB$
3     $M \leftarrow TN' \bmod R$
4     $T \leftarrow T+MN$
5   **return** T/R

Improvement 1. Instead of computing all of M at once, let us compute one digit m_i at a time, add $m_i N r^i$ to T, and repeat. The resulting T may not be the same as in the original algorithm but the effect of adding multiples of N will be: namely, to make T a multiple of R. This is essentially the approach Montgomery gives for multiple-precision integers. We note that this change allows us to compute $n_0' = N^{-1} \bmod b$ instead of N'.

MONTGOMERY-PRODUCT$'(A,B,N,R)$
1     $n_0' \leftarrow -n_0^{-1} \bmod b$
2     $T \leftarrow AB$
3   for $i \leftarrow 0$ to $n-1$
4   do $m_i \leftarrow t_i n_0' \bmod b$
5                   $T \leftarrow T + m_i N b^i$
6   return T/R

Improvement 2. Now let us interleave multiplication and reduction. We note that Montgomery reduction is intrinsically a right-to-left procedure. That is, m_i depends only on t_i. So we can begin adding this multiple to T as soon as we know t_i. This results in the following algorithm:

MONTGOMERY-PRODUCT$''(A,B,N,R)$
1     $n_0' \leftarrow -n_0^{-1} \bmod b$
2     $T \leftarrow 0$
3   for $i \leftarrow 0$ to $n-1$
4   do $T \leftarrow T + a_i B b^i$
5                   $m_i \leftarrow t_i n_0' \bmod b$
6                   $T \leftarrow T + m_i N b^i$
7   return T/R

Improvment 3. At this point we can begin to observe a potential difficulty for the DSP56000. The operation $T \leftarrow T + a_i B b^i$—the basic shift-and-add operation—is likely to break down into the following single-precision operations:

4.1   do $x \leftarrow t_i$
4.2   for $j \leftarrow 0$ to $n-1$
4.3   do $x \leftarrow x + a_i b_j$
4.4 $t_{i+j} \leftarrow x \bmod b$
4.5 $x \leftarrow x/b$ - right shift
4.6 $t_{i+n} \leftarrow x$ - (initial $t_{i+n} = 0$)

These operations involve not only n single-precision multiplications but also n right shifts. On many processors the "high part" and the "low part" of accumulators are separately addressable and the right shift can be accomplished with move instructions. This is true also on the DSP56000, but such shifting takes longer than a multiplication on the DSP56000, which motivates us to minimize the number of right shifts. Happily, there is a good way to avoid right shifts, and that is with the convolution-sum method of multiplication. In this method instead of performing operations like $T \leftarrow T + a_i B b^i$, we perform operations like $T \leftarrow T + (\sum_{0 \le i \le k} a_i b_{k-i}) b^k$. These involve $k+1$ multiplications but only one shift. The fact that Montgomery reduction is intrinsically right-to-left helps us again, and leads to our final algorithm.

MONTGOMERY-PRODUCT$'''(A,B,N,R)$
1     $n_0' \leftarrow -n_0^{-1} \bmod b$

2      $T \leftarrow 0$

3   **for** $k \leftarrow 0$ **to** $n-1$

4   **do** $T \leftarrow T + (\sum_{0 \leq i \leq k} a_i b_{k-i}) b^k + (\sum_{0 \leq i \leq k-1} m_i n_{k-i}) b^k$

5            $m_k \leftarrow t_k n_0' \bmod b$

6            $T \leftarrow T + m_k n_0 b^k$

7   **for** $k \leftarrow n$ **to** $2n-1$

8   **do** $T \leftarrow T + (\sum_{k-n+1 \leq i \leq n-1} a_i b_{k-i}) b^k + (\sum_{k-n+1 \leq i \leq n-1} m_i n_{k-i}) b^k$

9   **return** T/R

We expect that our final algorithm will generally be faster than the interleaved shift-and-add version on most processors, because our algorithm has fewer right shifts, $O(n)$ versus $O(n^2)$. It also has fewer stores, again $O(n)$ versus $O(n^2)$. (The number of other operations—fecthes, multiplies, and adds—is essentially the same for both algorithms.) However, we note that our algorithm has more complex loop control. We also note that our algorithm accumulates intermediate results that are a factor of $2n$ larger in magnitude than those in the shift-and-add algorithm, so we need a larger accumulator and addition instructions that can handle the larger accumulator.

On most processors we can implement the larger accumulator with multiple registers and the additions involving it with add-with-carry instructions. The DSP56000 is especially well suited since its accumulator is eight bits longer than the largest product its ALU can produce. Thus even without multiple registers or add-with-carry instructions the DSP56000 can handle the intermediate results for n up to 128.

The extent to which our algorithm is faster depends mostly on the relative speeds of multiplication and shifting. If multiplication is relatively slow then changes in the number of shifts will have an insignificant effect on total execution time. For example, on the Intel 80386 multiplication is an order of magnitude slower than shifting and we have observed what appears to be at best a 10 percent improvement in execution time. But on the DSP56000 the improvement is manyfold.

We conclude with a couple of remarks. First, we can derive a Montgomery squaring algorithm MONTGOMERY-SQUARE(A,N,R) in the usual way that is asymptotically 25 percent faster than the alternative MONTGOMERY-PRODUCT(A,A,N,R).

Second, we can precompute $n_0' = -n_0^{-1} \bmod b$ once during a Montgomery exponentiation since it depends only on the modulus N. Computing n_0' by a general modular inverse algorithm such as extended Euclidean GCD would not be all that expensive, since b is small. We have found instead (or rediscovered?) a very nice way to compute the modular inverse in the special case that n_0 is odd and b is a power of 2:

MODULAR-INVERSE(x,b)

1   - Computes $x^{-1} \bmod b$ for x odd and b a power of 2.

2      $y_1 \leftarrow 1$

3   **for** $i \leftarrow 2$ **to** $\lg b$

4   **do**   **if** $x y_{i-1} < 2^{i-1} \bmod 2^i$

```
5                          then $y_i \leftarrow y_{i-1}$
6                          else $y_i \leftarrow y_{i-1} + 2^{i-1}$
7        return $y_{\lg b}$
```

The correctness of MODULAR-INVERSE can be verified by induction with the hypothesis $xy_i \equiv 1 \bmod 2^i$.

3.3 Representation of multiple-precision integers

We have not yet defined "base b representation" for the DSP56000, so we do so now. The DSP56000 has a signed multiply instruction that multiplies two 24-bit two's-complement integers and adds their product to a 56-bit accumulator. Thus the logical choices for "base b representation" are a sequence of 24-bit signed digits and a sequence of 23-bit unsigned digits. A sequence of 24-bit *unsigned* digits is rather awkward with a signed multiply instruction. Given that the 23-bit unsigned representation of an integer would generally be longer than the 24-bit signed representation, we chose the signed representation.

Thus the digits a_i in Eq. 2 satisfy $-2^{23} \leq a_i \leq 2^{23}-1$.

We now prove our claim that MONTGOMERY-PRODUCT need not adjust its result to the range $[0,N-1]$ by showing that the redundant range $[-N,N-1]$ can be maintained through all intermediate calculations.

Theorem 2

Let $R = b^n$ where $b, n > 0$, and let A, B and N be n-digit, multiple-precision integers in the signed representation, where $N > 0$. If A and B are in the range $[-N,N-1]$ then for all n-digit multiple-precision integers M, $(AB+MN)/R$ is in the range $[-N,N-1]$.

Proof We begin by proving two identities:

$$N + M < R$$
$$-N + M > -R$$

The first follows from the observation that the largest positive n-digit integer in the signed representation is less than $R/2$. The second follows from the fact that the largest positive n-digit integer and the smallest negative n-digit integer differ by less than R.

The theorem follows, since

$$(AB+MN)/R \leq (N+M)N/R < N$$
$$(AB+MN)/R > (-N+M)N/R > -N$$

∎

We note that a similar property holds in the unsigned representation, but it requires the further condition that $N < R/4$.

4. DES algorithm

Our implementation of DES follows the paper of Davio *et. al.* [9]. We first recall the definition of DES, then describe how we implemented the improvement.

By way of review, DES consists of 16 nonlinear *rounds* that transform a 64-bit block according to a 48-bit round key. The 16 round keys are computed from a 56-bit key according to a DES key schedule that we do not describe further. The bits in the 64-bit block are permuted in a fixed way before the first round and after the last. This is summarized in the following algorithm:

DES(M,K)
  - Encrypts message M under key K with DES.
  $\langle K_1,...,K_{16} \rangle \leftarrow$ DES-KEY-SCHEDULE(K)
  $\langle L,R \rangle \leftarrow IP(M)$
  for $i \leftarrow 1$ **to** 16
    do $\langle L,R \rangle \leftarrow$ DES-ROUND($\langle L,R \rangle,K_i$)
  return $IP^{-1}(\langle R,L \rangle)$

DES-ROUND($\langle L,R \rangle,K$)
  $\langle x_1,...,x_8 \rangle \leftarrow E(R) \oplus K$
  for $i \leftarrow 1$ **to** 8
    do $y_i \leftarrow S_i(x_i)$
  return $\langle R,L \oplus P(\langle y_1,...,y_8 \rangle) \rangle$

Here E is a linear 32-to-48-bit mapping, P is a 32-bit permutation, and S_1, ..., S_8 are nonlinear six-to-four-bit mappings (the "S boxes"). IP is a 64-bit permutation and IP^{-1} is its inverse.

The primary difficulty with a direct implementation of DES is the expense of applying E and P. Davio *et al* observed that the linearity of E and P makes it possible to remove E and P entirely from the DES round. We can do this by modifying the S boxes to incorporate the permutation P that would follow in one round and the mapping E that would follow in the next. That is, we define S boxes S_1', ..., S_8' as

$$S_1'(x) = E(P(\langle S_1(x),0,...,0 \rangle))$$
$$S_2'(x) = E(P(\langle 0,S_2(x),0,...,0 \rangle))$$

$$\cdot$$
$$\cdot$$
$$\cdot$$

$$S_8'(x) = E(P(\langle 0,...,0,S_8(x) \rangle))$$

We also change the main algorithm to apply E at the beginning and E^{-1} at the end. The mapping E^{-1} can be any 48-to-32-bit mapping be any that satisfies $E^{-1}(E(X)) = X$ for all X. This leads to our algorithm.

DES'(M,K)
  - Encrypts message M under key K with DES.
  $\langle K_1,...,K_{16} \rangle \leftarrow$ DES-KEY-SCHEDULE(K)
  $\langle L,R \rangle \leftarrow IP(M)$
  $L' \leftarrow E(L)$
  $R' \leftarrow E(R)$
  for $i \leftarrow 1$ to 16
    do $\langle L',R' \rangle \leftarrow$ DES-ROUND'($\langle L',R' \rangle, K_i$)
  $L \leftarrow E^{-1}(L')$
  $R \leftarrow E^{-1}(R')$
  return $IP^{-1}(\langle R,L \rangle)$

DES-ROUND'($\langle R,L \rangle,K$)
  $\langle x_1,...,x_8 \rangle \leftarrow R \oplus K$
  for $i \leftarrow 1$ to 8
    do $y_i \leftarrow S_i'(x_i)$
  return $\langle R,L \oplus \langle y_1,...,y_8 \rangle \rangle$

Davio *et al* observed that this algorithm would be especially good on a processor with 48-bit words. We almost have this on the DSP56000, which can fetch two 24-bit words in one instruction. We note finally that speedups such as Davio *et al*'s have been adapted on many types of processor [10][17] and are quite common.

5. A "crypto-accelerator card" for the IBM PC

We now describe a working deployment of the DSP56000: our crypto-accelerator card. The card is a 3/4-length card occupying one expansion slot of an IBM PC, XT, AT or compatible. It uses the +5, +12, and -12 volt DC power supplies from the PC. The card is comprised of four major components: processor, optional DES chip, PC interface, and noise circuit. The total cost to us for the card is $400.

5.1 Processor

At the heart of the card is Motorola's XSP56001ZL20, a RAM-based member of the DSP56000. It is clocked at 20 MHz. The DSP consists of an arithmetic logic unit, address generation unit, and a program controller as well as internal program and data memory. There is extensive I/O support including a dedicated host interface port, an external memory/peripheral port, and two serial ports.

For our design, the DSP is attached to two banks of external 24-bit memory. The P program memory bank can hold 8K or 32K of RAM or ROM and the X data memory bank can hold 8K or 32K of RAM. An external power-up reset circuit holds the DSP in a reset state until the PC activates the DSP. This avoids the execution of spurious code upon power-up which could damage the DSP [15].

5.2 DES chip

The optional DES chip is Western Digital's WD20C03. It is clocked at 10 MHz. It can perform the ECB and CBC modes of DES [11]. The DES chip is attached to the DSP external memory/peripheral port and is mapped into the Y data memory bank's external I/O space. It transmits and receives data under program control of the DSP.

5.3 PC interface

The card communicates with the host PC via a multi-function interface. The interface has three main components: DSP external memory interface, direct DSP interface, and control and status.

DSP external memory interface. The DSP memory words are 24 bits wide and do not directly map to the PC's 8-bit memory space. For this reason, we designed special-purpose external memory support with these features:

- a *bank select* that specifies DSP external memory bank (P or X)

- an *address generation unit (AGU)* with a PC-loadable counter that indexes through the bytes of a 24-bit DSP word and then from word to word

- a *load mode* that indicates whether the address generation unit is to index DSP words in low-medium-high byte order or low byte only

- external memory *bus request/bus grant signals* programmable by the PC

To take advantage of fast block move instructions on the PC, the memory interface responds to any address in a large range in the PC's memory. Nevertheless, the AGU and not the particular PC memory address determines which DSP external memory byte is selected.

Direct DSP interface. The PC accesses the DSP56000's internal registers through the DSP's bidirectional host interface port. The port is mapped to eight locations in the PC's I/O address space. The port allows direct control of DSP software functions and gives access to software status and completion flags.

Control and status registers. Control and status is achieved through a series of latches and registers. These are mapped to eight locations in the PC's I/O address space. The registers support such functions as the loading of the AGU, DSP bus request/bus grant, location of the PC reserved memory, and DSP chip reset.

5.4 Noise circuit

The noise circuit consists of a noise diode, some amplification and an analog-to-digital converter. The noise diode is a CND6002A diode from Koep Precision Standards which produces a minimum noise output of 7 μv per root Hz over the range 10 Hz–100 kHz. Two op-amp stages amplify the noise and a precision comparator with TTL output converts the result into bits. The circuit is connected to the DSP and the status registers.

Since the primary purpose of the noise circuit is to generate random cryptographic keys, special care is taken to decouple the noise circuit from other computer components. Coupling would not only introduce unwanted noise but might also make the output of the noise circuit predictable.

6. Performance

We now summarize the performance of the cryptographic library.

Our RSA implementation on the DSP56000, as expected, achieves hardware speed: 11.6K bits/s for 512-bit exponentiation with the Chinese remainder theorem and 4.6K bits/s without. (This is for full or "private key" exponentiation, i.e., where the exponent is the same length as the modulus; "public key" exponentiation is faster.) The implementation is more than an order of magnitude faster than our software on a fast PC. We do not lose much performance in the overhead of interfacing the crypto-accelerator card to a PC.

The DES implementation runs at 350K bits/s in CBC mode for large blocks. ECB mode is a little faster. This is comparable to our software on a fast PC. In this case the interfacing overhead affects performance somewhat, such that it is faster for a fast PC to perform DES in software for blocks less than about 600 bytes than to interface to the crypto-accelerator card. A card with the optional DES chip runs at 3.8M bits/s and is faster than PC software for almost any block size.

Our RSA-MD2[1] implementation performs much like DES: for small blocks, fast PC software would be faster. The implementation achieves 190K bits/s, which is twice as fast as our software on a fast PC for large blocks.

---

[1]To be accurate, we have only implemented RSA-MD1, a proprietary message digest algorithm almost identical to RSA-MD2. We expect to include RSA-MD2 in later versions of the library.

We note that the lack of significant speedup in RSA-MD2 does not mean that our efforts to accelerate cryptography have failed. Although we have not made those algorithms "as fast as hardware," we have offloaded them from the PC, leaving the PC to perform such operations as I/O in parallel with the cryptography. We also note that there is room for improvement in all results. One factor significantly affecting performance is the limitation on internal memory. We do not store any of our program in internal memory, though we put some data there, so we expect a fair improvement just as a result of copying procedures to internal memory before executing them. There are other improvements as well.

Our results are summarized in Figs. 1–3 where we compare performances across the following hardware/software configurations:

- DSP56000 (the one stated in Sec. 5.1 with zero-wait-state memory): software alone, or with optional DES chip (the one stated in Sec. 5.1)

- fast PC (20 MHz Intel 80386): software alone, with crypto-accelerator card, or with crypto-accelerator card and optional DES chip

- slow PC (6 MHz Intel 80286): software alone, with crypto-accelerator card, or with crypto-accelerator card and optional DES chip

Our PC software is admittedly not the fastest in the world; a better baseline for comparison is probably Laurichesse's software [21].

7. Conclusion

We have described a flexible and fast cryptographic tool based on the Motorola DSP56000. Among the techniques we used are an algorithm for modular multiplication that interleaves multiplication with Montgomery modular reduction to give a very fast implementation of RSA, and the 48-bit model of DES due to Davio *et al*.

Since the time we began developing our cryptographic library, Motorola started producing a 27 MHz version of the DSP56000 [13]. Consequently our implementations can potentially be made 35 percent faster with no further investment on our part.

One of the areas of further interest is how fast RSA can be implemented on other digital signal processors. If we make the assumption that the speed of modular multiplication (Montgomery or otherwise) is proportional in speed to one FIR filter tap and to the square of the word size, we find that the processors likely to provide the best performance are AT&T's DSP16A (40M taps/s, 256 bits^2) and TI's TMS320C50 (29M taps/s, 256 bits^2) [28]. The DSP56000 measures in at 13M taps/s and 576 bits^2. We will probably explore the DSP16A and the TMS320C50 next.

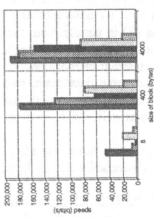

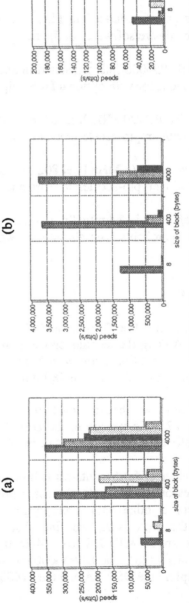

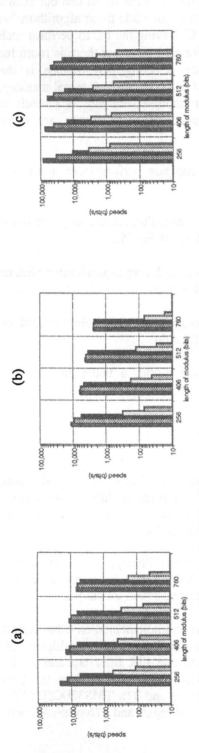

Figure 1 (a) Modular exponentiation speeds with the Chinese remainder theorem. (b) Modular exponentiation speeds without the Chinese remainder theorem. (c) Modular exponentiation speeds for 15-bit exponents. From left to right, bars represent: DSP56000 alone; fast PC with DSP; slow PC with DSP; fast PC alone; and slow PC alone.

Figure 2 (a) DES speeds in CBC mode without DES chip. From left to right, bars represent: DSP56000 alone; fast PC with DSP; slow PC with DSP; fast PC alone; slow PC alone. (b) DES speeds with DES chip. Bars represent: DSP56000 with DES chip; fast PC with DSP and DES chip; and slow PC with DSP and DES chip.

Figure 3 RSA-MD2 speeds. From left to right, bars represent: DSP56000 alone; fast PC with DSP; slow PC with DSP; fast PC alone; and slow PC alone.

Acknowledgements

We would like to thank our colleague Jeff Thompson for assisting in the implementation of the crypto-accelerator card and for preparing the timings described in Sec. 6. We also thank Jim Bidzos, Tom Knight, Ron Rivest, Ralph Sweitzer and Michael Wiener for their contributions.

References

[1]   Paul Barrett. Implementing the Rivest Shamir and Adleman public key encryption algorithm on a standard digital signal processor. In A.M. Odlyzko, editor, *Advances in Cryptology - CRYPTO '86 Proceedings*, volume 263 of *Lecture Notes in Computer Science*, pages 311–323. Springer-Verlag, 1987.

[2]   Eli Biham and Adi Shamir. Differential analysis of DES-like cryptosystems (preprint). *Proceedings of CRYPTO '90* (Santa Barbara, CA, August 12–15, 1990), to appear.

[3]   Jurjen Bos and Matthijs Coster. Addition chain heuristics. In G. Brassard, editor, *Advances in Cryptology - CRYPTO '89 Proceedings*, volume 435 of *Lecture Notes in Computer Science*, pages 400–407. Springer-Verlag, 1990.

[4]   Ernest F. Brickell. A survey of hardware implementations of RSA. In G. Brassard, editor, *Advances in Cryptology - CRYPTO '89 Proceedings*, volume 435 of *Lecture Notes in Computer Science*, pages 368–370. Springer-Verlag, 1990.

[5]   Duncan A. Buell and Robert L. Ward. A multiprecise integer arithmetic package. *The Journal of Supercomputing*, 3:89–107, 1989.

[6]   Computer data authentication. Federal Information Processing Standards Publication 113, National Bureau of Standards, U.S. Department of Commerce, 1985.

[7]   Data encryption standard. Federal Information Processing Standards Publication 46-1, National Bureau of Standards, U.S. Department of Commerce, 1977.

[8]   D.W. Davies and W.L. Price. The application of digital signatures based on public-key cryptosystems. In *Proceedings of the Fifth International Computer Communications Conference*, pages 525–530, 1980.

[9]   M. Davio, Y. Desmedt, M. Fosseprez, R. Govaerts, J. Hulsbosch, P. Neutjens, P. Piret, J.-J. Quisquater, J. Vandewalle and P. Wouters. Analytical characteristics of the DES. In D. Chaum, editor, *Advances in Cryptology: Proceedings of Crypto '83*, pages 171–202. Plenum Press, 1984.

[10]   Marc Davio, Yvo Desmedt, Jo Goubert, Frank Hoornaert and Jean-Jacques Quisquater. Efficient hardware and software implementations for the DES. In G.R. Blakley and D. Chaum, editors, *Advances in Cryptology: Proceedings of CRYPTO 84*, volume 196 of *Lecture Notes in Computer Science*, pages 144–146. Springer-Verlag, 1985.

[11]   DES modes of operation. Federal Information Processing Standards Publication 81, National Bureau of Standards, U.S. Department of Commerce, 1980.

[12]   W. Diffie and M.E. Hellman. New directions in cryptography. *IEEE Transactions on Information Theory*, IT-22(6):644–654, 1976.

[13]   *Digital Signal Processors Quarter 3, 1989*. Motorola, 1989.

[14]   *DSP56000/DSP56001 Digital Signal Processor User's Manual*. Motorola, 1990.

[15]   *DSP56001 Advance Information*. Motorola, 1988.

[16]   T. ElGamal. A public key cryptosystem and a signature scheme based on discrete logarithms. *IEEE Transactions on Information Theory*, IT-31:469–472, 1985.

[17]   David C. Feldmeier and Philip R. Karn. UNIX password security - ten years later. In G. Brassard, editor, *Advances in Cryptology - CRYPTO '89 Proceedings*, volume 435 of *Lecture Notes in Computer Science*, pages 44–63. Springer-Verlag, 1990.

[18]   A. Fiat and A. Shamir. How to prove yourself: Practical solutions to identification and signature problems. In A.M. Odlyzko, editor, *Advances in Cryptology - CRYPTO '86 Proceedings*, volume 263 of *Lecture Notes in Computer Science*, pages 186–194. Springer-Verlag, 1987.

[19]   L.S. Guillou and J.-J. Quisquater. A practical zero-knowledge protocol fitted to security microprocessor minimizing both transmission and memory. In C.G. Gunther, editor, *Advances in Cryptology - EUROCRYPT '88 Proceedings*, volume 330 of *Lecture Notes in Computer Science*, pages 123–128. Springer-Verlag, 1988.

[20]   Donald E. Knuth. *Seminumerical algorithms*, volume 2 of *The Art of Computer Programming*. Addison-Wesley, second edition, 1981.

[21]   Denis Laurichesse. Mise en oeuvre optimisee du chiffre RSA. Rapport Laas No. 90052, Laboratoire d'Automatique et D'Analyse des Systemes, 1990.

[22]   John Linn. Privacy enhancement for Internet electronic mail: Part III: Algorithms, modes, and identifiers. RFC 1115, Internet Activities Board Privacy Task Force, 1989.

[23]   Peter L. Montgomery. Modular multiplication without trial division. *Mathematics of Computation*, 44(170):519–521, 1985.

[24]   J.-J. Quisquater and C. Couvreur. Fast decipherment algorithms for RSA public-key cryptosystem. *Electronics Letters*, 18(21):905–907, 1982.

[25]   Ronald L. Rivest. The MD4 message digest algorithm (preprint). *Proceedings of CRYPTO '90* (Santa Barbara, CA, August 12–15, 1990), to appear.

[26]   Ronald L. Rivest, Adi Shamir and Leonard M. Adleman. A method for obtaining digital signatures and public-key cryptosystems. *Communications of the ACM*, 21(2):120–126, 1978.

[27]   A. Selby and C. Mitchell. Algorithms for software implementations of RSA. *IEE Proceedings*, 136 part E(3):166–170, 1989.

[28]   Michael K. Stauffer and Michael Slater. General-purpose digital signal processors. *Microprocessor Report*, 3(10):25–29, 1989.

[29]   M. Shand, P. Bertin and J. Vuillemin. Hardware speedups in long integer multiplication. *Proceedings of the Second ACM Symposium on Parallel Algorithms and Architectures* (Crete, July 2–6, 1990), to appear.

[30]   Michael Wiener. Personal communication, 1990.

VICTOR
an efficient RSA hardware implementation

Holger Orup
Erik Svendsen
Erik Andreasen

Computer Science Department, Aarhus University
Ny Munkegade 116
DK-8000 Aarhus C, DENMARK
e-mail: orup@daimi.aau.dk

Abstract

The latest improvements of RSA chips are based on progress in implementation technology and strategy i.e. smaller circuits and higher clock frequencies. There has been no improvements in efficiency of the algorithms. The efficiency is here defined as the number of bits produced pr. 1000 clock cycles.

We present algorithms which improve the efficiency by 300%–400%. The main strategy is multiple bit scan and parallel execution of two multiplications. Using these algorithms and the presented hardware architecture a bit rate greater than 90 Kbit/sec. can be achieved encrypting 512 bit blocks.

1   Introduction

Several implementations or suggestions of how to implement the RSA protocol in hardware have been presented in the past. Brickell made an overview of existing RSA chips. The three implementations with the highest bit rate, when the length of an encryption block is 512 bits, are shown in table 1 [Bri89] [EMI88].

We have defined the efficiency as the number of bits produced pr. 1000 clock cycles. Note that the efficiency of the three implementations are approximately the same, and the difference in bit rate is due to the difference in clock speed. The efficiency as defined here is a performance measure of the algorithm used. On the other hand, the clock rate is a rough performance measure of the technology and methods used for realizing the algorithm in hardware.

	Ω	Bit rate pr. 512 bits	efficiency
Thorn EMI	24 MHz	29.0 K	1.21
AT & T	15 MHz	19.0 K	1.27
Cryptech	14 MHz	17.0 K	1.21

Table 1: *The three fastest RSA chips to date*

Apparently there has been no development of more efficient algorithms suited for hardware implementation. In the following, we will present algorithms for exponentiation and multiplication which result in a higher efficiency than the above mentioned.

2 Exponentiation

The main operation in the RSA protocol is $M^E \bmod N$, where the length of each operand is at least 500 bits. Therefore it is essential to have an efficient exponentiation algorithm. The most commonly used algorithm is named *Russian Peasant* [Knu69]. Below is shown a variant in which E is read from the least significant bit. The i'th bit of E is denoted e_i.

```
Algorithm:    Modulo exponentiation.
Stimulation:  E, M, N, where E ≥ 0 and 0 ≤ M < N.
Response:     X = M^E mod N
Method:       i := 0
              X := 1;
              WHILE i < n DO
                  IF e_i = 1 THEN X := (X · M) mod N END;
                  M := (M · M) mod N;
                  i := i + 1;
              END;
```

Algorithm 1: *Variant of Russian Peasant for modulo exponentiation*

If we denote the length of M, E and N by n the time complexity is:

$$T[\mathrm{Exp}, n] = \frac{3}{2} n T[\mathrm{Mult}, n]$$

Assuming it is possible to perform *two* multiplications in parallel, this is indeed possible because the three statements in the loop do not depend on each other, the complexity is:

$$T[\mathrm{Exp}, n] = n T[\mathrm{Mult}, n]$$

This gives a 33% time reduction compared with the variant with one multiplication unit.

3 Multiplication

To implement the exponentiation algorithm mentioned above, we need an efficient way to perform modulo multiplication. Several algorithms have been presented [HDVG88] [Bar86] [ORSP86] [Bla83]. None of them are able to carry out a multiplication with fewer than n full additions of n-bit words.

The usual way of multiplying is by scanning the multiplier one bit at a time and conditionally accumulating the multiplicand parallelly. Assume we scan the multiplier k bits at a time, corresponding to base 2^k, we can express the serial-parallel multiplication scheme as in algorithm 2. In this algorithm S is the accumulator, n' is the number of

$$
\begin{aligned}
&S := 0;\ i := n' - 1; \\
&\text{WHILE } i \geq 0 \text{ DO} \\
&\qquad S := (2^k S + a_i B) \bmod N; \\
&\qquad i := i - 1; \\
&\text{END};
\end{aligned}
$$

Algorithm 2: *Serial-parallel multiplication with integrated modulo reduction*

digits in base 2^k, a_i is digit number i of the multiplier, B the multiplicand and N the modulus. The multiplier is scanned from the most significant digit.

The modulo reduction can be carried out by subtracting N from S until $S \in [0; N[$. The maximal number of subtractions will be $2^k + 2^k - 2$, because $a_i \in [0; 2^k - 1]$ and $B \in [0; N]$. Even though the number of subtractions is limited, this method is rather slow. Instead we can estimate the quotient, S div N, belonging to $[0; 2^{k+1} - 2]$ and carry out the reduction in *one* subtraction. This is shown in algorithm 3. Note that this method

$$
\begin{aligned}
&S := 0;\ i := n' - 1; \\
&\text{WHILE } i \geq 0 \text{ DO} \\
&\qquad q := \text{estimate}(S \text{ div } N); \\
&\qquad S := 2^k S + a_i B - 2^k q N; \\
&\qquad i := i - 1; \\
&\text{END}; \\
&\text{Correction of } S;
\end{aligned}
$$

Algorithm 3: *Modulo multiplication with quotient estimation*

is only feasible if we are able to generate the products $a_i B$ and qN rapidly. We could for example precalculate all the possible values of $a_i B$ and qN and save them in a table. According to Barrett [Bar86], the quotient estimate can be found by multiplying a few of the most significant bits of the dividend S and the reciprocal of the divisor N. We assume that the necessary amount of bits of $\frac{1}{N}$ is part of the input to the chip. If q is to have an accuracy of x bits, then by using $x + 2$ bits from S and $\frac{1}{N}$ we get an estimate which at the most is one less than the exact quotient.

In algorithm 3 the accumulator S does not necessarily belong to the interval $[0; N[$ after each iteration. This does not matter, as long as S belongs to the same residue as $S \bmod N$, and S does not diverge. But after the loop S has to be corrected by subtracting N until S belongs to the correct interval. It is proven in [OS90] that $S \in [0; (3 \cdot 2^k - 1)N]$ after each iteration. The range of q is therefore $[0; 3 \cdot 2^k - 1]$. As we shall see later, we can construct hardware that generates qN efficiently if the range of q belongs to $[0; 42]$, this means the scan factor k is limited to 3.

We are able to reduce the range of q, the idea is as follows: if we estimate $\frac{S}{2}$ div N instead of S div N, the range of the quotient is apparently halved. The modulo reduction is performed by subtracting $2 \cdot 2^k qN$ instead of $2^k qN$. However, the accuracy of the quotient estimate is hereby reduced, implying an increase of the range of S. A closer analysis [OSA90] shows that if we estimate $\frac{S}{2^r}$ div N we get a minimal range of q when $k \leq r$: $[0; 2^{k+1}]$. This means that the scan factor can be increased to 4. The final algorithm for modulo multiplication is shown as algorithm 4. Note that the final corrections can be made by iterating two extra times while setting $a_i = 0$ and further more assuming $r = k$.

Algorithm: Modulo multiplication.

Stimulation: $A = a_{n'-1}a_{n'-2}\cdots a_0 a_{-1}a_{-2}$,
where $a_i \in [0; 2^k - 1]$, $a_{-1} = a_{-2} = 0$ and $n' = \lceil \frac{n+1}{k} \rceil$;
B, where $B \in [0; 2N[$;
N, where $N \in]2^{n-1}; 2^n[$;
k, where $k \geq 3$;
r, where $r = k$.

Response: S div $2^{2k} \equiv_N AB$ and
S div $2^{2k} \in [0; 2N[$.

Method: $S := 0$; $i := n' - 1$;
WHILE $i \geq -2$ DO
       $q :=$ estimate(S div $2^r, N$);
       $S := 2^k S + a_i B - 2^{k+r} qN$;
       $i := i - 1$;
END;

Algorithm 4: *Modulo multiplication*

The final result is read from S discarding the $2k$ least significant bits. Note that this result belongs to the interval $[0; 2N[$. A further reduction is not necessary. When the exponentiation algorithm terminates, the result will also belong to $[0; 2N[$, here a reduction is necessary. This reduction is easily carried out while outputting the result serially. The correctness of the algorithm is proven in [OS90]. The time complexity is:

$$T[\text{Mult}, n] = (\lceil \frac{n+1}{k} \rceil + 2)T[\text{loop}]$$

In the rest of this paper we will describe how to perform the central operation of the loop: $S := 2^k S + a_i B - 2^{k+r} qN$. To do this we have to take a closer look at the hardware architecture of the multiplication unit.

3.1 Hardware architecture

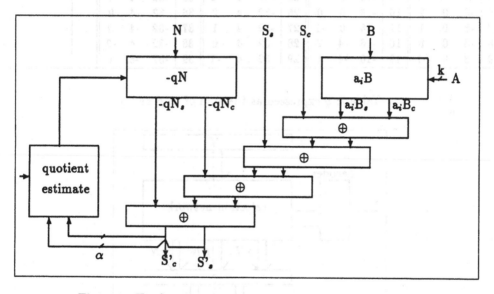

Figure 5: Hardware architecture of the multiplication unit

The multiplication unit consists of circuits for generating the values $-qN$ and $a_i B$. Each circuit returns the result represented in two words, i.e. the value $a_i B$ is represented as $a_i B_s$ and $a_i B_c$, where $a_i B_s + a_i B_c = a_i B$. Similarly the accumulator S is represented in two words S_s and S_c. The main task of the loop is now to add six words together and represent the sum as two words S'_s and S'_c. Using the carry save addition technique this is easily done with four rows of fulladders as shown in figure 5. The critical path of the multiplication unit is calculating the quotient estimate, generating $-qN$ followed by the delay of two fulladders. To be able to use the parallel version of the exponentiation algorithm we have to perform two multiplications i parallel. This can be achieved by pipelining the circuit and adding an extra A register. It is not necessary to duplicate the B register since the two multiplications always have a common operand. See algorithm 1.

3.2 Generating $a_i B$ and $-qN$

To compute $-qN$ we again use the carry save technique. Observe that all numbers in $[0; 42]$ can be expressed as a sum of three powers of two. Table 2 shows which values, α,

β and γ, are needed to compute $-qN = \alpha N + \beta N + \gamma N$. The values $\alpha N, \beta N$ and γN are generated through a selection network, and added through a row of fulladders as shown in figure 6. Here again the result is represented in two words $-qN_c$ and $-qN_s$.

q	α	β	γ	q	α	β	γ	q	α	β	γ	q	α	β	γ	q	α	β	γ
0	0	0	0	10	-16	4	2	20	-16	-4	0	30	-32	0	2	40	-32	-8	0
1	0	0	-1	11	-16	4	1	21	-16	-4	-1	31	-32	0	1	41	-32	-8	-1
2	0	-4	2	12	-16	4	0	22	-32	8	2	32	-32	0	0	42	-32	-8	-2
3	0	-4	1	13	-16	4	-1	23	-32	8	1	33	-32	0	-1				
4	0	-4	0	14	-16	0	2	24	-32	8	0	34	-32	0	-2				
5	0	-4	-1	15	-16	0	1	25	-32	8	-1	35	-32	-4	1				
6	-8	0	2	16	-16	0	0	26	-32	4	2	36	-32	-4	0				
7	-8	0	1	17	-16	0	-1	27	-32	4	1	37	-32	-4	-1				
8	-8	0	0	18	-16	-4	2	28	-32	4	0	38	-32	-4	-2				
9	-8	0	-1	19	-16	-4	1	29	-32	4	-1	39	-32	-8	1				

Table 2: q expressed as the sum of α, β and γ

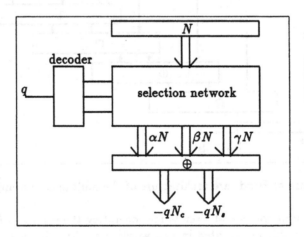

Figure 6: Unit for generating $-qN$

The computation of $a_i B$ is performed following the same principle.

3.3 Quotient estimation

The quotient estimate q is calculated by adding the δ most significant bits of S'_c and S'_s and then multiplying the sum with the ε most significant bits of $\frac{1}{N}$. The quotient can then be found by discarding the $\delta + \varepsilon - (k+2)$ least significant bits of the product, where k is the scan factor.

Earlier we have given an upper bound of $q = 2^{k+1}$, and this restricted $k \le 4$. In [OS90] we have investigated the interdependency of $q, \delta, \varepsilon, k$ and found an expression that gives a smaller upper bound for q:

$$q_{max} = \frac{2^k(2^{k+3-\delta} + 1 + 2^{1-k})}{1 - 2^{k-\varepsilon}}$$

Table 3 shows that we can achieve an upper bound equal to 42 with $k = 5$ by selecting $\delta = 10$ and $\varepsilon = 11$. This scan factor is optimal for the presented hardware architecture because the maximal value of a_i will exceed 42 if k is greater than 5. Simulations indicate

k	δ	ε	$\lfloor q_{max} \rfloor$
4	8	8	27
4	8	9	26
4	9	9	22
4	9	10	22
5	9	9	51
5	9	10	50
5	10	10	43
5	10	11	42

Table 3: Upper bounds for the quotient estimate

that an even lesser upper bound for q can be found, which means that δ and ε can be reduced, giving a simpler circuit.

4 Performance

The critical path of the multiplication unit has been designed in a 2μ process. Simulations show that a loop in the multiplication unit takes less than 85 ns. In a pipelined version each loop takes two clock cycles, thereby giving a clock period less than 50 ns., corresponding to a clock frequency of more than 20 MHz. The layout shows high regularity and the area is estimated at approximately 100 mm^2.

The efficiency of the algorithms is:

$$\frac{n \cdot 1000\text{bits}}{2 \cdot n(\frac{n+1}{k} + 2)\text{cycles}}$$

For $n = 512$ we achieve an efficiency of 3.8 for $k = 4$ and 4.8 for $k = 5$. The bit rates for a clock frequency of 20 MHz are 78 Kbit/sec and 97 Kbit/sec respectively.

5 Conclusion

We have presented a way to speed up a well known exponentiation algorithm by performing two multiplications in parallel, and we have shown how these multiplications can be performed efficiently using multiple bit scan. Further more we have developed a highly regular hardware architecture, based on the carry save addition technique, implementing the multiplication algorithm with a scan factor of up to 5.

Currently a prototype is being developed at the Computer Science Department, Aarhus University.

References

[Bar86]   Paul Barrett. Implementing the Riverst Shamir and Adleman public key encryption system on a standard digital signal processor. In *Advances in Cryptology - CRYPTO '86*, pages 311–323, 1986.

[Bla83]   G.R. Blakely. A Computer Algorithm for Calculating the Product AB Modulo M. *IEEE Trans. Computers*, C-32:497–500, 1983.

[Bri89]   Ernest F. Brickell. A Survey of Hardware Implementations of RSA. In *CRYPTO '89*, 1989.

[EMI88]   THORN EMI. RSA Evaluation Board. Technical Report 10, Thorn EMI Central Reasearch Laboratories, 1988.

[HDVG88] Frank Hoornaert, Marc Decroos, Joos Vandewalle, and René Govaerts. Fast RSA-Hardware: Dream or Reality ? In *Advances in Cryptology - EURO-CRYPT '88*, pages 257–264, 1988.

[Knu69]   Donald E. Knuth. *The Art of Computer Programming - Seminumerical Algorithms*, volume 2. Addison-Westley, 1969.

[ORSP86] G.A. Orton, M.P. Roy, P.A. Scott, and L.E. Peppard. VLSI implementation of public-key encryption algorithms. In *Advances in Cryptology - CRYPTO '86*, pages 277–301, 1986.

[OS90]   Holger Orup and Erik Svendsen. VICTOR. Forbedringer og videreudviklinger af VICTOR - en integreret kreds til understøttelse af RSA-kryptosystemer. Computer Science Department of Aarhus University - Internal report, 1990.

[OSA90]   Holger Orup, Erik Svendsen, and Erik Andreasen. VICTOR - Teoretiske og eksperimentelle undersøgelser af algoritmer til understøttelse af RSA-kryptosystemer med henblik på VLSI design. Master's thesis, Computer Science Department of Aarhus University, 1990.

Experimental Quantum Cryptography

Charles H. Bennett   François Bessette[†]   Gilles Brassard[§]

IBM Research[*]   *Université de Montréal*[‡]   *Université de Montréal*[‡]

Louis Salvail   John Smolin[¶]

Université de Montréal[‡]   *UCLA*[**]

Abstract

We describe initial results from an apparatus and protocol designed to implement *quantum public key distribution*, by which two users, who share no secret information initially: 1) exchange a random quantum transmission, consisting of very faint flashes of polarized light; 2) by subsequent public discussion of the sent and received versions of this transmission estimate the extent of eavesdropping that might have taken place on it, and finally 3) if this estimate is small enough, can distill from the sent and received versions a smaller body of shared random information (key), which is certifiably secret in the sense that any third party's expected information on it is an exponentially small fraction of one bit. Because the system depends on the uncertainty principle of quantum physics, instead of usual mathematical assumptions such as the difficulty of factoring, it remains secure against an adversary with unlimited computing power.

[*] Yorktown Heights, New York, NY 10598, USA.
[†] Supported in part by an NSERC Postgraduate Scholarship.
[‡] Département IRO, Université de Montréal, C.P. 6128, succursale "A", Montréal (Québec), Canada H3C 3J7.
[§] Supported in part by NSERC under grant A4107.
[¶] This work was performed while this author was visiting IBM Research.
[**] Physics Department, University of California at Los Angeles, Los Angeles, CA 90024, USA.

1 Introduction and History

Quantum cryptography has recently entered the experimental era. The first convincingly successful quantum exchange took place in October 1989. After a short historical review of quantum cryptography, we report on the new apparatus and the results obtained with it. These results extend the first report on the apparatus, which appeared in *SIGACT News* [4].

Quantum cryptography was born in the late sixties when Stephen Wiesner wrote "Conjugate Coding". Unfortunately, this highly innovative paper was unpublished at the time and it went mostly unnoticed. There Wiesner explained how quantum physics could be used in principle to produce bank notes that would be impossible to counterfeit and how to implement what he called a "multiplexing channel", a notion strikingly similar to what Rabin was to put forward more than ten years later under the name of "oblivious transfer" (in our opinion, it would be fair to give at least equal credit to Wiesner for the concept of oblivious transfer).

Fortunately, Charles H. Bennett knew Wiesner quite well and heard about his idea from the horse's mouth. Nevertheless, it was only when he met Gilles Brassard that quantum cryptography was revived. This happened on the occasion of the 20th IEEE Symposium on the Foundations of Computer Science, held in Puerto Rico in October 1979. Following our discussion of Wiesner's idea, we discovered how to incorporate the (almost new at the time) notion of public key cryptography, resulting in a CRYPTO '82 paper [6]. This brought Wiesner's paper back to life, and it was subsequently published in *SIGACT News* [17], together with a selection of papers from the earlier CRYPTO '81 workshop (for which "real" proceedings were not published).

Initially, quantum cryptography was thought of by everyone (including ourselves) mostly as a work of science-fiction because the technology required to implement it was out of reach (for instance, quantum bank notes [6] require the ability to store a single polarized photon or spin-1/2 particle for days without significant absorption or loss of polarization). Unfortunately, the impact of the CRYPTO '82 conference had left most people under the impression that everything having to do with quantum cryptography was doomed from the start to being unrealistic.

The main breakthrough came when Bennett and Brassard realized that photons were never meant to *store* information, but rather to *transmit* it (although it should be said that half of Wiesner's original paper dealt precisely with the use of quantum physics for the transmission of information). This lead initially to the *self-winding reusable one-time pad* [5] which was still not very practical. Later, Bennett thought of the *quantum public key distribution channel* and Brassard designed the somewhat less realistic *quantum coin-tossing protocol* [1, 2]. Quantum cryptography was also picked up by other researchers. For instance, Crépeau and Kilian showed how the quantum channel could be used to implement oblivious transfer in a strong way (Wiesner's original multiplexing channel could leak information on both channels), zero-knowledge protocols, and secure two-party computation [12, 11].

The principle of quantum cryptography has been described in major popular magazines such as *Scientific American* [15], *The Economist* [14], and *New Scientist* [13]. Also, Brickell and Odlyzko close their very thorough survey of recent (1988) results in cryptanalysis with these words: "If such systems [quantum cryptography] become feasible, the cryptanalytic tools discussed here [in their paper] will be of no use" [10].

In this paper, we report on the first experimental quantum public key distribution channel ever designed and actually put together. Although we assume that the reader is already familiar with the principles of quantum cryptography, the following section should provide sufficient background. (A good description of the quantum channel itself can be found in chapter 6 of [9].)

2 Quantum Public Key Distribution

The purpose of public key distribution is for two users "Alice" and "Bob", who share no secret information initially, to agree on a random key, which remains secret from an adversary "Eve", who eavesdrops on their communications. In conventional cryptography and information theory it is taken for granted that digital communications can always be passively monitored, so that the eavesdropper learns their entire contents, without the sender or receiver being aware that any eavesdropping has taken place. By contrast, when digital information is encoded in non-orthogonal states of an elementary quantum system, such as single photons with polarization directions 0, 45, 90 and 135 degrees, one obtains a communications channel with the property that its transmissions cannot in principle be reliably read or copied by an eavesdropper ignorant of certain key information used in forming the transmission. The eavesdropper cannot even gain partial information about such a transmission without disturbing it in a random and uncontrollable way likely to be detected by the channel's legitimate users.

The protocol we describe here is secure even against an enemy possessing unlimited computing power (even if $P = \mathcal{NP}$!), under any attack in which she is limited to measuring photons (or in the subsequent generalization, light pulses) one at a time, and combining the classical results of these measurements with information subsequently overheard during the public discussion. The formalism of quantum mechanics allows a more general kind of measurement, completely infeasible at present or in the foreseeable future. Such a measurement would treat the entire sequence of n photons sent during a key-distribution session as a single 2^n–state quantum system, cause it to interact coherently with an intermediate quantum system of comparable complexity, maintain the phase coherence of the intermediate system for an arbitrarily long time, then finally measure the intermediate system in a way depending on the information overheard during the public discussion. It is not known whether the protocol is secure against such an attack.

We first review the original quantum public key distribution (QPKD) protocol of [2], which illustrates the method most plainly. Then, we describe subsequent modifications of the protocol [7, 8, 3], which give it the ability, necessary in practice, to function despite partial information leakage to the eavesdropper and partial corruption of the quantum transmissions by noise. Finally, we describe the physical apparatus by which QPKD has actually been carried out. The essential quantum property involved, a manifestation of the uncertainty principle, is the fact that any measurement of a single photon's rectilinear (0 vs 90 degree) polarization randomizes its diagonal (45 vs 135 degree) polarization, and vice versa.

The basic QPKD protocol begins with Alice sending a random sequence of the four kinds of polarized photons to Bob. Bob then chooses randomly and independently for each photon (and independently of the choices made by Alice, of course, since these choices are unknown to him at this point) whether to measure the photon's rectilinear or diagonal polarization. Bob then announces publicly which kind of measurement he made (but not the result of the measurement), and Alice tells him, again publicly, whether he made the correct measurement (i.e. rectilinear for a 0 or 90 degree photon, diagonal for a 45 or 135 degree photon). Alice and Bob then agree publicly to discard all bit positions for which Bob performed the wrong measurement. Similarly, they agree to discard bit positions where Bob's detectors failed to detect the photon at all — a fairly common event with existing detectors at optical wavelengths. The polarizations of the remaining photons should be shared secret information between Alice and Bob, provided that no eavesdropping on the quantum channel has taken place. In the basic protocol, Alice and Bob next test for eavesdropping by publicly comparing polarizations of a random subset of the photons on which they think they should agree. In [2] it is shown that any measurement the eavesdropper makes on one of these photons while it is in transit from Alice to Bob has a 1/4 chance of inducing a discrepancy when the data of Bob and Alice are compared, assuming that this photon is detected in the correct basis by Bob (otherwise, this photon is lost to all parties). If Alice and Bob find no discrepancies, they may safely conclude that there are few or no errors in the remaining uncompared data, and that little or none of it is known to any eavesdropper.

The elementary protocol described above is inadequate in practice for two reasons:

1. Realistic detectors have some noise; therefore, Alice's and Bob's data will differ even in the absence of eavesdropping. Accordingly, they must be able to recover from a reasonably small error frequency.

2. It is technically difficult to produce a light pulse containing exactly one photon. It is much easier to produce a coherent pulse, which may be regarded as a superposition of quantum states with $0, 1, 2...$ photons; or an incoherent pulse, which may be regarded as a statistical mixture of coherent states. In either case, let μ be the expected number of photons per pulse. If μ is small (i.e. significantly less than 1), there is a probability approximately $\mu^2/2$ that an eavesdropper will be able to split a pulse into two or more photons, reading one and allowing the other(s) to go to Bob.

Below we briefly describe a practical protocol that allows these difficulties to be overcome. Further details may be found in [3, 7, 8].

The first task is for Alice and Bob to exchange public messages enabling them to reconcile the differences between their data, while revealing to Eve as little information as possible. We assume throughout that Eve listens to all the public messages between Bob and Alice.

An effective way for Alice and Bob to do this is for them first to agree on a random permutation of the bit positions in their strings (to randomize the locations of errors), then partition the permuted strings into blocks of size k such that single blocks are believed to be unlikely to contain more than one error. For each such block, Alice and Bob compare the block's parity. Blocks with matching parity are tentatively accepted as correct, while those of discordant parity are subject to a bisective search, disclosing $\log(k)$ further parities of sub-blocks, until the error is found and corrected. If the initial block size was much too large or too small, due to a bad a priori guess of the error rate, that fact will become apparent, and the procedure can be repeated with a more suitable block size. In order to avoid leaking information to Eve during the reconciliation process, Alice and Bob agree to discard the last bit of each block or sub-block whose parity they have disclosed. It is easy to see that Eve cannot know more information about the remaining truncated block than she did about the whole block before disclosure of its parity. However, because the string gets shorter, Eve's *proportion* of known information increases.

Of course, even with an appropriate block size, some errors will typically remain undetected, having occurred in blocks or sub-blocks with an even number of errors. To remove additional errors, the random permutation and block parity disclosure is repeated several more times, with increasing block sizes, until Alice and Bob estimate that very few errors remain in the data as a whole. At this point a different strategy is adopted to eliminate any errors that may remain and to verify, with high probability, that they have in fact been eliminated.

In each iteration of this strategy, Alice and Bob compare parities of a publicly chosen random subset of the bit positions in their entire respective data strings. If the data strings are not identical, then the random-subset parities will disagree with probability 1/2. If a disagreement is found, Alice and Bob undertake a bisective search, similar to that described above, to find and remove the error. As in the preceding block-parity stage of the reconciliation, the last bit of each compared subset is discarded to avoid leaking any information to Eve. Each subsequent random subset parity is, of course, computed with a new independent random subset of bit positions in the remaining string.

At some point, all errors will have been removed, but Alice and Bob will not yet be aware of their success. When this occurs, subsequent random subset parities will of course always agree. After the last detected error, Alice and Bob continue comparing random subset parities until sufficiently many consecutive agreements (say 20) have been found to assure them that their strings are indeed identical, with a negligible probability of not detecting the existence of remaining errors.

Alice and Bob are now in the possession of a string that is almost certainly shared, but only partly secret. As described in the next section, they can find a conservative upper bound on Eve's partial information on their string from the detected error frequency, the optical pulse intensity, and the 0/1 ratio of the received string. If their reconciled string x has length n, and if they estimate that Eve knows at most k bits about it (these need not be k physical bits, but any k bits of information about the string), it is shown in [8] that for any security parameter $s > 0$, a hash function h randomly and publicly chosen from an appropriate class of functions $\{0,1\}^n \rightarrow \{0,1\}^{n-k-s}$ will map their string into a value $h(x)$ about which Eve's expected information is less than $2^{-s}/\ln 2$ bit. An adequate hash function for this purpose can be obtained by continuing to compute $n - k - s$ additional independent random subset parities, but now keeping their values secret instead of revealing them. The class of hash functions thus realized is essentially the strongly-universal$_2$ class H3 discussed by Wegman and Carter [16].

3 Physical Apparatus

The apparatus occupies an optical bench approximately one meter long inside a light-tight box measuring approximately $1.5 \times .5 \times .5$ meters. It is controlled by a program running on an IBM PC computer, which contains separate software representations of the sender Alice, who controls the sending apparatus, the receiver Bob, who controls the receiving apparatus, and optionally an eavesdropper Eve. The program can also run in simulation mode, without the attached experimental apparatus. Even though they reside in the same computer, no direct communication is allowed between the software Alice and the software Bob, except the public channel communication called for by the protocol.

Alice's light source, at the left end of the optical bench, consists of a green light-emitting diode (Stanley type HBG5566X) as the source of incoherent light, a 50 micron pinhole and 25 mm focal length lens to form a collimated beam, a 500 ± 20 nm interference filter (Ealing type 45-5040) to reduce the intensity and spectral width of the light and select a portion of the spectrum at which the photomultipliers have relatively high quantum efficiency, and finally a Polaroid filter (i.e. a dichroic sheet polarizer) to polarize the beam horizontally. The LED is driven by current pulses (about 10^{-7} coulombs in 50 nanoseconds) yielding, after collimation, filtration and polarization, an intensity of about $0.1 - 0.2$ photon per pulse. The low intensity serves to minimize the chance that an eavesdropper will be able to split any one pulse into two or more photons.

Alice modulates the polarization of the beam by means of two Pockels cells (IN-RAD type 102-020), operated at $+$ or $-$ the quarter-wave voltage (about 800 volts), so as to be able to choose among the four polarization states {horizontal, vertical, left-circular, or right-circular} (circular polarizations are used instead of diagonal because they require only half the Pockels cell voltage). High voltage NPN transistors (type BU-205), in series with 200K ohm pull-up resistors, are used to switch the high voltage for the Pockels cells under control of low voltage TTL signals on output lines of the PC's parallel port (5.1 volt Zener diodes protect the computer from exposure to high voltage in case of transistor failure).

The quantum channel itself is a free air optical path of approximately 32 centimeters.

Bob's receiving apparatus, at the right end of the optical bench, consists of another Pockels cell and a calcite Wollaston prism (Melles-Griot type 03PPW001/C), oriented so as to split the beam into vertically and horizontally polarized beams, which are directed into two photomultiplier tubes (Hamamatsu type R1463-01) with integral preamplifiers and voltage dividers in the sockets (Hamamatsu type C716-05). Bob's Pockels cell is also operated at quarter wave voltage, allowing him to use the same Wollaston prism to make a measurement of either rectilinear or circular polarization, depending on whether the voltage is off or on.

The timing for each experiment is controlled by a timing and detection unit, which also contains the hardware for handling asynchronous communication with the PC's parallel port, and two potentiometers for setting the discrimination levels for rejecting small pulses from each photomultiplier preamplifier (no rejection of large pulses is necessary, owing to their infrequency). The pulse-height discrimination is carried out by fast ECL voltage comparators (Plessey type SP9687).

Upon receiving a "start" signal on one of the PC parallel port's output lines, the timing unit waits 60 μsec for the Pockels cell voltages to settle, turns the LED on for about 75 ns, gates the photomultiplier detection logic on for about 100 ns, and sets two input lines of the parallel port according to the result (for each photomultiplier, whether a count was detected during the gate interval). When it has done all this, the timing unit turns on another of the parallel port's input lines to signify "done", and begins waiting for the next start signal. When the computer sees the done signal it knows it can read the results of the present experiment and thereafter safely start the next experiment.

Alice's choice of polarization and Bob's choice of reading basis are made randomly (not pseudorandomly) using a large file of random bits supplied to the computer on a diskette. Of course, Alice and Bob feed on different bits from this diskette (recall that although they live on the same computer, they do not communicate or otherwise share information that is not called for by the public channel discussion). These random bits had been previously generated using the same experimental apparatus, by taking the physically random output of one of the photomultipliers, illuminated by an auxiliary nearby LED of intensity such as to yield a count in about 1/2 the time windows, removing the 0/1 bias by von Neumann's trick (i.e. in each consecutive pair of tosses taking HT=1, TH=0, ignoring HH and TT), and XORing the resulting bits with pseudorandom bits from the computer to hide any residual deviations from randomness caused by time-variation of the photomultiplier and pulse-detection circuit. The same file is used to supply additional random bits as needed by Alice and Bob during the data reconciliation and privacy amplification protocols described in the previous section.

The photomultipliers had quantum efficiency approximately 9 %, with dark count rates of about 1500 per second, or about 0.00015 per 100 ns time window. When using pulses of 0.15 expected photons per pulse, this dark count rate would yield a bit error rate of approximately 1 %; the actual error rate, about 5 %, was due primarily to imperfect alignment of the Pockels cells.

The driver program on the PC provides the ability to simulate two principal kinds of eavesdropping: intercept/resend and beamsplitting, by a hypothetical adversary "Eve" who has detectors of 100 % quantum efficiency.

Recall that μ is the expected number of photons per light pulse. If μ is sufficiently smaller than 1, it is approximately also the probability that a pulse would be detected by a perfectly efficient detector. In intercept/resend, Eve intercepts a light pulse and reads it in a basis of her choosing (she cannot be sure of choosing the correct basis, which has not yet been announced). If, with probability approximately μ, she is successful in detecting a photon, she fabricates and sends to Bob a pulse of the same polarization as she detected. It can be shown that the canonical bases, rectilinear or circular, are optimal for Eve to use in this attack, yielding an expected information 1/2 bit per intercepted photon, and inducing an error with probability 1/4 if the fabricated pulse is later detected in the correct basis by Bob. Other bases for Eve yield less information, induce more errors, or both. To avoid suspicion, Eve's fabricated pulses should be of such intensity (slightly higher than one expected photon per pulse) as to yield the same net rate of pulse detection by Bob as if no eavesdropping were taking place.

No additional hardware is needed to simulate this attack: when the software Eve wishes to intercept a pulse, she borrows the real receiving apparatus from Bob; when she wishes to resend to Bob, she borrows the sending apparatus from Alice. While Eve is borrowing the receiving apparatus, Alice obliges her by repeating the same transmission $1/q$ times, where q is the quantum efficiency of the actual detectors. This allows the software Eve to obtain a count with the same probability μ as a physical eavesdropper with perfectly efficient detectors.

The other attack, beamsplitting, would be technically easy for a real Eve, and depends on the fact that the transmitted light pulses are not pure single-photon states. To carry out this attack Eve would use a partly-silvered mirror or equivalent device to divert a fraction f of the original beam's intensity to detectors of her own, letting the remainder pass undisturbed to Bob. With probability approximately $f\mu/2$, Eve will succeed in detecting a photon, and will have by good luck measured it in the correct basis. This attack induces no errors, but does attenuate the intensity reaching Bob by a factor $1 - f$. If Eve is in control of the channel between Alice and Bob, and if this channel has significant attenuation, she can conceal the attenuation due to her beamsplitting by substituting a more transparent channel. Assuming conservatively that she can do this, she will divert most of the beam to herself, and learn a fraction roughly $\mu/2$ of the polarizations later correctly measured by Bob. An Eve with superior technology might be able to store her portion of the split beam and delay measuring it until after the correct bases were announced, thereby doubling her information yield. On the other hand, if Alice and Bob suspected Eve of having this capability, they could send and receive all the pulses first, wait an arbitrarily long time for Eve's stored beam to decay, and only then announce all the bases.

A dramatic but harmless variant on the beamsplitting attack would be for Eve to attempt to detect enough photons in the incoming pulse to determine its polarization uniquely, even without knowing the correct basis. An example of such a measurement would be for Eve to further split the intercepted portion of beam into two beams of intensity $f\mu/2$, and measure the rectilinear polarization of one and the circular polarization of the other. If, by extreme good luck, this measurement yielded three photons with polarizations vertical, horizontal, and right circular, Eve would know that the original pulse's polarization was definitely right-circular, and she could capitalize on this knowledge by sending Bob such a bright pulse of right-circular light that he would be sure to detect it. Fortunately this attack succeeds so rarely (roughly with probability $\mu^3/32$) that it is a less serious threat than simple 2-photon beamsplitting.

The driver program simulates beamsplitting simply by having the software Alice disclose directly to the software Eve the correct polarizations of a fraction $\mu/2$ of the pulses.

The expected information leaked to Eve through both kinds of eavesdropping is bounded above by

$$k = N(\mu/2 + 2p) \text{ bits,} \tag{1}$$

where N is the number of Alice's pulses received by Bob in the correct basis, μ is the pulse intensity at the upstream end of the channel, and p is the bit error rate. This estimate assumes that Eve has been able to manipulate the channel attenuation as described above to maximize her share f, that she does not have the superior technology required to delay measurement until after announcement of the correct bases (if she did, the first term above would be increased from $\mu/2$ to μ), and that intercept/resend eavesdropping is the only cause of transmission errors. These assumptions will in most cases be excessively conservative: e.g. in our case, the channel has negligible attenuation and many of the bit errors can be confidently attributed to causes other than eavesdropping. In our experiments, Eve has a small additional source of information: the excess 0:1 ratio (about 62/38) in the received data resulting from an inequality of quantum efficiency between Bob's two photomultipliers. This imbalance gives Eve a small additional amount of information (about 0.03N bits) on Alice's and Bob's string. This leakage could have been prevented by using photomultipliers of equal sensitivity, or by having Bob randomly permute the photomultipliers between measurements.

The present apparatus is only an experimental prototype. In a more realistic demonstration, the error rate could be reduced several orders of magnitude by better optical alignment and cooling the photomultipliers to reduce dark current, the quantum channel could be made much longer (e.g. a few km of optical fiber), and the protagonists Alice, Bob, and Eve could reside in separate buildings [3]. The feasible distance over which a QPKD system can operate depends on the noise and quantum efficiency of the detectors and especially on the attenuation of the optical channel: the weak signal entering the channel must still be recognizable above background upon leaving the channel.

4 Sample Data from the Apparatus

Here we give examples of data actually transmitted through the quantum channel, the subsequent public discussion, and the shared secret key ultimately distilled. The first batch of data is from a run in which there was in fact no eavesdropping, but the eavesdropper's potential information was nevertheless conservatively estimated as described above, from the known pulse intensity and error rate. The second batch of data illustrates the ability to distill a small amount of shared secret key from a run with significant amounts of both kinds of eavesdropping.

Here is raw data obtained from the quantum channel on Friday, April 13, 1990.

Alice
```
0101010010 1010001011 1100010100 0110001100 0001111001 1001001100 0001100000 1101000100 0001000100 0010100000
1011011001 0001000100 0000000100 1010001011 0011010101 0101010010 0000100011 1110000001 0101000001 0011010000
0111000100 0000011100 1100110100 0000101011 0000010001 1110000001 1100010000 0000100010 0010010110 0110000101
0111000000 1010011110 1101100111 0000000000 0010010100 0000000001 0000110110 0010010001 0001011110 1101100101
0101011000 1100100001 0000000100 1111111110 0010011010 0011000010 0111110000 0011000000 0000001010 1100010110
1010001010 0101000010 1110011000 0011111000 1100100011 1000100000 0001000101 1000101100 1101010111 0111011010
1001100101 0010000010 1000001100 0001110100
```

Bob
```
0101010010 1011001011 1100010100 0110001100 0001111001 1001000100 0001100000 1101000100 0001000100 0010100000
1011011001 0001000000 0000000100 1010001010 1011000101 0101010010 1000100011 1110000001 0101000011 0011000010
0111000100 0000111100 1100110100 0000101011 0000000000 1110000001 1100010000 0000110010 0010010100 0110000101
0111010000 1000011110 1101100111 0000000000 0010010100 0000000001 0000110110 0010010101 0001011110 1101100101
0101011000 1100100001 0000000100 1011101110 0010011010 0011000010 0111110000 0011000000 0100001010 0100010110
1010001010 0101000010 1110011000 0011111000 1100100011 1000100000 0001000101 1000101100 1100010101 0101011010
1001100101 0010100010 1000001000 0001110100
```

In this first example, out of about 85,000 light pulses of intensity 0.17 sent by Alice, 640 were received in the correct basis by Bob. Alice's corresponding string contained 242 ones out of 640 bits. Bob's string contained 28 errors, an error frequency of 4.375 %.

A random permutation and block parity comparison was performed with block size 10, reducing the string length to 509 bits with 8 undiscovered errors.

A second random permutation and block parity comparison was performed with block size 20, reducing the string length to 457 bits with 2 undiscovered errors.

Random subset parity comparison was then begun, revealing an error on the first attempt. Removal of this error by bisective search reduced the string length to 448 bits.

Another random subset parity was computed, revealing another error. Removal of this error reduced the string length to 439 bits with no undiscovered errors.

Twenty more random subset parities were compared and found to agree, confirming to Bob and Alice that with high probability their remaining strings, now 419 bits long, were identical.

From the 28 errors corrected during reconciliation, Bob and Alice estimated that the original error rate was 4.50 %. Estimated potential information leakage to Eve was 140 bits, including

- 58 bits from intercept/resend,

- 54 bits from beamsplitting, based on pulse intensity $\mu = 0.17$, and

- 28 bits from redundancy due to 0:1 imbalance (398/242) of initial string.

Therefore, allowing 60 bits excess compression for safety (e.g. in case Eve was especially lucky in her eavesdropping, and obtained several standard deviations more information than expected), it was decided to compress the string 200 bits by random subset hashing, leaving 219 bits of shared secret key distilled from 640 original bits.

The resulting secret key, the same for Alice and Bob, was

```
0000101010 1101100010 0100101100 0110010010 1000010100
1110011101 1001000011 1111101010 0000111010 0011111100
1100101000 1101011111 1110001101 0001100100 1000110011
0101110110 0011110110 1010100100 1011111010 0101111101
0110000000 010000101.
```

In the second example, out of another approximately 85,000 light pulses of intensity $\mu = 0.17$ sent by Alice, 640 were received in the correct basis by Bob. Alice's corresponding string contained 239 ones out of 640 bits. Bob's string contained 59 errors, an error frequency of 9.219 %. Through attempting to beamsplit all the pulses, and intercept/resending one sixth of them, the simulated Eve learned 100 individual bits of Alice's data as well as knowing 30 bits of distributed information about the string as a whole due to its 0/1 imbalance. (Her total information was actually slightly less than 130 bits, because of the correlation between these two kinds of information. As remarked earlier, Eve's absolute amount of information does not increase during reconciliation).

A random permutation and block parity comparison was performed with block size 5, reducing the string length to 399 bits with 13 undiscovered errors. (They start with a block size smaller than in the previous example because of reason (1) given at the end of this section.) Eve's information about the remaining string was still less than 130 bits, and included knowledge of 61 individual bits.

A second random permutation and block parity comparison was performed with block size 10, reducing the string length to 337 bits with 6 undiscovered errors. Eve's information about the remaining string was still less than 130 bits, and included knowledge of 51 individual bits.

A third random permutation and block parity comparison was performed with block size 20, reducing the string length to 292 bits and leaving no undiscovered errors. Eve's information about the remaining string was still less than 130 bits, and included knowledge of 43 individual bits.

Random subset parity comparison was then begun. Twenty consecutive successful comparisons with no failures convinced Alice and Bob that their strings, now consisting of 272 bits, were very probably identical. Eve's knowledge about the remaining string was still less than 130 bits, and included knowledge of 38 individual bits.

From the errors found and corrected, Alice and Bob estimated the error probability had been 9.59 % in the original string. From this error rate and from the known pulse intensity $\mu = 0.17$, Alice and Bob computed an upper bound of 207 bits on Eve's probable information, including:

- 123 bits from intercept/resend,

- 54 bits from beamsplitting,

- 30 bits from redundancy due to 0:1 imbalance.

Therefore, allowing 20 bits excess compression (about all we can afford), it was decided to compress the string 227 bits by random subset hashing, leaving 45 bits of shared secret key, distilled from 640 original bits. Since Eve's actual information was less than 130 bits, this amount of compression left Eve with an utterly negligible (less than 10^{-29} bit) expected information about the output of the hash function.

The resulting secret key, the same for Alice and Bob, was

0001110110 0011101001 1000100011 1111000010 10010.

The 640-bit batch size used above for illustrative purposes is far from optimal. In production use, a larger batch size (at least 10,000 bits) should be used for two reasons: 1) It would allow the users, by preliminary sampling, to get a good estimate of the bit error rate and so optimize the choice of block sizes used in the reconciliation stage; and 2) by reducing the statistical uncertainty in estimating Eve's possible information, it would reduce the proportional amount of compression needed in the privacy amplification stage to assure a given level of security.

Acknowledgements

We wish to thank Manuel Blum, Claude Crépeau, David Deutsch, Myron Mandel, and Stephen Wiesner for many helpful discussions.

References

[1] Bennett, C. H. and G. Brassard, "An update on quantum cryptography", *Advances in Cryptology: Proceedings of Crypto '84*, August 1984, Springer–Verlag, pp. 475–480.

[2] Bennett, C. H. and G. Brassard, "Quantum cryptography: Public key distribution and coin tossing", *Proceedings of IEEE International Conference on Computers, Systems, and Signal Processing*, Bangalore, India, December 1984, pp. 175–179.

[3] Bennett, C. H. and G. Brassard, "Quantum public key distribution system", *IBM Technical Disclosure Bulletin*, Vol. 28, 1985, pp. 3153–3163.

[4] Bennett, C. H. and G. Brassard, "The dawn of a new era for quantum cryptography: The experimental prototype is working!", *SIGACT News*, Vol. 20, no. 4, Fall 1989, pp. 78–82.

[5] Bennett, C. H., G. Brassard and S. Breidbart, "Quantum cryptography II: How to re-use a one-time pad safely even if $P = NP$", unpublished manuscript available from the authors, November 1982.

[6] Bennett, C. H., G. Brassard, S. Breidbart and S. Wiesner, "Quantum cryptography, or unforgeable subway tokens", *Advances in Cryptology: Proceedings of Crypto '82*, August 1982, Plenum Press, pp. 267–275.

[7] Bennett, C. H., G. Brassard and J.-M. Robert, "How to reduce your enemy's information", *Advances in Cryptology — Crypto '85 Proceedings*, August 1985, Springer–Verlag, pp. 468–476.

[8] Bennett, C. H., G. Brassard and J.-M. Robert, "Privacy amplification by public discussion", *SIAM Journal on Computing*, Vol. 17, no. 2, April 1988, pp. 210–229.

[9] Brassard, G., *Modern Cryptology: A Tutorial*, Lecture Notes in Computer Science, Vol. 325, Springer–Verlag, Heidelberg, 1988.

[10] Brickell, E. F. and A. M. Odlyzko, "Cryptanalysis: A survey of recent results", *Proceedings of the IEEE*, Vol. 76, no. 5, May 1988, pp. 578–593.

[11] Crépeau, C., "Correct and private reductions among oblivious transfers", PhD Thesis, Department of Electrical Engineering and Computer Science, Massachusetts Institute of Technology, February 1990.

[12] Crépeau, C. and J. Kilian, "Achieving oblivious transfer using weakened security assumptions", *Proceedings of 29th IEEE Symposium on the Foundations of Computer Science*, White Plains, New York, October 1988, pp. 42–52.

[13] Deutsch, D., *New Scientist*, December 9, 1989, pp. 25–26.

[14] Gottlieb, A., "Conjugal secrets — The untappable quantum telephone", *The Economist*, Vol. 311, no. 7599, 22 April 1989, p. 81.

[15] Wallich, P., "Quantum cryptography", *Scientific American*, Vol. 260, no. 5, May 1989, pp. 28–30.

[16] Wegman, M. N. and J. L. Carter, "New hash functions and their use in authentication and set equality", *Journal of Computer and System Sciences*, Vol. 22, 1981, pp. 265–279.

[17] Wiesner, S., "Conjugate Coding", manuscript written *circa* 1970, unpublished until it appeared in *SIGACT News*, Vol. 15, no. 1, 1983, pp. 78–88.

A Protocol to Set Up Shared Secret Schemes
Without the Assistance of a Mutually Trusted Party*

Ingemar Ingemarsson
Linköping University
Department of Electrical Engineering
S-58183 Linköping, Sweden

Gustavus J. Simmons
Sandia National Laboratories
Albuquerque, New Mexico 87185, USA

Introduction

All shared secret or shared control schemes devised thus far are autocratic in the sense that they depend in their realization on the existence of a single party—which may be either an individual or a device—that is unconditionally trusted by all the participants in the scheme [5,6]. The function of this trusted party is to first choose the secret (piece of information) and then to construct and distribute in secret to each of the participants the private pieces of information which are their shares in the shared secret or control scheme. The private pieces of information are constructed in such a way that any authorized concurrence (subset) of the participants will jointly have sufficient information about the secret to reconstruct it while no unauthorized collection of them will be able to do so. For many applications, though, there is no one who is trusted by all of the participants, and in the extreme case, no one who is trusted by anyone else. In the absence of a trusted party or authority, no one can be trusted to know the secret and hence—until now—it has appeared to be impossible to construct and distribute the private pieces of information needed to realize a shared control scheme. It is worth noting that in commercial and/or internation(al) applications, this situation is more nearly the norm than the exception.

* This work performed in part at Sandia National Laboratories supported by the U. S. Department of Energy under contract no. DE-AC04-76DP00789.

The single exception is that a way has been known (and used) for several years to ensure unanimous consent before a controlled action can be initiated [7]. For example, if it is desired that a specific two persons (controllers) must concur in order for a vault to be opened (or a weapon enabled or a missile fired) then each of these controllers could—during the initialization of the locking mechanism in the vault door—enter a randomly chosen k-digit number whose value is kept secret by the controller who chose it. The mod 10 sum of these two private and secret k-digit numbers would be the secret k-digit combination needed to open the vault. The subsequent entry of any pair of k-digit numbers whose mod 10 sum is equal to the secret combination determined by the two controllers would open the vault door. Clearly the probability that an outsider or either of the two insiders (controllers) alone being able to open the vault on the first try would be 10^{-k}. In this control scheme, two controllers are involved, and both must (in probability) concur in order for the controlled event to be initiated. For anything other than a unanimous consent scheme, however, a functional dependence must exist between the participants' private pieces of information reflecting the structure of the authorized concurrences—even though the participants don't trust each other so that cooperation in achieving such a dependence can't be assumed. In this paper we present a protocol to set up shared control schemes without the assistance of any trusted party, in which participants need only act in their own self interest.

The problem of setting up shared secret schemes in the absence of a trusted third party has been largely ignored by researchers in this area—with the single exception of a paper by Meadows [4]. In it she discusses this problem and at even greater length the twin questions of how new participants can be enrolled in an already existing shared control scheme and of how previously enrolled participants can be cut out. To accomplish this, she uses a construction which she calls a rigid linear threshold scheme that makes it possible for a predetermined number of the existing participants to delegate their capability to a new member—essentially to vote him into membership. Her constructions do not appear to be related to the approach to be presented here, espe-cially so since her primary proposal depends on a secure (unconditionally trustworthy) black box to replace the services of an unconditionally trustworthy key distribution center. Meadows attributes the question of whether a shared secret scheme can be set up without the assistance of a trusted key distribution center to Chaum, however the paper of his that she cites—"Some Open Questions"—did not appear in the Proceedings for Crypto'84 where she references it. The important point, though, is that her work appears to be the only prior reference to the problem of how a shared control scheme can be set up without the assistance of a trusted key (share) distribution center.

Democratic Shared Control Schemes

The essential notion to our protocol is that the secret (piece of information) will be jointly determined in a unanimous consent scheme from inputs made privately by each of the participants. Each of these inputs is to be equally influential in determining the value of the secret and will itself be kept secret by its contributor. Shared control schemes of this sort in which each participant has an equal influence on the determination of the secret (i.e., the information equivalent of the democratic principal of "one man-one vote") will be referred to as democratic (schemes) as contrasted to autocratic schemes in which the participants have no input to the initial determination of the secret (information). Once the secret has been determined, each participant may, if he wishes, devise private shared control schemes with which to distribute among the other participants private pieces of information that would make it possible for some groupings of them to reconstruct his contribution to the determination of the secret.

To summarize, the general protocol with which a group of mutually distrustful participants can set up a democratic shared control scheme that they must logically trust—without the assistance of any outside party is:

1. The participants first set up a democratic unanimous consent scheme; i.e., one in which

   a) they each contribute equally to the determination of the secret (piece of information), and

   b) all of their private inputs (contributions) must be made available in order for the secret to be reconstructed.

2. After the unanimous consent scheme is in place, any of the
    participants who trust some concurrence(s), i.e., subsets of
    the other participants, to faithfully represent their interests
    can then create private autocratic shared secret schemes to
    distribute information about the private (and secret) contri-
    bution they made to the determination of the overall secret
    among the members of those concurrences.

Step 1 doesn't require that anyone trust anyone else. After step 2 has
been completed, any concurrence of participants who were intrusted with
another participant's share in the unanimous consent scheme can act in
the stead of that participant—and no collection of the participants
that doesn't include one of these concurrences can do so. In setting up
these private shared secret schemes each participant is acting as his
own "trusted authority" to protect his own interests, so that he need
trust no one else insofar as the delegation of the capability to act in
his stead is concerned. In this way, each participant can guarantee
that only concurrences that either include him as a member or else that
include a subset of the other participants whom he trusts to represent
his interests will be able to initiate the controlled action or to
recover the shared secret. Since each participant acts similarly, the
net result is that democratic shared control schemes of arbitrary complexity (of control) can be established which accurately reflect the
placement of trust (or lack of it) by the participants in each other
[1]. In other words, every shared control scheme that would be acceptable to the participants and which could be set up by a mutually trusted
authority, can also be set up as a democratic scheme by the participants
themselves without anyone having to accept a greater risk of their
interests being abused than they would have had to accept in order for a
trusted authority to set up the scheme instead.

Implementation

Given this protocol, the first question is how the initial unanimous
consent scheme can be set up. We have only found two—inequivalent—
ways this can be done; either way, of course, can serve as the starting
point for setting up more complex schemes using the protocol described

above. The first is simply a generalization of the example given
earlier.

1. In a space whose cardinality is adequate for the concealment of
   the secret, i.e., in which the probability of selecting a randomly chosen secret (point) in a subsequent random drawing provides an acceptable level of security for the controlled action,
   each participant chooses at random a point as his contribution
   to the unanimous consent control scheme. During the initialization of the mechanism that implements the shared control, each
   of the participants secretly enters the point he has selected
   and the sum (vector, modular, exclusive-or, etc.) of all of the
   points becomes the jointly defined secret value. Since this
   procedure is an obvious generalization of Vernam encryption, the
   secret is unconditionally secure from discovery (or recovery) by
   any concurrence of fewer than all of the participants so long as
   the sum operation is an entropy preserving mapping. To see that
   this is true, consider the worst case scenario in which all but
   one of the participants conspire in an attempt to initiate the
   controlled action without the cooperation of the single missing
   participant. They can calculate the point which is the sum of
   all of their contributions, however, every point in the space is
   still equally likely to be the secret point depending on the
   point chosen by the missing participant. In other words, even
   in this worst case scenario, the best that the would-be cheaters
   can do is to "guess" at the value of the secret using a uniform
   probability distribution on all of the points in the space.
   Clearly, this is the best that can be achieved.

2. In an n-dimensional finite space, where n is the total number of
   participants in the scheme and the cardinality of the space is
   chosen such that the probability of randomly choosing a particular point (the secret) out of all of the points in a hyperplane
   of the space provides an adequate concealment for the secret,
   each participant randomly chooses a hyperplane as his private
   contribution to the determination of the secret. The secret in
   this case is the point defined by the intersection of the n
   hyperplanes. With virtual certainty (as the number of points in

a hyperplane increases) the n independently chosen hyperplanes will intersect in only a single point. For example, consider the case of two lines in PG(2,q) or of three planes in PG(3,q). A pair of lines in PG(2,q) either intersect in a point or else they are coincident. The probability of this later occurring is $O(1/q^2)$. Similarly, there are only two ways three planes in PG(3,1) can have more than a point in common: either they are all three coincident or else they form a pencil of planes on a common line of intersection. The probability of these occurrences is $O(1/q^6)$ and $O(1/q^2)$, respectively.

The important point is that n hyperplanes in an n-dimensional space almost certainly (with q) intersect in only a single point, so that the protocol described here will almost certainly define a unique value for the secret. In the (unlikely) event that they do not for a particular choice of hyperplanes by the participants, this would be detected during the initialization phase of setting up the shared control scheme and the participants would then have to make another (random) choice of inputs.

We will refer to these two ways of realizing unanimous consent schemes as point and plane protocols, respectively. It might at first appear that the point and plane protocols are in some sense simply two versions of a single scheme; especially so in view of the geometric duality of points and hyperplanes and the fact that these objects are the private choices of inputs in the two protocols. It is easy, however, to show that this cannot be the case.

In the point protocol, the uncertainty about the secret is the same for an outsider as it is for every combination of fewer than all of the participants: namely it is equally likely to be any point in the containing space. Furthermore there is no relationship between the dimension of the space in which the secret is concealed and the number of participants in the shared control scheme. The only requirement is that the number of points in the space be large enough that the probability of choosing the secret (one) at random will be sufficiently small. In other words, a k-out-of-k scheme could be implemented in a 1-dimensional space as well as any other, even if the dimension of the containing space is greater than k.

On the other hand, in the plane protocol the dimension of the space must equal the total number of participants, say n. Otherwise the intersection of the n randomly and independently chosen hyperplanes will almost certainly over or under-determine a point; i.e., the hyperplanes will either not have a common point of intersection or else will intersect in a subspace of higher dimension. More importantly, though, outsiders and all proper subsets of the insiders will be faced with substantial differences in uncertainty about the secret. An outsider knows only that p is some point in $PG(n,q)$, where all points are equally likely, i.e., an uncertainty about the secret of $O(q^{-n})$. Any single one of the participants, however, knows that p must be a point in the hyperplane he chose, i.e., a point in an $(n-1)$-dimensional subspace which is an uncertainty about p of only $O(q^{-(n-1)})$. Similarly, any pair of participants together could reduce the uncertainty about p to being a point in the $(n-2)$-dimensional flat which is the intersection of the two hyperplanes they chose, etc.

There are other, geometrical, arguments to show the inequivalence of these two protocols for setting up unanimous consent schemes in the absence of trust, but none so easy to see as this information based argument.

Given that the participants have set up a unanimous consent scheme (using either the point or the plane protocol) the next question is how each participant can then distribute shares in his input among concurrences of the other participants whom he trusts—if any exist. If the point protocol was used, so that each participant's input was a point, then conventional secret sharing schemes [5,6,7]—all of which are designed to control the recovery of a secret point—can be used. If the plane protocol was used, however, standard secret sharing schemes are not suitable, since the geometric object whose identification is to be shared is a hyperplane: not a point. To share a hyperplane requires that sets of subspaces (or varieties in general) be constructed such that the subsets of these held by trusted concurrences will suffice to determine the hyperplane, and no other subsets will do so. This may be easy. For example, if the plane protocol is used and the private hyperplanes are 3-dimensional, then it is easy to devise simple k-out-of-ℓ threshold schemes for k = 2, 3 or 4; the shares being a set of pairwise skew lines, a pencil of lines on a point—no three of which are coplanar, or a set of points—no four of which are coplanar, respectively. Note that only the first two cases are of interest, since the total number of participants in this case can only be four. On the other hand, it may be a difficult geometric problem. For example, if there are eight participants in all so that the private inputs are 7-dimensional hyperplanes, and one of the participants wishes to share his input with the others in such a way that a concurrence of any pair out of a particular subset of three of them or any concurrence of three participants will be able to act in his stead, it is difficult to see what the shares should be. It is easy to realize either a simple 2-out-of-3 or a 3-out-of-7 threshold scheme. In the first case, the shares could be three pairwise skew 3-spaces while in the second, the shares could be taken to be a pencil of 3-spaces chosen so that every pair intersect in a line and no three of which lie in a 6-dimensional subspace. The difficulty lies in constructing the three 3-spaces that are to be used for the shares for the three participants who are more trusted than the others so that they also function as shares in the 3-out-of-7 scheme. The only general solution we have found for problems of this sort, i.e., for trust schemes more complex than simple threshold schemes, makes use of the geometric dual to conventional shared secret schemes.

Points and hyperplanes are dual objects in any space. Conventional shared secret schemes are designed to conceal (and reveal) points, hence if we take the geometric dual of a conventional shared secret scheme replacing geometric unions by intersections and vice versa, we realize a scheme with the same control characteristics but in which the controlled object is a hyperplane instead of a point. Since the private pieces of information are points in the companion shared secret scheme, the private pieces of information will be hyperplanes in the dual scheme. We remark that while these dual constructions guarantee the existence of democratic shared secret schemes whenever shared secret schemes exist, the constructions that result may be unnecessarily complicated, as the examples will show.

While the discussion of shared secret schemes given here is purely geometric, there are other ways of looking at such schemes—which even if they prove to be equivalent may be very useful since they can draw on related disciplines in constructing such schemes. One of the more promising of these alternative formulations is based on the relationship between the reconstruction of a piece of information from partial information and the error correcting properties of error detecting and correcting codes, and in particular to maximum distance separable (MDS) codes.

The connection between shared secret schemes and Reed-Solomon codes has previously been observed by McEliece and Sarwate [3]. Reed-Solomon codes are special cases of Maximum Distance Separable Codes—MDS codes [2, Chapter 11]. These are block codes over some finite field GF(q) with block length n and q^k codewords. We will use the customary notation (n,k)-code. They have a property which is important in this context: the codeword can be reconstructed from any k of the n components of the codeword and if less than k components are known the remaining components are completely undecidable. Using the terminology of error correcting codes we say that the code is capable of correcting n-k erasures.

The use of MDS codes to construct autocratic secret shared schemes, i.e., with the assistance of a trusted authority, is straightforward: the authority randomly selects a codeword from an (n,k) MDS code. The selected codeword, or a part of it (say it's first component) may be regarded as the secret. Remaining components are distributed along with their position numbers in the codeword to the participants in the scheme. From the property of MDS codes described above it is clear that any k of the participants can reconstruct the codeword. A more general case, where some of the shares may be in error, is treated in [3].

In the absence of a mutually trusted party we modify this scheme to fit the protocol for setting up democratic shared control schemes described earlier. Each of the participants selects an (n,k) MDS code. Note that the choice of n and k may be different for each participant. Each participant then randomly selects a codeword in the code that he chose. These codewords are their shares in a unanimous consent scheme. The secret is the vector sum over GF(q) of their shares, which may require padding with trailing zeroes to obtain the same length.

Each participant now distributes distinct components, with position numbers, of his randomly selected codeword to a selected subset of the other participants whom he trusts to act in his stead. Due to the property of MDS codes any k (where k is the indicidual choice of the distributing particiapnt) of the members of the subset can reconstruct the participant's codeword.

In the simplest case of the above scheme a common (n,k) MDS code (where n is at least as large as the number, ℓ, of participants) is chosen beforehand. Each participant then, as his share in the unanimous consent scheme, randomly selects a codeword in the given code. In geo-metrical terms the shares are points on a k-dimensional hyperplane (through the origin) in the n-dimentional space over GF(q).

Each participant then distributes components and position numbers to all the other participants. Since the parameter k now is common to all the participants any k of them determine the shares of all the partici-pants and thus the joint secret. We thus have realized a simple k-out-of-ℓ threshold scheme without the assistance of a mutually trusted party.

Examples

The smallest example that fully illustrates the protocol is a 2-out-of-3 threshold scheme. Such a scheme might be used to allow any two out of three vice-presidents at a bank to open the vault door but to insure that no one of them alone could do so. We will show how the three vice-presidents can set up such a scheme using either the point or plane protocol described above. In either case, the vault combina-tion (the secret) can be thought of as a point in some suitable space.

For the first unanimous consent scheme the secret can be taken to be any point, p, on a line, say $\ell = PG(1,q)$. Each of the three partici-pants secretly and randomly chooses a point, p_i, on ℓ. p is defined to be the field sum of the three points;

$$p = \sum_{i=1}^{3} p_i . \tag{1}$$

Clearly $\sum$ satisfies the definition of an entropy preserving sum, since as any single summand, p_i, ranges over all q+1 possible values, with the other two points remaining fixed, p also ranges over all of the points on the line.

The inescapable conclusion that follows from the acceptability of a 2-out-of-3 threshold scheme is that each participant is willing to trust the other two to only initiate the controlled action (i.e., to open the vault door in the present example) when they should. By the same token, the need for a 2-out-of-3 concurrance presupposes a lack of confidence in what a single individual might do. Consequently each participant (in this example) must logically be willing to share his private input to the secret between the other two participants in such a way that they could jointly reconstruct his contribution, but in which they are individually totally uncertain of it. To do this, each participant constructs a private 2-out-of-2 scheme of the sort described earlier, i.e., he randomly chooses a pair of points whose sum is his contribution to the democratic shared secret scheme, and gives (in secret) each of the other participants a different one of these points. A convenient way to represent this implementation of the protocol is:

$$
\begin{array}{c c c c}
 & 1 & 2 & 3 \\
1 & p_1 & p_{21} & p_{31} \\
2 & p_{12} & p_2 & p_{32} \\
3 & p_{13} & p_{23} & p_3
\end{array}
$$

where the three points in column i are all chosen by participant i— subject to the condition that $p_i = \sum_{j \neq i} p_{ij}$. The three entries in row j are known to participant j: the entry on the diagonal because he chose it and the off diagonal entries because they are the private pieces of information (points) given to him by the other participants. Clearly, any two participants have between them all the information needed to compute $p = \sum p_i$, while any one of them is totally uncertain as to the value of p. Although we have described the protocol starting with the establishment of the democratic unanimous consent scheme, the scheme would probably be implemented in reverse order. Participant i would choose at random the two points p_{ij}, $j \neq i$, and then calculate his input, p_i, to the unanimous consent scheme $p_i = \sum_{j \neq i} p_{ij}$, etc.

To set up the other type of unanimous consent scheme each participant chooses at random a plane, π_i, in a projective 3-space PG(3,q). As
was noted earlier, since there are three participants, the second type
of scheme is only possible in a 3-dimensional space. With virtual
certainty (with increasing size of q), the three randomly and independently chosen planes intersect in only a single point. This point, p,
is the jointly determined secret (combination) $p = \bigcap_{i=1}^{3} \pi_i$. This protocol defines a 3-out-of-3 unanimous concurrence scheme, since the three
vice-presidents acting together can cause the secret to be reconstructed
within the vault door mechanism at any time by reentering their private
pieces of information (planes).

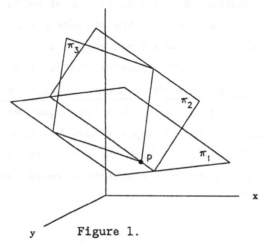

Figure 1.

As before, each participant also sets up a private 2-out-of-2 shared
secret scheme to distribute information about the plane he chose to the
other participants, constructed so that they can jointly reconstruct his
plane, but individually cannot do so. One way he could do this would be
to choose a pair of distinct lines lying in his plane and then give a
different one of these lines to each of the other participants (in private). Since the lines are distinct, they span the plane. Hence any
two vice-presidents have between them the capability to reconstruct all
three planes and thus redefine p. The private pieces of information for
each participant will be the plane he chose and the two lines given to
him by the other vice-presidents. Since the pair of lines are shares in
a perfect 2-out-of-2 scheme defining his secret plane, each vice-
president is assured that a successful concurrence must either include

him as a participant or else include both of the other vice-presidents.
A convenient way to represent this implementation of the protocol is

$$
\begin{array}{c|ccc}
 & 1 & 2 & 3 \\
\hline
1 & \pi_1 & \ell_{21} & \ell_{31} \\
2 & \ell_{12} & \pi_2 & \ell_{32} \\
3 & \ell_{13} & \ell_{23} & \pi_3
\end{array}
$$

where the lines (off diagonal) entries in column i are chosen by participant i—subject to the condition that they span the plane π_i,
$\pi_i = \bigcup_{j \neq i} \ell_{ij}$. The three entries in row j are known to participant j:
the entry on the diagonal because it is the plane he chose and the off
diagonal entries because they are the private pieces of information
(lines) given to him by the other participants.

This construction—for distributing shares of the participant's
private inputs—does not use the geometric dual of a companion shared
secret scheme. If we wish to use this technique, we must work in the
full 3-space since the dual of a point must be plane: the participant's
input which he wishes to share with the other participants. Consequently, the companion shared secret scheme would have to be of the form
shown in Figure 2.

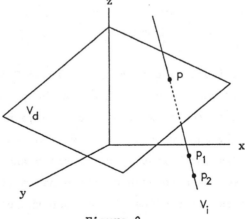

Figure 2.

The union (span) of the two points p_1 and p_2 (the shares of the point p
in this scheme) is the line V_i which intersects the publicly known plane
V_d in the (secret) point p. In the dual construction, Figure 3,

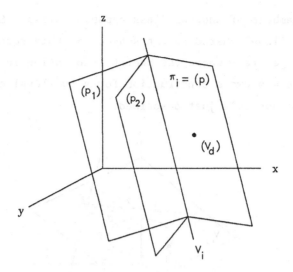

Figure 3.

the intersection of the two planes (p₁) and (p₂) (the shares of the
plane (p) in the dual scheme) is the line V_i whose union with the
publicly known point (V_d) is the secret plane (p). In other words, to
use a dual shared secret scheme to share a plane, π_i, each participant
would make public a point, (V_d), in the plane and then give to each of
the other participants one out of a pair of planes whose intersection is
a line, V_i, in π_i but not on the publicly known point (V_d). Clearly,
this is an alternative construction to the one described first for a
democratic 2-out-of-3 shared secret scheme, but equally clearly, not as
efficient. Each participant's private information now consists of three
planes: the one he chose and two given to him by the other participants. In addition, there are three publicly known points that are
essential to the shared secret scheme.

In view of the complexity of the dual geometric construction just
given, one might question whether the technique has any application. We
conclude by exhibiting an example in which (so far as we have been able
to determine) it is the only means of constructing a solution. We mentioned earlier a shared secret scheme in which a participant is willing
to trust any three of the other participants or any pair out of a specified subset of them to represent his interests. While it is easy to
realize efficient k-out-of-ℓ threshold schemes in general, it is very
difficult to realize schemes in which the members of one class can

function as members of another (less capable) class. Simmons [5,7] has
discussed multilevel shared secret schemes of this sort and shown how to
solve them in general—where the secret information is a point in some
space. Figure 4 shows a construction for a two-level control scheme
satisfying the controls just described.

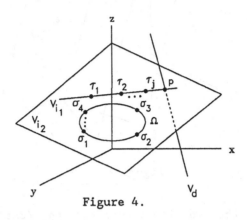

Figure 4.

The secret is the point, p, on the line V_d at which both the plane V_{i_2}
and the line V_{i_1} (in the plane) intersect it. The private pieces of
information held by the more capable class (the 2-out-of-ℓ_2 shared control scheme) are points τ_i one the line V_{i_1}; $\tau_i \neq p$. The private pieces
of information held by the less capable class (the 3-out-of-ℓ_3 shared
control scheme) are points σ_i on the oval Ω, where the points σ_i and τ_i
are chosen so that no pair of the points on Ω (used as private pieces of
information) are collinear with a point τ_i or with p on the line V_{i_1}.

The dual construction to this scheme is shown in Figure 5.

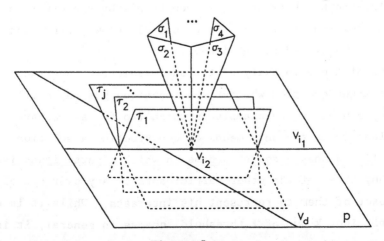

Figure 5.

The set of points σ_i on Ω, in Figure 4 no three of which are collinear and no pair of which are collinear with a point r_i on V_{i_1} (or with p), have been replaced in Figure 5 by the sheaf of planes π_i on the point V_{i_2} (dual to the plane V_{i_2} in Figure 4) no three of which intersect in a common line and none of which contain the line V_{i_1}. Similarly, the set of collinear points r_i on the line V_i in Figure 4 have been replaced by a pencil of planes on a common line V_{i_1} in Figure 5. The dual of the point p lying on the line V_d (and on the plane V_{i_2} and the line V_{i_1}) in Figure 4 is the plane p containing the point V_{i_2} and the line V_{i_1} in Figure 5. It is an easy matter to show that this dual configuration has the desired control characteristics. The line V_d (a unique line for each participant who wished to implement this sort of sharing of his secret input with the other participants) would be made public. Each participant would be (privately) given a plane as his share of the input. While it might be possible to devise a way of giving only lines and points in a plane as shares to the other participants to realize such a two-level control scheme, we have been unable to find such a construction. The directness of the geometric dual constructions may more than compensate for their lack of efficiency even if alternative constructions exist—which appears very doubtful for complex control schemes.

Conclusion

The protocol described here is so simple in principal that there is no question about its feasibility in practice—even though the private shared secret schemes may themselves require complex implementations to realize desired concurrences. The essential point is that the protocol insures that no participant can increase his capability (to contribute to the reconstruction of the secret) beyond what is acceptable to the other participants. However, an extended protocol is required if one also wishes to insure that a participant can't diminish the capability of other participants as well; i.e., to cause concurrences whose members believe they have the capability to recover the secret to not be able to do so. The authors plan to treat these extended (capability) protocols in a subsequent paper.

The bottom line is that the protocol described here permits democratic shared secret schemes, which must logically be trusted, to be set
up by mutually distrustful parties without outside assistance. In addition, no participant is required to accept a greater risk of the secret
information being misused than what he would have had to be willing to
accept if there had existed a trusted authority to set up the scheme
instead. Clearly this is the most that could be hoped for.

References

1. I. Ingemarsson and G. J. Simmons, "How Mutually Distrustful Parties
    Can Set Up a Mutually Trusted Shared Secret Scheme," *International
    Association for Cryptologic Research (IACR) Newsletter*, Vol. 7, No.
    1, January 1990, pp. 4-7.

2. F. J. MacWilliams and N.J.A. Sloane, "The Theory of Error Correct-
    ing Codes," North Holland, Amsterdam, 1981.

3. R. J. McEliece and D. V. Sarwate, "On Sharing Secrets and Reed-
    Solomon Codes," Communications of the ACM, Vol. 24, No. 9, Sept.
    1981, pp. 583-584.

4. C. Meadows, "Some Threshold Schemes Without Central Key Distrib-
    utors," *Congressus Numerantium*, 46, 1985, pp. 187-199.

5. G. J. Simmons, "How to (Really) Share a Secret," Crypto'88, Santa
    Barbara, CA, August 21-25, 1988, *Advances in Cryptology*, Vol. 403,
    Springer-Verlag, 1989, pp. 390-448.

6. G. J. Simmons, "Robust Shared Secret Schemes or 'How to be Sure You
    Have the Right Answer Even Though You Don't Know the Question',"
    18th Annual Conference on Numerical Mathematics and Computing,
    Sept. 29-Oct. 1, 1988, Winnipeg, Manitoba, Canada, *Congressus
    Numerantium*, Vol. 68, May 1989, pp. 215-248.

7. G. J. Simmons, "Prepositioned Shared Secret and/or Shared Control
    Schemes," Eurocrypt'89, Houthalen, Belgium, April 11-13, 1989,
    Advances in Cryptology, to appear.

LOWER BOUNDS FOR AUTHENTICATION CODES

WITH SPLITTING

Andrea Sgarro

Dept. of Mathematics and Computer Science
University of Udine, 33100 Udine, Italy

and: Dept. of Mathematical Sciences
University of Trieste, 34100 Trieste, Italy

Abstract. The role of non-deterministic authentication coding (coding with splitting) is discussed; a new substitution attack is put forward which is argued to be more relevant than usual substitution for codes with splitting. A reduction theorem is proved which allows to extend "abstract" bounds for impersonation (including the new JS-bound, which is shown to hold on the large class of "abstract" authentication codes) to the new substitution attack.

INTRODUCTION

In Simmons' model an authentication code is a finite random triple XYZ (random message, random codeword, random key; according to the original terminology: random source state, random authenticated message and random encoding rule). Further, the following is required: X and Z are independent; X = g(Y,Z) (decoding has to be deterministic). Instead, Y = f(X,Z) (deterministic encoding) is *not* required. If this happens the code is called deterministic, or *without splitting* . Below, as an example, we show an *encoding matrix* , the corresponding *decoding matrix*, and the binary matrix specifying *key-codeword admissibility;* this is obtained from the decoding matrix by writing ones instead of messages and zeroes instead of blanks, and tells which codewords are authenticated by which keys.

	x1	x2	x3
z1	y1	y2	y3
z2	y3	y4	y1

	y1	y2	y3	y4
z1	x1	x2	x3	
z2	x3		x1	x2

	y1	y2	y3	y4
z1	1	1	1	0
z2	1	0	1	1

To describe a code one may give the encoding matrix and specify the statistics of X and Z; however, in case of splitting many "homophones" can occupy the same entry; then one has also to specify the random selection rule of the homophones given the key and the message (for convenience, we rule out one-dimensional "objects", be they messages or keys, with zero probability; we assume $|X| \geq 2$). Decoding being deterministic, the same codeword appears *at most once* in each row of the encoding matrix.

Given a code XYZ we shall find it convenient to deal with the marginal couple YZ, forgetting about the random message X. We shall say that YZ is the *abstract code* derived from the *operational code* XYZ; actually we shall need all the abstract codes (random couples) YZ, even those which are not derivable from operational codes. In a way, we have the following inclusion: deterministic codes $\subseteq$ operational codes $\subseteq$ abstract codes.

THE JS-BOUND

So far our code might be a secrecy code, codewords being cryptograms, or even a source code, when only one key is there; we go now to authentication theory for good; cf /1,2/. In the *impersonation attack* a clever opponent, who ignores the key, chooses a codeword and sends it to the legal receiver in the hope that it will be accepted; the codeword is chosen so as to maximize the probability of fraud, P_I:

$$P_I = \max_y \ \text{Prob}(Z \in A_y), \quad \text{with } A_y = \{z: \text{Prob}(Y=y, Z=z) \neq 0\}$$

A_y is the set of keys which authenticate y and corresponds to the ones below y in the binary matrix (similarly, one defines A_z, the set of codewords authenticated by z, which corresponds to the ones on the right of z).

For P_I several lower bounds are known (below $I(Y;Z)=H(Y)+H(Z)-H(YZ)$ denotes mutual information in bits; $H(.)$ is Shannon entropy; bars denote size):

$$P_I \geq 2^{-I(Y;Z)} \qquad \text{Simmons bound} \qquad \text{Abstract}$$

$$P_I \geq \frac{|X|}{|Y|} \qquad \text{combinatorial bound} \qquad \text{Universal}$$

The first is abtract, that is it holds over the larger class of abstract codes; the second is universal, that is it is independent of (robust w.r.t.) source statistics. Actually the combinatorial bound can be easily generalized to

$$P_I \geq \frac{\min_z |A_z|}{|Y|}, \text{ which is both abstract and universal.}$$

For deterministic codes a standard manipulation of information quantities (cf /1/) shows that Simmons bound becomes: $P_I \geq 2^{H(X)-H(Y)}$; of course the combinatorial bound can be always written as $P_I \geq 2^{\log|X|-\log|Y|}$. Typically $|Y|>>|X|$, so Simmons bound is better; however, for this silly binary code below the combinatorial bound is better: $H(X)\cong 0$, $H(Z)=1$, $Y=X\oplus Z$ (then $H(Y)=1$). Simmons bound gives approximately $\frac{1}{2}$, the combinatorial bound gives correctly 1. Recently, Johannesson and this author /3/ have put forward a new bound, which is a sort of "universal strengthening" of Simmons bound; it improves also on the combinatorial bound. Below we shortly rederive the JS-bound, so as to show that it holds over the large class of abstract codes.

The starting point is this: P_I is defined through A_y, which is defined through joint probabilities codeword-key which differ from zero: it doesn't matter how much they differ from zero! Actually P_I depends *only* on distr(Z) and on the binary matrix for key-codeword admissibility, and so, in particular, P_I is itself "universal". (There is more to it, since P_I does not even depend on the possible correlation between X and Z; cf /3/).

Given an abstract code YZ consider the stochastic matrix $W=Y|Z$ of the conditional probabilities codeword-given-key. Observe that if one "binarizes" W by writing ones instead of its non-zero entries one re-obtains the binary matrix! Take Y*Z* s.t. Z* has the same distribution as Z, $W*=Y*|Z*$ has the same zeroes as W. Then $A_y=A_y*$. Then $P_I=P_I*$. So, the following chain holds:

for any $W*\sim W$ $P_I \geq 2^{-I(Y*,Z*)}$

$P_I \geq 2^{-\inf I(Y*,Z*)}$ inf w.r.t. all $W*\sim W$

$P_I \geq 2^{-\min I(Y*,Z*)}$ min w.r.t. all $W*\leq W$

(W*~W means that the stochastic matrix W* has exactly the same zeroes as W; W*≤W means that W* has at least the same zeroes as W; in the last line we simply closed the open minimization set: there are examples where the minimizing W* has more zeroes than W, and so lies on the boundary; cf /3/). Observe that the minimum in the exponent of the JS-bound can be interpreted as a suitable rate-distortion function evaluated at zero: so, the very efficient algorithm available for the latter can be used to compute it. (An intriguing question: why did a rate-distortion function show up in this authentication-theoretic context?)

For W*~W with uniform rows:

$$I(Y^*,Z^*) = H(Y^*) - E_Z \log |A_Z| \leq \log|Y| - \min \log |A_Z|$$

and one re-obtains the (abstract) combinatorial bound.

A GENERALIZED ATTACK

After this preliminary material on impersonation, we still have to pause on a boring insert devoted to a new "formal" attack: *conditional-constrained impersonation* . Let E and F be non-void sets of codewords and set:

$$P_{I(E,F)} = \max_{y\in F} \text{Prob}\{Z \in A_y | Y \in E\}$$

The opposer is constrained to choose his codeword inside F and can use the information $Y \in E$: this attack generalizes both impersonation and *substitution* of codeword c; in the latter case $F=\{y: y \neq c\}$, $E=\{c\}$. We shall prove a "reduction theorem": from YZ a new *abstract* code Y*Z* will be constructed, such that, under mild regularity assumptions:

$$P_{I(E,F)}(YZ) = P_I(Y^*Z^*)$$

Once one is able to "reduce" the fraud probability for the new attack to the fraud probability in an impersonation attack, all the (abstract) lower bounds obtained for impersonation are usable for the new attack! The construction of the new code follows. For Z* take Z|Y∈E (throw away zero-probability keys); Y* takes its values in F; to obtain Y*|Z*=z pump up the probabilities for Y|Z=z so as to make them sum to 1 (provided this pumping is feasible: this is the regularity assumption, which however will turn out to be trivially met when we shall need it). In symbols

$$\Pr\{Z^*=z\} = \Pr\{Z=z|Y\in E\}, \quad z\in \cup_{y\in E}A_y$$

$$\Pr\{Y^*=y|Z^*=z\} = \alpha^{-1}\Pr\{Y=y|Z=z\}, \quad y\in F$$

$$\text{with} \quad \alpha = \alpha(F) = \sum_{y\in F}\Pr\{Y=y|Z=z\}$$

The regularity condition is simply that α should be strictly positive for all z's in the range of Z*; in other terms, any key which authenticates at least a codeword in E should authenticate at least a codeword in F. Since this reduction theorem is an unimaginative generalization of a result given in /4/ for the case of substitution attacks details of the proof are omitted.

CODES WITH SPLITTING

Let's turn to (operational) codes with splitting. In a pure impersonation model they are *useless* (if one throws away homophones, that is ones in the binary matrix, P_I does not increase). However, authentication codes with splitting cannot be disposed of for very serious reasons, two of which follow. First: homophony (splitting) is a brilliant idea for secrecy ciphers and so splitting can be good in a mixed authentication-secrecy model, even if authentication is restricted to impersonation. Second: as shown by Simmons, splitting is essential in the mixed impersonation-substitution model which he has proposed, where the opposer is free to play either impersonation or substitution, according to which one pays better. Even so, however, a difficulty arises: we shall argue below that usual substitution is *not* a relevant attack to mount in the case of codes with splitting.

We shall make a distinction between two attacks: *codeword substitution* (substitution as defined usually) versus *message substitution*. These are genuinely different attacks when there is splitting. The point is the following. If the opposer substitutes the legal codeword by one of its homophones the system is safe: so, why declare his substitution successful? In the attack of message substitution we demand not only that the codeword is successfully substituted, but also that it is decoded to a message different from the legal message, so that havoc is brought about in the system. The probability of success is then:

$$P_{MS}(c) = \max_{y(\neq c)} \text{Prob}\{Z \in A_y - H_{y,c} | Y=c\}, \text{ with } H_{y,c}=\{z: g(y,z)=g(c,z)\}$$

By averaging with respect to $\text{Pr}\{Y=c\}$ one has the overall probability of message substitution $P_{MS}=\sum_y \text{Pr}\{Y=c\} \ P_{MS}(c)$; (the probabilities of codeword substitution are defined similarly, only omitting the conditions $Z \notin H_{y,c}$; observe that, unlike codeword substitution, message substitution does not make sense for abstract codes, since it explicitly involves messages).

Examples. Beside the case of impersonation and codeword or message substitution, below we consider also two probabilities of *deception* : $P_d=\max(P_I,P_{CS})$, as defined by Massey /2/, and $P_\delta=\max(P_I,P_{MS})$, which is a natural analogue of P_d in the case of message substitution. In general $P_\delta \leq P_d$; equality holds for deterministic codes (and for some probabilistic codes). (Deception as defined by Simmons is a more complicated game-theoretic notion; in his case the maximum is only a lower bound to P_d). The following examples /5/ show that, unlike in the case of pure impersonation, deleting homophones can be detrimental to the performance of the code in the case of substitution or deception.

Consider the three codes C_1, C_2 and C_3 specified by the encoding matrices below. The message is a fair coin, while the key probabilities are $\frac{3}{7}$, $\frac{3}{7}$ and $\frac{1}{7}$, top to bottom. C_2 and C_3 are obtained from C_1 by adding a codeword (by adding a one in the admissibility matrix). The homophones are equiprobable.

C1		C2		C3	
y1	y2	y1,y4	y2	y1	y2
y3	y4	y3	y4	y3,y5	y4
y5	y3	y5	y3	y5	y3

In the table below results are given in 56-ths to help comparisons; the easy computations are omitted.

	P_I	P_{CS}	P_d	P_{MS}	P_δ
C_1	32	52	52	52	52
C_2	48	51	51	51	51
C_3	32	56	56	48	48

In the case of C_2 splitting helps whenever substitution is involved, but it is harmful for impersonation; there is no practical difference between codeword and message substitution. In the case of C_3 the proposed splitting is catastrophic for *codeword* substitution; however the same splitting is advantageous for *message* substitution; it does no harm even to impersonation taken by itself, and so C_3 should be definitely preferred to both C_1 and C_2.

These examples show that deterministic coding can be "pointwise" improved by splitting, but they do not answer a deeper question: are there cases when (asymptotically) optimal encoding is necessarily probabilistic? Actually, at the moment such a question is not even well-defined, since a Shannon-theoretic framework for authentication theory is still in the make (cf /6/).

A REDUCTION THEOREM FOR MESSAGE SUBSTITUTION

Now we prove a reduction theorem for message substitution, which reduces it to *abstract* impersonation. From XYZ construct a new operational code XY'Z (only the random codeword changes) with an extra codeword d which takes the place of the homophones of c in the encoding matrix. Set $E=\{c\}$, $F=\{y: y\neq c,d\}$. Then:

$$P_{MS}(c) = P_{I(E,F)}(XY'Z) = P_I(Y_c Z_c)$$

where $Y_c Z_c$ is obtained from Y'Z in the same way as Y*Z* was obtained from YZ in the boring insert.

Proof. The first equality follows from the following obvious facts: Y'=c iff Y=c, $Z\in A'y$ iff $Z\in Ay-Hy,c$. The second follows from the insert; $|X|\geq 2$

ensures that the regularity assumption is met (recall that d is a homophone of c and so occupies the same entry of the encoding matrix).

Consequently, abstract lower bounds for impersonation can be recycled to bounds for message substitution. From each of them, one learns (necessary but not sufficient) conditions a code should meet to be a good code. Simmons bound recycled tells that Y and Z should be strongly correlated given c, but correlation due to homophones does not count. The JS-bound recycled improves on this and tells that only "deterministic" correlation (as measured by the infimum mutual information) matters:

$$P_{MS}(c) \geq 2^{-\inf I(Y^*;Z_c)}$$

where Y^*Z_c is constrained to have the same admissibility matrix as Y_cZ_c.

The abstract combinatorial bound yields instead:

$$P_{MS}(c) \geq \frac{|X|-1}{|Y|-1}, \quad \text{and therefore} \quad P_{MS} \geq \frac{|X|-1}{|Y|-1}.$$

Proof. Going from XYZ to XY'Z the number of codewords increases at most by 1; going from Y'Z to Y_cZ_c it decreases at least by 2; so $|Y_c| \leq |Y|-1$. Assume that c decodes to x under key z in XYZ, or XY'Z; at least $|X|-1$ more codewords are needed for key z to decode to the remaining $|X|-1$ messages; none of these can be equal to d, since $g(d,z)=g(c,z)$. So $|A^c_z| \leq |X|-1$ for any z.

A similar combinatorial bound was well-known for codeword substitution; it does not implies ours, though, as $P_{CS}(c)$ can be strictly greater than $P_{MS}(c)$.

We think that the foregoing vindicates the role of abstract codes in Simmons theory of authentication.

REFERENCES

/1/ G. J. Simmons "A survey of information authentication", Proceedings of the IEEE, may 1988, 603-620

/2/ J. Massey "An introduction to contemporary cryptology", Proceedings of the IEEE, may 1988, 533-549

/3/ R. Johannesson, A. Sgarro "Strengthening Simmons' bound on impersonation", submitted

/4/ A. Sgarro "Informational-divergence bounds for authentication codes", Proceedings of Eurocrypt 89, Houthalen, April 1989

/5/ M.-G. Croatto "Codifica probabilistica nei sistemi di segretezza e autenticazione", tesi di laurea, Università di Udine, marzo 1990

/6/ A. Sgarro "Towards a coding theorem in authentication theory", IEEE Information Theory Workshop, Eindhoven, June 1990

Essentially ℓ-fold secure authentication systems

Albrecht Beutelspacher
Justus Liebig-Universität Gießen
Mathematisches Institut
Arndtstr. 2
D-6300 Gießen

Ute Rosenbaum
Siemens AG
ZFE IS SOF 4
Otto Hahn-Ring 6
D-8000 München 83

Abstract

In this paper we first introduce the notion of essentially ℓ-fold secure authentication systems; these are authentication systems in which the Bad Guy's chance to cheat after having observed ℓ messages is – up to a constant – best possible. Then we shall construct classes of essentially ℓ-fold secure authentication systems; these systems are based on finite geometries, in particular spreads and quadrics in finite projective spaces.

1. Introduction

An **authentication system** A consists of an set S of **source states**, a set K of **keys**, and a set M of authenticated **messages** along with a mapping e: S × K → M. The mapping e together with the keys defines a set E of different **encoding rules**.
For a subset M' of M we denote by E(M') the set of encoding rules which have all m ∈ M' as possible messages.

In the authentication systems which we are going to deal with every message uniquely determines its source state. These authentication systems offer no secrecy. They were first studied by Gilbert, MacWilliams and Sloane [5].

In this paper we investigate the security of authentication system with respect to spoofing attacks of order ℓ. This means that a Bad Guy has observed ℓ messages authenticated with the same key. He is looking for the maximum probability in order to substitute another message.

We define p to be the maximum of the probabilities $p_0, ..., p_\ell$, where p_i is the probability of success when the Bad Guy has observed i different messages. Furthermore, we define the real number n by $n = 1/p$. We call n the ℓ-order of the authentication system **A**.

We assume throughout that the source states and the keys are equally distributed and independent.

Fåk [5] has proved the following result. *Let* **A** *be an authentication system without secrecy, in which the source states are uniformly distributed. If ℓ messages are observed, then*

$$p := \max\{p_0, ..., p_\ell\} \geq |E|^{-1/(\ell+1)} ;$$

equality holds if and only if

$$|E(M')| = |E|^{(\ell+1-i)/(\ell+1)}$$

for any $M' \subset M$, $|M'| = i$, $i = 0, ..., \ell + 1$, *where the* $m \in M'$ *belong to different source states.*

Definition. The authentication system **A** is called ℓ-**fold secure** if equality holds in the above result. A 1-fold secure system is also called **perfect** according to Simmons [8] (see also [4]).

In an ℓ-fold secure authentication system one has $p = |E|^{-1/(\ell+1)}$ and so its ℓ-order can be computed as $n = |E|^{1/(\ell+1)}$. Moreover, it follows that n is an integer. (*In fact,* $n = |E|^{1/(\ell+1)} = |E(M')|$ *for any ℓ-subset M' of M.*)

Definition. Let **A** be an authentication system and m be a fixed message. Then the **derivation** A_m of **A** with respect to m is defined as:
The set S_m of *source states* of A_m consists of the source states of **A** except for the source state belonging to m. The set E_m of *encoding rules* consists of the encoding rules of **A** under which m is a possible message, i.e. $E_m = E(m)$. The set M_m of *messages* consists of the messages of **A** which are possible under an encoding rule of E_m and a source state of S_m.

1.1 Lemma. *If m is a message in an ℓ-fold secure authentication system* **A***, then* A_m *is an $(\ell-1)$-fold secure authentication system.*

The proof is a direct consequence of the definition.□

1.2 Lemma. *The number σ of source states in a ℓ-fold secure authentication system of ℓ-order n is at most $n + \ell$.*

Proof. In [2] it was proved that the number of source states in a 1-fold secure authentication system is at most $1/\sqrt{|E|} + 1 = n + 1$. Thus, the assertion follows in view of Lemma 1. $\square$

For $\ell = 1$ and $\ell = 2$ there exist examples satisfying $\sigma = n + \ell$. We shall show that equality in the case $\ell = 2$ implies that n is an *even* integer. Consequently, if in an ℓ-fold secure authentication system one has $\sigma = n + \ell$ and $\ell \geq 2$, then n must be an even integer. For $\ell > 2$ one can construct examples with $\sigma = n + 1$. Recently Mitchell, Walker and Wild [7] have investigated ℓ-fold secure authentication systems with maximum number of source states.

ℓ-fold secure authentication systems have two disadvantages: They have very few source states compared to the number of keys. Also they seem to be very rare. For this reasons we introduce the notion of an essentially ℓ-fold authentication system.

Definition. Let C be a class of infinitely many authentication systems such that any two members of C have a different number of keys. Thus we can use the number of keys as index for the elements of C:

$$C = \{A_k \mid k = \text{number of keys in } A_k\}.$$

We say that C consists of **essentially ℓ-fold secure authentication systems**, if there is a constant c such that

$$p \leq c \cdot k^{-1/(\ell+1)}$$

for any authentication system A_k in C. We shall call a single authentication system **essentially ℓ-fold secure** if it is clear from the context in which essentially ℓ-fold secure class it is contained.

We believe that the above definition is quite useful, since it covers classes of authentication systems which are still unconditional secure, but for which there is a much larger flexiblity to construct them than ℓ-fold secure authentication systems in the strong sense.

Note, that we use the term "essentially secure" in a stronger sense as in [1,2,3].

3.1 Lemma. *Let* $C = \{A_q \mid q \text{ some index}\}$ *be a class of essentially ℓ-fold secure authentication systems with* $p \leq c^* \cdot 1/q$ *for any authentication system* A_q. *Then for each* A_q *we have for the number of keys k*

$$k \geq a \cdot q^{\ell+1}$$

with a constant a independent from q and k.

Proof. By Fåk [5] the probability of deception p is larger than $k^{-1/(\ell+1)}$ Hence

$$k^{-1/(\ell+1)} \leqq p \leqq c^*\cdot 1/q .$$

So, $k \geqq c^{*\cdot(\ell+1)}\cdot q^{\ell+1}$. $\square$

We present three classes of examples for essentially secure authentication systems. These examples are based on partial spreads and quadrics in finite projective spaces. Finally we shall deal with implementational aspects.

2. The maximum number of source states in a 2-fold secure authentication system

2.1 Theorem. *Let* **A** *be a 2-fold secure authentication system, denote by* σ *its number of source states. Then* $\sigma \leqq n + 2$. *If* $\sigma = n + 2$, *then* n *must be even.*

Sketch of the proof. It is easy to see that $\sigma \leqq n + 2$. Suppose therefore that $\sigma = n + 2$. One first shows certain regularity conditions: Every source state is on exactly n messages, every message is on exactly n^2 keys, etc.. Then it is possible to prove that any two keys share either 0 or exactly 2 messages. Using this, it follows that n must be even. $\square$

Example of a 2-fold secure authentication system with $n + 2$ messages (cf. [2]). Fix a point P_0 in a 3-dimensional projective space **P** of even order n. Since n is even, there is a set S of $n + 2$ lines through P_0 such that no three of which are in a common plane. Define S to be the source states, where the keys are the n^3 planes not through P_0 and the messages are the points $\neq P_0$ on the lines of S. This yields a 2-fold secure authentication system.

Remark. If, in the above example, we choose *all* lines through P_0 as source states, then the resulting authentication system is not essentially 2-fold secure. Indeed, if the Bad Guy has observed two messages P, Q (belonging to the source states P_0P and P_0Q), he knows that the key is a plane through the line PQ. Thus he is able to authenticate any source state L which is a line through P_0 intersecting PQ in a point $R \neq P, Q$. Then R is the message belonging to the source state L.

3. Some classes of essentially ℓ-fold secure authentication systems

Example 1 (cf. [2]). Consider a hyperbolic quadric Q (ruled quadric) in **P** = PG(3,q), the 3-dimensional projective space of odd order q. The following facts are well known (cf. [6]):

- Q is covered by two sets R and R', each consisting of $q + 1$ skew lines. The set R is called a **regulus**.

- Every plane of P intersects Q either in two lines or in an **oval** (that is a set of $q + 1$ points no three of which are collinear).

- There are sets S of q^2-q skew lines such that $S \cup R$ is a **spread** that is a set of skew lines covering all points of P.

We shall construct an authentication system $A(q)$ for any prime power q. Denote by $S \cup R$ a spread in $P = PG(3,q)$, where R is a regulus. The *source states* are the lines of S, the *keys* are the points on the lines of R, and the *messages* are the planes through the source states.

3.1 Theorem. *Let* $A(q)$ *be the above defined authentication system.*

(a) $A(q)$ *has* q^2-q *source states and* $(q + 1)^2$ *keys.*

(b) $p_0 = 1/(q + 1)$, $p_1 = 2/(q + 1)$.

(c) $A(q)$ *is essentially 1-fold secure.*

Proof. (b) Since each of the $q + 1$ planes through a source state contains the same number $q + 1$ of keys, we have $p_0 = 1/(q + 1)$. Assume that the Bad Guy knows a message m, that is a plane through a source state S_m. This plane intersects the set of keys in an oval, and the actual key is one point of the oval. Assume now that the Bad Guy wants to authenticate a source state $S^* \neq S_m$. The line S^* intersects m in a point X. For the Bad Guy it is sufficient to find a line L through X in m containing the actual key. (Then $<S^*,L>$ is the message.) Since any line of m through X contains at most two keys, his chance p_1 of success is at most $2/(q + 1)$.

(c) follows with $c = 2$. $\Box$

Note, that in $A(q)$ the number of source states has the same order of magnitude as the number of keys.

Now we change roles. Let Q be as above. Define $A'(q)$ as follows. The *source states* are the lines of R, the *keys* are the points outside R, and the *messages* are the planes containing a line of Q.

3.2 Theorem. *Let* $A'(q)$ *be the above defined authentication system.*

(a) $A'(q)$ *has* $q + 1$ *source states and* q^3-q *keys.*

(b) $p_0 = 1/(q + 1)$, $p_1 = 1/q$, $p_2 = 1/(q-1)$.

(c) $A'(q)$ *is essentially 2-fold secure.*

Proof. (b) The fact that $p_0 = 1/(q + 1)$ follows as above. Assume now that the Bad Guy knows a message m, that is a plane through a line S_m of R. This plane intersects the points of the lines of R in two lines S_m and T. The line T contains exactly one point of any source state. Suppose that the Bad Guy wants to authenticate $S^* \neq S_m$, which intersects m in a point $X \neq S_m \cap T$ of T. Observe, that T contains no key, but any of the q lines $\neq$ T through X contains exactly q–1 keys. Thus, p_1 = 1/q.

Finally, assume that the Bad Guy has observed two messages m and m' with source states S_m and $S_{m'}$. The planes m and m' intersect in a line L. Since the actual key is a point of L, L must contain at least one, hence exactly q–1 keys. The remaining two points of L are points of S_m and $S_{m'}$, respectively. It follows that any other source state S* is a line skew to L. Thus, in order to authenticate S*, the Bad Guy has to choose one of the q–1 keys on L, which gives q–1 distinct planes through S*. Hence p_2 = 1/(q–1).

(c) For $c \geq 6^{1/3}$ it follows

$$1/(q-1) \leq c \cdot (q^3-q)^{-1/3}$$

for any $q \geq 2$. $\square$

Remark. We have not only proved that **A'(q)** is essentially 2-fold secure; it satisfies the following stronger condition: Given *any* two messages, the Bad Guy has only a chance of 1/(q–1) to authenticate any third source state. (Note that "essentially 2-fold secure" only means, that, given a *randomly chosen* pair of messages, the Bad Guy has a certain chance to cheat.)

We will review the examples constructed in [3] using quadrics. It will turn out that these examples are essentially ℓ-fold secure authentication systems with large ℓ.

Example 2 (cf. [3]). Consider the quadrics Q in P = PG(d,q), q even, with the following properties:
–   Q does not contain the point N = (1,0,...,0),
–   any line through N intersects Q in exactly one point.

It is easy to check that these are exactly those quadrics satisfying an equation of the following type

$$f(x) = x_0^2 + \sum_{1 \leq i \leq j \leq d} a_{ij} \cdot x_i \cdot x_j = 0$$

with $a_{ij} \in GF(q)$.

Now we can define an authentication system $A_d(q)$ as follows. The *source states* are all lines through N and the *keys* are all quadrics of the above defined form. The *message* belonging to a source state S and a key K is the unique point of intersection of the line S with the quadric K. This authentication system contains $q^{d(d+1)/2}$ keys and $q^{d-1} + q^{d-2} + ... + q + 1$ source states; it is essentially [d(d + 1)/2 – 1]-fold secure. We shall first deal with the case d = 2.

3.3 Theorem. *Let* $A_2(q)$ *be the above defined authentication system for* $d = 2$. *Then* $A_2(q)$ *has* $q + 1$ *source states and* $k = q^3$ *keys. It is* 2-*fold secure.*

Proof. Since $d = 2$ these quadrics have the equation

$$f(x) = x_0^2 + a_{11} \cdot x_1^2 + a_{22} \cdot x_2^2 + a_{12} \cdot x_1 \cdot x_2 = 0, \text{ where } a_{ij} \in GF(q).$$

Each of the a_{ij} could be any of the q elements of $GF(q)$, so the number of keys (which are all possible quadrics of the above form) is q^3. The message, belonging to a source state S, could be any of the q points $\neq N$ on S. This is also true, if one message, that is a point $\neq N$ of P, is known. Thus, $p_0 = 1/q$ and $p_1 = 1/q$. Suppose, two messages P and Q, belonging to the source states S_P and S_Q, are known. If S^* is an arbitrary source state different from S_P and S_Q, then there exists exactly one quadric through each point $\neq N$ of S^*, which yields $p_2 = 1/q$.

We have $p := \max\{p_0, p_1, p_2\} = 1/q = k^{-1/3}$, so $A_2(q)$ is 2-fold secure in the strong sense.

3.4 Theorem. *Let* $A_d(q)$ *be the above defined authentication system.*
(a) $A_d(q)$ *has* $q^{d-1} + q^{d-2} + \ldots + q + 1$ *source states and* $q^{(d+1)d/2}$ *keys.*
(b) $A_d(q)$ *is essentially* $[d(d+1)/2 - 1]$-*fold secure.*

Proof. (a) Each set of different a_{ij}'s, $1 \leq i \leq j \leq d$, yields different quadrics of the type

$$f(x) = x_0^2 + \sum_{1 \leq i \leq j \leq d}' a_{ij} \cdot x_i \cdot x_j = 0.$$

Since the number of points i,j with $1 \leq i \leq j \leq d$ is $(d+1)d/2$ and each a_{ij} can take q different values, the number of keys of $A_d(q)$ is $q^{(d+1)d/2}$.

In $P = PG(d,q)$ there are $q^{d-1} + q^{d-2} + \ldots + q + 1$ lines through a point, so the number of source states is $q^{d-1} + q^{d-2} + \ldots + q + 1$.

(b) As for $A_2(q)$ it follows that $p_0 = 1/q$, $p_1 = 1/q$ and $p_2 = 1/q$. Let us first compute p_3.

Suppose that two messages P and Q belonging to the source states S_P and S_Q are known. Denote by E the plane through S_P and S_Q. It intersects Q in a quadric Q'. Since in E the quadric Q' has similar properties as Q in P, Q' is determined by any three of its points. If the Bad Guy happens to observe three messages P, Q and R, where the associated source states are in one plane, he can authenticate each of the other $q-2$ other source states in this plane.

To calculate p_3, we have to look at the definition of p_i. The probability p_i is defined as the probability that the Bad Guy succeeds in substituting a message belonging to a source state, which wasn't sent, given that he has observed i different messages. Thus

$$p_i = \sum_{M' \subseteq M, |M'| = i} p(M') \cdot \text{payoff}(M'),$$

where **M** is the set of messages, **M'** a subset of order i, and payoff(**M'**) the probability that the Bad Guy succeeds given that he has observed the messages in **M'**.

The probabilty to observe a set of messages depends only on the probability of the set of source states. Also payoff(**M'**) is constant for all **M'** that belong to the same set of source states S'. Hence in $A_d(q)$,

$$p_i = \sum_{S' \subseteq S, |S'| = i} p(S') \cdot \text{payoff}(S').$$

With this, we can calculate p_3. If the Bad Guy observes the messages P, Q, and R, where the corresponding source states are in a common plane, he can cheat with probability 1. In all other cases, his probability of success is only 1/q. Thus

$$
\begin{aligned}
p_3 &= p(S_R \subseteq \, <S_P, S_Q>) \cdot 1 + p(S_R \not\subseteq \, <S_P, S_Q>) \cdot 1/q \\
&= (q-1)/(q^{d-1} + \ldots + q - 1) + (q^{d-1} + \ldots + q - 1 - (q-1))/(q^{d-1} + \ldots + q - 1) \cdot 1/q \\
&= (q^{d-2} + \ldots + 2q - 1)/(q^{d-1} + \ldots + q - 1) \\
&\leq 2 \cdot 1/q.
\end{aligned}
$$

Now we proceed to the general case and show that $p_i \leq c \cdot 1/q$ for $i \leq d(d+1)/2$.

If the Bad Guy can determine the intersection of the key (quadric) Q with some subspace U of PG(d,q), then each point of the quadric $Q \cap U$ is a valid authenticator. If $Q \cap U$ contains at least one message not already observed he can cheat with probability 1. In all other case the probability of cheating is 1/q.

Hence, if p^* denotes the probability that the Bad Guy cannot determine the intersection of the key with some subspace containing unobserved messages, then

$$p_i = (1 - p^*) \cdot 1 + p^* \cdot 1/q \leq (1 - p^*) + 1/q.$$

It remains to show that

(*)
$$p^* \geq \frac{q^n + \sum\limits_{i=0}^{n-1} b_i \, q^i}{q^n + \sum\limits_{i=0}^{n-1} a_i \, q^i}$$

for $a_i, b_i \in N_0$, $n \in N$, all independent from q.

(Then

$$p^* \geq \frac{q^n + \sum\limits_{i=0}^{n-1} b_i\, q^i}{q^n + \sum\limits_{i=0}^{n-1} a_i\, q^i} \geq 1 - \frac{\sum\limits_{i=0}^{n-1} (a_i - b_i)\, q^i}{q^n + \sum\limits_{i=0}^{n-1} a_i\, q^i} \geq 1 - c \cdot \frac{1}{q}$$

for a suitable c independent of q. So $p_i \leq (1 - p^*) + 1/q \leq (c+1) \cdot 1/q$.)

We have to calculate the probability p^* that the Bad Guy cannot determine the intersection of the quadric Q with some subspace U of $PG(d,q)$ containing at least one unobserved message.

The Bad Guy knows i different messages m_j, $1 \leq j \leq i$, i.e. points of the quadric Q. In a t-dimensional subspace U, $t \geq 1$, of $<m_1,...,m_i>$ he can determine the quadric $Q \cap U$ if he knows at least n_u messages contained in U, where $n_u = (t+2) \cdot (t+1)/2 - 1$ if $N \notin U$ and $n_u = (t+1)t/2$ if $N \in U$. For this he needs at least $(t+2) \cdot (t+1)/2 - 1$ source states in a $(t+1)$-dimensional subspace or $(t+1)t/2$ source states in a t-dimensional subspace. For cheating with probability 1 the quadric $Q \cap U$ must contain at least one unobserved message, in particular we must have $t \geq 2$.

Thus,

$p^* = $ prob (Bad Guy cannot determine the intersection of the quadric Q with some subspace U of $PG(d,q)$ such that $Q \cap U$ contains at least one unobserved message)

$\geq$ prob (no $(t+1)t/2$ source states are in a t-dimensional and no $(t+2)(t+1)/1 - 1$ source states are in a $(t+1)$-dimensional subspace of $PG(d,q)$, $t \geq 2$)

$=$ prob (among i points of $PG(d-1,q)$ there are no $(t+1)t/2$ in a $(t-1)$-dimensional and no $(t+2)(t+1)/2-1$ in a t-dimensional subspace for all t, $2 \leq t < d$)

$\geq$ prob (among i points of $PG(d-1,q)$ there are no $(t+1)$ in a $(t-1)$-dimensional subspace for all t, $2 \leq t \leq d$)

$=: p'$

because $t + 1 \leq (t+1)t/2$ and $t + 2 \leq (t+2)(t+1)/2 - 1$ for $t \geq 2$.

Denote by s the minimum of $d-1$ and i. Then

$$p' = \prod_{j=2}^{s} \text{prob (the } (j+1)\text{-th point is not contained in the subspace generated by the first } j \text{ points)}$$

$$\cdot \prod_{j=s+1}^{i-1} \text{prob (the } (j+1)\text{-th point is not contained in any of the } \binom{j}{d-1}$$

$$(d-2)\text{-dimensional subspaces generated by the first } j \text{ points)}$$

$$= \prod_{j=2}^{s} \frac{\#\text{points in } PG(d-1,q) - \#\text{points in a } j-\text{dimensional space}}{\#\text{points in } PG(d-1,q) - j}$$

$$\cdot \prod_{j=s+1}^{i-1} \frac{\#\text{points in } PG(d-1,q) - \#\text{points on the} \binom{j}{d-1} (d-2)-\text{dimensional subspace}}{\#\text{points in } PG(d-1,q)}$$

$$\geqq \prod_{j=2}^{i-1} \frac{\#\text{points in } PG(d-1,q) - \binom{j}{d-1} \#\text{points in } PG(d-2,q)}{\#\text{points in } PG(d-1,q)}$$

$$= \prod_{j=2}^{i-1} \frac{q^{d-1}+q^{d-2}+\ldots+1 - \binom{j}{d-1}(q^{d-2}+q^{d-3}+\ldots+1)}{q^{d-1}+q^{d-2}+\ldots+1}$$

Because i is independent of q, we have shown (*).

So, $A_d(q)$ is essentially $[d(d+1)/2-1]$-fold secure. $\square$

Example 3 (cf. [3]). Consider the quadrics Q in $P = PG(d,q)$, with the following properties:

- Q contains the point $N = (1,0,...,0)$,
- $x_1 = 0$ is tangential hyperplane in N.

It is easy to verify that these are exactly those quadrics satisfying an equation of the following type

$$f(x) = x_0 \cdot x_1 + \sum_{1 \leq i \leq j \leq d} a_{ij} \cdot x_i \cdot x_j = 0$$

with $a_{ij} \in GF(q)$.

Now we can define the authentication system $A'_d(q)$ as follows. The *source states* are all lines through N not in the hyperplane $x_1 = 0$ and the *keys* are all quadrics of the above defined form. The *message* belonging to a source state S and a key K is the unique point of intersection of the line S with the quadric K different from N.

3.5 Theorem. Let $A'_d(q)$ be the above defined authentication system.

(a) $A'_d(q)$ has q^{d-1} source states and $q^{(d+1)d/2}$ keys.

(b) $A'_d(q)$ is essentially $[d(d+1)/2-1]$-fold secure.

The *proof* is similar to the proof of Theorem 3.4. $\square$

This example can be generalized in the following way.

In $PG(d,q)$ the equation

$$f(x) = x_0 \cdot x_1^{n-1} + \phi^n(x_1,...,x_d) = 0,$$

where ϕ^n is a homogeneous polynomial of degree n, represents a hypersurface T.

T has multiplicity $n-1$ at $N = (1,0,...,0)$. The tangent cone in N has the equation $x_1^{n-1} = 0$, and the number of undetermined coefficients of the above equation is $\binom{d+n-1}{n}$.

Now the authentication system can be defined as follows. The *source states* are the q^{d-1} lines of $PG(d,q)$ through N, which do not lie on the hyperplane $x_1 = 0$. The *keys* are the hypersurfaces with an equation of the above form. The *message* belonging to a source state S and a key K is the unique point of intersection different from N of the line S with the hypersurface K.

The authentication system is essentially ℓ-fold secure with $\ell = \binom{d+n-1}{n} - 1$.

For $n = 2$ the hypersurfaces are the quadrics of Theorem 3.5.

4. Implementation

We discuss a possible implementation of the example 1 in section 3.

We represent a *point* of $P = PG(3,q)$ by its homogeneous coordinates $(x_0,x_1,x_2,x_3) \neq (0,0,0,0)$, where the x_i are in $GF(q)$, the field with q elements. Two such 4-tuples (x_0,x_1,x_2,x_3), (y_0,y_1,y_2,y_3) represent the same point if there exists an $h \in GF(q)$ with $(x_0,x_1,x_2,x_3) = h \cdot (y_0,y_1,y_2,y_3)$.

The *planes* are the sets of all points (x_0,x_1,x_2,x_3) satisfying an equation

$$a_0 \cdot x_0 + a_1 \cdot x_1 + a_2 \cdot x_2 + a_3 \cdot x_3 = 0 \quad (a_i \in GF(q), \text{ not all } a_i = 0).$$

Thus, we can represent any plane by a 4-tuple $[a_0,a_1,a_2,a_3]$, this 4-tuple being unique up to a factor $k \neq 0$.

The *lines* of P through the distinct points (x_0,x_1,x_2,x_3) and (y_0,y_1,y_2,y_3) consist of all the points with homogeneous coordinates

$$a \cdot (x_0,x_1,x_2,x_3) + b \cdot (y_0,y_1,y_2,y_3), \text{ with } a, b \in GF(q);$$

We represent this line by $<(x_0,x_1,x_2,x_3), (y_0,y_1,y_2,y_3)>$.

Given a line g through the points (a_0,a_1,a_2,a_3) and (b_0,b_1,b_2,b_3), and a point $P = (x_0,x_1,x_2,x_3)$ not on g, then the plane $m = [m_0,m_1,m_2,m_3]$ through g and P is obtained by solving the following system of linear equations:

$$a_0 \cdot m_0 + a_1 \cdot m_1 + a_2 \cdot m_2 + a_3 \cdot m_3 = 0$$
$$b_0 \cdot m_0 + b_1 \cdot m_1 + b_2 \cdot m_2 + b_3 \cdot m_3 = 0$$
$$p_0 \cdot m_0 + p_1 \cdot m_1 + p_2 \cdot m_2 + p_3 \cdot m_3 = 0$$

As the set S of source states we shall take a "regular spread" in PG(3,q) with $q \equiv$ 3 mod 4. This has the advantage that the coding of the "real" source states into the lines of S is easily performed:

Such a regular spread S can be described as follows. It consists of the lines

$$g_{h,k} = \;<(h,k,1,0), (-k,h,0,1)> \qquad h,k \in GF(q)$$

and the line

$$g_\infty = \;<(1,0,0,0), (0,1,0,0)>.$$

Moreover

$$R = \{g_\infty\} \cup \{g_{k,0}|\; k \in GF(q)\}$$

is a regulus contained in S.

Now, if q is a prime, we can represent the source states by a tuple (h,k) of integers with $0 \leqq h < q, 0 < k < q$. The tuple (h,k) is will be encoded by the source state $g_{h,k}$. The keys are represented by a tuple (k_1,k_2) with $0 \leqq k_1, k_2 \leqq q$, where k_1 identifies the line of R and k_2 is a point of this line. A message is represented by the homogeneous coordinates of the plane, i.e. a tuple (m_0,m_1,m_2,m_3) with $0 \leqq m_i < q$, $i = 0,...,3$.

To authenticate a source state with a key and to determine the source state from a message, one has only to solve a system of three linear equations.

References

[1]   Beutelspacher, A., Perfect and essentially perfect authentication systems. Proc. Eurocrypt 87, Lecture notes in Computer Science **304**, p. 167-170, 1988.

[2]   Beutelspacher, A., Rosenbaum U., Geometric Authentication Systems. Ratio Math. 1 (1990), 39-50.

[3]   Beutelspacher, A.,Tallini, G., Zanella C., Examples of essentially s-fold secure geometric authentication systems with large s. To appear in Rend. mat. Roma.

[4]   Gilbert, E.N., MacWilliams, F.J., Sloane, N.J.A., Codes which detect Deception. The Bell System Technical Journal, vol. **53**, no. 3, pp. 405-424, March 1974.

[5]   Fåk, V., Repeated use of Codes which Detect Deception. IEEE Transactions in Information Theory, vol. IT-25, no. 2, pp. 233-234, March 1979.

[6]   J.W.P. Hirschfeld, Finite projective spaces of three dimensions. Clarendon Press, Oxford 1985.

[7]   Mitchell, C., Walker, M., Wild, P., The combinatorics of perfect authentication schemes. To appear.

[8]   Simmons, G.J., Authentication Theory / Coding Theory. Proc. of Crypto 84, Lecture Notes in Computer Science **196**, pp. 411-432, 1985.

On the construction of authentication codes with secrecy and codes withstanding spoofing attacks of order $L \geq 2$

Ben Smeets, Peter Vanroose and Zhe-xian Wan

Department of Information Theory
University of Lund
Box 118
S-221 00 Lund, Sweden

Abstract

We present an analysis of some known cartesian authentication codes and their modification into authentication codes with secrecy, with transmission rate $R = r/n$, where $n = 2, 3, \ldots$, and $1 \leq r \leq n - 1$ using $(n - r)(r + 1)$ q-ary key digits. For this purpose we use a grouping technique.

Essentially the same key grouping technique is used for the construction of codes that withstand spoofing attacks of order $L \geq 2$. The information rate of this scheme is also r/n, and it requires $(L + r)(n - r)$ q-ary key digits. Moreover these codes allow that previously transmitted source states can be reused.

1 Introduction

Authentication codes are codes which allow two trusting parties to communicate information to each other in the presence of an opponent who might submit false messages and/or substitute legally transmitted messages by false ones that will misinform the receiver. One of the first constructions of such codes were the codes given by Gilbert, MacWilliams and Sloane [1]. Their constructions stem from the projective spaces over finite fields.

In this paper we address two different problems. The first problem is the construction of authentication codes which also achieve perfect secrecy. That is, given an observed transmitted message, the opponent obtains no information about the source state that was transmitted by the legal sender. Recently various authors have presented some constructions most of them stemming from combinatorial designs [2,3]. The other problem is the construction of codes for multiple authentication. The latter type of codes have to withstand the so-called spoofing attacks of order $L \geq 2$. Most researchers investigated this problem under the ad-hoc assumption that previously transmitted source states were excluded from subsequent transmissions. For a discussion of the origin of this assumption and its effects we refer to [4]. In this paper we extend the concept of spoofing attack [5] so that it allows the transmitter to repeat previously used source states.

In [1] a modification of the code is presented allowing a simpler logic circuitry than for the orginal code. In Section 2, we analyse this construction in more detail and introduce a grouping technique of the encoding rules. This grouping is then used in the code constructions given in the subsequent sections.

In Section 3 we use the grouping technique to obtain codes that achieve perfect secrecy and in Section 4 we discuss how to use it to obtain codes for multiple authentication.

Finally, Section 5 contains a description of a generalization of the construction allowing a tradeoff between the security requirements and the desire to achieve high information rates.

2 The Modified GMS Scheme and a Grouping Technique

Let q be a prime power. Gilbert, MacWilliams and Sloane [1] constructed an authentication code from the projective plane $\mathbf{PG}(2, \mathbf{F}_q)$ over the field $\mathbf{F}_q$ with q elements as follows. Fix a line ℓ in $\mathbf{PG}(2, \mathbf{F}_q)$. The points on ℓ are regarded as the source states, the points not lying on ℓ are regarded as the encoding rules and the lines different from ℓ are regarded as the messages. Given a source state s and an encoding rule e, there is a unique line passing through s and e, which will be called m, i.e., $m = \mathfrak{L}(s, e)$. The source state s will now be encoded into the message m by the encoding rule e. This is an authentication code with $q + 1$ source states, q^2 encoding rules and $q^2 + q$ messages.

The probabilities of successfull impersonation and substitution attack of this code are

$$P_I = P_S = \frac{1}{q}.$$

This scheme will be referred to as the GMS scheme and the corresponding code is a so-called cartesian code because each message conveys the transmitted source state. For $q = 2$, the GMS scheme can be represented by the following encoding table:

	a	b	c	d	e	f
e_0	0			1	2	
e_1		0	1		2	
e_2	0		1			2
e_3		0		1		2

Table 1: The GMS Scheme derived from $\mathbf{PG}(2, \mathbf{F}_2)$

where 0, 1, 2 are the 3 source states, being all points of a given line ℓ, e_0, e_1, e_2, e_3 are the encoding rules, being the 4 other points, and a, b, c, d, e, f are the messages, being the lines different from ℓ.

In Section V of reference [1] the authors also gave a modification of their scheme. For this modified scheme one chooses in addition to the line ℓ also a fixed point P on ℓ. The source states (they are q in number) are then all the points on ℓ different from P, the encoding rules are all the points not on ℓ and the messages are all the lines not passing through P. We obtain a code with q source states, q^2 encoding rules and q^2 messages. This scheme will be referred to as the Modified GMS scheme. For this code we also have $P_I = P_S = 1/q$. In Table 2 we illustrate the modification with the case $q = 2$.

	a	b	c	d	
e_0	0		1		P
e_1	0			1	P
e_2		0	1		P
e_3		0		1	P

Table 2: The Modified GMS scheme derived from $\mathbf{PG}(2, \mathbf{F}_2)$.

For this example, we observe that the four encoding rules can be grouped into two groups, $\{e_0, e_3\}$ and $\{e_1, e_2\}$, such that each message contains exactly one encoding rule from each group [4]. This kind of *grouping* holds in the general case. The groups are the lines through P.

Theorem 1: In the Modified GMS scheme, if we group the q^2 encoding rules into q groups, such that each group consists of the q rules lying on a line through P, then each message contains exactly one encoding rule from each group.

Proof: This follows from the property of the projective plane that any two lines meet in exactly one point. □

3 Authentication codes with perfect secrecy

We continue with the example of the foregoing section:

	Cartesian						Perfect Secrecy				
	a	b	c	d			a	b	c	d	
e_0	0		1		P		e_0	0		1		P
e_1	0			1	P		e_1	1			0	P
e_2		0	1		P		e_2		1	0		P
e_3		0		1	P		e_3		0		1	P

Table 3:The Construction of a code with perfect secrecy from the Modified GMS scheme derived from $\mathbf{PG}(2, \mathbf{F}_2)$.

Above we interchanged 0 and 1 in the rows of the second group of encoding rules $\{e_1, e_2\}$. In the new code, each message contains two source states, one 0 and one 1. Thus perfect secrecy is achieved. This is the well-known "simplest" example of an authentication code with secrecy, see [7].

For the general case, we label the source states of the Modified GMS scheme with $0, 1, \ldots, q - 1$ and the groups of encoding rules also with $0, 1, \ldots, q - 1$. We change the encoding table of the Modified GMS scheme according to the following procedure. Replace source state s lying in the row of the encoding rule e and the column of the message m in the encoding table, where e belongs to the group i, by $s - i(\bmod q)$. That is, m will be used to transmit the source state $s - i(\bmod q)$ under the encoding rule e. We can prove the following:

Theorem 2: Each message in the new authentication code contains each of the q source states $0, 1, \ldots, q - 1$ exactly once, thus we have perfect secrecy for this code.

Proof: In the encoding table of the original Modified GMS construction, each column, i.e., each message, contains one source state q times. Theorem 1 guarantees that these are replaced by q different source states. □

Finally we note that the information rate of this code is 1/2, i.e., the same as the rate for the Modified GMS scheme. The key information required to specify the encoding rule used is two q-ary digits (considering the case that the encoding rules are selected according to a uniform probability distribution).

4 Authentication codes withstanding spoofing attacks of order $L \geq 2$

The previous geometrical scheme (with reduced source state set and with grouping of keys) can also be turned into a code that can be used L consecutive times, having probability of deception $1/q$ at each use. The information rate of this scheme is also 1/2, and it requires $2 + (L - 1)$ q-ary key digits. But we can no longer guarantee the source state to be secret from the second use on.

At the first time slot, the scheme of the previous paragraph (with or without secrecy) is used, consuming the first 2 q-ary key digits. At any of the next $L - 1$ time slots, the group label of the key used at the previous slot is incremented by one (modulo q), and 1 extra q-ary key digit is used to determine the key inside this new group to be used for this time slot, using the same geometrical scheme as at the first time slot. The number of key digits required per transmission is asymptotically optimal ($L \to \infty$), i.e., one q-ary digit per transmission. These codes have a probability $1/q$ of deception at every transmission.

Group $i - 1$ Group i Group $i + 1$

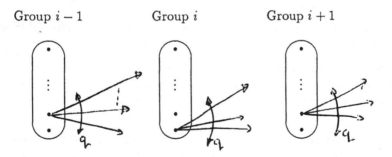

Figure 1: Illustration of the paths of possible encoding rules

5 Generalizations

The modification technique of Gilbert, MacWilliams and Sloane and our grouping technique can be applied to several generalizations of the GMS scheme. For instance, we can use some authentication codes given by Beutelspacher [6]. We shall present a new generalization below, to which both the modification technique and the grouping technique can be applied. It will give us a generalization of the two dimensional schemes of the previous sections to the n-dimensional projective space.

Fix an r-flat $\mathcal{L}$ of the n-dimensional projective space $\mathbf{PG}(n, \mathbf{F}_q)$, where $1 \le r \le n-1$. The source states are all the t-flats, $0 \le t < r$, contained in $\mathcal{L}$. Encoding rules are all $(n-r-1)$-flats which have empty intersection with $\mathcal{L}$. The messages are the $(n-r+t)$-flats intersecting $\mathcal{L}$ in a t-flat. The message corresponding to source state s and encoding rule e is the unique $(n-1)$-flat containing s and e. For this code we have:

The number of source states: $\begin{bmatrix} r+1 \\ t+1 \end{bmatrix}_q$ [1)]

The number of encoding rules: $q^{(n-r)(r+1)}$

The number of messages: $q^{(n-r)(r-t)} \begin{bmatrix} r+1 \\ t+1 \end{bmatrix}_q$

The probability of impersonation: $P_I = q^{-(n-r)(r-t)}$

The probability of substitution: $P_S = q^{-(n-r)}$

The information rate: $\ge \frac{t+1}{n-r+t+1}$ (for large q).

Substitution is done with the highest probability of success by replacing the legal message by a message having an as large as possible intersection with the legal one, i.e., an $(n-r+t-1)$-flat. For $r = n-1$ and $t = n-2$, this is the generalization given in [1].

Now take $t = r - 1$. The modification of the code is carried out as follows. Fix a point P on $\mathcal{L}$. Source states are all $(r-1)$-flats of $\mathcal{L}$ *not containing* P. Encoding rules are the same as above and the messages are the $(n-1)$-flats which do not contain P. For the modified code we have:

The number of source states: q^r

The number of encoding rules: $q^{(n-r)(r+1)}$

The number of messages: q^n

The probability of impersonation: $P_I = \frac{1}{q^{n-r}}$

The probability of substitution: $P_S = P_I$

The information rate: $\frac{r}{n}$.

The grouping of the encoding rules is done as follows. Consider all possible $(n-r)$-flats through P, intersecting $\mathcal{L}$ only in P. The encoding rules are grouped such that encoding rules lying in a common $(n-r)$-flat of the above type belong to the same group. This results in the partitioning of the $q^{(n-r)(r+1)}$ encoding rules into $q^{(n-r)r}$ groups of q^{n-r} encoding rules such that each message contains exactly one encoding rule from each group. Thus from this modified code we can construct an authentication code with

---

[1)] A Gaussian binomial coefficient $\begin{bmatrix} r+1 \\ t+1 \end{bmatrix}_q \stackrel{\text{def}}{=} \frac{(q^{r+1}-1)\dots(q^{r-t+1}-1)}{(q^{t+1}-1)\dots(q-1)}$

perfect secrecy as well as an authentication code withstanding spoofing attacks of order $L \geq 2$ as before. But in the case of codes for multiple authentication we have to spend additional $(n - r)$ q-ary digits per additional use of the code. That is $(L + r)(n - r)$ q-ary key digits in total. Hence, the asymptotic number of key digits per transmission is $n - r$.

References

[1] E.N. Gilbert, F.J. MacWilliams and N.J.A. Sloane, "Codes which detect deception", *Bell Syst. Techn. J.* **53**, pp. 405–424, 1974.

[2] M. De Soete, "Some constructions for authentication - secrecy codes", Eurocrypt'88, Davos, Switzerland, May 25-27, 1988, in *Advances in Cryptology — Eurocrypt '88*, Ed. C.G. Günther, Springer-Verlag, Berlin, pp. 469-472, 1988.

[3] D.R. Stinson, "A construction for authentication codes/secrecy codes from certain combinatorial designs", J. Cryptology, **1**, pp.119-127, 1988.

[4] G. Simmons, B. Smeets, "A paradoxical result in unconditionally secure authentication codes – and an explanation", *IMA Conference on Cryptography and Coding*, Dec. 18-20, 1989, Cirencester, England. to appear.

[5] J.L. Massey, "Cryptography – A selective survey", in *Digital Communications*, pp.3-21. 1986.

[6] A. Beutelspacher, "Perfect and essentially perfect authentication schemes", *Advances in Cryptology — EUROCRYPT '87*, **LNCS 304**, Springer-Verlag, Berlin, pp. 167–170, 1987.

[7] J.L. Massey, "An introduction to contemporary cryptology", *Proc. IEEE* **76**, pp. 533–549, 1988.

Cryptanalysis of a public-key cryptosystem based on approximations by rational numbers *

Jacques Stern
Équipe de Logique
Université de Paris 7
et
Département de mathématiques et informatique
École Normale Supérieure

Philippe Toffin
Département de mathématiques
Université de Caen

Abstract

At the Eurocrypt meeting, a public-key cryptosystem based on rational numbers has been proposed [2]. We show that this system is not secure. Our attack uses the LLL algorithm. Numerical computations confirm that it is successful.

1 The proposed cryptosystem

We briefly review the article of H.Isselhorst [2], in which the system was described. The secret key consists of a large prime number p, (a size of 250 decimal digits is suggested), together with a (small) integer k and a (k, k)-matrix A, with an inverse A^{-1} mod p.

The public part of the system essentially consists of a matrix

$$C = (c_{i,j})_{1 \leq i \leq k \ \& \ 1 \leq j \leq k}$$

where $c_{i,j}$ is computed from A and a fixed public integer t, $1 \leq t < p$, by truncating the decimal expansion of $ta_{i,j}/p$ after n digits, which we write

$$c_{i,j} = \text{Float}(ta_{i,j}/p, n)$$

Also included in the public key data are integers z and m satisfying inequalities which will be given later on

The plaintext is a vector X, with k coordinates, all of them being positive integers bounded by m. The encryption is as follows:

*Research supported by the PRC mathématiques et informatique

- Compute $U = C.X$
- Set $V = U \bmod t$, where the mod function is applied coordinatewise
- Output $Y = \text{Float}(V, z)$, where the Float function is applied coordinatewise

We now explain why and how the ciphertext can be decoded. If u_i (resp. v_i) is the ith coordinate of U (resp. V), we can write:

$$\frac{pv_i}{t} = \frac{pu_i}{t} \bmod p$$

and using the definition of Y,

$$\frac{py_i}{t} = \frac{pv_i}{t} - \frac{pe_i}{t} \text{ with } 0 \le e_i < 10^{-z}$$

thus, for some integers α_i, we get

$$\frac{py_i}{t} = \alpha_i p + \frac{pv_i}{t} - \frac{pe_i}{t}$$

which gives

$$\frac{py_i}{t} = \alpha_i p + \frac{p}{t} \sum_j c_{i,j} x_j - \frac{pe_i}{t}$$

$$= \alpha_i p + \frac{p}{t} \sum_j \left(\frac{ta_{i,j}}{p} - r_{i,j} \right) x_j - \frac{pe_i}{t}$$

$$= \alpha_i p + \sum_j a_{i,j} x_j - \frac{p}{t} \sum_j r_{i,j} x_j - \frac{pe_i}{t}$$

Noting that

$$0 \le r_{i,j} < 10^{-n}$$

we get that the sum of the last two terms is bounded by

$$\frac{p}{t} 10^{-n} km + \frac{p}{t} 10^{-z}$$

Now, if both terms are bounded by 1/4, then the real value of

$$\sum_j a_{i,j} x_j$$

can be easily recovered from p, by rounding py_i/t and reducing mod p. From these values, the original message is obtained via A^{-1}

The inequalities that are needed to carry through the above argument are easy consequences of thoses which are proposed in the paper [2], namely

$$10^{70} \le m \le p/10$$

$$4kpm/t \le 10^n < p^2 10^{-50}/t$$

$$10^{z-1} \le 4p/t < 10^z$$

Presumably, the other inequalities have been added to ensure security.

2 A cryptanalytic attack

Our attack uses the LLL algorithm [3], very much in the same way as known attacks against the knapsack-based cryptosystem (see [1]). It can be described as follows

• Pick four distincts values c_1, c_2, c_3, c_4 among the k^2 possible $c_{i,j}$'s, included the largest one c_1. Set

$$\gamma_i = 10^n c_i \quad 1 \leq i \leq 4$$

Note that the γ_i's are integers.

• Apply the LLL algorithm to the 4-dimensional lattice generated by the columns of the matrix

$$\begin{pmatrix} 1 & 0 & 0 & 0 \\ -\gamma_2 & \gamma_1 & 0 & 0 \\ -\gamma_3 & 0 & \gamma_1 & 0 \\ -\gamma_4 & 0 & 0 & \gamma_1 \end{pmatrix}$$

• Output the first coordinate a_1 of the first vector of the reduced basis of L, obtained through LLL.

We claim that a_1 is precisely the original value $a_{i,j}$, corresponding to the largest of the $c_{i,j}$, which was choosen as c_1. We will give a heuristic justification of this fact. The argument can actually be put on a firmer theoretical basis by a precise probabilistic analysis. Anyhow, as will be seen in section 3, the success of the attack is confirmed by numerical experiments.

First observe that, for $i = 1, ..4$, if we denote by $a_1, ..a_4$ the values of $a_{i,j}$ corresponding to $c_1, ..c_4$, we have

$$0 \leq \frac{t a_i}{p} - c_i < 10^{-n}$$

which gives , by linear combination

$$|a_1 c_i - a_i c_1| \leq 10^{-n} p, \quad i = 2, 3, 4$$

multiplying by 10^n, we get

$$|a_1 \gamma_i - a_i \gamma_1| \leq p, \quad i = 2, 3, 4$$

together with the inequality $1 \leq a_1 < p$, this shows that the integers a_1, a_2, a_3, a_4 provide a linear combination V of the columns of the matrix of L, whose coordinates are bounded by p. Now, the determinant of L is γ_1^3. Because c_1 is the largest of the $c_{i,j}$'s, it is presumably close to t; thus γ_1 is close to $t 10^n$ so that the expected size of the coordinates of a short vector is about $3(n + \log_{10} t)/4$ digits. Letting $m = p^\alpha$, and using the fact that

$$4kpm/t \leq 10^n$$

we see that the expected value of the coordinates of a short vector of L should be bounded from below by $p^{3(1+\alpha)/4}$. If α is significantly greater than $1/3$, $p^{3(1+\alpha)/4}$ is definitely greater than p, so that the LLL algorithm will actually disclose the very short vector V, defined above, whose first coordinate is precisely a_1.

If we consider the size suggested in [2], namely 250 digits, we see that our attack is presumably successful when the size of the coded messages m is 85 digits or more. Of corse, for a smaller choice of m, it is possible to apply an analogous method, provided one chooses more than 4 values c_i and one apply the LLL algorithm in a larger dimension. For example, the 6D-version of the attack works as soon as α is significantly greater than $1/5$.

Once a_1 has been correctly recovered, p can be computed by rounding ta_1/c_1; this because of the inequality

$$\left| \frac{ta_1}{c_1} - p \right| \le \frac{p}{c_1 10^n} = \frac{p}{\gamma_1}$$

Similarly, the correct value of each $a_{i,j}$ is obtained by rounding $a_1 c_{i,j}/c_1$. This is because of the following inequality

$$\left| a_{i,j} - \frac{a_1}{c_1} c_{i,j} \right| \le \frac{p}{c_1 10^n} = \frac{p}{\gamma_1}$$

3 Numerical experiments

For numerical experiments, we used the Symbolic Computation System Maple. In all our experiments, we restricted ourselves to the case $k = 2$, which involves a (4,4)-matrix Mat to be reduced by LLL.

3.1 Main part of our program

In order to test our cryptanalytic attack, we first choose the 3 following parameters: t, n , whose role is explained in the previous sections and nb_of_digits which is the number of digits of the prime number p. We then choose randomly 4 nb_of_digits-long integers a_i, a_1 beeing the largest, and p a prime number greater than these 4 numbers. The c_i are the values of $Float(ta_i/p, n)$ and are the public key. The γ_i and the matrix Mat are then built as in section 2 and we obtain by LLL-reduction a new matrix new_Mat. We may assume that $new\_Mat_{1,1}$ is positive. Our algorithm fails if $new\_Mat_{1,1}$ is different of a_1, and if not, we let $new\_a_1 = new\_Mat_{1,1}$. We then get a value

$$new\_p = \text{closest\_integer}(t * 10^n \, new\_a_1/\gamma_1)$$

Again the attack fails if $new\_p \ne p$, and if not, we let

$$new\_a_i = \text{closest\_integer}(new\_a_1 * \gamma_i/\gamma_1)$$

We reach complete success if for each i, we have $new\_a_i = a_i$.

3.2 Results

We made 4 different trials under different values of the parameters.

- $t = 1$, $nb\_of\_digits = 20$, $n = 30$: 10 different runs reached complete success.
- The smallest bound for m suggested by Isselhorst beeing 70, when $t = 1$, we took $nb\_of\_digits = 121$, and $n = 192$: 2 different runs reached complete success.
- $nb\_of\_digits = 121$, and t is a randomly choosen 50-digits integer. As is clear from section 2, this allows a lower value for n. We set $n = 141$, which corresponds to messages m of length 70. 2 different runs reached complete success.
- The runs with the largest figures:

$t = 1$, $nb\_of\_digits = 250$, $n = 336$, which allows messages m of length 85. 4 different runs reached complete success.

3.3 Remarks

All our trials gave values which can be saved and can be used again. We also checked that for $t = 1$, $nb\_of\_digits = 20$, we get a failure as soon as $n \leq 27$. This is in accordance with the analysis os section 2.

3.4 Final conclusion

All these complete success justify the basic claim of our theoretical analysis above: The cryptosystem proposed by Isselhorst is not secure.

References

[1] E.F.Brickell, The cryptanalysis of knapsack cryptosystems. *Proceedings of the third SIAM discrete mathematics conference.*

[2] H.Isselhorst, The use of fractions in public-key cryptosystems. *Proceedings Eurocrypt'89.*

[3] A.K.Lenstra, H.W.Lenstra, L.Lovász, Factoring polynomials with rational coefficients. *Math. Annalen* 261 (1982) 515-534.

A Known-Plaintext Attack on Two-Key Triple Encryption

Paul C. van Oorschot   *Michael J. Wiener*

BNR
P.O. Box 3511 Station C
Ottawa, Ontario, Canada, K1Y 4H7

Abstract. A chosen-plaintext attack on two-key triple encryption noted by Merkle and Hellman is extended to a known-plaintext attack. The known-plaintext attack has lower memory requirements than the chosen-plaintext attack, but has a greater running time. The new attack is a significant improvement over a known-plaintext brute-force attack, but is still not seen as a serious threat to two-key triple encryption.

Key Words. triple encryption, cryptanalysis, DES.

1. Introduction

Due to questions raised (e.g., see [Diff77]) regarding the adequacy of security by the 56-bit key in the Data Encryption Standard (DES) [FIPS46], several varieties of multiple encryption have been considered. Given a few plaintext-ciphertext pairs, an exhaustive search defeats (single) DES in on the order of 2^{56} operations. Double DES encryption, using two independent 56-bit keys (see Figure 1), requires on the order of 2^{112} operations to attack by this naive approach. This may be reduced to on the order of 2^{56} operations and 2^{56} words of memory using a simple "meet-in-the-middle" attack [Diff77].

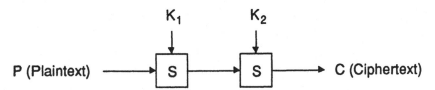

S is a private-key cryptosystem such as DES.

Figure 1: Double Encryption

Two-key triple DES (see Figure 2) can be defeated by the naive approach in on the order of 2^{112} operations. This may be reduced to on the order of 2^{56} operations and 2^{56} words of memory using a chosen-plaintext attack due to Merkle and Hellman which requires 2^{56} chosen-plaintext plaintext-ciphertext pairs [Merk81]. This latter attack, although impractical, is of interest in that it exhibits what Merkle and Hellman refer to as a "certificational" weakness in two-key triple encryption.

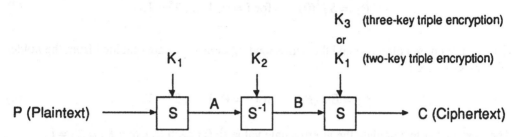

Figure 2: Triple Encryption

This paper presents a *known*-plaintext attack on two-key triple encryption. The Merkle-Hellman attack is first reviewed in §2. The new attack is presented in §3 and briefly analyzed in §4, showing it to require a running time on the order of $2^{120-log_2 n}$ operations and n words of memory, where n is the number of available plaintext-ciphertext pairs. This is the best known-plaintext attack on two-key triple DES that the authors are aware of. In §5, we consider a hardware implementation of the new attack using $n = 2^{32}$.

As with the Merkle-Hellman attack, the new attack poses no serious threat to two-key triple encryption in practice. However, it is of interest in that it may be used to both reduce the memory requirements and relax the chosen-plaintext condition in the Merkle-Hellman attack, and may lead to further advances. It is also highly amenable to parallel implementation. As with the Merkle-Hellman attack, the ideas discussed in this paper are not restricted to DES, but apply to any similar cipher.

2. The Merkle-Hellman Attack on Two-Key Triple Encryption

Let $C = S_K(P)$ denote that the plaintext P, enciphered using key K, results in ciphertext C. Then as in [Merk81], denote two-key triple encryption by the function $Enc()$:

$$C = Enc(P) = S_{K_1}(S_{K_2}^{-1}(S_{K_1}(P))). \tag{1}$$

Let A and B be the intermediate values in $Enc(P)$:

$$A = S_{K_1}(P) \quad \text{and} \quad B = S_{K_2}^{-1}(A). \tag{2}$$

The Merkle-Hellman attack finds the desired two keys $K_1 = \kappa_1$, $K_2 = \kappa_2$ by finding the plaintext-ciphertext pair such that intermediate value A is 0. The first step is to create a list of all of the plaintexts that could give $A = 0$:

$$P_i = S_i^{-1}(0) \qquad \text{for } i = 0, 1, ..., 2^{56}\text{-}1. \tag{3}$$

Each P_i is a chosen plaintext and the corresponding ciphertexts are obtained from the holder of keys κ_1 and κ_2:

$$C_i = Enc(P_i) \qquad \text{for } i = 0, 1, ..., 2^{56}\text{-}1. \tag{4}$$

The next step is to calculate the intermediate value B_i for each C_i using $K_3 = K_1 = i$.

$$B_i = S_i^{-1}(C_i) \qquad \text{for } i = 0, 1, ..., 2^{56}\text{-}1. \tag{5}$$

A table of triples of the following form is constructed:

$$(P_i \text{ or } B_i, i, flag),$$

where *flag* indicates either a P_i-type or B_i-type triple. Note that the 2^{56} values P_i from equation (3) are also potentially intermediate values B, by equation (2). All P_i and B_i values from equations (3) and (5) are placed in this table, and the table is sorted on the first entry in each triple, and then searched in order to find consecutive P and B values such that $B_i = P_j$. If $B_i = P_j$, then i, j is a candidate for the desired pair of keys κ_1, κ_2. This fact is illustrated in the two-key triple encryption depicted in Figure 3.

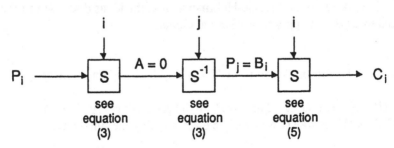

Figure 3: Two-Key Triple Encryption with a Candidate Pair of Keys

Because $C_i = Enc(P_i)$ for both the candidate pair of keys i, j and the desired keys κ_1, κ_2, it is reasonable to expect that the two pairs of keys might be equal. Each candidate pair of keys found from the sorted table is tested on a few other plaintext-ciphertext pairs to filter out "false alarms". The reason the attack succeeds is that a match $P_j = B_i$ is found in the table with $i = \kappa_1$; this is that i for which $S_{\kappa_1}(P_i) = 0$. Testing all candidate pairs guarantees that κ_1 and κ_2 will be found [Merk81].

3. Known-Plaintext Extension of the Merkle-Hellman Attack

Because the Merkle-Hellman algorithm computes a table based on the fixed value $A = 0$, and it is not known *a priori* which plaintext P results in the intermediate value $A = 0$, it is necessary to test all 2^{56} possibilities (i.e., $S_i^{-1}(0)$ for all possible keys i). Also, the attacker must request that each of these plaintexts be enciphered for him by his adversary. This makes the Merkle-Hellman attack far from practical. The idea for extending the algorithm is to remove the reliance on a single, fixed value of A; rather, we choose values for A at random, and for each choice, carry out a tabulation. We continue until a "lucky" choice of A is made which results in the success of the algorithm. As the attacker, we no longer require access to the adversary's $Enc()$ function. Instead, we assume that we are given n plaintext-ciphertext pairs.

The new algorithm proceeds as follows. Tabulate the (P, C) pairs, sorted or hashed on the plaintext values (see Table 1 in Figure 4). Table 1 is independent of A and requires $O(n)$ words of storage. Now randomly select and fix (for this stage of computation) a value a for A, and create a second table (see Table 2 in Figure 4) as follows. For each of the 2^{56} possible keys $K_1 = i$, calculate what the plaintext value would be if i were used for K_1:

$$P_i = S_i^{-1}(a).$$

Next, look up P_i in Table 1. If P_i is found in the first column of Table 1, take the corresponding ciphertext value C and compute the intermediate value

$$B = S_i^{-1}(C).$$

Place this value of B along with the key i into Table 2. Table 2 is sorted or hashed on the B values.

Each entry in Table 2 consists of an intermediate B value and corresponding key i which is a candidate for κ_1; as described above, each (B, i) pair is associated with a (P, C) pair from Table 1 which satisfies $S_i(P) = a$. The remaining task is to search for the desired value of K_2. For each of the 2^{56} candidate keys $K_2 = j$, calculate what the intermediate B value would be if j were used for K_2:

$$B_j = S_j^{-1}(a).$$

Next, look up B_j in Table 2. For each appearance of B_j (if any), the corresponding key i along with key j is a candidate for the desired pair of keys κ_1, κ_2. (To handle the rare case that a given B-value appears more than once in Table 2, a few bits could be added in Table 2 entries to indicate the multiplicity of each B-value.) Each candidate pair of keys (i, j) is tested on a few other plaintext-ciphertext pairs. If all of these additional (P, C) pairs have P mapped to C by the key pair (i, j), then $(i, j) = (\kappa_1, \kappa_2)$ and the task is complete.

This algorithm will find κ_1 and κ_2 the first time *any one* of the available (P, C) pairs has a first intermediate value $(S_{\kappa_1}(P))$ that is equal to a chosen a. If the algorithm does not succeed for a given a, the process is repeated for another value of A until ultimately the desired keys κ_1, κ_2 are found.

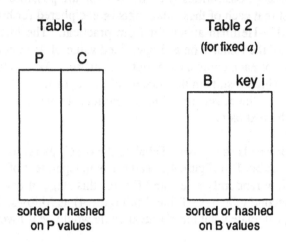

Figure 4: Tables used in the Known-Plaintext Attack

4. Time and Space Analysis

In this section, we briefly summarize the running time and memory requirements of the known-plaintext attack.

The time required for building and hashing Table 1 is the time required to hash n items. This time is dominated by other computations required in the attack, for $n < 2^{56}$. The space required for Table 1 is $O(n)$.

For each value of A that is tried, the time required to build Table 2 is on the order of 2^{56}, assuming that Table 1 is hashed on the plaintext values so that lookups take constant time. Because only 2^{56} out of 2^{64} possible texts are searched for in Table 1, the expected number of entries in Table 2 is $n/2^8$. This space is reusable across different values of A. The time required to work with Table 2 to find candidate pairs of keys is on the order of 2^{56}.

The probability of selecting a value of A that leads to success is $n/2^{64}$. The expected number of draws required to draw one red ball out of a bin containing n red balls and $N - n$ green balls is $(N + 1)/(n + 1)$ if the balls are not replaced. Therefore, assuming that one does not try the same value of a more than once, the expected number of values of a that must be tried is

$$(2^{64} + 1)/(n + 1) \approx 2^{64}/n \quad \text{for } n \text{ large.}$$

Thus, the expected running time for the attack is on the order of $(2^{56})(2^{64}/n) = 2^{120-log_2 n}$, and the space required is $O(n)$.

5. Parallel Hardware Implementation

In this section we present one possible parallel hardware implementation of the known-plaintext attack on two-key triple DES, assuming that $n = 2^{32}$ plaintext-ciphertext pairs are available. Given a number of assumptions concerning the cost of components and the performance that can be achieved by present-day technology, the illustrated implementation of the attack is shown to be four orders of magnitude faster (for an attacker with fixed resources) than a brute-force known-plaintext attack. This is the best known-plaintext attack the authors are aware of, but this attack is still not feasible. We conclude that two-key triple DES is currently not vulnerable to attack in practice.

The following hardware implementation is suitable for an attacker with a large amount of resources. We will assume that the attacker has 1 billion (10^9) dollars and $n = 2^{32}$ plaintext-ciphertext pairs available to him. Note that the execution time is not particularly sensitive to n (provided that n is not too small) because as n increases, the number of operations required for the attack ($2^{120-log_2 n}$) decreases, but memory requirements increase, and the number of machines that can be built with a fixed amount of money decreases.

Each machine for attacking two-key triple DES (see Figure 5) consists of a central component containing Table 1, and 512 peripheral components each containing its own version of Table 2 (for distinct sets of values for A).

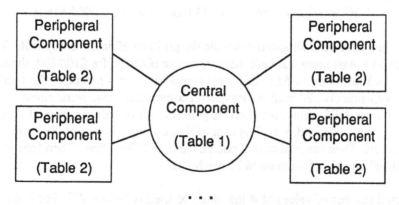

512 peripheral components in all

Figure 5: A Single Machine for Attacking Two-Key Triple DES

The function of the central component is to service requests from the peripheral components for the ciphertexts (if any) which correspond to a specified plaintext. In order to service these requests quickly, Table 1 is hashed on the plaintext values. To reduce overhead during table lookup of hashed values caused by hashing collisions, the density of the hashing table is restricted to 50%. In this case, the total memory required for Table 1 is

$$2(2^{32} \text{ words})(64 + 64 = 128 \text{ bits per word}) = 2^{40} \text{ bits.}$$

Assuming that bulk memory can be obtained for \$10/Mbit, the cost of this memory is approximately \$10 million.

If each memory chip is 1M x 1-bit, then Table 1 is organized as approximately 8000 rows, with 128 chips in each row. These rows are independent and can be accessed in parallel. This makes it possible for the central component to service the requests from the peripheral components in parallel. Each request will be directed to one of the 8000 rows. There should be few collisions among 512 requests out of 8000 rows. We will assume that the cost of the complex routing and arbitration circuitry required to make this work will double the cost of the memory making the total cost of a central component \$20 million.

We will assume that the average time required to service a request from a peripheral component is 250 ns. This may seem slow considering the current speed of memories, but this figure takes into account delays caused by the routing and arbitration circuitry, delays due to collisions among the 512 requests, and delays due to hashing collisions which lead to extra probes into Table 1.

The expected number of words required for Table 2 is $n/2^8 = 2^{24}$. Again, restricting the density of Table 2 to 50%, the total memory required for Table 2 is

$$2(2^{24} \text{ words})(64 + 56 + 4 = 124 \text{ bits per word}) \approx 4000 \text{ Mbits.}$$

(Four extra bits have been allocated to handle the problem of possible duplicate B-values as indicated in §3.) Assuming that bulk memory can be obtained for \$10/Mbit, the cost of this memory is \$40 000. For all 512 peripheral components in a machine, the total memory costs are approximately \$20 million. Peripheral components have some circuitry other than memory, such as DES chips, but there is just enough of this circuitry that the 250 ns request rate is not slowed down. The cost of this circuitry is negligible compared to the cost of memory. Then the total cost of one machine is \$40 million. Therefore, the attacker who has \$1 billion can afford to build 25 machines.

The expected number of values of A that must be tried is $2^{64}/n = 2^{32}$. For each value a of A, 2^{56} accesses of Table 1 are required to build Table 2. Also 2^{56} accesses of Table 2 are required to find all candidate pairs of keys. Assuming that accesses of Table 2 also require 250 ns, the expected time required to find the desired pair of keys is

$$(2^{32})(2^{56} + 2^{56})(250 \text{ ns}) / (25 \times 512 \text{ peripheral components}) \approx 4 \times 10^8 \text{ years.}$$

Next, we consider a brute-force known-plaintext attack. Analysis indicates that a DES chip could be built in volume for about \$10/chip [BNR]. A similar chip with added comparison circuitry and modified input/output could be built for about the same cost and used for attacking DES. The cost of building a machine for attacking two-key triple DES would include overhead in addition to the cost of the DES chips; assume this overhead cost to be roughly equal to the total cost of the DES chips. Then for $1 billion, the attacker could afford to build a machine with 50 million DES chips. Using current technology, each DES chip could perform a DES operation in about 500 ns. One would expect to have to search through about half of the 2^{112} pairs of keys, and testing each pair of keys requires 3 DES operations. Therefore, the expected time required for a brute-force search is

$$(3)(0.5)(2^{112})(500 \text{ ns}) / (50 \times 10^6 \text{ DES chips}) \approx 2.5 \times 10^{12} \text{ years.}$$

Therefore, the known-plaintext attack is approximately four orders of magnitude faster than a brute-force search, based on the assumptions made in the preceding arguments. However, this is of little practical consequence unless new ideas improve the running time of the former by several more orders of magnitude.

6. Conclusion

The new attack presented in this paper demonstrates a known-plaintext variation of the chosen-plaintext Merkle-Hellman attack, with a decreased memory requirement. The penalty that is paid for these improvements is increased running time.

The new attack gives approximately four orders of magnitude improvement over a brute-force known-plaintext attack, provided that a sufficient number of plaintext-ciphertext pairs are available. Despite the improvement, for practical purposes, two-key triple encryption remains currently invulnerable to known-plaintext attacks.

The authors encourage others to pursue other known-plaintext attacks on two-key triple encryption, which further reduce the running time.

References

[Merk81]   Merkle, R. and M. Hellman, "On the Security of Multiple Encryption", *Communications of the ACM*, vol. 24, no. 7, pp. 465-467, July 1981. See also *Communications of the ACM*, vol. 24, no. 11, p. 776, November 1981.

[Diff77]   Diffie, W. and M. Hellman, "Exhaustive Cryptanalysis of the NBS Data Encryption Standard", *Computer*, vol. 10, no. 6, pp. 74-84, June 1977.

[FIPS46]   "Data Encryption Standard", National Bureau of Standards (U.S.), Federal Information Processing Standards Publication (FIPS PUB) 46, National Technical Information Service, Springfield VA, 1977.

[BNR]   Internal study, BNR, Ottawa, 1989.

Confirmation that Some Hash Functions Are Not Collision Free

Shoji Miyaguchi   *Kazuo Ohta*   *Masahiko Iwata*

NTT Communications and Information Processing Laboratories

Nippon Telegraph and Telephone Corporation

1-2356, Take, Yokosuka-shi, Kanagawa, 238-03 Japan

Abstract: Hash functions are used to compress messages into digital signatures. A hash function has to be collision free; i.e., it must be computationally infeasible to construct different messages which output the same hash-value. This paper shows that five hash functions are not collision free, including the assumptions that an attacker can modify an initial value of the hash function. These hash functions are analyzed from the standpoints of their structure, the complementation property and the weak keys of the block ciphers used in them. As a result, it is clear that many pairs of messages can be created to generate the same hash-values. Therefore, users desiring to use these hash functions should be notified of their weakness.

1 Introduction

Digital signature techniques are methods that can confirm the contents of a communication message and its origin [1]. It is recommended to sign a compressed form of the message, the hash-value, in order to enhance digital signature efficiency. Hash functions are normally used for this purpose, especially for long messages. The hash function used to generate the hash-value has to be collision free; i.e., it must be computationally infeasible to construct different messages which output the same hash-value. When a hash function is not collision free, an attacker or malicious sender first signs a message M, and then later changes M into a false message M' that yields the same hash-value, and hence the same signature, as message M. Therefore, senders can abuse the signature system.

An n-bit hash function is a hash function that outputs n-bit hash-values. Some existing schemes [3, 4] are 64-bit hash functions, i.e., $n=64$, which include a 64-bit block cipher algorithm [9, 12]. However the birthday attack method [2, 6] can attack any n-bit hash function and about $2^{n/2}$ attacks have a success probability of about 0.5. Consequently, the collision free property requires that the length of the hash-value should be at least about 100-bits (128-bits to ensure a sufficient safety margin) to attain the comparable security of a 64-bit block cipher algorithm. Therefore, several $2n$-bit hash functions based on n-bit block cipher algorithms were recently proposed [5, 7].

In this paper, five hash functions are analyzed from the standpoints of their structure, the complementation property and the weak keys of the block ciphers used in them. We show that they are not collision free.

A hash-value is calculated from a message and an initial value. Although there are several ways of sharing the initial value between sender and verifier, we will consider the collision free property of the hash functions in the worst case scenario. That is, attacker might be able to modify both the message and the initial value. It is clarified under what conditions the hash functions are collision prone.

2 Notation

Hereafter, we use the following notation. Hash function H,
$H = h(M, I)$, is represented as follows:

$H_0 = I$ (Initial value)

$H_i = \theta(M_i, H_{i-1})$ for i from 1 to N

$H = h(M, I) = H_N$ (Hash-value)

where, $M = M_1\|\ldots\|M_i\|\ldots\|M_N$: message

|| : concatenation, ϕ : n-zeros

θ : the iterative function of the hash function

$eK(P)$: ciphertext block of plaintext block P, enciphered by an n-bit block cipher using key K

$dK(C)$: plaintext block of ciphertext block C, deciphered by an n-bit block cipher using key K

$X \oplus Y$: bitwise exclusive-or of data X and Y

$X\|Y$: concatenation of data X and Y

$\sim X$: complement of X (All bit inverse of X)

3 Meyer-Matyas hash function [8]

3.1 Algorithm

This is described as:

$H_0 = I$ (Initial value)

$H_i = \theta_1(M_i, H_{i-1}) = eH_{i-1}(M_i) \oplus M_i$ for i from 1 to N

$H = h_1(M, I) = H_N$

M_i , H_i , I : n-bit blocks

3.2 Consideration of collision prone property

We will describe four kinds of attacks. Three attacks are deterministic and the other one is probabilistic.

3.2.1 Attack 1

We assume that an n-bit block cipher satisfies the complementation property below:

$$e \sim K(\sim P) = \sim eK(P) \tag{1}$$

Then,

$$\theta_1(\sim M_i, \sim H_{i-1}) = e \sim H_{i-1}(\sim M_i) \oplus \sim M_i$$
$$= \sim eH_{i-1}(M_i) \oplus \sim M_i$$
$$= eH_{i-1}(M_i) \oplus M_i$$
$$= \theta_1(M_i, H_{i-1})$$

holds. Therefore, $h_1(M, I) = h_1(M', \sim I)$ for any given M and I, where $M = M_1\|M_2\| \ldots \|M_N$, $M' = (\sim M_1)\|M_2\| \ldots \|M_N$.

The DES [9] cipher has the complementation property shown in Equation(1). Thus, the Meyer-Matyas hash function is not collision free if the DES cipher is used in its iterative function.

3.2.2 Attack 2

When this hash function includes the DES cipher in its iterative function, the substantial bit length of H_i is 56. Therefore, an example of the collision prone property is found with 2^{28} attacks where the success probability is about 0.5 using the birthday attack [2, 6].

For example, define a message ξ and a mapping Ψ as follows:

$$\xi = M_1\|M_2\| \ldots \|M_{k-1}$$
$$\Psi(\xi) = K_k,$$

where K_k is the substantial bits of $H_{k-1}(k \geq 2)$. Keep both I and M_k constant, apply the birthday attack to Ψ in order to find ξ and ξ' as:

$$\Psi(\xi) = \Psi(\xi'), \xi \neq \xi'$$

Since K_k is 56-bits long, a pair of ξ and ξ' is found with 2^{28} attacks where the success probability is about 0.5. Define M and M' as:

$$M = \xi\|M_k\|M_{k+1}\| \ldots \|M_N , \quad M' = \xi'\|M_k\|M_{k+1}\| \ldots \|M_N.$$

Then, $h_1(M, I) = h_1(M', I)$ holds for any given I and $M_k\| \ldots \|M_N$ ($k \geq 2$).

3.2.3 Attack 3

We assume that an n-bit block cipher has the weak keys, K_1 and K_2, satisfying Equation(2) for an arbitrary n-bit block K.

$$eK_2(eK_1(K)) = K \tag{2}$$

Then,

$\theta_1(K, K_1) = eK_1(K) \oplus K$

$\theta_1(eK_1(K), K_2) = eK_2(eK_1(K)) \oplus eK_1(K) = eK_1(K) \oplus K$

The *DES* cipher has the weak keys. Thus, the Meyer-Matyas hash function is collision prone if the *DES* cipher is used in its iterative function. This attack can work well with the *Attack 1*.

3.2.4 Attack 4

We assume that an n-bit block cipher has the collision keys, K_x and K_y as:

$eK_x(P) = eK_y(P)$, $K_x \neq K_y$ for some data block P

Then,

$\theta_1(P, K_x) = eK_x(P) \oplus P = eK_y(P) \oplus P = \theta_1(P, K_y)$

The *DES* cipher has the collision keys. An example of collision keys, K_x and K_y is given [6, 11] below in hex.

$P = 04\ 04\ 04\ 04\ 04\ 04\ 04\ 04$

$K_x = \text{CE}\ 80\ 6\text{E}\ \text{EE}\ 7\text{C}\ \text{FC}\ \text{D2}\ \text{EC}$

$K_y = \text{AE}\ 88\ 38\ 90\ 48\ 74\ \text{C6}\ 06$

$eK_x(P) = eK_y(P) = 15\ 0\text{E}\ 0\text{B}\ 6\text{F}\ \text{F3}\ 5\text{B}\ 4\text{F}\ 0\text{E}$

Then, the Meyer-Matyas hash function is collision prone. This attack can work smoothly with the *Attack 1*.

4   Davies-Price hash function [4]

4.1 Algorithm

This is described as:

$$H_0 = I \text{ (Initial value)}$$
$$H_i = \theta_2(M_i, H_{i-1})$$
$$= eM_i(H_{i-1}) \oplus H_{i-1} \quad \text{for } i \text{ from 1 to } N$$
$$H = h_2(M, I) = H_N$$
$$M_i, H_i, I : n\text{-bit blocks}$$

4.2 Consideration of collision prone property

The basic idea to find collision prone examples will be described below. Find an n-bit block H and a message K so as to satisfy:

$$H = \theta_2(K, H) \tag{3}$$

Then, the following relation holds and the hash function is proven to be collision prone:

$$h_2(M, H) = h_2(M', H) = H,$$

where $M = K$, $M' = K \| \ldots \| K$.

Furthermore, find an initial value I and a message K^* so as to satisfy:

$$H = \theta_2(K^*, I),$$

where H is the value obtained above. The following relation holds:

$$h_2(M, I) = h_2(M', I) = H,$$

where $M = K^*$, $M' = K^* \| K \| \ldots \| K$.

4.2.1 Attack 5

Define an n-bit block H as follows:

$$H = dK(\phi) \quad \text{for arbitrary } K$$

Then Equation(3) holds. Thus the following relation holds:

$$h_2(M, dK(\phi)) = h_2(M', dK(\phi)) = dK(\phi),$$

where $M = K$, $M' = K \| \ldots \| K$.

This attack becomes so powerful because it combines the meet-in-the-middle attack [3, 8, 10] that it is used when an initial value is given.

Let I be any given initial value. A pair of K^* and K can be found with 2^{32} attacks so as to satisfy the following:

$$eK^*(I) \oplus I = dK(\phi)$$

Then,

$$h_2(M, I) = h_2(M', I) = dK(\phi) \quad \text{for any given } I,$$

where $M = K^*$, $M' = K^*\|K\|\ldots\|K$.

4.2.2 Attack 6

We assume that an n-bit block cipher satisfies the complementation property of Equation(1). Define an n-bit block H as follows:

$$H =\sim d \sim K(\sim \phi) \quad \text{for arbitrary } K$$

Then Equation(3) holds. Thus the following relation holds:

$$h_2(M, \sim d \sim K(\sim \phi)) = h_2(M', \sim d \sim K(\sim \phi))$$
$$=\sim d \sim K(\sim \phi),$$

where $M = K$, $M' = K\|\ldots\|K$.

This attack can be extended to the case an initial value is given as well as *Attack 5*.

Let I be any given initial value. A pair of K^* and K can be found with 2^{32} attacks so as to satisfy the following [3, 8, 10]:

$$eK^*(I) \oplus I =\sim d \sim K(\sim \phi)$$

Then,

$$h_2(M, I) = h_2(M', I) =\sim d \sim K(\sim \phi),$$

where $M = K^*$, $M' = K^*\|K\|\ldots\|K$.

4.2.3 Attack 7

We assume that there is a pair of keys, K_1 and K_2, in a block cipher satisfying Equation(4).

$$eK_2(eK_1(\phi)) = \phi \tag{4}$$

Note that the weak keys satisfy this equation. Then,

$$\theta_2(K_1, \phi) = eK_1(\phi) \oplus \phi = eK_1(\phi)$$
$$\theta_2(K_2, eK_1(\phi)) = eK_2(eK_1(\phi)) \oplus eK_1(\phi) = eK_1(\phi)$$

Thus, $h_2(M, \phi) = h_2(M', eK_1(\phi)) = eK_1(\phi)$,
where $M = K_1\|K_2\|\ldots\|K_2$, $M' = K_2\|\ldots\|K_2$.

4.2.4 Attack 8

We assume that there is a pair of keys, K_1 and K_2, in a block cipher satisfying Equation(5) and (6).

$$eK_2(eK_1(\sim \phi)) =\sim \phi \tag{5}$$

$$e \sim K_2(\sim eK_1(\sim \phi)) =\sim eK_2(eK_1(\sim \phi)) \tag{6}$$

Note that the weak keys satisfy Equation(5), and if the block cipher has the complementation property of Equation(1), the weak keys also satisfy Equation(6). Then,

$$\theta_2(K_1, \sim \phi) = eK_1(\sim \phi) \oplus \sim \phi =\sim eK_1(\sim \phi)$$
$$\theta_2(\sim K_2, \sim eK_1(\sim \phi)) = e \sim K_2(\sim eK_1(\sim \phi)) \oplus \sim eK_1(\sim \phi)$$
$$=\sim eK_1(\sim \phi)$$

Therefore,

$$h_2(M, \sim \phi) = h_2(M', \sim eK_1(\sim \phi)) =\sim eK_1(\sim \phi),$$
where $M = K_1\| \sim K_2\|\ldots\| \sim K_2$, $M' =\sim K_2\|\ldots\| \sim K_2$.

4.2.5 Attack 9

This attack is similar to *Attack 1*, that is, the complementation property of Equation(1) is assumed. Then,

$$\theta_2(\sim K_i, \sim H_{i-1}) = e \sim K_i(\sim H_{i-1}) \oplus \sim H_{i-1}$$
$$=\sim eK_i(H_{i-1}) \oplus \sim H_{i-1} = \theta_2(K_i, H_{i-1})$$

holds. Therefore, $h_2(M, I) = h_2(M', \sim I)$ for any given I, where $M = K_1\|K_2\|\ldots\|K_N$, $M' = (\sim K_1)\|K_2\|\ldots\|K_N$.

4.3 Summary

The above mentioned results are summarized in *Table 1*. The *DES* cipher has the complementation property of Equation(1) and its weak keys satisfy Equations(4) and (5). If the Davies-Price hash function includes the *DES* cipher, all the examples listed in *Table 1* exist.

5 Quisquater-Girault hash function (April version) [5]

In this section, the $2n$-bit hash function presented at Eurocrypt '89 [5] by Quisquater and Girault will be analyzed.

5.1 Algorithm

Figure 1 shows the iterative function, θ_3, of this $2n$-bit hash function. Its procedures are shown below, where H_i and M_i are $2n$-bit blocks, while the others are n-bit blocks.

$H_0 = b_{-1} \| b_0$ (Initial value)

$H_i = \theta_3(M_i, H_{i-1})$ for i from 1 to N

where

$M_i = m_{2i-1} \| m_{2i}$

$w_i = em_{2i-1}(b_{2i-3} \oplus m_{2i}) \oplus m_{2i} \oplus b_{2i-2}$

$b_{2i-1} = em_{2i}(w_i \oplus m_{2i-1}) \oplus m_{2i-1} \oplus b_{2i-3} \oplus b_{2i-2}$

$b_{2i} = w_i \oplus b_{2i-3}$

$H_i = b_{2i-1} \| b_{2i}$

$H = h_3(M, I) = H_N$ (Hash-value)

5.2 Consideration of collision prone property

Considering a state transition of hash-values for this hash function yields *Figure 2*. Typical examples are as follows:

Assume the key K in a block cipher satisfies Equation(7) and (8):

$$eK(eK(K)) = K \tag{7}$$

$$e \sim K(\sim eK(K)) = \sim eK(eK(K)) \tag{8}$$

Note that the weak keys satisfy Equation(7), and if the block cipher has the complementation property of Equation(1), the weak keys also satisfy Equation(8).

Define $H_0 = \sim \phi\|\phi$ and $M_1 = K\| \sim K$, then

$$H_1 = \theta_3(M_1, H_0) = \phi\|eK(K) \oplus K \qquad (9)$$

holds, where Equations(7) and (8) are used.

This state transition is indicated from the state $H_0(=\sim \phi\|\phi)$ to the state $H_1(= \phi\|eK(K) \oplus K)$ using the input $M_1(= K\| \sim K)$. The arrow from H_0 to H_1 drawn with a straight line means the weak keys and complementation property are necessary.

Define $M_2 = K\|K$, then

$$H_2 = \theta_3(M_2, H_1) = \phi\|\phi \qquad (10)$$

holds, where no condition is necessary.

Define $M_3 = K\|K$, then

$$H_3 = \theta_3(M_3, H_2) = \phi\|eK(K) \oplus K \qquad (11)$$

holds, where Equation(7) is used.

Here, because $H_3 = H_1$ holds, the two states, H_1 and H_2, are connected and form a loop. There are many initial values other than $\sim \phi\|\phi$ but all reduce to the state, H_1 or H_2, as shown in *Figure 2*.

Moreover, consider the complementation of an arbitrary message block M_i. Then

$$H_i = \theta_3(\sim M_i, H_{i-1})$$
$$= \theta_3(\sim m_{2i-1}\| \sim m_{2i}, H_{i-1}) = \theta_3(m_{2i-1}\|m_{2i}, H_{i-1})$$
$$= \theta_3(M_i, H_{i-1})$$

holds, where the complementation property of Equation(1) is used. Therefore, there are two invoking messages between each state in *Figure 2*. One message is always the complement of the other.

5.2.1 Attack 10

For any block cipher, the following relation holds:

$$h_3(M, \phi \| eK_a(K_a) \oplus K_a) = h_3(M', \phi \| eK_b(K_b) \oplus K_b) = \phi \| \phi,$$

where $M = K_a \| K_a$, $M' = K_b \| K_b$, K_a and K_b are arbitrary n-bit blocks, $K_a \neq K_b$.

5.2.2 Attack 11

This attack can modify any message component M_i to $\sim M_i$ without changing the initial value and the hash-value. The complementation property of Equation(1) is assumed. Then

$$\theta_3(\sim M_i, H_{i-1}) = \theta_3(M_i, H_{i-1})$$

holds for an arbitrary message block M_i. Thus the following relation holds:

$$h_3(M, I) = h_3(M', I) \text{ for any given } M \text{ and } I,$$

where $M = M_1 \| \ldots \| M_i \| \ldots \| M_N$,

$M' = M_1 \| \ldots \| M_{i-1} \| \sim M_i \| M_{i+1} \| \ldots \| M_N$.

5.2.3 Attack 12

The state transition diagram shown in *Figure 2* allows us to construct many examples of the collision prone property.

For example, when we use the three state transitions, Equations (9), (10), (11), the following relation holds:

$$h_3(M, \sim \phi \| \phi) = h_3(M', \sim \phi \| \phi) = \phi \| \phi,$$

where

$$M = K \| \sim K \| K \| K,$$

$$M' = K \| \sim K \| K \| K \| (K \| K \| K \| K) \| \ldots \| (K \| K \| K \| K).$$

If this hash function includes the *DES* cipher in its iterative function, all the collision prone examples obtained from *Figure 2* exist, because the *DES* cipher has the complementation property and the weak keys.

6 Quisquater- Girault hash function (October version) [7]

6.1 Algorithm

This scheme was proposed as a modification of the Quisquater-Girault hash function (April version) by originators [7]. In this algorithm, a new message block, M_{N+1}, is introduced as a pseudo message block.

$M_1, \ldots, M_N$: real message blocks, where $M_i = m_{2i-1} \| m_{2i}$

M_{N+1}: a pseudo message block, where $M_{N+1} = m_{2N+1} \| m_{2N+2}$

$$m_{2N+1} = m_1 \oplus m_2 \oplus \ldots \oplus m_{2N-1} \oplus m_{2N}$$

$$m_{2N+2} = m_1 + m_2 + \ldots + m_{2N-1} + m_{2N} \bmod (2^n - 1)$$

H_i, M_i, I : $2n$-bit blocks, others : n-bit blocks

$H_0 = b_{-1} \| b_0$ (Initial value)

$H_i = \theta_4(M_i, H_{i-1})$ for i from 1 to $N + 1$

where

$$w_i = em_{2i-1}(b_{2i-3}) \oplus b_{2i-2}$$

$$b_{2i-1} = em_{2i}(w_i) \oplus b_{2i-3} \oplus b_{2i-2}$$

$$b_{2i} = w_i \oplus b_{2i-3}$$

$$H_i = b_{2i-1} \| b_{2i}$$

$$H = h_4(M, I) = H_{N+1} \quad (\text{ Hash-value })$$

6.2 Attack 13

For any block cipher,

$$\theta_4(m_1 \| m_2, a \| em_1(a) \oplus a) = \theta_4(m_2 \| m_1, a \| em_2(a) \oplus a)$$
$$= em_1(a) \oplus em_2(a) \| \phi$$

holds, where a is an arbitrary n-bit block. Furthermore,

$$m_1 \oplus m_2 \oplus \ldots \oplus m_{2N-1} \oplus m_{2N}$$
$$= m_2 \oplus m_1 \oplus \ldots \oplus m_{2N-1} \oplus m_{2N},$$

and

$$m_1 + m_2 + \ldots + m_{2N-1} + m_{2N} \bmod (2^n - 1)$$

$$= m_2 + m_1 + \ldots + m_{2N-1} + m_{2N} \bmod (2^n - 1)$$

hold. Thus the following relation holds:

$$h_4(M, a \| em_1(a) \oplus a) = h_4(M', a \| em_2(a) \oplus a)$$

where $M = (m_1 \| m_2) \| M_2 \| \ldots \| M_N \| M_{N+1}$,

$$M' = (m_2 \| m_1) \| M_2 \| \ldots \| M_N \| M_{N+1}.$$

7 2n-bit hash funtion [8]

A $2n$-bit hash function has been proposed that includes an n-bit block cipher as the parallel processing element.

7.1 Algorithm

This is described as:

$H_0 \| H_0' = I \| I'$ (Initial value)

$H_i \| H_i' = \theta_5(M_i, H_{i-1} \| H_{i-1}')$ for i from 1 to N

where

$T_i = eK_i(M_i) \oplus M_i$ where $K_i = \text{Adj10 } (H_{i-1})$

$T_i' = eK_i'(M_i) \oplus M_i$ where $K_i' = \text{Adj01 } (H_{i-1}')$

$H_i = T_i[\text{left}] \| T_i'[\text{right}]$ $H_i' = T_i'[\text{left}] \| T_i[\text{right}]$

$T_i[\text{left}]$: Left half of T_i $T_i[\text{right}]$: Right half of T_i

$T_i'[\text{left}]$: Left half of T_i' $T_i'[\text{right}]$: Right half of T_i'

$H = h_5(M, I) = H_N \| H_N'$ (Hash-value)

H_i, H_i', M_i, I, I' : n-bit blocks

Here, $\text{Adj10}(X)$ means that: bit positions 2,3 of X are set to '10', and other bits (i.e., bit positions 1(MSB), 4,5,...,64(LSB)) remain unchanged. $\text{Adj01}(X)$ means that: bit positions 2,3 of X are set to '01', and other bits (i.e., bit positions 1,4,5,...,64) remain unchanged.

Paper[8] has the following comment about the initial values:

> *These* (initial values) *can be standardized, or randomly generated by the authenticator. If static origin keys* (initial values) *are used, they are defined here* $I = 5252525252525252$ *in hex,* $I' = 2525252525252525$ *in hex.*

7.2 Attack 14

Find collision keys, K_x and K_y, using the birthday attack [2, 6] such that:

$eK_x(P) = eK_y(P), K_x \neq K_y$, for some data block P

bit positions 2,3 of K_x are 1,0,   bit positions 2,3 of K_y are 0,1.

Then, the following equations hold:

Adj10 $(K_x) = K_x$,   Adj01 $(K_y) = K_y$

Adj10 $(\sim K_y) =\sim K_y$,   Adj01 $(\sim K_x) =\sim K_x$

We assume the complementation property of Equation(1). Then,

$$H_1\|H_1 = \theta_5(P, K_x\|K_y) = \theta_5(\sim P, \sim K_y\| \sim K_x).$$

The *DES* cipher has collision keys [11]. Thus, users of the $2n$-bit hash function have to adopt at least the initial values as specified by paper[8].

8   Conclusion

The collision free property of hash functions has been analyzed considering hash function structures, the complementation property and the weak keys of the block cipher used in their iterative functions. This paper has shown that five hash functions are not collision free. One of our assumptions is that an attacker can modify the initial value of the hash function. Thus, users desiring to use these hash functions should be notified of their weakness. The authors recommend that the initial value should be standardized to decrease collision-prone cases. We have proposed a new 128-bit hash function, *N-Hash*[13], which seems to be collision free.

Acknowledgement

The authors would like to thank Dr. Marc Girault with SEPT for various comments on our previous version, and Dr. D. W. Davies for informing us of the initial value requirements of the 2n-bit hash function [8].

References

[1] Davies,D.W. :"Applying the RSA digital signature to electronic mail, " IEEE Computer, 16, 2, pp. 55-62 (Feb. 1983)

[2] Yuval,G. :"How to swindle Rabin," Cryptologia, 3, 3, pp.187-190 (July 1979)

[3] Meyer,C.H. and Matyas,S.M. :"Cryptography : A new dimension in Computer data security," John Willy and Sons, Inc. (1982)

[4] Davies,D.W. and Price,W.L. :"Digital Signatures - An Update," 7th Int. Conf. on Computer Communication, pp.845-849 (1984)

[5] Quisquater, J.J. and Girault,M. :"2n-bit Hash- Functions Using n-bit Symmetric Block Cipher Algorithms," Eurocrypt'89 (Abstract)

[6] Quisquater,J.J. and Delescaille,J.P. :"How easy is collision search? Application to DES," Eurocrypt'89 (Abstract)

[7] Quisquater, J.J. and Girault,M. : Manuscript of "2n-bit Hash-Functions Using n-bit Symmetric Block Cipher Algorithms," Euro-crypt'89 (for Lecture Note) (October 1989, private correspondence)

[8] Meyer,C.H. and Schilling,M. :"Secure Program Load with Modi-fication Detection Code," Proc. of SECURICOM 88, pp.111-130 (1988)

[9] "Data Encryption Standard," FIPS Pub.46, NBS(1977)

[10] Ohta,K. and Koyama,K. : "A Meet-in-the-Middle Attack against Mixed-Type Digital Signature Methods," Trans. IEICE Japan, J70-D, 2, pp.415-422 (Feb.1987)

[11] Quisquater,J.J. and Delescaille,J.P. :"How easy is collision search? New results and application to DES," Crypto'89 (Abstract)

[12] Miyaguchi,S., Shiraishi,S. and Shimizu,S. :"Fast Data Encipher-ment Algorithm FEAL-8," Review of the Electrical Communication Laboratories, 36, 4, pp.433-437 (1988)

[13] Miyaguchi,S., Ohta,K. and Iwata,M. :"128-bit Hash Function (N-Hash)," Proc. of SECURICOM 90, pp.123-137 (Mar.1990)

Table 1. Examples of collision messages in the Davies–Price hash function

Initial value	Message chain	Hash-value	Note	Attack
$dK(\phi)$	$K\|\|K\|\|K\|\|...$	$dK(\phi)$	1)	5
given I	$K^*\|\|K\|\|K\|\|...$	$dK(\phi)$	2)	
$\sim d\sim K(\sim\phi)$	$K\|\|K\|\|K\|\|...$	$\sim d\sim K(\sim\phi)$	1),4)	6
given I	$K^*\|\|K\|\|K\|\|...$	$\sim d\sim K(\sim\phi)$	3),4)	
$d\sim K(\sim\phi)$	$\sim K\|\|K\|\|K\|\|...$	$\sim d\sim K(\sim\phi)$	1),4)	6, 9
ϕ	$K_1\|\|K_2\|\|K_2\|\|...$	$eK_1(\phi)$	5)	7
$eK_1(\phi)$	$K_2\|\|K_2\|\|K_2\|\|...$	$eK_1(\phi)$		
$\sim eK_1(\phi)$	$\sim K_2\|\|K_2\|\|K_2\|\|...$	$eK_1(\phi)$	6)	7, 9
$\sim\phi$	$K_1\|\|\sim K_2\|\|\sim K_2\|\|...$	$\sim eK_1(\sim\phi)$	7)	8
$\sim eK_1(\sim\phi)$	$\sim K_2\|\|\sim K_2\|\|\sim K_2\|\|...$	$\sim eK_1(\sim\phi)$		
$eK_1(\sim\phi)$	$K_2\|\|\sim K_2\|\|\sim K_2\|\|...$	$\sim eK_1(\sim\phi)$	7)	8, 9
given I and $\sim$I	$K_1\|\|K_2\|\|..\|\|K_N$ and $(\sim K_1)\|\|K_2\|\|..\|\|K_N$	H_N	4)	9

Notation:

$H_0 = I$ (initial value), C_i : message chain

$H_i = eC_i(H_{i-1}) \oplus H_{i-1}$, from $i=1$ to n

$C = eK(P)$, $P = dK(C)$: e:enciphering, d:deciphering

K : key, C : ciphertext, P : plaintext, ϕ : null data block

$\sim x$: all bit inverse of x, $\|$: concatenation

Notes:

1) K is arbitrary 64–bit data block.

2) K^* and K are 64–bit data blocks that satisfy $eK^*(I) \oplus I = dK(\phi)$. A pair of K^* and K can be found by the meet–in–the–middle attack using any given initial value I.

3) K^* and K are 64–bit data blocks that satisfy $eK^*(I) \oplus I = \sim d\sim K(\sim\phi)$. A pair of K^* and K can be found by the meet–in–the–middle attack using any given initial value I.

4) The encipherment algorithm should satisfy the following for massage block K:
$$e\sim K(\sim P) = \sim eK(P)$$

5) The encipherment algorithm should satisfy the following for a pair of keys, K_1 and K_2:
$$eK_2(eK_1(\phi)) = \phi$$

6) The encipherment algorithm should satisfy the following for a pair of keys, K_1 and K_2:
$$eK_2(eK_1(\phi)) = \phi \text{ and } e\sim K_2(\sim eK_1(\phi)) = \sim eK_2(eK_1(\phi))$$

7) The encipherment algorithm should satisfy the following for a pair of keys, K_1 and K_2:
$$eK_2(eK_1(\sim\phi)) = \sim\phi \text{ and } e\sim K_2(\sim eK_1(\sim\phi)) = \sim eK_2(eK_1(\sim\phi))$$

The DES cipher has the properties listed in notes 4)–7).

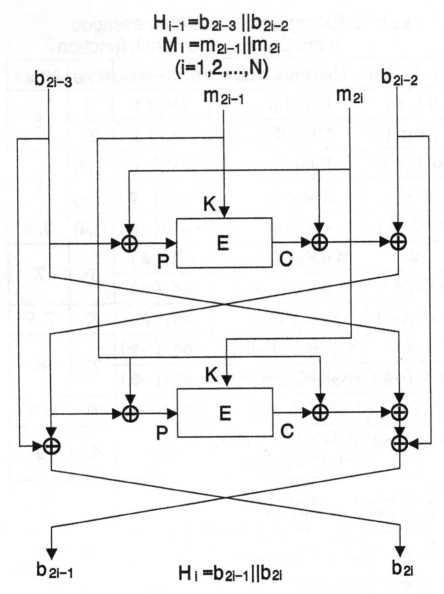

K : Key input, P : Plaintext input, C : Ciphertext output
E : n−bit block cipher

Figure 1. Iterative function of the Quisquater–Girault
hash function (April 1989 version)[5]

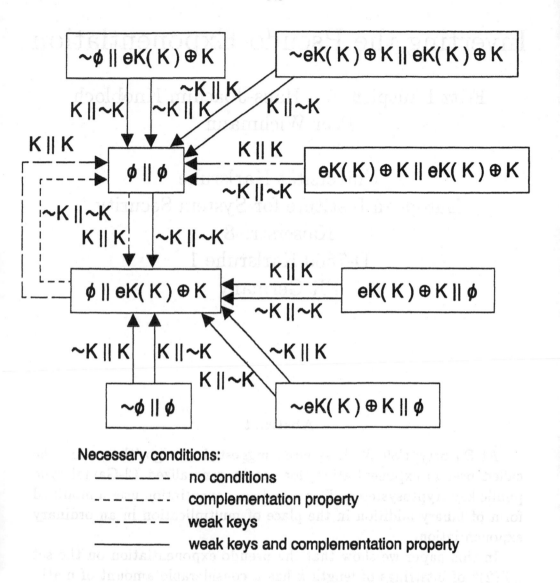

Figure 2. State transition of the Quisquater–Girault hash function (April 1989 version)

Inverting the Pseudo Exponentiation

Fritz Bauspieß   Hans-Joachim Knobloch
Peer Wichmann

Universität Karlsruhe
European Institute for System Security
Kaiserstr. 8
D-7500 Karlsruhe 1
FR Germany

Abstract

At Eurocrypt'89 W. J. Jaburek suggested an algorithm, which he called pseudo exponentiation, for use in generalized El-Gamal type public key cryptosystems. This pseudo exponentiation uses a modified form of binary addition in the place of multiplication in an ordinary exponentiation.

In this paper we show that the pseudo exponentiation on the set $GF(2)^k$ of bitstrings of length k has a considerable amount of mathematical structure. Using this structure we present an algorithm for inverting pseudo exponentiation that has a running time polynomial in k.

1 Introduction

In their famous work [2] Diffie and Hellman presented a protocol for public key exchange based on the discrete log problem. Then, in 1985 El-Gamal presented a public key cryptosystem [3] based on this probloem. Some years later Beth found a corresponding zero-knowledge identification scheme [1].

These cryptosystems are based on an exponential structure over a set **G**, usually a group. There a generalized exponentiation x^e is the application of the $(e-1)$–fold composition of an associative function $f : \mathbf{G} \times \mathbf{G} \rightarrow \mathbf{G}$ on $x \in \mathbf{G}$, namely

$$x^e = \underbrace{f(\dots \ \underbrace{f(x,x), \dots x)}_{e \text{ times}}}_{e-1 \text{ times}}$$

The cryptosystems mentioned above make use of the associativity of f:

$$f \text{ is associative} \Rightarrow \begin{cases} (x^a)^b = x^{ab} = (x^b)^a \\ f(x^a, x^b) = x^{a+b} \end{cases}$$

Furthermore if f is associative (and can be computed in polynomial time), x^e can be efficiently computed in polynomial time (with complexity parameter $\varepsilon = \log e$) using the square and multiply algorithm, whereas it is hoped that in general computing its inverse is much harder.

Now consider the order $< x >$ of any element $x \in \mathbf{G}$, defined as

$$< x >= \max\{n \in \mathbf{N} \ : \ \forall \nu : 1 < \nu \leq n \Rightarrow x^\nu \neq x\}$$

Because **G** is finite the repeated application of f on x will lead into a cycle of f-images. Therefore $< x >$ is either infinite or $< x > \leq |\mathbf{G}|$. From a cryptographical point of view, given $x, y \in \mathbf{G}$, as much as possible uncertainty about the integer e with $x^e = y$ is required. In terms of the order of elements this means that most of the elements of **G** have an order of roughly $|\mathbf{G}|$. For complexity theoretic analysis this allows for the simplification that $e = O(|\mathbf{G}|)$.

Some well-known f-functions are the multiplication in the multiplicative group of a finite field $GF(q)$ and the addition of points on an elliptic curve ([5]).

2 Reviewing the Pseudo Exponentiation

To achieve efficient implementations of public key cryptosystems, in [4] Jaburek suggests a modified binary addition as as suitable function f, there called *pseudo addition*. It is defined over the vector space $GF(2)^k$ of bitstrings of length k. The resulting exponential structure is called *pseudo exponentiation*. For the rest of the paper we shall denote pseudo addition with † and pseudo exponentiation with ‡.

Let A be a $k \times k$ matrix over $GF(2)$, satisfying the conditions

$$a_{m,m} = 0 \tag{1}$$

$$a_{l,m} = 1 \ \Rightarrow \ \forall \lambda \neq l : a_{\lambda,m} = 0 \tag{2}$$

Pseudo addition of two vectors x and y is performed using Algorithm 1.

Algorithm 1 (Pseudo addition)

```
Variables are a, b, c
a ← x
b ← y
While b ≠ (0,...,0) Do
        c ← (a₁ · b₁,..., aₖ · bₖ)
        a ← a + b
        b ← c · A
EndWhile
x † y ← a
```

The pseudo addition so defined is associative and commutative. Note that

$$x \dagger x = \sum_{\substack{n=0 \\ x_n=1}}^{k-1} a_n \tag{3}$$

where a_n denotes the n-th row vector of A, so that together with (2) in particular

$$(\forall n : x_n = 1 \Rightarrow a_n = 0) \Leftrightarrow x\ddagger^2 = 0 \tag{4}$$

Furthermore it is worth noting that the $k \times k$ matix

$$Z_k = \begin{pmatrix} 0 & 1 & 0 & \cdots & 0 \\ \vdots & \ddots & \ddots & \ddots & \vdots \\ \vdots & & \ddots & \ddots & 0 \\ 0 & & & \ddots & 1 \\ 1 & 0 & \cdots & \cdots & 0 \end{pmatrix} \tag{5}$$

describes the usual addition in $\mathbf{Z}_{2^k-1}$.

From a technical viewpoint, pseudo addition can be looked upon as a series of k full adders with widely scattered carry lines. Each carry output can be connected to one or more carry inputs or can be ignored. Each carry input can be connected to one carry output or to zero. These connections are given by the matrix A:

$$a_{l,m} = \begin{cases} 1 & \text{if the carry output of the } l\text{-th full adder is connected} \\ & \text{to the carry input of the } m\text{-th full adder} \\ 0 & \text{otherwise} \end{cases}$$

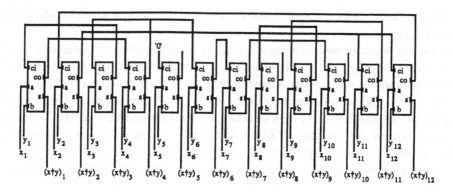

Figure 1: Full adder representation of the pseudo addition defined by Å

An example of this hardware representation for the matrix

$$\text{Å} = \begin{pmatrix} 0 & 0 & 0 & 1 & 0 & 0 & 0 & 0 & 0 & 0 & 0 & 0 \\ 0 & 0 & 0 & 0 & 0 & 0 & 0 & 0 & 0 & 0 & 1 & 1 \\ 1 & 0 & 0 & 0 & 0 & 0 & 0 & 0 & 0 & 0 & 0 & 0 \\ 0 & 0 & 1 & 0 & 0 & 1 & 0 & 0 & 0 & 0 & 0 & 0 \\ 0 & 0 & 0 & 0 & 0 & 0 & 0 & 0 & 0 & 0 & 0 & 0 \\ 0 & 0 & 0 & 0 & 0 & 0 & 1 & 0 & 0 & 0 & 0 & 0 \\ 0 & 0 & 0 & 0 & 0 & 0 & 0 & 0 & 1 & 0 & 0 \\ 0 & \wedge & 0 & 0 & 0 & 0 & 0 & 0 & 0 & 0 & 0 & 0 \\ 0 & 0 & 0 & 0 & 0 & 0 & 0 & 1 & 0 & 0 & 0 & 0 \\ 0 & 0 & 0 & 0 & 0 & 0 & 0 & 0 & 0 & 0 & 0 & 0 \\ 0 & 0 & 0 & 0 & 0 & 0 & 0 & 0 & 1 & 0 & 0 & 0 \\ 0 & 1 & 0 & 0 & 0 & 0 & 0 & 0 & 0 & 0 & 0 & 0 \end{pmatrix}$$

is given in fig. 1.

3 Identifying the Mathematical Structure

To see the mathematical properties of pseudo addition and exponentiation more clearly we introduce the notion of a *cluster*. A cluster C_i is a subset of the vector space $GF(2)^k$ that is defined as follows:

$$C_i = \{x \in GF(2)^k : x\ddagger^{2^{i-1}} \neq 0 \,,\; x\ddagger^{2^i} = 0\} \qquad i = 1, 2, \ldots, m \qquad (6)$$

$$C_\infty = \{x \in GF(2)^k : x\ddagger^{2^i} \neq 0 \,,\; i = 1, 2, \ldots\} \qquad (7)$$

for some integer $0 \leq m \leq k$. We call i the order of cluster C_i.

Equation (4) directly implies that any member x of C_1 is nonzero at most in those coordinates x_n, for which the corresponding row vectors a_n of matrix A are zero. We will identify these coordinates of x and corresponding rows and columns of A with C_1. In an analog way we will identify those row vectors a_n of A (and their corresponding columns of

A and vector coordinates x_n) with cluster C_i, $i = 2, 3, \ldots, m$, who are themselves members of cluster C_{i-1}.

For ease of notation we will assume from now on that the coordinates of x and hence the rows and columns of A are ordered in a way that all bits belonging to the same cluster are adjacent and the clusters are sorted according to their order, so that x_0 belongs to the cluster with highest order and x_k to the cluster with lowest order. Such an ordering can easily be obtained by applying a suitable permutation matrix P at the beginning and the inverse matrix P^{-1} at the end of the algorithm. In the hardware representation of the pseudo addition this is equivalent to permuting the order of the full adders.

According to this definition the matrix $A' = PA$ shows the following structure:

$$A' = \begin{pmatrix} A'_\infty & ? & \cdots & \cdots & ? \\ 0 & A'_m & \ddots & & \vdots \\ \vdots & \ddots & \ddots & \ddots & \vdots \\ \vdots & & \ddots & A'_2 & ? \\ 0 & \cdots & \cdots & 0 & A'_1 \end{pmatrix} \tag{8}$$

where the $A'_\infty, A'_m \ldots, A'_1$ are square sub-matrices, not necessarily equal in size associated with clusters $C_\infty, C_m, \ldots, C_1$ respectively. Note that this is the most general case and that either $A_1, \ldots, A_m$ or A_∞ need not exist for some of the possible matrices definig a pseudo addition.

Considering the definition of a cluster above it is obvious, that

$$A'_1 = A'_2 = \cdots = A'_m = 0. \tag{9}$$

The sub-matrix A'_∞ must contain at most one 1 per column (because of the definition of A, cf. equation (2)). On the other hand it must contain at least one 1 per row, else the specific row would belong either to one of the existing clusters $C_m, \ldots, C_1$ or to a new cluster C_{m+1}. Therefore A'_∞ has exactly one 1 per row and column (i. e. A'_∞ is a permutation matrix), none of them on the main diagonal. So A'_∞ can be permuted to a form

$$A''_\infty = \begin{pmatrix} Z_j & 0 & \cdots & 0 \\ 0 & \ddots & \ddots & \vdots \\ \vdots & \ddots & \ddots & 0 \\ 0 & \cdots & 0 & Z_i \end{pmatrix} \tag{10}$$

with cyclic permutation matrices $Z_i, \ldots, Z_j$ as defined in equation (5). For simplicity we will assume that P is suitably chosen so that $A'_\infty = A''_\infty$

To obatin such an order for our example matrix $\mathring{A}$, we must exchange the rows and

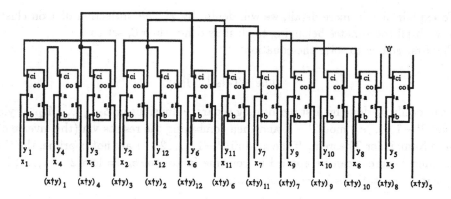

Figure 2: Sorted full adder representation of Å

columns 5 with 12, 8 with 11, 7 with 8 and 2 with 4 giving:

$$
\text{Å}' = \begin{pmatrix}
0 & 1 & 0 & 0 & 0 & 0 & 0 & 0 & 0 & 0 & 0 & 0 \\
0 & 0 & 1 & 0 & 0 & 1 & 0 & 0 & 0 & 0 & 0 & 0 \\
1 & 0 & 0 & 0 & 0 & 0 & 0 & 0 & 0 & 0 & 0 & 0 \\
0 & 0 & 0 & 0 & 1 & 0 & 1 & 0 & 0 & 0 & 0 & 0 \\
0 & 0 & 0 & 1 & 0 & 0 & 0 & 0 & 0 & 0 & 0 & 0 \\
0 & 0 & 0 & 0 & 0 & 0 & 0 & 1 & 0 & 0 & 0 & 0 \\
0 & 0 & 0 & 0 & 0 & 0 & 0 & 0 & 1 & 0 & 0 & 0 \\
0 & 0 & 0 & 0 & 0 & 0 & 0 & 0 & 0 & 1 & 0 & 0 \\
0 & 0 & 0 & 0 & 0 & 0 & 0 & 0 & 0 & 0 & 1 & 0 \\
0 & 0 & 0 & 0 & 0 & 0 & 0 & 0 & 0 & 0 & 0 & 0 \\
0 & 0 & 0 & 0 & 0 & 0 & 0 & 0 & 0 & 0 & 0 & 0 \\
0 & 0 & 0 & 0 & 0 & 0 & 0 & 0 & 0 & 0 & 0 & 0
\end{pmatrix}
$$

The example matrix thereby divides into four clusters C_∞, C_3, C_2 and C_1. The sub-matrix $\text{Å}'_\infty$ divides into two cyclic matrices Z_3 and Z_2. The corresponding sorted full adder chain is shown in fig. 2.

4 Computing the Pseudo Logarithm

We will call a solution e of the equation

$$
y = x \ddagger^e \tag{11}
$$

the pseudo logarithm of y (with respect to the base x).

The pseudo logarithm can be computed by first finding an exponent that yields a correct result for those bits belonging to cluster C_∞ using Chinese Remainder techniques. Then this exponent will then be successively adjusted so that it yields also the correct result for those bits belonging to clusters $C_m, \ldots, C_1$.

To explain this in more detail, we will denite by $x|_i$ the projection of x on cluster C_i, i.e. x with all coordinates belonging to clusters other than C_i set to 0. The first goal is to solve the equation

$$y|_\infty = (x|_\infty \ddagger^{e_\infty})|_\infty \qquad (12)$$

This can easily be done by first performing separate divisions in $\mathbf{Z}_{2^i-1}, \ldots, \mathbf{Z}_{2^j-1}$, yielding $e_\infty \bmod 2^i - 1, \ldots, e_\infty \bmod 2^j - 1$, and then combining the results via (the inverse of) the Chinese Remainder Theorem. If the moduli $2^i - 1, \ldots, 2^j - 1$ are not coprime, the Chinese Remainder Theorem can be applied to coprime factors of $z_\infty = \mathrm{lcm}(2^i - 1, \ldots, 2^j - 1)$. The solution e_∞ satisfies $e_\infty = e \bmod z_\infty$, i.e.

$$e = e_\infty + f z_\infty \qquad (13)$$

for some $f \in \mathbf{Z}$.

Multiples of z_∞ may now be used as correction terms to adjust the other bits of $x\ddagger^{e_\infty}$, because a direct consequence of the definition of z_∞ is that $(x\ddagger^{z_\infty})|_\infty = 0$.

There are, however, two distinct i-bit representations of $0 \in \mathbf{Z}_{2^i-1}$, namely i zeros and i ones. So, in general, $(x\ddagger^{z_\infty})|_\infty$ will consist of several runs of zeros and ones, all representing 0 in one of $\mathbf{Z}_{2^i-1}, \ldots, \mathbf{Z}_{2^j-1}$. It can nevertheless been shown that this fact has no influence on the algorithm for computing pseudo logarithms.

Thus the remaining task of determining f in equation (13) can be completely done by evaluating the clusters of finite order. Algorithm 2 computes f bit by bit, starting with the least significant bit. Each cluster contributes exactly one bit to f, because definition (6) implies that $x \ddagger^{2^i z_\infty}|_m = \cdots = x \ddagger^{2^i z_\infty}|_{m-i+1} = 0$.

Algorithm 2 (Pseudo logarithm for clusters of finite order)

```
Variables are g, i, r, s, a, b
g ← 0
i ← 0
Repeat
        r ← x ‡^e∞ † x‡^{gz∞}
        s ← x ‡^e∞ † x‡^{(2^i+g)z∞}
        a ← max{j : r|_j ≠ y|_j} (or 0 if the set is empty)
        b ← max{j : s|_j ≠ y|_j} (or 0 if the set is empty)
        If a > b Then
            g ← 2^i + g
        EndIf
        i ← i + 1
Until a = b
f ← g
```

If y was the result of a pseudo exponentiation, not an arbitrary vector, algorithm 2 will terminate with $a = b = 0$ and yield a correct result. In the case that C_∞ does not exist for a given A, set e to 0 and z_∞ to 1 in order to compute the pseudo logarithm.

5 Computational Complexity

The permutation matrix P can be found in at most $O(k^3)$ steps. The same holds for the permuation from A_∞ to A'_∞.

Division in $\mathbf{Z}_{2^i-1}$ can be done using the extended Euclidean algorithm. Such a division is needed at most $O(k)$ times. Finding coprime factors of z_∞, if necessary, can be done by applying the Euclidean algorithm on the moduli $2^i - 1, \ldots, 2^j - 1$ at most $O(k^2)$ times. The inverse of the Chinese Remainder Theorem is needed only once. Both the Euclidean algorithm and the inverse Chinese Remainder Theorem are well known polynomial time algorithms.

The subsequent computation of f consists of at most k steps, whereby the computational cost of each step is dominated by a pseudo exponentiation.

Thus the presented algorithm prohibits the cryptographic application of the pseudo exponentiation.

6 Remarks

The interested reader may find that in the original publication [4] the matrix A is described with the main diagonal from top right to bottom left.

An implementation of pseudo exponentiation and pseudo logarithm for values of $k \approx 200$ showed that computing the pseudo log is almost as fast as the pseudo exponentiation.

It is also feasible to solve equation (11) for x given y and e to compute 'pseudo roots'.

References

[1] T. Beth, *Efficient Zero-Knowledge Identification Scheme for Smart Cards*, Adv. in Cryptology - EUROCRYPT '88, Springer, Berlin 1988, pp. 77-84.

[2] W. Diffie, M. Hellman, *New Directions in Cryptography*, IEEE Trans. on Information Theory 22, 1976, pp. 644-654.

[3] T. El-Gamal, *A Public Key Cryptosystem and a Signature Scheme Based on Discrete Logarithms*, IEEE Trans. on Information Theory 31, 1985, pp. 469-472.

[4] W. J. Jaburek, *A generalization of El-Gamal's public key cryptposystem*, Adv. in Cryptology - EUROCRYPT '89, to appear.

[5] N. Koblitz, *Elliptic Curve Cryptosystems*, Math. of Computation 48, 1985.

Cryptosystem for Group Oriented Cryptography

Tzonelih Hwang

National Cheng Kung University

Institute of Information Engineering

Tainan, Taiwan, R. O. C.

Abstract *A practical non—interactive scheme is proposed to simultaneously solve several open problems in group oriented cryptography. The sender of the information is allowed to determine the encryption/decryption keys as well as the information destination without any coordination with the receiving group. The encrypted message is broadcasted to the receiving group and the receivers may authenticate themselves for legitimacy of the information directly from the ciphertext. The security of the scheme can be shown to be equivalent to the difficulty of solving the discrete logarithm problem.*

Key Words: Group Oriented Cryptography, Threshold Scheme, Chinese Remainder Theorem, Diffie—Hellman Key Distribution Scheme, Lagrange Interpolating Polynomial.

1. Introduction

When messages are intended for a group oriented society (or a company), there are several needs for the information sender/receiver depending on the nature of the information. The information maybe so important that it is readable only when a set of authorized receivers agree to decipher it. It may be so urgent that anyone of the authorized receivers could decipher it, whereas the unauthorized user is not allowed to do so. The message can also be transmitted in private to a particular user as usual. These problems and other related ones have been addressed in [Desmedt 88 , Frankel 89 , Desmedt 89 , Hwang 89].

Desmedt has proposed solutions to these problems [Desmedt 88] based on [

Goldreich 87] but are impractical and interactive [Desmedt 89]. Frankel has proposed a protocol to solve some of these problems [Frankel 89]. However, his protocol requires the use of trusted clerks or tamperfree modulars to distribute the encrypted message, thus may be impractical for use in large group oriented networks. In their recent paper, Desmedt and Frankel propose a non–interactive scheme based on the idea of threshold scheme discussing mainly the problem of deciphering the message by a group of people [Desmedt 89]. In their scheme, a trusted center has to distribute the shadows of the deciphering key in private to the authorized receivers. It is not very convenient if the deciphering key is renewed or if the number of the authorized receivers work together to decipher the message is changed.

In this paper, we propose a scheme based on the Diffie–Hellman key distribution scheme [Diffie 76] and Shamir's secret sharing scheme [Shamir 79] to solve these open problems simultaneously.

The sender, depending on the nature of the information, may broadcast the encrypted message to the destination company in such a way that either the ciphertext is decipherable only when a group of authorized receivers work together or it can be deciphered by anyone inside the authorized group, or it is decipherable only by a particular member. There are no assumptions of the existence of tamperfree modulars and trusted clerks or centers.

2. The Protocol

Assume that each member A_i inside the company A holds a secret $x_{A_i} \in \{ 1, \ldots, p-1 \}$ and publishes the value

$$Y_{A_i} = g^{x_{A_i}} \pmod{p}$$

where p is a large prime and g is a fixed primitive element in GF(p) [Diffie 76]. Each member A_i is also assigned a public prime number N_i ($N_i > p$). Note, $N_i \neq N_j$ if $i \neq j$.

2.1 Messages for a Group of Receivers

The sender may send a message M to a group G of n members inside A in such a way that M is readable only when any subset of t members ($t \leq n$) from G agree to decipher the message.

[The Sender]:

(1) Obtain the public values g, p, N_i and Y_{A_i} ($1 \leq i \leq n$, $A_i \in G$) from the public directory of A .

(2) Generate a secret random number $x_s \in \{ 1, \dots , p{-}1 \}$.
Compute

$$Y_s = g^{x_s} \pmod{p}$$

$$K_{sA_i} = g^{x_{A_i} x_s} \pmod{p}, \quad 1 \leq i \leq n .$$

Repeat this step if $K_{sA_i} = K_{sA_j}$ for all $i \neq j$.

(3) Construct a polynomial h(x) of degree t−1 with random coefficients over GF(P),

$$h(x) = a_{t-1} x^{t-1} + \dots + a_1 x + K \pmod{p}.$$

K will serve as the encryption key later.

(4) Encipher M into C_1 using the key K

$$C_1 = E_K (M),$$

where E denotes the predetermined encryption algorithm and D is the corresponding decryption algorithm.

(5) Compute n shadows $W_i = h(K_{sA_i})$ (mod p), $1 \leq i \leq n$.

(6) Compute a common solution C_2 using Chinese Remainder Theorem (CRT) from the following system of equations:

$$X = W_i \pmod{N_i}, \quad 1 \leq i \leq n.$$

(7) Broadcast the ciphertext C

$$C = (C_1, C_2, N, Y_s)$$

where $N = N_1 N_2 \dots N_n$ is the product of all N_i' s.

The purpose of introducing CRT here is to compute a common solution (C_2) of all shadows so that the ciphertext (C) can be broadcasted to the receiving group and every authorized receiver in the receiving group may compute his own part. Alternatively, the sender may send W_i directly to the member A_i. In this case, these N_i's and CRT are no more required.

[The Receiver]:

(1) The authorized user A_i can authenticate himself as a legal receiver by verifying

$$N_i \mid N.$$

(2) A_i computes his shadow W_i by

$$C_2 = W_i \pmod{N_i}.$$

Then he computes

$$K_{sA_i} = (Y_s)^{x_{A_i}} \pmod{p},$$

(3) When t authorized receivers (assume that, without loss of generality, they are $A_1, A_2, \dots, A_t$) work together, the key K can be computed by using Shamir's (n,t) threshold scheme [Shamir 79].

(4) $M = D_K(C_1)$.

2.2 Message for Anyone in the Group

The sender may send M to G in the company A such that anyone in G can recover M. However, anyone outside the group G may not be able to recover M. Here, we extended the idea of the conference key distribution scheme in [Laih 88] to solve this problem.

[The Sender]:

(1) The sender first performs the steps (1), (2), (3), (4) and (5) as in section 2.1 .

(2) Compute $t-1$ extra shadows such that

$$W_i' = h(i) \quad (\bmod\ p), \quad 1 \leq i \leq t-1 .$$

Assume that $K_{sA_i} > t$ for all i.

(3) Compute the common solution C_2 using CRT from the following equations

$$X = W_i' \quad \bmod\ p_i , \quad 1 \leq i \leq t-1$$

$$X = W_j \quad \bmod\ N_j , \quad 1 \leq j \leq n$$

where p_i's are distinct public primes ($p_i > p$) and $p_i \neq N_j$ for all i and j.

(4) Broadcast $C = (C_1, C_2, N, Y_s)$, where $N = p_1\, p_2 \cdots p_{t-1}\, N_1\, N_2 \cdots N_n$.

[The authorized Receiver A_j]

(1) Compute the $t-1$ extra shadows by

$$C_2 = W_i' \quad \bmod\ p_i , \quad 1 \leq i \leq t-1 .$$

(2) Compute $K_{sA_j} = (Y_s)^{x_{A_j}} \pmod{p}$.

(3) Obtain K using Shamir's (n,t) threshold scheme.

(4) $M = D_k(C_1)$.

Notice that for the message intended to everyone in the group, it will be advantageous to use a polynomial $h(x)$ of degree one (i.e., $t=2$). In this case, only one extra shadow is required.

It is clear that the sender can communicate with a particular member i inside the company A in private by using the common key K_{sA_i} to encipher/decipher the message.

3. Discussion & Security Analysis

Shamir's threshold scheme is applied to solving the group–oriented secret sharing problem. It is obvious that the encryption key K cannot be obtained easily even if $t-1$ authorized receivers are acting in collusion.

If the cryptanalyst tries to compute X_{A_i} from Y_{A_i}, he has to solve the discrete logarithm problem [Diffie 76].

To obtain K_{sA_i} from W_i ($= C_2 \mod N_i$), the conspirators has to solve X from the polynomial [Purdy 74, Denning 82]

$$W_i = a_{t-1} X^{t-1} + ... + a_1 X + K \quad (\mod p),$$

with unknown coefficients. Therefore, this attack won't be successful.

4. Conclusion

We have proposed a scheme to simultaneously solve several open problems in group oriented cryptography. The new scheme is particularly useful in the case that the information sender has the authority to decide the destination of the information. It can also be modified to solve the case that the receiving group decides the destination of the information. In this scheme, since only one ciphertext is needed, the ciphertext can be broadcasted to the destination.

The scheme works without the use of trusted clerks, centers or tamperfree modulars. The encryption/decryption key and the number of receivers that have to work together to recover the plaintext can be renewed easily by the sender without any coordination with the destination. Furthermore, both the conventional or public–key cryptosystems are applicable to this scheme. The security of this scheme depends on the difficulty of computing the discrete logarithm problem.

[Acknowledgment]

The author would like to thank the referee of this paper for the comments and suggestions. This work was supported by the National Science Council of the Republic of China as research project NSC 80–0408–E–006–02.

[References]

<1> [Denning 82] D. E. R. Denning. *Cryptography and Data Security.* Addison – Wesley, Reading, Mass., 1982.

<2> [Diffie 76] W. Diffie and M. E. Hellman. "New directions in cryptography". *IEEE Trans. Inform. Theory,* IT–22(6):644–654, November 1976.

<3> [Desmedt 88] Y. Desmedt. "Society and group oriented cryptography : a new

concept". In C. Pomerance, editor, Advances in Cryptology, *proc. of Crypto'87* (Lecture Notes in Computer Science 293),pages 120–127. Springer–Verlag,1988. Santa Barbara, California, U.S.A., August 16–20.

<4> [Desmedt 89] Y. Desmedt and Y. Frankel, "Threshold Cryptosystems". presented at *Crypto'89*. Santa Barbara, California, U.S.A. Aug. 20–24.

<5> [Frankel 89] Y. Frankel. "Practical Protocol for Large Group Oriented Networks". Presented at *Eurocrypt'89*, Houthalen, Belgium, to appear in Advances in Cryptology. Proc. of *Eurocrypt'89* (Lecture Notes in Computer Science), Springer–Verlag,April 1989.

<6> [Goldreich 87] O. Goldreich, S. Micali, and A. Wigderson. "How to play any mental game". In Proceedings of the *Nineteenth ACM symp.* Theory of Computing, STOC, pages 218–239, May 25–27, 1987.

<7> [Hwang 89] T. Hwang, "On the Secure Communications of Group Oriented Socienties", 1990 *IEEE International Symposium on Info. Theory*.

<8> [Laih 88] C. S. Laih, L. Harn, and J. Y. Lee, "A new threshold scheme and its application on designing the conference key distribution cryptosystem.", *Info. processing letters*, North–Holland Vol. 32, No. 3, 24 Aug. 1989.

<9> [Purdy 74] G. p. Purdy, " A High Security Log–in Procedure", *Commun. ACM* vol. 17(8), pp. 442–445 Aug. 1974.

<10> [Shamir 79] A. Shamir. How to share a secret. *Commun. ACM*, 22:612–613, November 1979.

A Provably-Secure Strongly-Randomized Cipher

Ueli M. Maurer

Institute for Signal and Information Processing
Swiss Federal Institute of Technology
CH-8092 Zurich [1]

Abstract. Shannon's pessimistic theorem, which states that a cipher can be perfect only when the entropy of the secret key is at least as great as that of the plaintext, is relativized by the demonstration of a randomized cipher in which the secret key is short but the plaintext can be very long. This cipher is shown to be "perfect with high probability". More precisely, the enemy is unable to obtain any information about the plaintext when a certain security event occurs, and the probability of this event is shown to be arbitrarily close to one unless the enemy performs an infeasible computation. This cipher exploits the existence of a publicly-accessible string of random bits whose length is much greater than that of all the plaintext to be encrypted before the secret key and the randomizer itself are changed. Two modifications of this cipher are discussed that may lead to practical provably-secure ciphers based on either of two assumptions that appear to be novel in cryptography, viz., the (sole) assumption that the enemy's memory capacity (but not his computing power) is restricted and the assumption that an explicit function is, in a specified sense, controllably-difficult to compute, but not necessarily one-way.

---

[1]The author is presently with the Dept. of Computer Science, Princeton University, Princeton, NJ 08540.

1. Introduction

One of the most important practical and theoretical open problems in cryptography is to devise a cipher that is both provably-secure and practical. The significance of a result on provable security crucially depends on the definition of security used, on the assumptions about the enemy's knowledge and resources, and on the practicality of the cipher. Excluding approaches that are based on an unproven hypothesis such as the intractability of a certain problem (e.g., factoring), one observes that every approach to provable security that has previously been proposed is either impractical or is based on a generally unrealistic assumption about the enemy's *a priori* and/or obtainable knowledge. To list a few examples: the one-time pad [7] is, because of its large key size, impractical in most applications; perfect local randomizers [4] are based on the generally unrealistic assumption that an enemy can only obtain a small number of ciphertext bits; Wyner's wire-tap channel [8] is based on the generally unrealistic assumption that the enemy's channel is noisier than the main channel; and the Rip van Winkle cipher proposed by Massey and Ingemarsson [1,2] is completely impractical since the legitimate receiver's deciphering delay is on the order of the square of the time the enemy must spend in order to break the cipher. Finally, the result that a cascade of additive stream ciphers is at least as secure as any of its component ciphers [5] yields provably-secure ciphers only when a set of additive stream ciphers can be constructed that provably contains at least one computationally secure cipher.

In this paper, we present a new approach to provable security that was motivated by [2] and is based on the availability of a very large publicly-accessible string of random bits. The need for this public randomizer is the only (but serious) detriment to the practicality of the proposed cipher. The randomizer could, for instance, be stored on a high-density storage medium, copies of which are publicly available, or it could be broadcast by a satellite.

The enemy's computational effort needed to break the cipher is measured in terms of the number of randomizer bits that he must examine. A very general way of modeling algorithms is by execution trees, where each branch corresponds to one or more operations and where the branching points correspond to decisions to be made during the execution of the algorithm. Because every examination of a randomizer bit corresponds to a branching point, the average depth of the execution tree, which is a lower bound on the average number of operations performed, is lower bounded by the average number of examined bits.

The basic idea of our approach is to prove that, even if he uses an optimal strategy for examining randomizer bits, an enemy obtains no information in Shannon's sense about the plaintext with probability very close to one unless he accesses a substantial fraction of all the randomizer bits. More precisely, we prove that if a certain event occurs, then the enemy's entire observation, consisting of the cryptogram and the examined randomizer bits, is statistically independent of the plaintext. The probability of this event is lower bounded by a quantity that depends on the number of bits examined by the enemy, and it is very close to one unless the enemy examines a substantial fraction (e.g., 2/3) of the entire randomizer. It is obviously impossible to prove that the number of bits that the enemy must examine is greater than the total number of randomizer bits, and thus our result is close to optimum within our framework of provable security. Note that we prove that the size of the necessary input of every algorithm breaking the cipher is infeasibly large rather than that the enemy must perform any operation on the input in addition to examining it.

Since the effort to examine a random bit is in current technology roughly equal to that required to generate one, our lower bound on the enemy's computational effort appears to be on the same order as the effort needed to generate the randomizer. Therefore, our strongly-randomized cipher is truly practical only when either an existing source of randomness can be used (for example, a deep-space radio source or the surface of the moon) or should a much easier way of generating large amounts of random data be discovered (e.g., by generating identical copies of a very complicated quasi-crystal). It is not the purpose of this paper to discuss further the technical problem of generating a huge amount of publicly available random data. Rather, our interest is in exploring the question whether provable security is possible in such a model. Note, however, that when the randomizer is broadcast before the transmission of the actual cryptogram, an enemy must store essentially the whole randomizer if his chances of receiving any information about the plaintext from the succeeding ciphertext are to be non-negligible. Therefore, the amount of random data needed to achieve an acceptable level of security, even when the enemy has infinite computing power, is only somewhat larger than the enemy's memory capacity. This "broadcast" version of our cipher may be more practical than the original one.

The results of this paper appear to be somewhat surprising for two reasons. First, they demonstrate that, although perfect secrecy can be achieved only when the entropy of the secret key is at least equal to that of the plaintext (see [6]), relaxing the notion of perfectness only slightly allows one to build a provably-secure cipher whose secret key is very short compared to the length of the plaintext. Second, although information-theoretic security usually implies that the enemy has infinite computing power, our proposed cipher is secure for an information-theoretic notion of security only when the enemy is computationally restricted.

In Section 2, our model of a cipher with public randomizer is introduced, and a particular randomized cipher is presented. After describing a general model of attacks against randomized ciphers, a proof of security of our cipher against all feasible attacks is given in Section 3. In Section 4, techniques are suggested for basing the (provable) security of ciphers on either one of two assumptions, viz., that the enemy's memory capacity is restricted or that a certain function is difficult to compute in a specified sense, but not necessarily one-way.

2. Description of the Randomized Cipher

Throughout this paper, random variables are denoted by capital letters, whereas the corresponding small letters denote specific values taken on by these random variables. Underlined capital letters or superscripted capital letters denote random vectors. Our model of a strongly-randomized cipher is as follows. As in a conventional symmetric cryptosystem, the communicating parties share a short randomly-selected secret key. The randomizer R is a binary random string of length L, whose bits can be read in a random-access manner by the legitimate parties as well as by all potential opponents, i.e., R is assumed to be publicly accessible. The cryptogram is a function of the plaintext, the secret key and the randomizer such that, given the cryptogram, the key and the randomizer, the plaintext is uniquely determined. The goal of the design of a randomized cipher is to devise an encryption transformation such that the cryptogram depends on only a few randomizer bits whose positions in turn depend on the secret key in such a manner that without the secret key it is impossible to determine any of the plaintext without examining a very large number of randomizer bits.

We now describe our specific strongly-randomized cipher. It is a binary additive stream cipher in which the plaintext $X = [X_1, \ldots, X_N]$, the cryptogram $Y = [Y_1, \ldots, Y_N]$ and the keystream $W = [W_1, \ldots, W_N]$ are binary sequences of length N. The cryptogram Y is obtained by adding X and W bitwise modulo 2:

$$Y_n = X_n \oplus W_n \quad \text{for } 1 \leq n \leq N.$$

The publicly-accessible binary random string R consists of K blocks of length T and thus has total length $L = KT$ bits. These blocks are denoted by $R[k,0], \ldots, R[k, T-1]$ for $1 \leq k \leq K$, i.e., the randomizer can be viewed as a two-dimensional array of binary random variables (see Figure 1). The secret key $Z = [Z_1, \ldots, Z_K]$, where $Z_k \in \{0, \ldots, T-1\}$ for $1 \leq k \leq K$, specifies a position within each block of R, and is chosen to be uniformly distributed over the key space $S_Z = \{0, \ldots, T-1\}^K$. Thus the number of bits needed to represent the key is $K \log_2 T$.

$$
\begin{array}{cccc}
R[1,0] & R[1,1] & \cdots & R[1,T-1] \\
R[2,0] & R[2,1] & \cdots & R[2,T-1] \\
\cdot & \cdot & & \cdot \\
\cdot & \cdot & & \cdot \\
R[K,0] & R[K,1] & \cdots & R[K,T-1]
\end{array}
$$

Figure 1. The randomizer $\underline{R}$, viewed as a two-dimensional array

The keystream $\underline{W}$, which is a function of the secret key $\underline{Z}$ and the randomizer $\underline{R}$, is the bitwise modulo 2 sum of the K subsequences of length N within the randomizer starting at the positions specified by the key, where each block (row) of $\underline{R}$ is considered to be extended cyclically, i.e., the second index is reduced modulo T:

$$
W_n = \sum_{k=1}^{K} R[k, (n-1+Z_k) \bmod T] \tag{1}
$$

for $1 \leq n \leq N$, where $\sum$ denotes summation modulo 2. The sub-array of the randomizer that determines $\underline{W}$ is denoted by $R^{\underline{Z}}$ and is depicted in Figure 2. A diagram of the sending site of the cipher system is shown in Figure 3. Note that the legitimate receiver who knows the secret key needs to examine only KN of the L random bits, i.e., a very small fraction N/T of all bits when $T \gg N$ as we shall assume.

$$
\begin{array}{cccc}
R[1,Z_1] & R[1,Z_1+1] & \cdots & R[1,Z_1+N-1] \\
R[2,Z_2] & R[2,Z_2+1] & \cdots & R[2,Z_2+N-1] \\
\cdot & \cdot & & \cdot \\
\cdot & \cdot & & \cdot \\
R[K,Z_K] & R[K,Z_K+1] & \cdots & R[K,Z_K+N-1]
\end{array}
$$

Figure 2. The sub-array $R^{\underline{Z}}$ of the randomizer $\underline{R}$ is selected by the secret key $\underline{Z}$. All second indices are to be reduced modulo T. The keystream $\underline{W} = [W_1,\ldots,W_N]$ is formed by adding the K rows of $R^{\underline{Z}}$ bitwise.

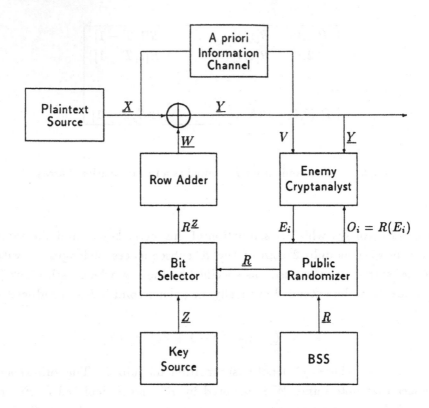

Figure 3. A block diagram of the specific strongly-randomized cipher investigated in Section 3. The public randomizer $\underline{R}$ is an array of independent and completely random binary random variables. The keystream $\underline{W}$ is formed by letting the key $\underline{Z}$ select the sub-array $R^{\underline{Z}}$ of bits of $\underline{R}$ consisting of K rows of length N, and adding these rows bitwise modulo 2. The enemy uses an arbitrary, possibly probabilistic, sequential strategy to determine the addresses $E_1, E_2, \ldots$ of the randomizer bits $O_1, O_2, \ldots$ that he examines.

3. Model of Attacks and Main Results

An enemy trying to break the cipher may have (possibly partial) knowledge of the plaintext statistics and may also have some other *a priori* information about the plaintext. Let $P_{\underline{X}}$ be the probability distribution of the plaintext and let V be a random variable, jointly distributed with $\underline{X}$ according to $P_{\underline{X}V}$, that summarizes the enemy's other *a priori* information about $\underline{X}$. Since precise knowledge of $P_{\underline{X}V}$ and thus also of $P_{\underline{X}}$ can only help the enemy and because we assume that he precisely knows these distributions, our proof of security remains valid when the enemy actually has only partial knowledge about $P_{\underline{X}V}$.

Our model of the enemy's *attack* is described in the sequel. We allow the enemy to use an arbitrary, possibly probabilistic, sequential strategy for selecting the positions of the randomizer bits that he examines. At each step of the attack, the enemy can make use of the entire available information, i.e., the cryptogram $\underline{Y}$, the side-information V, and the positionss and values of the bits observed so far. Let $E_i = [A_i, B_i]$ denote the address of the i-th randomizer bit examined by the enemy, where A_i and B_i satisfy $1 \leq A_i \leq K$ and $0 \leq B_i \leq T-1$ for $i = 1, 2, \ldots$. Let further $O_i = R(E_i) = R[A_i, B_i]$ denote the observed value of the randomizer bit at address E_i that is examined by the enemy at the i-th step of his attack. Note that the randomizer bit O_i is a binary random variable whose address E_i is a random variable rather than a constant. However, we will make use several times of the fact that, given $E_i = e_i$, O_i corresponds to the randomizer bit $R(e_i)$ at the specific address e_i. We use the notation $E^m = [E_1, \ldots, E_m]$ and $O^m = [O_1, \ldots, O_m]$ for all $m \geq 1$. For a particular sequence $e^m = [e_1, \ldots, e_m]$ of m bit addresses, where $e_i = [a_i, b_i]$ with $1 \leq a_i \leq K$ and $0 \leq b_i \leq T-1$ for $1 \leq i \leq m$, $R(e^m) = [R(e_1), \ldots, R(e_m)]$ denotes the corresponding sequence of randomizer bits. Correspondingly, we have $O^m = R(E^m)$ for $m \geq 1$.

For $m \geq 1$, the bit position E_m is determined by the enemy as a (possibly randomized) function of the entire information he possesses at this time, i.e., the cryptogram $\underline{Y}$, the values O^{m-1} of all previously examined bits together with their addresses E^{m-1}, and the a priori information V. The enemy's strategy is hence completely specified by the sequence of conditional probability distributions $P_{E_1|\underline{Y}V}$, $P_{E_2|\underline{Y}VE_1O_1}$, $P_{E_3|\underline{Y}VE_1E_2O_1O_2}$, etc.. The following theorem is the main result of this paper.

Theorem: *There exists an event $\mathcal{E}$ such that, for all joint probability distributions $P_{\underline{X}V}$ and for all (possibly probabilistic) strategies for examining bits $O_1, \ldots, O_M$ of $\underline{R}$ at addresses $E_1, \ldots, E_M$,*

$$I\left(\underline{X}; \underline{Y}E^M O^M \mid V, \mathcal{E}\right) = 0 \quad \text{and} \quad P(\mathcal{E}) \geq 1 - N\delta^K,$$

where $\delta = M/KT$ is the fraction of randomizer bits examined by the enemy.

Here $I(\underline{X}; \underline{Y}E^M O^M \mid V, \mathcal{E})$ denotes the (mutual) information that $\underline{Y}, E^M$ and O^M together give about $\underline{X}$, given that V is known and given that the event $\mathcal{E}$ occurs. The theorem states that if the event $\mathcal{E}$ occurs, then the enemy's total observation $[\underline{Y}, E^M, O^M]$ gives no information about the plaintext $\underline{X}$ beyond the information already provided by V. Clearly, if the enemy knew the value of a random variable V that uniquely determines $\underline{X}$, i.e., such that $H(\underline{X}|V) = 0$, it would make little sense to use a cipher at all. But the point is that, no matter what *a priori* information about the plaintext the enemy has, this does not help him to obtain any *additional* information. For instance, even if the enemy knew all but one bit of the plaintext, he would still get no information about this remaining bit if $\mathcal{E}$ occurs, and the probability of $\mathcal{E}$ could not be reduced by exploiting his virtually complete knowledge about the plaintext. Note that the theorem asserts the existence of a high-probability event $\mathcal{E}$, but does not specify it. However, in the proof we will specify such an event.

Example: Assume $K = 50$, $T = 10^{20}$ and let the plaintext be one gigabit, i.e., $N = 2^{30} \approx 10^9$. The key size of this cipher is $50 \cdot \log_2 10^{20} \approx 3320$ bits. The legitimate users need to examine only 50 randomizer bits per plaintext bit. An enemy, however, even if he used an optimal strategy for examining a fraction $\delta = 1/4$ of all bits, i.e., $M = KT/4 = 1.25 \cdot 10^{21}$ bits in total or $1.16 \cdot 10^{12}$ bits per plaintext bit, would have a chance of obtaining any new information about the plaintext not greater than $2^{30} \cdot (1/4)^{50} < 10^{-21}$.

The proof of the above theorem is divided into a sequence of four lemmas. The complete proofs of Lemmas 1 to 3 are given in [3].

Definition: The sequence $e^M = [e_1, \ldots, e_M]$ of $M \geq 1$ bit positions yields a *consistency check* for the key $\underline{z} = [z_1, \ldots, z_K]$ if and only if there exists an interger $n \in [1, N]$ and a subset $\{[1, t_1], [2, t_2], \ldots, [K, t_K]\}$ of $\{e_1, \ldots, e_M\}$ such that

$$t_k - z_k \equiv n - 1 \pmod{T} \quad \text{for } 1 \leq k \leq K.$$

In other words, e^M yields a consistency check for $\underline{z}$ if and only if $R(e^M)$ and $\underline{Z}$ together determine at least one (the n-th) bit of the keystream $\underline{W} = [W_1, \ldots, W_N]$ or, equivalently, if and only if $R(e^M)$ completely determines at least one column of $R^{\underline{Z}}$ (cf. Figure 2). Furthermore, let $\mathcal{Z}(e^M) \subseteq S_{\underline{Z}}$ denote the set of keys for which e^M yields at least one consistency check, i.e.,

$$\mathcal{Z}(e^M) = \left\{ \underline{z} \in S_{\underline{Z}} : e^M \text{ yields at least one consistency check for } \underline{z} \right\}.$$

The idea behind this definition is that if the enemy knew the plaintext (and hence also the keystream because he knows the ciphertext) and the set $R(e^M)$ of randomizer bits, then, for every key $\underline{z} \in \mathcal{Z}(e^M)$, he could perform one consistency check per keystream bit that he could compute from $R(e^M)$, by comparing the computed keystream bit for the key $\underline{z}$ with the actual keystream bit. If all computed (for key $\underline{z}$) keystream bits agree with the actual keystream bits, the key $\underline{z}$ is still a possible candidate, but if any of the computed keystream bits differs from the corresponding actual keystream bit, then z cannot be the actual key. Note that when e^M consists of one bit in each block of $\underline{R}$, then e^M yields exactly one consistency check for N different keys. In general, if e^M consists of m_k bits in the k-th block for $1 \leq k \leq K$, then e^M yields a total number $N \prod_{k=1}^{K} m_k$ of consistency checks, but in general several of these checks will be for the same key. The event $\mathcal{E}$ introduced in the main theorem will later be defined as the event that the actual key does not belong to the set of keys for which the enemy's set E^M of observed bits yields a consistency check.

Lemma 1: *For all joint probability distributions $P_{\underline{X}V}$, for every sequence $e^M = [e_1, \ldots, e_M]$ of $M \geq 1$ bit positions, and for all $\underline{x}, v, \underline{y}, r^M \in \{0,1\}^M$ and $\underline{z} \notin \mathcal{Z}(e^M)$ such that the conditioning event has non-zero probability,*

$$P\left[\underline{X} = \underline{x} \mid V = v, \underline{Y} = \underline{y}, E^M = e^M, O^M = r^M, \underline{Z} = \underline{z}\right] = P\left[\underline{X} = \underline{x} \mid V = v\right].$$

Idea of proof: (A formal proof is given in [3].) Every bit of the keystream $\underline{W}$ is the sum of K randomizer bits (see equation (1)). The crucial observation is that when $\underline{Z} \notin \mathcal{Z}(e^M)$, then, for every n satisfying $1 \leq n \leq N$, at least one of the K randomizer bits contributing to W_n is not contained in the sequence $R(e^M)$ of randomizer bits. Therefore, given that the event $\underline{Z} \notin \mathcal{Z}(E^M)$ occurs, the keystream $\underline{W}$ is completely random and statistically independent of $\underline{X}, V, O^M = R(E^M)$ and $\underline{Z}$. Thus, also the plaintext $\underline{X}$ is statistically independent of $\underline{Y}, E^M, O^M$ and $\underline{Z}$. $\quad\square$

Lemma 2: *For all probability distributions $P_{\underline{X}V}$ and for all (possibly probabilistic) strategies for examining $M \geq 1$ bits $O_1, \ldots, O_M$ of $\underline{R}$ at addresses $E_1, \ldots, E_M$, we have*

$$I\left(\underline{X}; \underline{Y} E^M O^M \underline{Z} \mid V, \underline{Z} \notin \mathcal{Z}(E^M)\right) = 0.$$

This lemma establishes the first part of the main theorem when $\mathcal{E}$ is defined as the event that $\underline{Z} \notin \mathcal{Z}(E^M)$. It states that if the enemy does not succeed in choosing the bit positions E^M such that $\underline{Z} \in \mathcal{Z}(E^M)$, then he does not obtain any information whatsoever about the plaintext beyond the information already conveyed by V, even if an oracle would give him the key $\underline{Z}$ for free after he has finished his observation. Note that it is crucial, however, that the enemy does not know the key while selecting bits.

Proof: The conditional mutual information of Lemma 2 can be written as a difference of conditional uncertainties:

$$I\left(\underline{X};\underline{Y}E^M O^M \underline{Z} \mid V, \underline{Z} \notin \mathcal{Z}(E^M)\right)$$

$$= H\left(\underline{X} \mid V, \underline{Z} \notin \mathcal{Z}(E^M)\right) - H\left(\underline{X} \mid V \underline{Y} E^M O^M \underline{Z}, \underline{Z} \notin \mathcal{Z}(E^M)\right).$$

It is an immediate consequence of Lemma 1 that both uncertainties are equal. $\square$

It remains to prove the second part of the theorem, i.e., to lower bound the probability of the event $\mathcal{E}$ that $\underline{Z} \notin \mathcal{Z}(E^M)$. Let $|S|$ denote the cardinality of the set S.

Lemma 3: *For all probability distributions $P_{\underline{X}V}$ and for all (possibly probabilistic) strategies for examining $M \geq 1$ bits $O_1, \ldots, O_M$ of $\underline{R}$ at addresses $E_1, \ldots, E_M$,*

$$P\left[\underline{Z} \notin \mathcal{Z}(E^M)\right] \geq 1 - \frac{\max_{e^M} |\mathcal{Z}(e^M)|}{T^K}.$$

Idea of proof: The proof is based on the observation that no matter which bits the enemy examines, all keys $\underline{z}$ that are not in the set $\mathcal{Z}(E^M)$ for which the enemy's sequence of observed bit positions yields a consistency check, are still equally likely candidates. More precisely, it is proved in [3] that

$$P\left[\underline{Z} = \underline{z} \mid \underline{Y} = \underline{y}, V = v, E^M = e^M, O^M = r^M\right] = T^{-K} \tag{2}$$

for all $e^M, \underline{y}, v, r^M$ and $\underline{z} \notin \mathcal{Z}(e^M)$. This result appears to be somewhat counter-intuitive, since it states that the *a posteriori* probabilities of the keys $\underline{z} \notin \mathcal{Z}(e^M)$ are equal to the *a priori* probabilities even when there exists a key $\underline{z} \in \mathcal{Z}(e^M)$ that satisfies many consistency checks and therefore appears to be the correct key. Equation (2) implies that

$$P[\underline{Z} = \underline{z} \mid E^M = e^M] = T^{-K}$$

for all e^M and $\underline{z} \notin \mathcal{Z}(e^M)$. Summing these probabilities over all keys $\underline{z} \notin \mathcal{Z}(e^M)$, i.e., over $T^K - |\mathcal{Z}(e^M)|$ terms, we obtain

$$P\left[\underline{Z} \notin \mathcal{Z}(E^M) \mid E^M = e^M\right] = \sum_{\underline{z} \notin \mathcal{Z}(e^M)} P\left[\underline{Z} = \underline{z} \mid E^M = e^M\right]$$

$$= 1 - \frac{|\mathcal{Z}(e^M)|}{T^K}. \tag{3}$$

Since $P[\underline{Z} \notin \mathcal{Z}(E^M)]$ is equal to the average of $P[\underline{Z} \notin \mathcal{Z}(E^M) \mid E^M = e^M]$ over all values of e^M, we immediately have

$$P\left[\underline{Z} \notin \mathcal{Z}(E^M)\right] \geq 1 - \frac{\max_{e^M} |\mathcal{Z}(e^M)|}{T^K}. \quad \square$$

Equation (3) demonstrates that the enemy's optimal strategy for making the event $\mathcal{E}$ that $\underline{Z} \notin \mathcal{Z}(E^M)$ as unlikely to occur as possible is simply to make the set $\mathcal{Z}(E^M)$ as large as possible. Notice that, surprisingly, this strategy is independent of $\underline{Y}, O^M$ and V. In other words, letting the selected bit positions $E_1, \ldots, E_M$ depend on the observed bits $O_1, \ldots, O_M$, the cryptogram $\underline{Y}$ and on the *a priori* information V cannot help the enemy in reducing the probability of the event $\mathcal{E}$. However, to base the strategy on $\underline{Y}, O^M$ and V can increase the amount of information that the enemy gets about the plaintext in case that $\mathcal{E}$ does not occur, i.e., in case that $\underline{Z} \in \mathcal{Z}(E^M)$. Note that although $P[\underline{Z} \notin \mathcal{Z}(E^M) | E^M = e^M]$ equals the number of keys that are not in $\mathcal{Z}(e^M)$ divided by the total number of keys, equation (3) is non-trivial because E^M is a random variable that, because it depends on $\underline{Y}$, also depends on $\underline{Z}$.

Lemma 4: *For every sequence* $e^M = [e_1, \ldots, e_M]$ *of* $M \geq 1$ *bit positions,*

$$|\mathcal{Z}(e^M)| \leq N \left(\frac{M}{K} \right)^K.$$

Proof: Let m_k, for $1 \leq k \leq K$, be the number of randomizer bits specified by e^M that belong to the k-th block of $\underline{R}$, i.e., whose first address component is equal to k. Every subset of elements of e^M of the form $\{[1, t_1], [2, t_2], \ldots, [K, t_K]\}$ yields a consistency check for exactly N keys, namely for the keys $\underline{z} = [(t_1 - x) \bmod T, (t_2 - x) \bmod T, \ldots, (t_K - x) \bmod T]$ for $0 \leq x \leq N - 1$. There are exactly $\prod_{k=1}^{K} m_k$ different subsets of the described form and hence there are at most $N \prod_{k=1}^{K} m_k$ keys for which e^M yields a consistency check. $\prod_{k=1}^{K} m_k$ is maximized for real m_k under the restriction $\sum_{k=1}^{K} m_k = M$ by the choice $m_1 = \cdots = m_K = M/K$ for which $\prod_{k=1}^{K} m_k = (M/K)^K$. Clearly, this maximum is also an upper bound on $\prod_{k=1}^{K} m_k$ under the restriction that $m_1, \ldots, m_K$ must be integers satisfying $\sum_{k=1}^{K} m_k = M$. $\square$

Proof of the Theorem: Lemma 2 shows that if we define $\mathcal{E}$ as the event that $\underline{Z} \notin \mathcal{Z}(E^M)$, then $I(\underline{X}; \underline{Y}E^M O^M \underline{Z} | V, \mathcal{E}) = 0$ and therefore also $I(\underline{X}; \underline{Y}E^M O^M | V, \mathcal{E}) = 0$. This last step follows from the two basic facts that mutual information is always non-negative and that giving additional random variables (here $\underline{Z}$) to the information-giving set cannot reduce the information about $\underline{X}$. Lemmas 3 and 4 finally give

$$P[\mathcal{E}] \geq 1 - \frac{\max_{e^M} |\mathcal{Z}(e^M)|}{T^K} \geq 1 - N \left(\frac{M}{KT} \right)^K = 1 - N\delta^K. \quad \square$$

4. Conclusions

In this section, we suggest two modifications of the randomized cipher presented in Section 2 that are more practical in that the size of the public randomizer required to achieve a sufficient level of security is much smaller. A rigorous proof of security for the first suggested modification would lead to the first cipher that is provably-secure under the sole assumption that the enemy's memory capacity, but not necessarily his computing power, is restricted. The second suggested modification has the potential of leading to an existence proof for secure cryptosystems without necessarily leading to a specific realization.

We first discuss a version of our strongly-randomized cipher in which instead of having the randomizer *stored* in a publicly-accessible way, it is *broadcast* by a sender (e.g., a satellite), i.e., the randomizer evolves in time rather than in space. There may exist natural sources of randomness, such as a deep-space radio source, that could be used. Alternatively, the randomizer could be transmitted as a burst of random data over the (insecure) communication channel prior to the transmission of the actual cryptogram. Because in this version of our cipher, the randomizer is not available at the time that the ciphertext is transmitted, an enemy must not only examine but also *store* a substantial fraction of the randomizer in order to be able to obtain any information about the plaintext from the ciphertext. Thus, if the enemy's memory capacity is not more than δ times the number of randomizer bits, then there exists no strategy for storing randomizer bits such that these will later be of any use to the enemy with probability more than $N\delta^K$, where N is the length of the plaintext. Note, however, that in general an enemy is not restricted to storing randomizer bits. Rather, he can store the values of boolean functions applied to the randomizer. We conjecture that a result similar to the above theorem holds even for such an extended model of the enemy's attack.

A second modification of our cipher is based on the observation that the size of the randomizer can be greatly reduced if the bit access operation can be made more difficult. In [3], a version of our cipher is discussed that is based on a function that is difficult to compute in a specified sense, but not necessarily one-way.

Finally, we would like to point out that randomization techniques similar to those presented in this paper may be useful for the construction of practical ciphers, even when the randomizer is not sufficiently long to guarantee a reasonable lower bound on the enemy's computational effort required to break the cipher or when the randomizer is replaced by a pseudo-random sequence.

Acknowledgement

I would like to thank Jim Massey for motivating this research and for many helpful discussions.

References

[1] J.L. Massey, *An introduction to contemporary cryptology*, Proceedings of the IEEE, vol. 76, no. 5, pp. 533-549, May 1988.

[2] J.L. Massey and I. Ingemarsson, *The Rip van Winkle cipher - a simple and provably computationally secure cipher with a finite key*, in IEEE Int. Symp. Info. Th., Brighton, England, (Abstracts), p. 146, June 24-28, 1985.

[3] U.M. Maurer, *Conditionally-perfect secrecy and a provably-secure randomized cipher*, to appear in Journal of Cryptology, special issue EUROCRYPT'90.

[4] U.M. Maurer and J.L. Massey, *Local randomness in pseudo-random sequences*, to appear in Journal of Cryptology, special issue CRYPTO'89.

[5] U.M. Maurer and J.L. Massey, *Cascade ciphers: the importance of being first*, presented at the 1990 IEEE Int. Symp. Inform. Theory, San Diego, CA, Jan. 14-19, 1990 (submitted to J. of Cryptology).

[6] C.E. Shannon, *Communication theory of secrecy systems*, Bell Syst. Tech. J., vol. 28, pp. 656-715, Oct. 1949.

[7] G.S. Vernam, *Cipher printing telegraph systems for secret wire and radio telegraphic communications*, J. American Inst. Elec. Eng., vol. 55, pp. 109-115, 1926.

[8] A. Wyner, *The wire-tap channel*, Bell Systems Technical Journal, vol. 54, no. 8, pp. 1355-1387, Oct. 1975.

General public key residue cryptosystems and mental poker protocols

Kaoru KUROSAWA   Yutaka KATAYAMA   Wakaha OGATA

Shigeo TSUJII

Department of Electrical and Electronic Engineering

Faculty of Engineering

Tokyo Institute of Technology

2-12-1, Ookayama, Meguro-ku, Tokyo 152, JAPAN

Tel.   +81-3-726-1111(Ext. 2577)

Fax   +81-3-729-0685

E-mail   kkurosaw@ss.titech.ac.jp or

kkurosaw%ss.titech.ac.jp@relay.cs.net

Abstract

This paper presents a general method how to construct public key cryptosystems based on the r-th residue problem. Based on the proposed method, we present the first mental poker protocol which can shuffle any set of cards. Its fault tolerant version is given, too. An efficient zero knowledge interactive proof system for quadratic non-residuosity is also shown.

1   Introduction

Goldwasser and Micali presented a probabilistic encryption scheme based on the quadratic residue problem[GM]. Cohen generalized this binary system to r valued systems for prime r, and applied it to a secret voting system [CF][BY]. Zheng showed a sufficient condition which the parameters must satisfy for odd r[ZMI]. Since such cryptosystems have a nice additive homomorphic property, they have many cryptographic applications.

This paper shows a general method how to construct a public key cryptosystem based on the r-th residue problem for any r (both odd and even). The necessary and sufficient conditions are presented which the parameters must satisfy.

We apply our result to mental poker protocols (Note that the number of cards is 52, which is even.) Previous mental poker protocols[C][MUS] are not that realistic. They cannot shuffle only discarded cards, for example. We propose the first mental poker protocol which can shuffle any set of cards. The difference is in the way of card expression. In the previous protocols, the k-th card is given by the composition of each player's secret permutation. In our protocol, card k is given by the sum of each player's secret random number. Its fault tolerant version is given, too.

The related zero knowledge interactive proof systems are also shown.

2 Preliminaries

2.1 Public key residue cryptosystems

(Secret key) two large prime numbers, p and q.

(Public key) $N(= pq)$ and y

(Plaintext) $m(0 \leq m < r)$

(Encryption) $E(m) = y^m x^r \bmod N$, where x is a random number.

This cryptosystem has the following homomorphic property.

$$E(m + n) = E(m)E(n)z^r \bmod N \text{ for some } z.$$

Under what condition, is any element of Z_n^* uniquely deciphered ?

The condition for $r = 2$ is [GM]

$$(y/p) = (y/q) = -1$$

The condition for prime r is [CF][BY]

(1) $r|p - 1$, $r \nmid q - 1$

(2) y is an r-th non-residue.

2.2 Some known lemmas

Let G be an abelian multiplicative group and H be a subgroup of G.

(Lemma A)[P]

Two elements g and g' of G are in the same coset of H if and only if $g^{-1}g'$ is an element of H.

(Lemma B)[P]

Every elememt of G is in one and only one coset of H.

(Lemma C)[P]

(Order of H)(index of G over H)=(order of G)

(Lemma D)[K]

In $GF(p)$, the number of r-th roots of unity is $gcd(r, p-1)$, where p is a prime number.

2.3 Notations

$$Z_N^* = \{x | 0 < x < N, gcd(x, N) = 1\}$$

$$Z_N^*(+1) = \{x | x \in Z_N^*, (x/N) = 1\}$$

$$B_N(r) = \{w | w = x^r \bmod N, x \in Z_N^*\}$$

p, q : two prime numbers.

3 How to construct residue cryptosystems

3.1 Conditions of the parameters

[Theorem 1]

In the public key residue cryptosystem in 2.1, any element of Z_n^* is uniquely deciphered if and only if eq.(1)~(6) are satisfied.

$$y^j \notin B_N(r) \quad 1 \leq j < r \tag{1}$$

$$gcd(p-1, r) = e_1 \tag{2}$$

$$gcd(q-1, r) = e_2 \tag{3}$$

$$r = \begin{cases} e_1 e_2 & \text{if } r \text{ is odd.} \\ (e_1 e_2)/2 & \text{if } r \text{ is even.} \end{cases} \tag{4}$$

$$gcd(e_1, e_2) = \begin{cases} 1 & \text{if } r \text{ is odd.} \\ 2 & \text{if } r \text{ is even.} \end{cases} \tag{5}$$

$$(y/N) = 1 \text{ if } r \text{ is even.} \tag{6}$$

(Definition)

We call y which satisfies the above conditions "a basic element".

(Decryption)

In mod p,

$$\{E(m)\}^{(p-1)/e_1} = (y^m x^r) y^{(p-1)/e_1}$$

$$= (y^{(p-1)/e_1})^m (x^{r/e_1})^{(p-1)}$$

$$= (y^{(p-1)/e_1})^m$$

Similarly,

$$\{E(m)\}^{(q-1)/e_2} = (y^{(q-1)/e_2})^m \bmod q$$

Therefore, for $0 \leq i < r$, just compare

$$\{E(m)\}^{(p-1)/e_1} \bmod p \quad \text{and} \quad \{E(m)\}^{(q-1)/e_2} \bmod q.$$

with

$$(y^{(p-1)/e_1})^i \quad \text{mod } p \quad \text{and} \quad (y^{(q-1)/e_2})^i \quad \text{mod } q.$$

(Sketch of proof of Theorem 1)

Observe that

$$E\{0\} = 1, 2^r, \cdots$$

$$E\{m\} = y, y2^r, \cdots, \quad (1 \le m < r)$$

Notice that $\{E(0)\}$ is a multiplicative group and $\{E(m)\}$ is a coset. Therefore, any element

of Z_n^* is uniquely deciphered if and only if

(1) y^m is the coset leader $(1 \le m < r)$.

(2) $\displaystyle\bigcup_{m=0}^{r-1} \{E(m)\} = Z_n^*$

Eq.(1) is the necessary and sufficient condition for (1) from Lemma A. It can be proved

that Eq.(2)~(6) are the necessary and sufficient conditions for (2).   Q.E.D.

3.2   Proof of Theorem 1

Theorem 1 is based on the following lemmas, which are derived from lemma A~D in 2.2.

We discuss only the case of $r = odd$ for the simplicity. The case of $r = even$ is similar.

[Lemma 1]

(1) Let w be an r-th residue mod p. Then, the number of r-th roots of w is e_1.

(2) Let $w \in B_N(r)$. Then, the number of r-th roots of w is $e_1 e_2$.

(3) Z_n^* is a multiplicative group and its order is $(p-1)(q-1)$.

(4) $B_N(r)$ is a subgroup of Z_n^* and its order is $(p-1)(q-1)/e_1 e_2$.

(5) The index of Z_n^* over $B_N(r)$ is $e_1 e_2$.

[Lemma 2]

(1) $x^{e_1} \in B_p(r)$ for any x.

(2) Let g be a primitive element of GF(p). Then,

$$g^j \notin B_p(r) \quad 1 \leq j < e_1$$

[Lemma 3]

Let $s = lcm(e_1, e_2)$. Then,

(1) $x^s \in B_N(r)$ for any x

(2) Let y be an integer which is a primitive element of GF(p) and GF(q). Then,

$$y^j \notin B_p(r) \quad 1 \leq j < s$$

(3) Let $r = e_1 e_2$. Then, there exists y such that eq.(1) holds if and only if eq.(5) holds.

(4) If $r \neq e_1 e_2$, there is no y such that any element of Z_n^* is uniquely deciphered.

3.3 Existency of basic elements

[Theorem 2]

(number of basic elements)$/|Z_N^*| = \phi(r)/r$ if r is odd.

(number of basic elements)$/|Z_N^*(+1)| = \phi(r)/r$ if r is even.

It is known that [HW]

$$r/\phi(r) = O(loglogr).$$

Therefore, the expected number of trials to choose a basic element is $O(loglogr)$.

[Theorem 3]

Let $r = \prod_{i=1}^{k} p_i^{j_i}$, where p_i is prime for $1 \leq i < k$. Then,

$$y^i \notin B_N(r) \quad 1 \leq i < r$$

if and only if

$$y^{(r/p_i)} \notin B_N(r) \quad 1 \leq i < k$$

[Lemma 4]

$$w \in B_N(r) \quad \text{if and only if}$$

$$w^{(p-1)/e_1} = 1 \mod p \quad \text{and} \quad w^{(q-1)/e_2} = 1 \mod q$$

4 Proposed mental poker protocols

We propose the first mental poker protocols which can shuffle any set of cards. In Poker 1, a distributed sum scheme is introduced for the card representation. Poker 2 is a fault tolerant version.

Suppose that $P_1, \cdots, P_n$ want to play poker. Each palyer constructs the r-th residue cryptosystem, where $r = 52$ in Poker 1 and $r = 53$ in Poker 2.

(1) P_i publicizes his public key (N_i, y_i).

(2) P_i proves that (N_i, y_i) satisfies the conditions in 3.1 by a zero knowledge proof system.

4.1 Poker 1

A DECK of cards is $\{0, \cdots, 51\}$. In Poker 1, card k in DECK is expressed by

$$\{E_1(f_1(k)), \cdots, E_n(f_n(k))\}$$

where $f_i(k)$ is P_i's secret random number such that

$$k = f_1(k) + \cdots + f_n(k) \bmod 52$$

We show how to assign such random numbers to each player in a distributed and trusted way.

(Putting cards upside down)

P_1 choose randomly $\{r_{i,j}\}$ such that

$$k = r_{1,k} + \cdots + r_{n,k} \bmod 52, \; 0 \le k \le 51$$

and publicizes them. At this stage, every card is open. P_1 then encrypts each $r_{i,k}$ by P_i's public key. He publicizes the result and the related random numbers which were used in the encryptions. Let

$$C_k = (t_{1,k}, \cdots, t_{n,k}), \quad 0 \le k \le 51$$

where $t_{i,k} = E_i(r_{i,k})$.

At this stage, each card has become the backside. However, everyone knows, for each card, what the front side is. So, next, each player shuffles the cards in turn.

(Shuffling cards in DECK)

Do the following for $h = 1, \cdots, n$.

P_h chooses random numbers $\{s_{i,j}\}$ such that

$$0 = s_{1,k} + \cdots + s_{n,k} \bmod 52, \quad 0 \le k \le 51$$

He then chooses a random permutaion π and computes

$$t'_{1,k} = t_{1,\pi(k)} \times E_i(s_{i,k}) \bmod N_i$$

Note that, for some π',

$$E_1^{-1}(t_{1,k})' + \cdots + E_n^{-1}(t'_{n,k}) = \pi'(k) \bmod 52, \quad 0 \leq k \leq 51$$

because of the homomorphic property of E_i and eq.(4.1). He sets $t_{1,k} = t'_{1,k}$ and publicizes

$$C_k = (t_{1,k}, \cdots, t_{n,k}), \quad 0 \leq k \leq 51$$

By a zero knowledge interactive proof system, he proves that he computed C_k according to the protocol.

Now, we have made a DECK.

(Getting a card from the DECK)

When P_i gets a card from the DECK, the other players open their plaintexts of $\{t_{i,k}\}$ and the related random numbers. Only P_i can obtain the card by computing

$$\pi'(k) = E_1^{-1}(t_{1,k}) + \cdots + E_n^{-1}(t_{n,k}) \bmod 52$$

The other players can compute the back side of this card as follows.

$$E_i(\pi'(k)) = t_{1,k} \times y_i^{E_1^{-1}(t_{1,k}) + \cdots, \overset{i}{v}, \cdots + E_n^{-1}(t_{n,k})}$$

(Shuffling Hand)

Let the cards in the hand of P_i be $E_i(c_1), \cdots, E_i(c_m)$. P_i shuffles them before he opens or discards one of them. For the suffling, he chooses a random permutaion π and publicizes

$$E_i(c_{\pi(1)})E_i(0), \cdots, E_i(c_{\pi(m)})E_i(0)$$

He proves that he followed the protocol by a zero knowledge interactive proof system.

(Openning a card)

When P_i opens $E_i(\pi'(k))$, he just opens $\pi'(k)$ and the related random number.

(Discarding a card)

When P_i discards $E_i(\pi'(k))$, he just declares that he discards $E_i(\pi'(k))$.

(Giving a card)

When P_i gives $E_i(\pi'(k))$ to P_j, he publicizes $E_j(\pi'(k))$. By a zero knowledge interactive proof system, P_i proves that the plaintexts of $E_i(\pi'(k))$ and $E_j(\pi'(k))$ are the same.

(Shuffling any set of cards)

Suppose that all players want to mix the cards in DECK with a discarded card, $E_i(\pi'(k))$, and reshuffle them. Note that

$$\pi'(k) = 0 + \cdots + \pi'(k) + \cdots + 0$$

and $E_i(\pi'(k))$ is given. Then, it is easy to see that we can use the same protocol as "shuffling cards in DECK".

(Remark)

The necessary zero knowledge interactive proof systems are easy to obtain from the homomorphic property of the cryptosystem.

4.2 Poker 2

This protocol is a fault tolerant version of Poker 1. When n=2t+1, at most t faulty players are allowed. Let g(x) be a random polynomial with degree t and with the constant term k. Card k is expressed as follows.

$$E_1(g(1)), \cdots, E_n(g(n))$$

The other part of the protocol is almost the same, except that P_j sends g(j) to P_i secretly, $j \neq i$, when P_i gets a card from the DECK.

5 ZKIP for public key residue cryptosystems

5.1 Knowledge of a plaintext

Suppose that z is given such that

$$z = y^m x^r \bmod N \tag{7}$$

A(lice) wants to convince B(ob) that she knows m and x.

(Protocol 1)

  Repeat the following n times, where $n = |N|$.

(step 1) A chooses m' and x' randomly and computes

$$z' = y^{m'} x'^r \bmod N$$

She sends z' to B.

(step 2) B sends a random bit e to A.

(step 3) A sends m'' and x'' to B such that

$$z^e z' = y^{m''} x''^r \bmod N$$

(step 4) B checks the above equation.

5.2 On the basic element

Let

$$
S = \begin{cases}
Z_n^* & \text{if } r \text{ is odd.} \\
\\
Z_n^*(+1) & \text{if } r \text{ is even.}
\end{cases}
$$

A wants to convince B that (N, y) satisfies the following two conditions.

(1) For any $z \in S$, there exists m and x such that eq.(7) holds, where $0 \le m < r$.

(2) The above m is unique.

[ZKIP for (1)]

Repeat the following n times, where $n = |N|$.

(step 1) B chooses $z \in S$ at random and sends it to A.

(step 2) A shows that she knows the m and x by protocol 1.

[ZKIP for (2)]

Repeat the following n times, where $n = |N|$.

(step 1) B chooses m and x randomly and computes eq.(7). He sends the z to A.

(step 2) B shows that he knows m and x by protocol 1.

(step 3) A computes the plaintext of z, $\hat{m}$, and send it to B.

(step 4) B checks that $\hat{m} = m$.

(Remarks) When $r = 2$,

(1) ZKIP for (1) is also a ZKIP for

$$
N = p^i q^j, \quad (y/p^i) = (y/q^j) = -1
$$

(2) ZKIP for (2) is also a ZKIP for that y is a quadratic non-residue. The number of bits communicated is a half of [GMR].

Acknowledgement

We are grateful to Dr.Itoh and Mr.Kishimoto for useful advice.

References

[BY]   Benaloh and Yung: "Distributing the power of a government to enhance the privacy of voters", Proc. 5th Annual Symp. on pronciples of distributed computing, ACM, pp.52-62 (1985)

[C]   Crepeau: "A zero knowledge poker protocol that achieves confidentiality of players' strategy, or how to achieve an electronic poker face", CRYPTO'86, pp.239-247 (1986)

[CF]   Cohen and Fischer: "A robust and verifiable cryptographically secure election scheme", Proc. 26th FOCS, pp.372-382 (1985)

[GM]   Goldwasser and Micali: "Probabilistic encription and how to play mental poker, keeping secret all partial information", 14th STOC, pp.365-377 (1982)

[GMR]   Goldwasser, Micali and Rackoff: "The knowledge complexity of interactive proof systems", SIAM J. on computing, vol.18, No.1, pp.186-208 (1989)

[HW]   Hardy and Wright: "An introduction to the theory and numbers", 5th ed., Oxford Univ. Press (1979)

[K]   Koblitz: "A course in number theory and cryptography", Springer-Verlag (1987)

[MUS]   Miyama, Uyematsu and Sakaniwa: "A mental poker protocol without later verifications", (in Japanese) (1987)

[P]   Peterson: "Error correcting codes", MIT Press (1961)

[ZMI]   Zheng, Matsumoto and Imai: "Residuosity problem and its application to cryptography", Trans, IEICE, vol.E71, No.8, pp.759-767 (1988)

A Proposal for a New Block Encryption Standard

Xuejia Lai   James L. Massey

Institute for Signal and Information Processing
Swiss Federal Institute of Technology
CH–8092 Zürich, Switzerland

Abstract

A new secret-key block cipher is proposed as a candidate for a new encryption standard. In the proposed cipher, the plaintext and the ciphertext are 64 bit blocks, while the secret key is 128 bit long. The cipher is based on the design concept of "mixing operations from different algebraic groups". The cipher structure was chosen to provide confusion and diffusion and to facilitate both hardware and software implementations.

1 Introduction

A new secret-key block cipher is proposed herein as a candidate for a new encryption standard. In the proposed cipher, the plaintext and the ciphertext are 64 bit blocks, while the secret key is 128 bit long. The cipher is based on the design concept of "mixing operations from different algebraic groups". The required confusion is achieved by successively using three different group operations on pairs of 16-bit subblocks and the cipher structure was chosen to provide the necessary diffusion. The cipher is so constructed that the deciphering process is the same as the enciphering process once the decryption key subblocks have been computed from the encryption key subblocks. The cipher structure was chosen to facilitate both hardware and software implementations.

The cipher is described in Section 2. Section 3 considers the relation of the three chosen different operations to one another and the effect of their "mixing". The design principles for the cipher are discussed in Section 4. Section 5 discusses the implementation of the cipher in software as well as in hardware. A C-language program of the cipher is given in Appendix B together with examples that can be used to test the correctness of implementations.

2   Description of the Proposed Cipher

The computational graph of the encryption process is shown in Fig.1. The process consists of 8 similar rounds followed by an output transformation. The complete first round and the output transformation are depicted in Fig.1.

In the encryption process shown in Fig.1, three different group operations on pairs of 16-bit subblocks are used, namely,

 – bit-by-bit exclusive-OR of two 16-bit subblocks, denoted as $\oplus$;

 – addition of integers modulo 2^{16} where the subblock is treated as the usual radix-two representation of an integer, the resulting operation is denoted as $\boxplus$;

 – multiplication of integers modulo $2^{16} + 1$ where the subblock is treated as the usual radix-two representation of an integer except that the all-zero subblock is treated as representing 2^{16}, and the resulting operation is denoted as $\odot$.

For example,
$$(0,...,0)\odot(1,0,...,0) = (1,0,...,0,1)$$
because
$$2^{16}2^{15} \bmod (2^{16} + 1) = 2^{15} + 1.$$

The 64-bit plaintext block is partitioned into four 16-bit subblocks, the i-th of which is denoted as X_i in Fig.1. The four plaintext subblocks are then transformed into four 16-bit ciphertext subblocks, Y_1, Y_2, Y_3 and Y_4, under the control of 52 key subblocks of 16 bits, where the six key subblocks used in the r-th (r=1,..,8) round are denoted as $Z_1^{(r)}, .., Z_6^{(r)}$ and where the four key subblocks used in the output transformation are denoted as $Z_1^{(9)}, Z_2^{(9)}, Z_3^{(9)}, Z_4^{(9)}$.

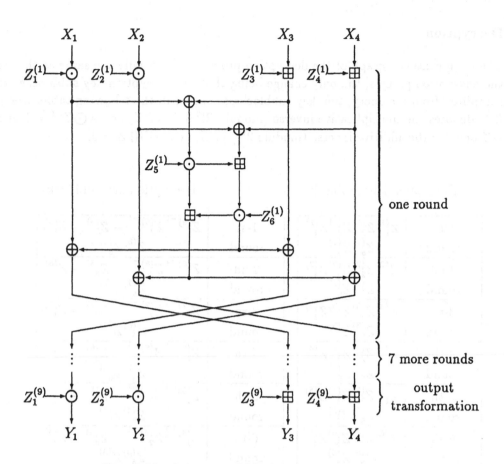

X_i : 16-bit plaintext subblock
Y_i : 16-bit ciphertext subblock
$Z_i^{(r)}$: 16-bit key subblock
$\oplus$: bit-by-bit exclusive-OR of 16-bit subblocks
$\boxplus$: addition modulo 2^{16} of 16-bit integers
$\odot$: multiplication modulo $2^{16} + 1$ of 16-bit integers
with the zero subblock corresponding to 2^n

Figure 1: Computational graph for encryption

Decryption

The computational graph of the decryption process is essentially the same as that of the encryption process, the only change being that the decryption key subblocks are computed from the encryption key subblocks as shown in the following table, where Z^{-1} denotes the multiplicative inverse (modulo $2^{16} + 1$) of Z, i.e., $Z \odot Z^{-1} = 1$ and $-Z$ denotes the additive inverse (modulo 2^{16}) of Z, i.e., $-Z \boxplus Z = 0$.

Encryption key subblocks		Decryption key subblocks	
1-st round	$Z_1^{(1)} Z_2^{(1)} Z_3^{(1)} Z_4^{(1)}$ $Z_5^{(1)} Z_6^{(1)}$	1-st round	$Z_1^{(9)^{-1}} Z_2^{(9)^{-1}} - Z_3^{(9)} - Z_4^{(9)}$ $Z_5^{(8)} Z_6^{(8)}$
2-nd round	$Z_1^{(2)} Z_2^{(2)} Z_3^{(2)} Z_4^{(2)}$ $Z_5^{(2)} Z_6^{(2)}$	2-nd round	$Z_1^{(8)^{-1}} Z_2^{(8)^{-1}} - Z_3^{(8)} - Z_4^{(8)}$ $Z_5^{(7)} Z_6^{(7)}$
3-rd round	$Z_1^{(3)} Z_2^{(3)} Z_3^{(3)} Z_4^{(3)}$ $Z_5^{(3)} Z_6^{(3)}$	3-rd round	$Z_1^{(7)^{-1}} Z_2^{(7)^{-1}} - Z_3^{(7)} - Z_4^{(7)}$ $Z_5^{(6)} Z_6^{(6)}$
4-th round	$Z_1^{(4)} Z_2^{(4)} Z_3^{(4)} Z_4^{(4)}$ $Z_5^{(4)} Z_6^{(4)}$	4-th round	$Z_1^{(6)^{-1}} Z_2^{(6)^{-1}} - Z_3^{(6)} - Z_4^{(6)}$ $Z_5^{(5)} Z_6^{(5)}$
5-th round	$Z_1^{(5)} Z_2^{(5)} Z_3^{(5)} Z_4^{(5)}$ $Z_5^{(5)} Z_6^{(5)}$	5-th round	$Z_1^{(5)^{-1}} Z_2^{(5)^{-1}} - Z_3^{(5)} - Z_4^{(5)}$ $Z_5^{(4)} Z_6^{(4)}$
6-th round	$Z_1^{(6)} Z_2^{(6)} Z_3^{(6)} Z_4^{(6)}$ $Z_5^{(6)} Z_6^{(6)}$	6-th round	$Z_1^{(4)^{-1}} Z_2^{(4)^{-1}} - Z_3^{(4)} - Z_4^{(4)}$ $Z_5^{(3)} Z_6^{(3)}$
7-th round	$Z_1^{(7)} Z_2^{(7)} Z_3^{(7)} Z_4^{(7)}$ $Z_5^{(7)} Z_6^{(7)}$	7-th round	$Z_1^{(3)^{-1}} Z_2^{(3)^{-1}} - Z_3^{(3)} - Z_4^{(3)}$ $Z_5^{(2)} Z_6^{(2)}$
8-th round	$Z_1^{(8)} Z_2^{(8)} Z_3^{(8)} Z_4^{(8)}$ $Z_5^{(8)} Z_6^{(8)}$	8-th round	$Z_1^{(2)^{-1}} Z_2^{(2)^{-1}} - Z_3^{(2)} - Z_4^{(2)}$ $Z_5^{(1)} Z_6^{(1)}$
output transform.	$Z_1^{(9)} Z_2^{(9)} Z_3^{(9)} Z_4^{(9)}$	output transform.	$Z_1^{(1)^{-1}} Z_2^{(1)^{-1}} - Z_3^{(1)} - Z_4^{(1)}$

The key schedule

The 52 key subblocks used in the encryption process are generated from the 128 bit user-selected key as follows:

The 128 bit user-selected key is partitioned into 8 subblocks that are directly used as the first eight key subblocks, where the ordering of key subblocks is as follows: $Z_1^{(1)}, Z_2^{(1)}, .., Z_6^{(1)}, Z_1^{(2)}, .., Z_6^{(2)}, .., Z_1^{(8)}, .., Z_6^{(8)}, Z_1^{(9)}, Z_2^{(9)}, Z_3^{(9)}, Z_4^{(9)}$.

The 128 bit user-selected key is then cyclic shifted to the left by 25 positions, after which the resulting 128 bit block is again partitioned into eight subblocks that are taken as the next eight key subblocks. The obtained 128 bit block is cyclic shifted again to the left by 25 positions to produce the next eight key subblocks, and this procedure is repeated until all 52 key subblocks have been generated.

3 Group Operations and Their Interaction

The cipher is based on the design concept of "mixing operations from different algebraic groups having the same number of elements". Group operations were chosen because the statistical relation of random variables X, Y, Z related by a group operation as $Y = X * Z$ has the desired "perfect secrecy" property, i.e., if one of the random variables is chosen equally likely to be any group element, then the other two random variables are statistically independent. The interaction of different group operations contributes to the "confusion" required for a secure cipher. In this section, the interaction of different operations will be considered in terms of isotopism of quasigroups and in terms of polynomial expressions.

The three operations as quasigroup operations

Def. Let S be a set and let $*$ denote an operation from pairs (a, b) of elements of S to an element $a * b$ of S. Then $(S, *)$ is said to be an *quasigroup* if, for any $a, b \in S$, the equations $a * x = b$ and $y * a = b$ both have exactly one solution in S. A *group* is a quasigroup in which the operation is associative, i.e., for which $a * (b * c) = (a * b) * c$ for all a, b and c in S. Quasigroups $(S_1, *_1), (S_2, *_2)$ are said to be *isotopic* if there are bijective mappings $\theta, \phi, \psi : S_1 \rightarrow S_2$, such that, for all $x, y \in S_1$, $\theta(x) *_2 \phi(y) = \psi(x *_1 y)$. Such a triple (θ, ϕ, ψ) is then called an *isotopism* of $(S_1, *_1)$ upon $(S_2, *_2)$. Two groups are said to be *isomorphic* if they are isotopic as quasigroups for which the isotopism is (θ, θ, θ).

It can be shown that two groups are isomorphic if and only if they are isotopic [3].

Let n be one of the integers 1,2,4,8 or 16 so that the integer $2^n + 1$ is a prime, and let Z_{2^n} denote the ring of integers modulo 2^n. Let $(Z^*_{2^n+1}, \cdot)$ denote the multiplicative group of the non-zero elements of the field Z_{2^n+1}, let $(Z_{2n}, +)$ denote the additive group of the ring Z_{2^n}, and let $(F_2^n, \oplus)$ denote the group of n–tuples over F_2 under the bitwise exclusive-or operation. Then the following theorem states some of the "incompatibility" properties of these groups.

Theorem 1   *For $n \in \{1, 2, 4, 8, 16\}$:*
1) *Quasigroups $(F_2^n, \oplus)$ and $(Z_{2n}, +)$ are not isotopic for $n \geq 2$.*
2) *Quasigroups $(Z^*_{2^n+1}, \cdot)$ and $(F_2^n, \oplus)$ are not isotopic for $n \geq 2$.*
3) *(θ, ϕ, ψ) is an isotopism of $(Z^*_{2^n+1}, \cdot)$ upon $(Z_{2n}, +)$ if and only if there exist constants $c_1, c_2 \in Z_{2n}$ and a primitive element a of the field $Z^*_{2^n+1}$ such that, for all x in Z_{2n},*

$$\theta(x) - c_1 = \phi(x) - c_2 = \psi(x) - (c_1 + c_2) = \log_a(x), \tag{1}$$

i.e., any isotopism between these groups is essentially the logarithm. Moreover, if (θ, ϕ, ψ) is an isotopism, none of these maps will be the "mixing mapping" m from $Z^*_{2^n+1}$ to Z_{2^n} defined by $m(i) = i$, for $i \neq 2^n$ and $m(2^n) = 0$ when $n \geq 2$.

Proof.

1) For $n \geq 2$, the groups $(F^n_2, \oplus)$ and $(Z_{2^n}, +)$ are not isomorphic because $(Z_{2^n}, +)$ is a cyclic group while $(F^n_2, \oplus)$ is not. Thus, they are not isotopic as quasigroups.

2) $(Z^*_{2^n+1}, \cdot)$ and $(Z_{2^n}, +)$ are isomorphic groups for $n = 1, 2, 4, 8, 16$ because both groups are cyclic. Thus, $(Z^*_{2^n+1}, \cdot)$ is isotopic to $(F^n_2, \oplus)$ if and only if $(Z_{2^n}, +)$ is isotopic to $(F^n_2, \oplus)$.

3) If (1) holds, then for any x, y in Z_{2^n+1},

$$\psi(x \cdot y) = \log_a(x \cdot y) + c_1 + c_2 = \log_a(x) + \log_a(y) + c_1 + c_2 = \theta(x) + \phi(y).$$

Now suppose that (θ, ϕ, ψ) is an isotopism. Then for all $x, y \in Z^*_{2^n+1}$, $\theta(x) + \phi(y) = \psi(x \cdot y)$. Let $\theta_1(x) = \theta(x) - \theta(1)$, $\phi_1(x) = \phi(x) - \phi(1)$ and $\psi_1(x) = \psi(x) - \psi(1)$, then $(\theta_1, \phi_1, \psi_1)$ is also an isotopism and $\psi_1(1) = \theta_1(1) = \phi_1(1) = 0$. In the equation $\theta_1(x) + \phi_1(y) = \psi_1(x \cdot y)$, setting first x and then y to 1 results in $\theta_1(y) = \phi_1(y) = \psi_1(y)$ so that $\psi_1(x \cdot y) = \psi_1(x) + \psi_1(y)$. Let a be the element of $Z^*_{2^n+1}$ such that $\psi_1(a) = 1$, then $\psi_1(a^i) = i$ for $i = 1, 2, ..2^n - 1$ and $\psi_1(a^{2^n}) = 0$. This implies that a is a primitive element of Z_{2^n+1}. Thus, for each $x \in Z^*_{2^n+1}$, there exists a $t \in Z_{2^n}$ such that,

$$\psi_1(x) = \psi_1(a^t) = t = t \log_a(a) = \log_a(a^t) = \log_a(x).$$

Letting $c_1 = \theta(1)$, $c_2 = \phi(1)$, we arrive at equation (1).

Finally, suppose that the mixing mapping m is an isotopism. Then $m(x) = \log_a(x) + c$ implies $1 = m(1) = \log_a(1) + c = c$, $2 = m(2) = \log_a(2) + 1$, which implies that $a = 2$, and $0 = m(2^n) = \log_2(2^n) + 1 = n + 1$, which implies that n=1.

Polynomial expressions for multiplication and addition

Under the mixing mapping m, multiplication modulo $2^n + 1$, which is a bilinear function over the field Z_{2^n+1}, is a two variable function over the ring Z_{2^n}, which we denote by $x \odot y$. Similarly, under the inverse mixing mapping m^{-1}, addition modulo 2^n, which is an affine function in each argument over the ring Z_{2^n}, is a two variable function over the field Z_{2^n+1}, which we denote by $F(X, Y)$. Here and hereafter in this section, we denote arguments with lower-case letters when we consider them to be elements of Z_{2^n} and with upper-case letters when we consider them to be elements of Z_{2^n+1}. For example, when $n = 1$, we have

$$x + y \bmod 2 \longleftrightarrow F(X, Y) = 2XY \bmod 3,$$

$$XY \bmod 3 \longleftrightarrow x \odot y = x + y + 1 \bmod 2.$$

Theorem 2 *For $n \in \{2, 4, 8, 16\}$:*

1. For any fixed $X \neq 2^n (i.e., x \neq 0)$, the function $F(X, Y)$, corresponding to addition $x + y \bmod 2^n$ in Z_{2^n}, is a polynomial in Y over Z_{2^n+1} with degree $2^n - 1$. Similarly, for any fixed $Y \neq 2^n$, $F(X, Y)$ is a polynomial in X over Z_{2^n+1} with degree $2^n - 1$.

2. For any fixed $x \neq 0, 1$ (i.e., $X \neq 2^n, 1$), the function $x \odot y$, corresponding to multiplication $XY \bmod 2^n + 1$ in Z_{2^n+1} cannot be written as a polynomial in y over Z_{2^n}. Similarly, for any fixed $y \neq 0, 1$, $x \odot y$ is not a polynomial in x over Z_{2^n}.

Example 1

For $n = 2$, in Z_5, the function $F(X, Y)$ corresponding to $x + y \bmod 4$ is

$$F(X, Y) = 3(X^3 Y^2 + X^2 Y^3) + 3(X^3 Y + XY^3) + 2X^2 Y^2 + 4(X^2 Y + XY^2).$$

Proof.

1. In any finite field $GF(q)$, we have

$$\prod_{j \neq i, \alpha_j \neq 0} (x - \alpha_j)(-\alpha_i) = \begin{cases} \prod_{j(j \neq i), \alpha_j \neq 0} (\alpha_i - \alpha_j)(-\alpha_i) = -\prod_{\alpha_i \neq 0} \alpha_i = 1 & x = \alpha_i \\ 0 & x \neq \alpha_i, 0. \end{cases}$$

Thus, every function $f(x)$ from the set $GF(q) - \{0\}$ to $GF(q) - \{0\}$ can be written as a polynomial over GF(q) of degree at most $q - 2$ as follows:

$$f(x) = \sum_{\alpha_i \in GF(q) - \{0\}} f(\alpha_i) \prod_{\alpha_j \neq \alpha_i, 0} (x - \alpha_j)(-\alpha_i). \tag{2}$$

The function $F(A, X)$ corresponding to $a + x \bmod 2^n$ is a function from $GF(2^n + 1) - \{0\}$ to $GF(2^n + 1) - \{0\}$ in X. According to (2), this function can be written as

$$
\begin{aligned}
F(A, X) &= \begin{cases} A + X & 1 \leq X \leq 2^n - A \\ A + X + 1 & 2^n - A < X \leq 2^n \end{cases} \\
&= \sum_{I=1}^{2^n - A} (A + I) \prod_{\substack{J \neq I \\ 1 \leq J \leq 2^n}} (X - J)(-I) + \sum_{I = 2^n - A + 1}^{2^n} (A + I + 1) \prod_{\substack{J \neq I \\ 1 \leq J \leq 2^n}} (X - J)(-I) \\
&= \sum_{I=1}^{2^n} (A + I) \prod_{\substack{J \neq I \\ 1 \leq J \leq 2^n}} (X - J)(-I) + \sum_{I = 2^n - A + 1}^{2^n} \prod_{\substack{J \neq I \\ 1 \leq J \leq 2^n}} (X - J)(-I).
\end{aligned}
$$

The coefficient of $X^{2^n - 1}$ in $F(A, X)$ is

$$
\begin{aligned}
&\sum_{I=1}^{2^n} (A + I)(-I) + \sum_{I = 2^n - A + 1}^{2^n} (-I) = -A \sum_{I=1}^{2^n} I - \sum_{I=1}^{2^n} I^2 + \sum_{I=-A}^{-1} (-I) \\
&= \sum_{I=1}^{A} I = \frac{A(A + 1)}{2},
\end{aligned}
$$

which is zero if and only if $A = 0$ or $A = -1 = 2^n$, which cases are excluded by the assumption. Thus, $\deg F(X, A) = 2^n - 1$.

2. We show first the following lemma:

Lemma 1 *If $f(x)$ is a polynomial over Z_{2^n}, then for all $x \in Z_{2^n}$,*

$$f(2x) \bmod 2 = f(0) \bmod 2.$$

Proof. Let $f(x) = a_k x^k + a_{k-1} x^{k-1} + \cdots + a_1 x + a_0$.
Then for all $x \in Z_{2^n}$,

$$f(2x) = a_k (2x)^k + a_{k-1}(2x)^{k-1} + \cdots + a_1 2x + a_0.$$

Taking both sides of this equation modulo 2^n, we have $f(2x) = 2e + a_0 \bmod 2^n$, where e is an element in Z_{2^n}. Thus, $f(2x) \bmod 2 = a_0 \bmod 2 = f(0) \bmod 2$.

Now let $n > 1$, then for every integer a, $1 < a < 2^n$, there exists an integer $x_0 \in \{1, 2, \ldots, 2^n\}$ such that $2^n + 1 < 2ax_0 < 2(2^n + 1)$ and $0 \le 2a(x_0 - 1) < 2^n + 1$. The first inequality is equivalent to that $0 < 2ax_0 - (2^n + 1) < 2^n + 1$ and $2ax_0 - (2^n + 1)$ is an odd integer. Hence the function $f_a(x) = a \odot x$ corresponding to $AX \bmod (2^n + 1)$ satisfies

$$f_a(2x_0) \bmod 2 = (2ax_0 - (2^n + 1)) \bmod 2 = 1.$$

On the other hand, the inequality $0 \le 2a(x_0 - 1) < 2^n + 1$ implies that $2a(x_0 - 1)$ is an even integer in $\{0, 1, .., 2^n\}$ so that

$$f_a(2(x_0 - 1)) \bmod 2 = 2a(x_0 - 1) \bmod 2 = 0.$$

By the Lemma, $f_a(x)$ is not a polynomial over Z_{2^n}.

4 Design Principles for the Proposed Cipher

Confusion

Confusion (see [1,2]) means that the ciphertext depends on the plaintext and key in a complicated and involved way.

The confusion is achieved by mixing three different group operations. In the computational graph of the encryption process, the three different group operations are so arranged that the output of an operation of one type is **never** used as the input to an operation of the same type.

The three operations are incompatible in the sense that:

1. No pair of the 3 operations satisfies a distributive law. For example,

$$a \boxplus (b \odot c) \ne (a \boxplus b) \odot (a \boxplus c).$$

2. No pair of the 3 operations satisfies an associative law. For example,

$$a \boxplus (b \oplus c) \ne (a \boxplus b) \oplus c.$$

3. The 3 operations are combined by the mixing mapping m, which inhibits isotopisms as was shown in Theorem 1. Thus, using any bijections on the operands, it is impossible to realize any one of the three operations by another operation.

4. Under the mixing mapping, multiplication modulo $2^{16} + 1$, which is a bilinear function over $Z_{2^{16}+1}$, corresponds a non-polynomial function over $Z_{2^{16}}$; Under the inverse mixing mapping, addition modulo 2^{16}, which is an affine function in each argument over $Z_{2^{16}}$, corresponds a two variable polynomial of degree $2^{16} - 1$ in each variable over $Z_{2^{16}+1}$.

Diffusion

The *diffusion* requirement on a cipher is that each plaintext bit should influence every ciphertext bit and each key bit should influence every ciphertext bit(see [1,2]).

For the proposed cipher, a check by computer has shown that the diffusion requirement is satisfied after the first round, i.e., each output bit of the first round depends on every bit of the plaintext and on every bit of the key used for that round.

Diffusion is provided by the transformation called the multiplication–addition (MA) structure. The computational graph of the MA structure is shown in Fig.2. The MA structure transforms two 16 bit subblocks into two 16 bit subblocks controlled by two 16 bit key subblocks. This structure has the following properties:

– for any choice of the key subblocks Z_5 and Z_6, $MA(\cdot, \cdot, Z_5, Z_6)$ is an invertible transformation; for any choice of U_1 and U_2, $MA(U_1, U_2, \cdot, \cdot)$ is also an invertible transformation;

– this structure has a "complete diffusion" effect in the sense that each output subblock depends on every input subblock, and

– this structure uses the least number of operations (four) required to achieve such complete diffusion (see Appendix A for the proof).

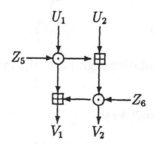

Figure 2: Computational graph of the MA structure

Similarity of encryption and decryption

The *similarity* of encryption and decryption means that decryption is essentially the same process as encryption, the only difference being that different key subblocks are used. This similarity results from

- using the output transformation in the encryption process so that the effect of (Z_1, Z_2, Z_3, Z_4) can be cancelled by using inverse key subblocks $(Z_1^{-1}, Z_2^{-1}, -Z_3, -Z_4)$ in the decryption process,

and from

- using an involution (i.e., a self-inverse function) with a 64 bit input and a 64 bit output controlled by a 32 bit key within the cipher. The involution used in the cipher is shown in Fig.3. The self-inverse property is a consequence of the fact that the exclusive-OR of (S_1, S_2) and (S_3, S_4) is equal to the exclusive-OR of (T_1, T_2) and (T_3, T_4); Thus, the input to the MA structure in Fig.3 is unchanged when S_1, S_2, S_3 and S_4 are replaced by T_1, T_2, T_3 and T_4. Thus, if T_1, T_2, T_3 and T_4 are the inputs to the involution, the left half of the output is

$$
\begin{aligned}
& (T_3, T_4) \oplus MA((T_1, T_2) \oplus (T_3, T_4), Z_5, Z_6) \\
= \ & (S_1, S_2) \oplus MA((S_1, S_2) \oplus (S_3, S_4), Z_5, Z_6) \oplus MA((S_1, S_2) \oplus (S_3, S_4), Z_5, Z_6) \\
= \ & (S_1, S_2).
\end{aligned}
$$

Similarly, the right half of the output is (S_3, S_4).

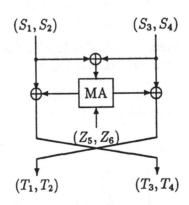

Figure 3: Computational graph of the involution

Perfect secrecy for a "one-time" key

Perfect secrecy in the sense of Shannon is obtained in each round of encryption if a "one-time" key is used. In fact, such perfect secrecy is achieved at the input transformation in the first round because each operation is a group operation. In addition, for every choice of (p_1, p_2, p_3, p_4) and of (q_1, q_2, q_3, q_4) in F_2^{64}, there are exactly 2^{32} different choices of the key subblocks $(Z_1, .., Z_6)$ such that the first round of the cipher transforms (p_1, p_2, p_3, p_4) into (q_1, q_2, q_3, q_4).

5   Implementations of the Cipher

The cipher structure was chosen to facilitate both hardware and software implementations. In the encryption process, a regular modular structure was chosen so that the cipher can be easily implemented in hardware. A VLSI implementation of the cipher is being carried out at the Integrated Systems Laboratory of the ETH, Zürich. The estimated data rate of this chip varies from 45 Mbits per second to 115 Mbits per second, depending on the complexity of the architecture chosen.

The cipher can also be easily implemented in software because only operations on pairs of 16-bit subblocks are used in the encryption process.

The most difficult part in the implementation, multiplication modulo $(2^{16} + 1)$, can be implemented in the way suggested by the following lemma.

Lemma 2 *Let a, b be two n-bit non-zero integers in Z_{2^n+1}, then*

$$ab \bmod (2^n+1) = \begin{cases} (ab \bmod 2^n) - (ab \operatorname{div} 2^n) & \text{if } (ab \bmod 2^n) \geq (ab \operatorname{div} 2^n) \\ (ab \bmod 2^n) - (ab \operatorname{div} 2^n) + 2^n + 1 & \text{if } (ab \bmod 2^n) < (ab \operatorname{div} 2^n) \end{cases}$$

where $(ab \operatorname{div} 2^n)$ denotes the quotient when ab is divided by 2^n.

Note that $(ab \bmod 2^n)$ corresponds to the n least significant bits of ab, and $(ab \operatorname{div} 2^n)$ is just the right-shift of ab by n bits. Note also that $(ab \bmod 2^n) = (ab \operatorname{div} 2^n)$ implies that $ab \bmod (2^n + 1) = 0$ and hence cannot occur when $2^n + 1$ is a prime.

Proof. For any non-zero a and b in Z_{2^n+1}, there exist unique integers q and r such that

$$ab = q(2^n + 1) + r, \qquad 0 \leq r < 2^n + 1, \ 0 \leq q < 2^n.$$

Moreover, $q + r < 2^{n+1}$. Note that $r = ab \bmod (2^n + 1)$. We have

$$(ab \operatorname{div} 2^n) = \begin{cases} q & \text{if } q + r < 2^n \\ q+1 & \text{if } q + r \geq 2^n \end{cases}$$

and

$$(ab \bmod 2^n) = \begin{cases} q+r & \text{if } q + r < 2^n \\ q+r-2^n & \text{if } q + r \geq 2^n. \end{cases}$$

Thus

$$r = \begin{cases} (ab \bmod 2^n) - (ab \operatorname{div} 2^n) & \text{if } q + r < 2^n \\ (ab \bmod 2^n) - (ab \operatorname{div} 2^n) + 2^n + 1 & \text{if } q + r \geq 2^n. \end{cases}$$

But $q + r < 2^n$ if and only if $(ab \bmod 2^n) \geq (ab \text{ div } 2^n)$. This proves the Lemma.

A C-program of the cipher together with some examples for checking the correctness of the implementation are given in Appendix B.

6 Conclusion

The cipher described above is proposed as a possible candidate for a new encryption standard. The cipher is based on the design concept of "mixing 3 different group operations" to achieve the required confusion and diffusion. Confusion is achieved by arranging the operations in a way that no pair of successive operations are of the same type and by the fact that operations of different types are incompatible. The structure of the cipher is so chosen that diffusion can be achieved using a small number of operations.

Enciphering and deciphering are essentially the same process, but with different key subblocks. Because of the use of 16-bit operations and a regular modular structure, the cipher can be implemented efficiently in both hardware and software. In particular, bit-level permutations are avoided in the encryption process because such permutations are difficult to implement in software.

In all of the statistical testings conducted up to now, we have not found any significant difference between the permutation of F_2^{64} determined by the cipher with a randomly chosen key and a permutation equiprobably chosen from all possible permutations of F_2^{64}.

The security of the proposed cipher needs further intensive investigation. The authors hereby invite interested parties to attack this proposed cipher and will be grateful to receive the results (positive or negative) of any such attacks.

References

[1] C. E. Shannon, "Communication Theory of Secrecy Systems", B.S.T.J., Vol. 28, pp.656-715, Oct. 1949.

[2] J. L. Massey, "An Introduction to Contemporary Cryptology", *Proc. IEEE*, Vol. 76, No. 5, pp. 533–549, May 1988.

[3] J.Dénes, A.D.Keedwell, *Latin squares and their applications*, Akadémiai Kiadó, Budapest 1974.

Appendix A: Number of Operations Required for Diffusion

An *operation* is a mapping from two variables to one variable. A *computational graph* is a directed graph in which the vertices are operations, the edges entering a vertex are the inputs to the operation, the edges leaving a vertex are the outputs of the operation, the edges entering no vertex are the graph outputs, and the edges leaving no vertex are the graph inputs. An algorithm to compute a function determines a computational graph where the graph inputs are the inputs of the algorithm and the graph outputs are the outputs of the algorithm.

Consider a function having the form

$$(Y_1, Y_2) = E(X_1, X_2, Z_1, Z_2), \qquad X_i, Y_i \in F_2^m, \quad X_i \in F_2^k \tag{3}$$

such that, for every choice of (Z_1, Z_2), $E(\cdot, \cdot, Z_1, Z_2)$ is invertible. Such a function will be called a *cipher function*. A cipher function is said to have *complete diffusion* if each of its output variable depends non-idly on every input variable.

Theorem 3 *If a cipher function of the form (3) has complete diffusion, then the computational graph of any algorithm that computes the function contains at least 4 operations.*

Proof. Let $Y_1 = E_1(X_1, X_2, Z_1, Z_2)$, and $Y_2 = E_2(X_1, X_2, Z_1, Z_2)$. Then if E_1 has complete diffusion, it contains at least 3 operations because there are four input variables. Suppose E_1 contains 3 operations. The invertibility of the cipher function implies that $E_2 \neq E_1$ and complete diffusion requires that E_2 not equal any intermediate result that appears in E_1. Thus, at least one operation not appearing in E_1 is required in E_2. This proves the theorem.

It is easy to check that the MA structure shown in Fig.2, which has four operations, is a cipher function with complete diffusion, i.e., that each of the output variables depends non-idly on all four input variables.

Appendix B: C-program of the Cipher and Sample Data

```
/* C - program  blockcipher */

# define maxim 65537
# define fuyi  65536
# define one   65535
# define round    8
void   cip(unsigned IN[5],unsigned OUT[5],unsigned Z[7][10]);
void   key( short unsigned uskey[9],unsigned Z[7][10] );
void   de_key(unsigned Z[7][10],unsigned DK[7][10]);
unsigned inv(unsigned xin);
unsigned mul(unsigned a, unsigned b);
```

```
main()
{
  unsigned  Z[7][10], DK[7][10], XX[5],TT[5], YY[5];
  short unsigned uskey[9];

/*                to generate encryption key blocks */
      key(uskey,Z);
/*                to generate decryption key blocks */
      de_key(Z,DK);
/*          to encipher plaintext XX to ciphertext YY by key blocks Z */
      cip(XX,YY,Z);
/*          to decipher ciphertext YY to  TT by decryption key blocks DK */
      cip(YY,TT,DK);
}

    /* encryption algorithm */
void  cip(unsigned IN[5],unsigned OUT[5],unsigned Z[7][10])
 {
     unsigned r, x1,x2,x3,x4,kk,t1,t2,a;
     x1=IN[1];  x2=IN[2]; x3= IN[3]; x4=IN[4];
     for (r= 1; r<= 8; r++) {
        x1 =mul(x1,Z[1][r]);
        x2 =mul(x2,Z[2][r]);
        x3 =( x3+ Z[3][r] ) & one;
        x4 =( x4 + Z[4][r] ) & one;
        kk = mul( Z[5][r], ( x1^x3 ) );
        t1 = mul( Z[6][r], (  kk+ ( x2^x4) )& one);
        t2 = ( kk+ t1 ) & one;
        a = x1^t1;       x1 = x3^t1;      x3 = a;
        a = x2^t2;       x2 = x4^t2;      x4 = a;
        }
     OUT[1] = mul( x1,Z[1][round+1] );
     OUT[2] = mul( x2,Z[2][round+1] );
     OUT[3] = ( x3 + Z[3][round +1] )& one;
     OUT[4] = ( x4 + Z[4][round+1] ) & one;
 }


   /* the multiplication */
unsigned mul(unsigned  a, unsigned  b)
{
  long int p;
  long unsigned q;
     if (a==0)    p = maxim-b;
     else if ( b==0 )  p = maxim-a;  else
       { q=a*b; p=( q & one) - (q>>16);  if (p<=0) p= p+maxim; }
  return (unsigned)(p & one);
}


          /*  compute multiplication inverse of integer xin
                  by Euclidean gcd algorithm */
unsigned inv(unsigned  xin)
{
     long n1,n2,q,r,b1,b2,t;
     if ( xin == 0 )  b2 = 0;
```

403

```
   else {
       n1=maxim; n2 = xin; b2= 1; b1= 0;
       do   { r = (n1 % n2);  q = (n1-r)/n2 ;
              if (r== 0) {if ( b2<0 )  b2 = maxim+b2; }
              else { n1= n2; n2= r; t = b2; b2= b1- q*b2; b1= t; }
             } while (r != 0);
   }
   return (unsigned)b2;
 }


   /*  generate key blocks Z[i][r] from user key uskey */
void   key( short unsigned uskey[9], unsigned  Z[7][10] )
{
   short unsigned S[55];
   int  i,j,r;
                   /*   shifts  */
   for (i = 1; i<9; i++)  S[i-1] = uskey[i];
   for (i = 8; i< 55; i++ ) {
       if ( (i+2)%8 == 0 )                  /* for S[14],S[22],.. */
        S[i] = ( S[i-7]<<9 )^( S[i-14]>>7 );
       else  if ( (i+1)%8 ==0 )             /* for S[15],S[23],.. */
        S[i] =( S[i-15]<<9 )^( S[i-14]>>7 );
       else                                 /* for other S[i]     */
        S[i] = ( S[i-7]<<9 )^( S[i-6]>>7 );
   }
             /* get endcryption key blocks    */
      for (r= 1; r<=round+1; r++)    for(j= 1;j<7; j++)
             Z[j][r] = S[6*(r-1) + j-1];
 }


   /* compute decryption key blocks DK[i][r]
          from encryption key blocks Z[i][r] */
void  de_key(unsigned Z[7][10],unsigned DK[7][10])
{
   int  j;
   for (j = 1; j<=round+1; j++) {
       DK[1][round-j+2] = inv(Z[1][j]);
       DK[2][round-j+2] = inv(Z[2][j]);
       DK[3][round-j+2] = ( fuyi-Z[3][j] ) & one;
       DK[4][round-j+2] = ( fuyi-Z[4][j] ) & one;
   }
   for (j= 1;j<=round+1;j++)
     { DK[5][round+1-j] = Z[5][j];  DK[6][round+1-j] = Z[6][j];}
 }
```

Some sample data for checking the correctness of implementations are given below. The numbers are 16-bit integers in the decimal form.

the eight 16-bit user key subblocks: uskey[i]

	1	2	3	4	5	6	7	8

encryption key subblocks Z[i][r]

	Z[1][r]	Z[2][r]	Z[3][r]	Z[4][r]	Z[5][r]	Z[6][r]
1-st round	1	2	3	4	5	6
2-nd round	7	8	1024	1536	2048	2560
3-rd round	3072	3584	4096	512	16	20
4-th round	24	28	32	4	8	12
5-th round	10240	12288	14336	16384	2048	4096
6-th round	6144	8192	112	128	16	32
7-th round	48	64	80	96	0	8192
8-th round	16384	24576	32768	40960	49152	57345
output transf	128	192	256	320	--	--

decryption key subblocks   DK[i][r]

1-st round	65025	43350	65280	65216	49152	57345
2-nd round	65533	21843	32768	24576	0	8192
3-rd round	42326	64513	65456	65440	16	32
4-th round	21835	65529	65424	65408	2048	4096
5-th round	13101	43686	51200	49152	8	12
6-th round	19115	53834	65504	65532	16	20
7-th round	43670	28069	61440	65024	2048	2560
8-th round	18725	57345	64512	64000	5	6
output transf	1	32769	65533	65532	--	--

plaintext XX	0	1	2	3
after 1-st round	177	202	180	207
after 2-nd round	5054	10696	5085	10583
after 3-rd round	42790	64040	25583	15559
after 4-th round	16281	58571	61463	33861
after 5-th round	62321	51187	1399	59053
after 6-th round	37668	1126	6125	42057
after 7-th round	49700	61227	19644	21245
after 8-th round	2688	12695	2372	1339
YY =cip(XX,Z)	16379	12571	2628	1659

TT=cip(YY,DK)	0	1	2	3

A new trapdoor in knapsacks

Valtteri Niemi

Mathematics Department

University of Turku

20500 Turku, Finland

Abstract. *A public key scheme with trapdoor based on a group of modular knapsacks is proposed. In parallel architecture encryption and decryption are very fast.*

1 Introduction

Recently Adi Shamir [4] proposed a new identification scheme that possesses two nice characteristic features. First, he got rid of huge number arithmetics thus making it easier to implement a smart card identification system. Secondly, the security of the new scheme does not depend on the difficulty of factoring.

In this paper we present a new public key cryptosystem based on the same two features as Shamir's identification scheme. We do not obtain as low levels of time and space complexity as in [4] (which is quite understandable since we must build a trapdoor). Nevertheless, we may use 8 bit numbers and at least in parallel architecture the operations can be carried out very fast.

The underlying difficult problem we try to imitate consists of a group of modular knapsack type equations. It is NP-complete in the strong sense which means that it is possible to use small numbers as coefficients in the equations.

Our trapdoor is built roughly as follows. Let us consider square matrices whose elements are very small numbers. We disguise them by matrix multiplication with an arbitrary square matrix. All operations are carried out in a finite field of a prime order. Note that if we originally have only one "small" square matrix this process causes no limitations on the disguised matrix obtained after multiplication (provided the original matrix has an inverse).

In our scheme, we begin with two "small" matrices and use the two disguised versions as a public key. (They constitute the coefficients in knapsack type equations.) The security of the trapdoor is thus based on the difficulty of determining the multiplier matrix after two repetitions of the disguising procedure.

2 The underlying hard problem

Let us consider two $n \times n$-square matrices A and B over a finite field $\mathbf{Z}_p$. Combine these matrices to one $n \times 2n$-matrix $E = (A \mid B)$. Let x be a column vector (i.e. $2n \times 1$-matrix) of 0's and 1's. Now we obtain an $n \times 1$-matrix c as a matrix product : $c = Ex$.

Modular knapsack group problem :
Given : an $n \times 2n$-matrix E, an n-vector c and a prime p.
Find : a $2n$-vector x whose elements belong to the set $\{0,1\}$ satisfying the equation $Ex \equiv c \pmod{p}$.

A quite straight-forward transformation from the EXACT 3-COVER problem (see [1]) shows that our problem is NP-complete even in the case where the elements of E and c are also 0's and 1's. Hence, the problem is NP-complete in the strong sense.

3 The trapdoor

Let us first define a notion of an *absolute value* in the field $\mathbf{Z}_p$: An absolute value $|a|$ of $a \in \mathbf{Z}_p$ is the minimum of the least *positive* remainders (modulo p) of the two integers a and $-a$.

Example. In $\mathbf{Z}_{19}$ $|3| = |16| = 3, |-8| = |8| = |11| = |30| = 8$ etc.

In general, $0 \leq |a| \leq \frac{p}{2}$. We say that a is k-small if $|a| \leq k$, and a is k-large if $|a| \geq \frac{p}{2} - k$. In the sequel, we speak shortly of small and large numbers thus leaving k unfixed. However, we make a general unrigorous assumption that k is relatively small compared with the moduli p. (The exact choice of k is discussed in the next section.)

Let us now fix $n \times n$-square matrices C, D, S of *small* numbers and an arbitrary $n \times n$-matrix R. Furthermore, fix a *diagonal* $n \times n$-matrix Δ of *large* numbers. Compute the two square matrices

$$A = R^{-1}(\Delta - SC) \quad \text{and} \quad B = -R^{-1}SD \tag{1}$$

(which exist provided R has a full rank.)
Now the following matrix identity holds :

$$(R \ \ S) \begin{pmatrix} A & B \\ C & D \end{pmatrix} = (\Delta \ \ 0).$$

Our cryptosystem can now be defined :

Public key : Matrix $E = (\ A \quad B\ )$.

Private key : Matrix R. (Matrices C, D, S and Δ are also private but they are not needed after the initial construction.)

Messages : $2n$-bit vectors x.

Encryption : an n-vector $c = Ex$.

Decryption : Compute an n-vector $l = Rc$. Now the first half of x can be found by the following rule :

$$\begin{cases} \text{If } l_i \text{ is small then } x_i = 0 \\ \text{If } l_i \text{ is large then } x_i = 1 \end{cases} \qquad (i = 1, ..., n).$$

The other half of x can be decrypted by elementary linear algebra (n equations, n variables).

Our decryption rule is valid by the following observations. First,

$$(\ R \quad S\ )\begin{pmatrix} A & B \\ C & D \end{pmatrix}(x) = (\ \Delta \quad 0\ )(x) = \begin{pmatrix} \Delta_1 x_1 \\ \vdots \\ \Delta_n x_n \end{pmatrix},$$

where Δ_i:s are Δ:s (large) diagonal elements.

On the other hand,

$$(\ R \quad S\ )\begin{pmatrix} A & B \\ C & D \end{pmatrix}(x) = (\ R \quad S\ )\begin{pmatrix} c \\ \alpha \end{pmatrix},$$

where α is a small n-vector, since C, D and x consist of small numbers (of course, the parameter k in the definition of smallness must be increased in the case of α).

Furthermore,

$$(\ R \quad S\ )\begin{pmatrix} c \\ \alpha \end{pmatrix} = Rc + S\alpha = l + S\alpha,$$

where $S\alpha$ is a small n-vector as S and α consist of small numbers only.

Thus, $\Delta_i x_i = l_i + (S\alpha)_i$ and since the last term is small, l_i determines whether the right-hand side is large or small. Similarly, x_i determines whether the left-hand side is large or small.

A toy example. Let us fix $n = 5, k = 1$ and $p = 23$. We choose matrices

C,D (whose elements are 1-small) and R at random :

$$C = \begin{pmatrix} 0 & 0 & -1 & 1 & -1 \\ -1 & 0 & -1 & 0 & 0 \\ 1 & 1 & 0 & 1 & 0 \\ -1 & 0 & -1 & 0 & 0 \\ 0 & 1 & 0 & 1 & 1 \end{pmatrix}, D = \begin{pmatrix} 0 & 1 & 1 & 0 & -1 \\ 1 & 1 & -1 & 0 & 0 \\ -1 & 0 & 0 & 1 & 1 \\ 0 & -1 & 0 & 0 & 1 \\ 1 & 0 & 0 & 0 & 1 \end{pmatrix},$$

$$R = \begin{pmatrix} 15 & 14 & 14 & 11 & 21 \\ 6 & 17 & 6 & 17 & 4 \\ 13 & 15 & 11 & 7 & 13 \\ 19 & 18 & 19 & 7 & 3 \\ 18 & 13 & 14 & 8 & 8 \end{pmatrix}.$$

Also, we choose $S = I$ (the identity matrix) and $\Delta = 11 \cdot I$. Now

$$R^{-1} = \begin{pmatrix} -7 & 3 & 17 & 4 & 5 \\ 1 & 1 & 4 & 11 & -8 \\ 11 & -5 & -1 & 2 & 9 \\ 17 & 4 & 13 & 9 & 18 \\ 21 & 5 & 13 & -2 & -6 \end{pmatrix}$$

and we may calculate the public key matrices by (1) :

$$A = \begin{pmatrix} 5 & 11 & 3 & 6 & 20 \\ -4 & 15 & 11 & 9 & 13 \\ 4 & 6 & -3 & 3 & 9 \\ 3 & 13 & 12 & 5 & 13 \\ 14 & 2 & 6 & -4 & 7 \end{pmatrix}, B = \begin{pmatrix} 9 & 8 & 10 & 6 & 13 \\ 11 & 9 & 0 & -4 & -6 \\ 18 & -4 & 7 & 1 & 1 \\ -9 & 11 & 10 & 10 & 0 \\ 14 & -5 & 7 & 10 & 16 \end{pmatrix}.$$

The message

$$x = (\ 1 \quad 1 \quad 0 \quad 1 \quad 0 \quad 1 \quad 0 \quad 1 \quad 1 \quad 1\ )$$

is encrypted as

$$c = (\ A \quad B\ ) \cdot x^T = (\ -9 \quad 21 \quad -6 \quad 9 \quad 13\ )^T.$$

To decrypt we first calculate

$$l = Rc = (\ 10 \quad 12 \quad 19 \quad 11 \quad 19\ )^T = (\ 10 \quad 12 \quad -4 \quad 11 \quad -4\ )^T$$

from which we can derive the first 5 bits of the message: first, second and fourth element are large, thus corresponding to 1's, while third and fifth element are (comparatively) small, corresponding to 0's. We skip here the second part of decyption (that uses methods of linear algebra).

4 Observations on security and complexity

As already noted in the previous section, it is sufficient to decrypt only half of the cryptotext bits, since the other half can be determined easily by the public key only. Of course, the cryptanalyst is also able to complete the decryption if she already knows half of the plaintext. Also, it is possible to derive dependencies between plaintext bits solely from the public information. For these reasons it is recommended to combine each n-bit plaintext block with a random padding of n bits.

From (1) in the previous section we see that the matrices A and B are obtained from two special-type matrices (i.e. $\Delta - SC$ and $-SD$) by multiplying them with the *same* matrix. In principle, this relation between A and B gives us a starting point for cryptanalysis. The multiplier R^{-1} can be eliminated, leading to the equation

$$-SD = (\Delta - SC)A^{-1}B.$$

Hence, we should be able to multiply the known matrix $A^{-1}B$ by some matrix whose nondiagonal elements are small and diagonal ones are large and the result of this operation should be a small matrix. It is easy to see that if we can fulfil these conditions, a suitable decryption matrix, say R', can be found.

This approach also shows that from the point of security the choices of S and Δ are quite unessential. We may, for instance, let S to be an identity matrix I and Δ to be $\lfloor \frac{p}{2} \rfloor \cdot I$. (Hence, S and Δ could be universal entities.) Now the critical requirement of the previous section, i.e. that $S\alpha$ should be small, can be stated exactly in the form of inequality :

$$|\alpha| < \lfloor \tfrac{p}{4} \rfloor.$$

On the other hand, since $\alpha = (C \quad D) \cdot (x)$, it follows from the triangle inequality (which clearly is valid also for this definition of absolute value) that

$$|\alpha| \leq 2n \cdot \max\{|c| : c \in C \cup D\} = 2nk \tag{2}$$

(where k refers to the definition of k-smallness). Hence, the critical demand can be restated as

$$nk < \frac{p}{8} \tag{3}$$

To exclude exhaustive search attacks we should choose n sufficiently large, e.g. $n = 100$ seems to be suitable. If $k = 1$, which means that the elements of C and D must be chosen from the set $\{-1, 0, 1\}$, the condition (3) gives a lower bound for the moduli $p : p > 800$. This would mean that 8 bit numbers are slightly too small for our purposes. However, in practice the value of $|\alpha|$ is considerably smaller than the theoretical upper bound $2nk$ derived in (2).

Indeed, the value $2nk$ could be reached only in the extreme case where all elements of C, D and x are equal to 1. Since the choice of C and D is free within the set $\{-1, 0, 1\}$, we can easily reduce even the theoretical upper bound to one third or even more. This gives us the possibility of using, e.g., a moduli $p = 251$ that is a 8 bit number.

In fact, a false decryption of some bits due to too large value of $|\alpha|$ does not cause serious problems, seen as follows. Let us suppose the receiver has decrypted a cryptogram and she has a proposal for the correct plaintext. She can easily check whether the proposal is valid by the encryption mechanism. If the result of checking is negative she must determine which bits are wrong. Fortunately, the receiver can use her proposal to calculate estimates for the $|\alpha|$-values. Bits corresponding to largest $|\alpha|$-values are best candidates as false ones, and the receiver can correct her guess. Of course, the process converges only if the portion of falsely decrypted bits is small.

Another question is the choice of k. If one does not trust on too small elements in C and D but rather chooses, e.g., $k = 128$ the size of the moduli respectively extends to 16 bits etc. On the other hand, it does not seem to be very likely that the security of the system would depend too heavily on the size of k.

In the case of $n = 100, k = 1$ we must, in average, execute 10 000 single-byte additions to encrypt a padded 100-bit message. Decryption takes the same number of single-byte multiplications. On the other hand, the mechanism is particularly suitable for parallel computers, since only matrix operations are needed. For instance, a special hardware with 100 processors needs only one single-byte addition to encrypt one bit of plaintext and one single-byte multiplication per decrypted bit. With more processors involved the operations are even faster.

The keys are, unfortunately, quite large. The public key consists of 20 kilobytes in the same case as above, while the private key needs 10 kB. The latter one is basically a random matrix, hence it can be stored in a form of a pseudorandom function but the same idea does not suit for the public matrix. In general it can be said that as regards time and space requirements this new system is in the same class as some known systems based on error-correcting codes (see e.g. [3] and [2].)

5 Some variations

Perhaps the most immediate variations of the basic scheme are found by changing the underlying field structure. For example, we could choose elements of R^{-1}, Δ, S, C and D from $\mathbf{Z}$ (which means that also A and B are $\mathbf{Z}$-matrices). In this case, the decryption matrix R will be a rational matrix, and obviously both keys will be larger than in the modular version. Finite fields of order p^k do not seem very suitable, since the ordering in the field is crucial to the scheme.

Another possible variation is to change the form of the matrices involved. Let A and B be general $n \times m$-matrices instead of square ones. Similarly, C and D are $n \times m$-matrices and, respectively, R and S are $m \times n$-matrices. Further, Δ is still a square (of order m). The length of plaintext blocks is $2m$, while cryptotext blocks are still n-vectors. This variant means that R cannot be chosen completely randomly in the case $m > n$.

The following variation deals with the problem of large keys. Recall that the private key R was chosen at random and the public key matrices were calculated by (1). We could as well choose A (or B) at random and calculate R and B (or R and A). Then half of the public key is random and could be stored in pseudorandom form, thus reducing the need of storage essentially by half.

Our last variation is in fact an addition to the disguising procedure. We can try to improve the security of the system by a standard way of permuting the columns of the encryption matrix E. As a result there is no trivial way of separating matrices A and B, hence the starting point of cryptanalysis must be based on the properties of single columns.

Unfortunately, we have no variants for the purpose of signatures. As usual in knapsack-type systems, the encryption function is not surjective.

References

[1] Garey, M. R. and Johnson, D. S., *Computers and intractability : A guide to the theory of NP-completeness*, 1979.

[2] MacEliece, R. J., A public-key cryptosystem based on algebraic coding theory, *DSN Progress Rep. 42-44, Jet Propulsion Laboratory*, 114-116, 1978.

[3] Niederreiter, H., Knapsack-type cryptosystems and algebraic coding theory, *Problems of Control and Information Theory*, **15**, 159-166, 1986.

[4] Shamir, A., An Efficient Identification Scheme Based on Permuted Kernels (extended abstract), 1989.

On the Design of Provably-Secure Cryptographic Hash Functions

Alfredo De Santis*

Dipartimento di Informatica ed Applicazioni

Università di Salerno

84081 Baronissi (Salerno), Italy

Moti Yung

IBM Research Division

T. J. Watson Research Center

Yorktown Heights, NY 10598

(extended summary)

Abstract

Recently, formal complexity-theoretic treatment of cryptographic hash functions was suggested. Two primitives of Collision-free hash functions and Universal one-way hash function families have been defined. The primitives have numerous applications in secure information compression, since their security implies that finding collisions is computationally hard. Most notably, Naor and Yung have shown that the most secure signature scheme can be reduced to the existence of universal one-way hash (this, in turn, gives the first trapdoor-less provably secure signature scheme).

In this work, we first present reductions from various one-way function families to universal one-way hash functions. Our reductions are general and quite efficient and show how to base universal one-way hash functions on any of the known concrete candidates for one-way functions. We then show equivalences among various definitions of hardness for collision-free hash functions.

1 Introduction

Cryptographic Hash Functions are important tools in secure information compression and as building blocks for other cryptographic procedures. A hash function is cryptographically strong if collision finding is computationally hard.

The usefulness of cryptographic hashing was used and known in practice [M1, Gi, M2] (and is already mentioned in the original Diffie-Hellman paper which introduced the notions of trapdoor and one-way functions and their applications). Nevertheless, only recently two formal complexity-theoretic definitions of cryptographic hash functions were given and implementations based on one-way functions were suggested.

---

*Part of this work was done while the author was visiting IBM Research Division, T. J. Watson Research Ctr, Yorktown Heights, NY 10598.

The first function family, suggested by Damgård, is the *Collision-Free Hash Functions* (CFHF) [D], which is based on *Claw-Free* functions of Goldwasser, Micali and Rivest [GMRi]. The second function family, suggested by Naor and Yung [NY], is the *Universal One-Way Hash Function* family (UOWHF). In a CFHF family, the function is given and then finding a colliding pair is hard, while in UOWHF definition is weaker: first an adversary chooses an input to compress and then the function is drawn at random; (the weaker definition may imply implementations based on weaker assumptions).

CFHF family can be based on any one-way homomorphism. Its applicability was shown as this family when having also a trapdoor property was used to implement a secure signature scheme [GMRi, D]; it was also shown to give efficient zero-knowledge proof-systems [NY]. On the other hand, [NY] showed that a secure signature scheme is reduces to the existence of UOWHF, and they show how to achieve UOWHF based on 1-1 one-way functions, giving the first provably secure signature scheme which is trapdoor-less (unlike all previous secure signature schemes which had followed the Diffie-Hellman model of basing signatures on a trapdoor property).

It is theoretically important to base cryptographic primitives and basic tools on reduced complexity assumptions, it is also practically important to give efficient implementations of such tools. In this work we investigate implementations of cryptographic hash functions on reduced complexity assumptions and investigate reductions among various definitions of hardness of collision finding. We would like the reductions to be efficient as well.

We first give a reduction from a 1-1 one-way function to a UOWHF (our proof is easier and the reduction is more efficient than the original construction in [NY]). We then show that a UOWHF can be based on a one-way function with the property that the expected size of the preimage of an element in the range is small (i.e., when an element in the domain is randomly chosen). We call this function *small expected preimage-size function*. We then show how to construct such a function if a regular function [GKL] is available; this function family includes a large number of concrete examples (see [GKL]). Even more generally, we show how to reduce a very general one-way function family to a small expected preimage-size family. The general property requirement is that: given an element in the range, an estimate on the size of the preimage set is almost always easily computable (where the estimate should only be polynomially close to the real size). We call such a function an *almost-known preimage-size function*. This requirement is a mild one since it implies some structure of the domain-range relationship which all concrete candidates for one-way functions (based on number theory, algebra, coding theory or combinatorics (subset-sum)) have. Then, we investigate various definitions of hardness of hash functions and we show reductions and equivalences among the various definitions; such relations may lead to finding CFHF based on reduced complexity assumptions or on concrete functions which are assumed to be one-way.

Recently, Rompel has come up with a construction of generating a UOWHF based on any one-way function [Ro]. This general construction does not rely on the approximate knowledge of the preimage size (the property mentioned above) and it is much more involved than ours. It is ingenious and theoretically optimal, as it shows that the Naor-Yung approach leads to a signature based on any one-way function (which is necessary by the work of Impagliazzo and Luby [IL]). However it is much less practical than the work presented here for all known concrete candidates for one-way functions.

The rest of the paper is organized as following. Section 2 present background on one-way functions, universal hash functions, UOWHF, and signature schemes. The reader familiar with these notions can skip this section. Section 3 gives the construction based on any 1-1 one-way function, while Section 4 gives the construction based on small expected preimage-size, regular, and the most general almost-known preimage-size families. In section 5 we present reductions among different notions of difficulty of cryptographic hash functions.

2 Notations and background

Let x be a string, then $|x|$ is the length of x. Let x and y be two strings, then $x \diamond y$ is the concatenation of x and y. By the symbol "$\circ$" we mean the composition of functions. Thus, let f and g be functions, $y = f \circ g(x)$ is the value $f(g(x))$.

Probability ensembles: A probability ensemble D is the set $\{D_n \mid n \in N^+\}$ where D_n is a distribution probability on $\{0,1\}^n$. For $x \in \{0,1\}^n$, $D[x]$ is the probability assigned to x by D_n. For $X \subseteq \{0,1\}^n$, $D[X]$ is the sum $\sum_{x \in X} D[x]$. By the notation $x \in_R B$, we mean that x has been chosen from the set B under the uniform distribution (i.e. each element in B has the same probability $1/|B|$ of being selected).

Accessible ensembles: An ensemble D is accessible if there exists a probabilistic polynomial time algorithm G, such that on input n, the probability distribution induced by the output of G (depending on its internal coin-flips) is D_n.

Functions: A function f is a collection $\{f_n: \{0,1\}^n \rightarrow \{0,1\}^{l(n)} \mid n \in N^+\}$ where $l(n)$ is the output length. Hereafter, for sake of brevity we often omit the subscript in f_n. All functions considered will be polynomial time computable, i.e. given an input n and an argument x, the value $f_n(x)$ can be computed in time polynomial in n.

Definition 1 [One-way function.] *f is one-way if for every polynomial time algorithm A, for all polynomials p and all sufficiently large n, the probability that A on input $f(x)$, when $x \in_R \{0,1\}^n$, outputs a y such that $f(y) = f(x)$ is*

$$Pr[f(x) = f(A(f(x))) \mid x \in_R \{0,1\}^n] < 1/p(n).$$

We remark that the above definition is of a *strong one-way function* which is implied by the existence of the weaker *somewhat one-way function* using Yao's amplification technique [Y]. A somewhat one-way function has the same definition as above, but the hardness of inversion is smaller, i.e. its probability is inverse polynomially away from 1. (In the above definition the probability is at most $1 - 1/q(n)$ for a given polynomial (instead of $1/p(n)$ for any polynomial above)).

Unless stated otherwise, f will have input length n and output length $l(n)$. The requirement that the range of f has a uniform length is without loss of generality, as we may use a variable to fixed length encoding. For instance, if f has output length less than or equal to m, then we can construct a function f' with output length at most $2m$ (or $m + 2\lceil\log m\rceil + 2$) by employing a suitably prefix encoding [E] explained below.

A prefix set of strings is a set S with the property that if any two strings $x, y \in S$ are such that $x = y \diamond w$ (where w is a string) then $x = y$ (and w is the empty string). A prefix encoding from a set S to a set R is a bijection from S to R where R is a prefix set.

In our case, the set is the range of a one-way function f (where $f = f_n$), we employ the following encoding $f'(x) = 0^{|f(x)|-1} \diamond 1 \diamond f(x)$. This has the property that it is easy to compute $f'(x)$ given $f(x)$ and vice-versa, even without knowing x. Since the range of f' is a prefix set, we can add dummy zeroes to make the range of the same length m. That is $f''(x) = f'(x) \diamond 0^{m-|f'(x)|}$.

2.1 Collision-Free Hash Functions

Let $\{n_{1_i}\}$ and $\{n_{0_i}\}$ be two increasing sequences such that for all i $n_{0_i} \leq n_{1_i}$, but $\exists q$, a polynomial such that $q(n_{0_i}) \geq n_{1_i}$ (we say that these sequences are polynomially related). Let H_k be a collection of functions such that for all $h \in H_k$, $h\colon \{0,1\}^{n_{1_k}} \mapsto \{0,1\}^{n_{0_k}}$ and let $U = \bigcup_k H_k$. Let A be a probabilistic polynomial time algorithm (A is a *collision adversary*) that given a random $h \in H_k$ attempts to find $x, y \in \{0,1\}^{n_{1_k}}$ such that $h(x) = h(y)$ but $x \neq y$. In other words, after getting a hash function it tries to find a collision pair.

Definition 2 *Such a U is called a* family of Collision-Free Hash Functions *(CFHF) if for all polynomials p, for all polynomial time probabilistic algorithms A, and all sufficiently large k the following holds:*

1. *$Pr[A(h) = (x, y), h(x) = h(y), y \neq x] < 1/p(n_{1_k})$ where the probability is taken over all $h \in H_k$ and the random choices of A.*

2. *$\forall h \in H_k$ there is a description of h of length polynomial in n_{1_k}, such that given h's description and x, $h(x)$ is computable in polynomial time.*

3. *H_k is accessible : there exists an algorithm G such that G on input k generates uniformly at random a description of $h \in H_k$.*

Based on the existence of claw-free permutations (as defined in [GMRi]) one can construct a CFHF [D]; also, based on any one-way function which is homomorphism, one can construct CFHF.

2.2 Universal Hash Functions

The following definition is from Carter and Wegman [CW].

Definition 3 [Universal$_2$ hash function.] *Let G be a family of functions from C to B. We say that G is a* universal$_2$ *if for any pair of inputs (a_1, a_2) and any pair of outputs (b_1, b_2), the number of functions that map a_1 to b_1 and a_2 to b_2 is $|G|/|B|^2$.*

Let $x \in C$, $S \subseteq C - \{x\}$, $g \in G$ and $\delta_g(x,S)$ be the number of $y \in S$ such that $g(x) = g(y)$. Then by [CW], the expected value μ of $\delta_g(x,S)$, for each fixed x and S and when g is uniformly chosen from G, is $\mu = |S|/|B|$. Markov's inequality tells us that when g is randomly chosen from G then for any $t > 1$:

$$Pr[\delta_g(x,S) > t \cdot \mu] < 1/t.$$

Definition 4 [NY] (extended): *A strongly universal$_2$ family G has the* collision accessibility *property if, given a requirement $g(x) = a_1$ and $g(y) = a_2$, it is possible to generate in polynomial time a function uniformly among all functions in G that obey the requirement.*

The above is an extended property which is necessary for our construction. A simple example of such a family (which we will use from now on, just for clarity) is the set $G_{n,m} = \{g_{a,b} \mid g_{a,b}(x) = chop(ax + b), a, b \in GF(2^n)\}$, where all computations are in $GF(2^n)$ and $chop : \{0,1\}^n \rightarrow \{0,1\}^m$ returns the first m bits of its n-bit argument. We will denote by G_n the set $G_{n,n-1}$. An interesting property of the family $G_{n,m}$ is that each function is a 2^{n-m}-1 function; i.e. exactly 2^{n-m} elements in the domain have the same value in the range. In particular, when $n = m + 1$ the functions are 2-1; and when $m = n$, $G_{n,n}$ is a permutation.

From now on, all the strong universal$_2$ families we consider are supposed to have the collision accessibility property.

The following simple lemma states that the composition of two universal$_2$ functions is still universal$_2$.

Lemma 1 [Composition.] *Let G_1 and G_2 be two universal$_2$ families from C_1 to C_2 and from C_2 to C_3, respectively. Then the set $G = \{g = g_2 \circ g_1 \mid g_1 \in G_1, g_2 \in G_2\}$ is a universal$_2$ family from C_1 to C_3.*

2.3 Universal One-way Hash Functions

In this subsection we review the definition and the important properties of universal one-way hash functions (UOWHF), as introduced and discussed in [NY].

Let $\{n_{1_i}\}$ and $\{n_{0_i}\}$ be two increasing sequences such that for all i $n_{0_i} \leq n_{1_i}$, but $\exists q$, a polynomial such that $q(n_{0_i}) \geq n_{1_i}$ (we say that these sequences are polynomially related). Let H_k be a collection of functions such that for all $h \in H_k$, $h: \{0,1\}^{n_{1_k}} \mapsto \{0,1\}^{n_{0_k}}$ and let $U = \bigcup_k H_k$. Let A be a probabilistic polynomial time algorithm (A is a *collision adversary*) that on input k outputs $x \in \{0,1\}^{n_{1_k}}$ which we call an *initial value*, then given a random $h \in H_k$ attempts to find $y \in \{0,1\}^{n_{1_k}}$ such that $h(x) = h(y)$ but $x \neq y$. In other words, after getting a hash function it tries to find a collision with the initial value.

Definition 5 *Such a U is called a* family of universal one-way hash functions *if for all polynomials p, for all polynomial time probabilistic algorithms A, and for all sufficiently large k the following holds:*

1. *If $x \in \{0,1\}^{n_{1_k}}$ is A's initial value, then $Pr[A(h,x) = y, h(x) = h(y), y \neq x] < 1/p(n_{1_k})$ where the probability is taken over all $h \in H_k$ and the random choices of A.*

2. *$\forall h \in H_k$ there is a description of h of length polynomial in n_{1_k}, such that given h's description and x, $h(x)$ is computable in polynomial time.*

3. *H_k is accessible : there exists an algorithm G such that G on input k generates uniformly at random a description of $h \in H_k$.*

Notice that H_k is actually a collection of descriptions of functions; two different descriptions might correspond to the same function.

In this definition the collision adversary A is a (uniform) algorithm. We can alternatively define UOWHF where A is a polynomial sized circuit (the non-uniform case). In this case, all our results still hold, but we require the one-way functions that we use to be one-way in the non-uniform setting as well.

An important property is the composition lemma: composing families of UOWHF yields a family of UOWHF. Because of this lemma it will be enough to prove the existence of UOWHF that compress one bit. Invoking the composition lemma allows us to construct a family for any input and output size that are polynomially related.

Let $H_1, H_2, ..., H_l$ be families of functions such that $\forall i$ and $\forall h_i \in H_i$, h_i: $\{0,1\}^{n_i} \mapsto \{0,1\}^{n_i - 1}$ and $n_i < n_{i+1}$. We call $H = \{h \mid h = h_1 \circ h_2 ... \circ h_l\}$ an l-composition of $H_1, H_2, ..., H_l$. H is a multiset; If $h_1 \circ h_2 ... \circ h_l = h_1' \circ h_2' ... \circ h_l'$ for different $(h_1, h_2, ..., h_l)$ and $(h_1', h_2', ..., h_l')$, both instances are members of H. (In other words, we use the set of concatenated functions and sample an element by sampling each H_i independently and uniformly).

Lemma 2 [NY] *Let H be an l-decomposition as above. If there exists an algorithm A that produces an initial value x and when given a uniformly random $h \in H$ $Pr[A(h,x) = y, h(x) = h(y), y \neq x] > \epsilon$, then there exists an $1 \leq i \leq l$ and an algorithm A' such that*

- *A' produces an initial value $x_i \in \{0,1\}^{n_i}$*

- *then on input $h_i \in H_i$ tries to find a y_i that collides with x_i.*

- *$Pr[A'(h_i, x_i) = y_i, h(x_i) = h(y_i), y_i \neq x_i] > \epsilon/l$ where the probabilities are taken over $h_i \in H_i$ and A''s random choices.*

- *The running time of A' is similar to that of A.*

We remark that an equivalent definition to the above is when the initial input x is chosen (in a more specific way) at random. For a given $h \in H$ and x chosen by an arbitrary way by A, one can come up with another UOWHF family $H' = G_{n,n} \circ H$ where $G_{n,n}$ is a universal$_2$ permutation family, which randomizes the initial value.

UOWHF can be successfully applied to solve various authentications problems. Signature schemes and public fingerprintings for files among the others [NY]. Next we briefly describe signature schemes, and how to base a trapdoor-less secure signature scheme on UOWHF.

2.4   Signature Schemes

In this subsection we review the definition of a signature scheme, its security and the relation of trapdoor-less signature schemes and UOWHF. For a more complete treatment, the reader is encouraged to consult the original paper [NY].

Digital signature is a primitive suggested right at the birth of modern (public-key) cryptography by Diffie and Hellman [DH]. The first implementations of their idea provided digital signature as well [MH, RSA, R]; these proposed signature schemes were based on trapdoor one-way functions, but lacked a precise notion of security. Following [DH], signature systems design has become an extensive field of research (see [GMRi]); we concentrate here only on provably secure systems.

The first scheme to deal formally with the notion of security of signature scheme was suggested by Goldwasser, Micali and Yao [GMY] who also pointed out flaws in the Diffie-Hellman scheme. They based their probabilistic scheme on the problem of factoring. Then, the strongest known definition of security was formalized by Goldwasser, Micali, and Rivest [GMRi]; they defined what it means for a system to be *existentially unforgeable under an adaptive chosen plaintext attack* (which we call "secure" in the rest of the paper). This is an attack by an adversary (forger) who initially computes a plaintext and receives from the signature algorithm a corresponding valid signatures; this is repeated in an adaptive fashion, polynomially many times. Then, the forger has to produce, without the cooperation of the signature algorithm, an extra signature for a message that was not previously signed. A secure system was designed under the assumption that factoring is hard, or a more general assumption that claw-free trapdoor permutations exist. Bellare and Micali [BeM] have shown how to construct secure signature system based on the assumption that trapdoor one-way permutations exist; this matches the original suggestion of Diffie and Hellman, but this time the system had a proof of security.

Naor and Yung [NY] were the first to conceive that the trapdoor property is not necessary for secure signature, (even one robust against the adaptive chosen plaintext attack). They proved that a one-way permutation is sufficient and invented the primitive of universal one-way hash family (UOWHF) to achieve (among other things) a secure signature.

2.4.1   Definition of a Signature Scheme and its Security

A signature scheme includes the following components:

1. A *security parameter* k which determines the size of keys, messages and other resources; all sizes and algorithms are polynomial in k.

2. A *message space* MS, we allow all messages of a given size polynomial in k.

3. A *key component* which includes a *key space* $KS(k)$ a family from which keys are being drawn and a *generation algorithm* KAL which chooses random keys.

4. A *signature bound* SB, a polynomial representing a bound on the number of messages signed; any polynomial should work.

5. A *system state* s which represents the state of the system; there is an initial state and execution states.

6. A *signing algorithm* SAL which is given a message, a system state, and a key, generates a signature and updates the system's state.

7. A *verification algorithm* VAL which is given a message, a signature and a system's state, checks the validity of the signature.

A signature system is a distributed system in which each user is a polynomial-time machine which initiates its instance of the signature scheme.

Next we describe attacks on signature schemes. The most general attack on a signature scheme ([GMRi]) has two phases. First, it allows a polynomial-time adversary F (a *forger*) to use the signature algorithm in an adaptive fashion, getting signatures to polynomially many plaintexts of its choice. Next, the attack has an existential nature, i.e., the forger itself has to come with a valid signature of a new message of its choice, in which case we say that it was successful. A scheme is *p-forgeable* if for a polynomial p there is a forger F which for infinitely many k's, succeeds in the attack with probability larger than $1/p(k)$, where the probability is taken over the random choices of keys by KAL, the choices of the signatures by SAL, and the coin flips of F itself. We say that a system is *secure* if it is not p-forgeable for any polynomial p.

2.4.2   A Signature Scheme based on UOWHF

Here we review the new approach to signature scheme as developed in [NY]. We briefly describe their reduction of signature to UOWHF.

Consider the Diffie-Lamport *tagging system* [La]. It consists of making public a one-way function f and a *window*, which is an ordered pair of values $< f(x_o), f(x_1) >$, for randomly chosen x_o, x_1 in the function domain. The user, then, is committed to the window and later on when it sends a bit b, it is done by publishing a tag x_b, an operation we call *opening half a window*. We say that the other half of the window remains *unused*. The construction can be extended to tag a message of length m-bits, by initially publishing (committing) to a *row of windows* $[< f(x_o^i), f(x_1^i) >, i = 1, ..., m]$ and then opening the halves corresponding to the bits of the message. Since f is one-way, only the committed user can open a tag, and no one else can tag a different message unless it can invert a random value of f, furthermore, anyone can verify tags; in this sense the system resembles a signature scheme.

The drawback of the above system is that the size of the initial commitment limits the number of bits which can be tagged; to eliminate it (and transform it into a signature scheme) [NY] suggests to use UOWHF. The general strategy of their system is to extend the tagging system, enhancing it with the capability of "regenerating rows of windows". The system is represented as a linked list, a system's state is a list consisting of nodes. Each node is associated with a message, i.e., it tags that message. The node is also connected to its successor in the list, i.e., it tags the successor node as well.

The node N_i contains three data fields: h_i a UOWHF, and two rows of windows rm_i and rs_i, the first will tag the next message M_{i+1} while the second one will tag the successor node in the list, N_{i+1}. The UOWHF family is publicly known, or is otherwise produced by each user.

Next we sketch the algorithms and the dynamic behavior of the system. The system has an initial state (state 0) in which a user deposits an initial (root) node N_0 in the public directory.

In a typical situation the system is in state s_{i-1} where there is a list of $i-1$ nodes and the *last-node* N_{i-1} is unused. The connection between nodes will be explained in the following sketch of the signature and the verification algorithms.

SAL: Each message signing changes the state of the system, the list is grown by a node which becomes the new last-node. At state s_{i-1} the user sends a message M_i and tags it using the row rm_{i-1}, furthermore a new node N_i is generated by algorithm KAL: $N_i = <h_i, rm_i, rs_i>$ where its components are chosen at random: h_i is a random element of the UOWHF family based on f, and the rows are encryption by f of random tag values.

In order to link the new node into the list, the user has to tag the new node by its predecessor. Notice that the new node as a string of random bits is larger than the tagging capabilities of the row rs_{i-1} which was given this tagging task! Here is where the UOWHF hash is needed in a non-trivial way. The algorithm first computes the hash value of the new node by evaluating $n_i = h_{i-1}(N_i)$, then the smaller string n_i is being tagged by opening the corresponding half-windows in rs_{i-1}. This defines a signature on M_i and a new valid state of the system s_i.

VAL: A verification of a validity of a message can be done by checking the tagging of the message M_i by rm_{i-1} and testing the validity of the system's state by checking that the tagging of $n_j = h_{j-1}(N_j)$ is a valid one, namely, it was done by a proper opening of rs_{j-1} for all $j = 1, \ldots, i-1$. This is done all the way to the root and if all checks are valid the user accepts the signature.

Since a UOWHF implies the existence of a one-way function [NY], we can state that:

Theorem 1 *[NY] If UOWHF exist, then the signature scheme described above is secure.*

It is also possible to improve the efficiency of the above scheme [Go, NY].

3   UOWHF Based on 1–1 One-way Functions

Naor and Yung [NY] showed how to construct UOWHF from any 1-1 one-way function. In this section we describe a construction different from them which is easier to prove and is more economical in the number of applications of one-way function used in the construction. Actually only a single application of a one-way function is used.

Let f be a 1–1 one-way function (i.e. a 1–1 function that is also one-way), with input length n and output length $l(n)$.

Define $H_n = \{h = g_n \circ g_{n+1} \circ \ldots \circ g_{l(n)} \circ f_n \mid g_i \in G_i\}$, where G_i is a strongly universal$_2$ family from i-bit strings to $(i-1)$-bit strings.

Theorem 2 $U = \bigcup_n H_n$ (based on 1-1 one-way function) is a UOWHF family.

Proof. The proof is by contradiction. If there is a polynomial time algorithm that can find collisions for a randomly chosen h, then it can be used to invert the one-way function f.

Suppose there is an algorithm A, that on input n produces an $x \in \{0,1\}^n$ and given a randomly chosen $h \in_R H_n$ outputs a y such that $h(x) = h(y)$ and $x \neq y$, with probability greater than ϵ (here the probability is taken over $h \in_R H_n$, and the random choices of A). Given x and $g_n,..., g_{l(n)}$, denote by COL_i, $n \leq i \leq l(n)$, the set of y such that $g_i \circ g_{i+1} \circ ... \circ g_{l(n)} \circ f_n(x) = g_i \circ g_{i+1} \circ ... \circ g_{l(n)} \circ f_n(y)$ and, for $i < l(n)$, $g_{i+1} \circ ... \circ g_{l(n)} \circ f_n(x) \neq g_{i+1} \circ ... \circ g_{l(n)} \circ f_n(y)$.

Let j, $n \leq j \leq l(n)$, be an integer such that with probability at least $\epsilon/(l(n) - n + 1)$ the algorithm A, on input x and $h \in_R H_n$, gives a value $A(x,h) \in COL_j$. Such a j must exist by the pigeonhole principle (wlog now we assume it is known).

Given x and $g_{j+1},..., g_{l(n)}$, let W be the set of $g_j \in G_j$ such that $|COL_j| \leq 4(l(n) - n + 1)/\epsilon$, i.e. the set of functions which have zero or more, but not too many, collisions in COL_j.

The probability $Pr[A(x,h) \in COL_j]$ that A on input x and $h \in_R H_n$, returns a value in COL_j can be written as

$$\sum_{g \in W} Pr[A(x,h) \in COL_j \mid g_j = g] Pr[g] + Pr[A(x,h) \in COL_j \text{ and } g_j \notin W],$$

where $Pr[g]$ is the probability of choosing g from G_j under the uniform distribution. Thus, for $g \in W$ we have $Pr[g] = Pr[W]/|W|$, where $Pr[W] = \sum_{g \in W} Pr[g]$, and hence $Pr[A(x,h) \in COL_j]$ is equal to

$$\sum_{g \in W} Pr[A(x,h) \in COL_j \mid g_j = g] Pr[W]/|W| + Pr[A(x,h) \in COL_j \text{ and } g_j \notin W].$$

Let $g \in W$ and u be such that $g \circ g_{j+1} \circ ... \circ f_n(x) = g \circ g_{j+1} \circ ... \circ f_n(u)$ and $g_{j+1} \circ ... \circ f_n(x) \neq g_{j+1} \circ ... \circ f_n(u)$. Since g is a 2-1 function, the number of elements in COL_j is equal to the number of collisions the composition of the first $l(n) - (j + 1) + 1$ universal$_2$ functions (that is in turn a universal$_2$ function), makes with u. The expected number of collisions for this composition, $\delta_{g_{j+1} \circ ... \circ g_{l(n)} \circ f_n}(u, \{y \neq u\})$, is equal to $(2^n - 1)/2^j$, which is less than 2. Thus, from the Markov's inequality it follows that given x if we randomly choose $g_{j+1} \in_R G_{j+1},..., g_{l(n)} \in_R G_{l(n)}$, and g in G_j, then the probability, $1 - Pr[W] = Pr[|COL_j| > 4(l(n) - n + 1)/\epsilon]$ is less than $(1/2)\epsilon/(l(n) - n + 1)$. Since $Pr[A(x,h) \in COL_j \text{ and } g_j \notin W] \leq Pr[|COL_j| > 4(l(n) - n + 1)/\epsilon]$ and because $Pr[A(x,h) \in COL_j] \geq \epsilon/(l(n) - n + 1)$, it follows

$$\sum_{g \in W} Pr[A(x,h) \in COL_j \mid g_j = g] Pr[W]/|W| \geq \frac{1}{2} \frac{\epsilon}{l(n) - n + 1}$$

and thus

$$\sum_{g \in W} Pr[A(x,h) \in COL_j \mid g_j = g] \frac{1}{|W|} \geq \frac{1}{2} \frac{\epsilon}{l(n) - n + 1} \left(1 - \frac{1}{2} \frac{\epsilon}{l(n) - n + 1} \right).$$

$$(1)$$

Now, consider the algorithm A' that on input $z = f_n(w)$ where $w \in_R \{0,1\}^n$,

1. runs A to produce x (if $f_n(x) = z$ stop successfully –this is, of course, negligible);

2. for $i = n, ..., j - 1, j + 1, ..., l(n)$, randomly chooses $g_i \in_R G_i$;

3. randomly chooses $g_j \in_R G_j$ such that $g_j \circ g_{j+1} \circ ... \circ g_{l(n)-n+1} \circ f_n(x) = g_j \circ g_{j+1} \circ ... \circ g_{l(n)-n+1}(z)$;

4. gets y by running A on input $h = g_n \circ ... \circ g_{l(n)} \circ f_n$ and x;

5. outputs y.

Notice that the probability that $g_{j+1} \circ ... \circ g_{l(n)} \circ f_n(x) = g_{j+1} \circ ... \circ g_{l(n)}(z)$ is $1/2^j$, which is negligible; in this case we say that A' fails and we can stop it. Otherwise, with probability $p_s = 1 - 1/2^j$, $f^{-1}(z)$ belongs to COL_j (by the forced collision). Next we compute the probability of inversion of z when A' does not stop. For the rest of the calculation p_s will be a multiplicative factor of the successful event.

Denote by D the distribution on G_j, according to which g_j is chosen at step 3. This is not an uniform distribution, not even among the functions in G_j that have at least one collision $g_j \circ g_{j+1} \circ ... \circ g_{l(n)} \circ f_n(x) = g_j \circ g_{j+1} \circ ... \circ g_{l(n)} \circ f_n(y)$, $y \neq x$. Indeed, for any two functions g', g'' we have $D[g'] = (|COL_j(g')|/|COL_j(g'')|)D[g'']$, where $COL_j(g)$ is the set COL_j when $g_j = g$. That is the probability $D[g]$ of a function g to be chosen at step 3 is proportional to the number of collisions there are in $COL_j(g)$, thus it is dependent on the previous choices of $g_{j+1}, ..., g_{l(n)}$.

The number of elements in COL_j is given by $\delta_{g_{j+1}\circ...\circ g_{l(n)}\circ f_n}(f^{-1}(z), \{y \neq f^{-1}(z)\})$, and its expected value when h has been chosen according to A''s algorithm, is equal to $(2^n - 1)/2^j$, which is less than 2. Thus, from Markov's inequality it follows that $D[W]$ is greater than or equal to $p_s\{1 - (1/2)\epsilon/(l(n) - n + 1)\}$. Let g' and g'' be two functions in W. From $D[g'] = (|COL_j(g')|/|COL_j(g'')|)D[g'']$, it follows that $D[g'] \geq p_s(1/|W|)(\epsilon/\{4(l(n) - n + 1)\})\{1 - (1/2)\epsilon/(l(n) - n + 1)\}$.

The probability $Pr[A'(x, h) \in COL_j]$ that A' on input x and h chosen by A' in H_n, returns a value in COL_j is at least

$$\sum_{g \in W} Pr[A(x, h) \in COL_j \mid g_j = g]D[g],$$

which is greater than or equal to

$$\sum_{g \in W} Pr[A(x, h) \in COL_j \mid g_j = g]\frac{p_s}{4|W|}\left(\frac{\epsilon}{l(n) - n + 1}\right)\left(1 - \frac{1}{2}\frac{\epsilon}{l(n) - n + 1}\right).$$

Making use of (1), it follows that $Pr[A'(x, h) \in COL_j]$ is greater than

$$\frac{p_s}{8}\left(\frac{\epsilon}{l(n) - n + 1}\right)^2\left(1 - \frac{1}{2}\frac{\epsilon}{l(n) - n + 1}\right)^2.$$

The probability that $g_j \in W$ when it is chosen accordingly to A''s algorithm is $D[W] \geq p_s\{1 - (1/2)(\epsilon/(l(n) - n + 1))\}$. When A' at step 5 returns an element $y \in COL_j$ that collides with x, then the probability that $f(y) = z$ is $1/|COL_j|$

(since z is not an input to A). Hence the probability that $y = A'(x, h)$ satisfies $f(y) = z$ is at least

$$\frac{p_s^2}{32} \left(\frac{\epsilon}{l(n) - n + 1} \right)^3 \left(1 - \frac{1}{2} \frac{\epsilon}{l(n) - n + 1} \right)^3,$$

which is polynomially related to ϵ (notice that, say, $p_s > 1/4$). □

4 Further Reducing Complexity Assumptions

Next, we show how to construct UOWHF using one-way functions that are more general. First, we show that a function with a small preimage size gives us a UOWHF as well. This will be followed by a more general result: a function with the property that the expected size (when an element in the domain is randomly chosen) of the preimage of an element in the range is small gives us a UOWHF. We show also how to construct such a function if a regular function [GKL] is available, or even when a function where given an element in the range an estimate (with polynomial uncertainty) on the size of the preimage set is easily computable.

4.1 UOWHF Based on One-way Functions with Small Expected Preimage Size

Here we describe how to construct a UOWHF if a one-way function that has at most only polynomially many collisions on the average is available. We first define formally what we mean by small preimage size and then by small expected preimage size. Roughly speaking, the latter is a function f with the property that for a randomly chosen x, the expected size of the preimage of $f(x)$ is small. Then, we discuss why the previous scheme does not work with such functions. We describe a scheme for these functions and prove its correctness. In the next subsection we show how to construct such a function when it is only required that there is a feasible algorithm that when given an element z in the range, it gives a relatively good estimate on the size of the preimage set $f^{-1}(z)$.

The property of a small preimage size is shared by all 1–1 one-way functions, but also include for example, one-way functions based on the generalized factoring assumption of composite with two or more primes, such as modular squaring (whose inverse is extracting square roots) (see [GKL]).

Definition 6 *Let $r(\cdot)$ be a function from $\mathcal{N}^+$ to $\mathcal{N}^+$. A one-way function has a $r(n)$-preimage size if for each $x \in \{0, 1\}^n$*

$$|f^{-1}(f(x))| \leq r(n).$$

Definition 7 *Let $r(\cdot)$ be a function from $\mathcal{N}^+$ to $\mathcal{N}^+$. A one-way function has an expected $r(n)$-preimage-size if when x is randomly chosen in $\{0, 1\}^n$ the expected size of $f^{-1}(f(x))$ is at most $r(n)$.*

Definition 8 *A one-way function f has a small expected preimage-size if there is a polynomial p such that f has an expected $p(n)$-preimage-size.*

Let f be a one-way function with expected $p(n)$-preimage-size. Denote by $\delta_f(x, \{y \neq x\})$ the number of elements such that $f(y) = f(x)$, then Markov's inequality implies that $Pr[\delta_f(x, \{y \neq x\}) > t \cdot p(n)] < 1/t$, where the probability is over the choices of $x \in_R \{0,1\}^n$. The Markov's inequality essentially states that there is only a negligible probability that there are more than polynomially many collisions. This is an important property for the proof of our scheme for UOWHF.

Why the previous scheme does not work with small preimage size functions
Suppose we use the same scheme described earlier to construct UOWHF based on 1–1 functions, but we plug in a $p(n)$-preimage size function as the underlying one-way function. So, h is constructed as $g_n \circ g_{n+1} \circ \ldots \circ g_{l(n)} \circ f_n$, where each g_i is a hash function that shrinks the input by one bit. To prove its correctness we should derive a contradiction with the difficulty of inverting f; i.e. the ability to easily find a collision for the h would imply the ability to invert the f. It is immediate that this approach is doomed to failure. Indeed, suppose that there is a poly-time algorithm that on input x outputs y such that $f(x) = f(y)$ and $y \neq x$ (this is not in contradiction with the difficulty of inverting f). Then, this latter algorithm can be used to find a collision for h. Squaring modulo a composite is such a function f, so is any one-way function which is independent of part of its input and just applies to the rest of the argument.

A provably secure scheme
We just saw the problem in dealing with functions that are not 1–1, as the difficulty of inverting does not rule out the possibility of easily finding collisions for the one-way function and may thus jeopardize the security of the h function that is based on it. Here we show how to deal with this problem.

Let $\delta > 0$ be a constant. Let $f = \{f_n: \{0,1\}^n \to \{0,1\}^{l(n)} \mid n \in N^+\}$ be a one-way function with the output length $l(n)$ and expected $p(n)$-preimage-size, where $p(\cdot)$ is a polynomial. Recall that $G_{n, \lfloor (\log n)^{1+\delta} \rfloor}$ and G_i are families of universal$_2$ functions from $\{0,1\}^n$ to $\{0,1\}^{\lfloor (\log n)^{1+\delta} \rfloor}$ and from $\{0,1\}^i$ to $\{0,1\}^{i-1}$, respectively. For a positive integer k, let $\widehat{H}_k$ be the set of functions $\widehat{h}_k: \{0,1\}^k \to \{0,1\}^{k-1}$ defined as

$$\widehat{h}_k(w) = g_k \circ g_{k+1} \circ \ldots \circ g_{l(k)} \circ f_k(w)$$

where $g_i \in G_i$, $i = k, \ldots, l(k)$. Let H_n be the set of functions $h: \{0,1\}^n \to \{0,1\}^{n-1}$ defined as

$$h(x) = g(x) \diamond (\widehat{h}_{n - \lfloor (\log n)^{1+\delta} \rfloor} \circ \ldots \circ \widehat{h}_n \circ h'(x))$$

where $g \in G_{n, \lfloor (\log n)^{1+\delta} \rfloor}$, $\widehat{h}_i \in \widehat{H}_i$, $i = n - \lfloor (\log n)^{1+\delta} \rfloor, \ldots, n$, and $h' \in G_{n,n}$ is a universal$_2$ permutation.

To randomly choose an element in H_n, we first randomly select $g \in_R G_{n, \lfloor (\log n)^{1+\delta} \rfloor}$, then $h' \in_R G_{n,n}$, and finally $\widehat{h}_i \in_R \widehat{H}_i$, $i = n - \lfloor (\log n)^{1+\delta} \rfloor, \ldots, n$, uniformly and independently from each other.

In the above scheme to compress c bits one needs to apply one-way functions only $\lfloor (\log n)^{1+\delta} \rfloor + c$ times.

Theorem 3 $U = \bigcup_n H_n$, based on small expected preimage functions as above, is a UOWHF family.

Proof. Suppose there is an algorithm A, that on input n produces an $x \in \{0,1\}^n$ and given a randomly chosen $h \in_R H_n$ outputs a y such that $h(x) = h(y)$ and $x \neq y$, with probability greater than ϵ. As for the previous proof of Theorem 2, we will describe a probabilistic poly-time algorithm A' that on input $z = f(w)$, where w has been randomly chosen, finds a u such that $f(u) = f(w)$, with probability polynomially related to ϵ.

It is unlikely that A on input x and $h \in_R H_n$ returns a value $y = A(x, h)$ such that $y \neq x$, $g(y) = g(x)$ and $\hat{h}_n(y) = \hat{h}_n(x)$. Indeed, when $\hat{h}_n \in_R \hat{H}_n$, $h' \in_R G_{n,n}$, and $g \in_R G_{n,\lfloor (\log n)^{1+\delta} \rfloor}$, the probability that there exists a y such that $f_n(h'(x)) \neq f_n(h'(y))$, $g(y) = g(x)$ and $\hat{h}_n(y) = \hat{h}_n(x)$ is at most $2^n/2^{n-1+\lfloor (\log n)^{1+\delta} \rfloor}$, which is negligible. Moreover, when $h' \in_R G_{n,n}$, and $g \in_R G_{n,\lfloor (\log n)^{1+\delta} \rfloor}$, the probability that there exists a y such that $f_n(h'(x)) = f_n(h'(y))$ and $g(y) = g(x)$, is at most $p(n)/2^{\lfloor (\log n)^{1+\delta} \rfloor}$ (by Markov's inequality) which is negligible. (This is indeed the reason why, in the definition, $\lfloor (\log n)^{1+\delta} \rfloor$ of the output bits of $h(x)$ have been chosen to be an hashed value of x through the universal$_2$ g.) Thus, we assume this is not the case (we can assume that for all sufficiently large n, this will happen with probability at least $1/2$, to be generous).

Given x and h, denote by COL_i, $n - \lfloor (\log n)^{1+\delta} \rfloor \leq i < n$, the set of y such that $g(x) = g(y)$, $\hat{h}_i \circ \ldots \circ \hat{h}_n(x) = \hat{h}_i \circ \ldots \circ \hat{h}_n(y)$ and $\hat{h}_{i+1} \circ \ldots \circ \hat{h}_n(x) \neq \hat{h}_{i+1} \circ \ldots \circ \hat{h}_n(y)$. Fix an integer j, $n - \lfloor (\log n)^{1+\delta} \rfloor \leq j < n$, such that with probability at least $(1/2)\epsilon/(\lfloor (\log n)^{1+\delta} \rfloor + 1)$ the algorithm A, on input x and $h \in_R H_n$, returns a value $A(x, h) \in COL_j$. Such a j must exist by the pigeonhole principle (wlog we assume now that j is known).

Let $\hat{h}_j$ be the composition of $g_j \circ g_{j+1} \circ \ldots \circ g_{l(j)} \circ f_j$, and let $s = \hat{h}_{j+1} \circ \ldots \circ \hat{h}_n \circ f_n(x)$. Denote by C_i, $i = j, \ldots, l(j)$, the set of $y \in COL_j$, such that $g_i \circ g_{i+1} \circ \ldots \circ g_{l(j)} \circ f_j(s) = g_i \circ g_{i+1} \circ \ldots \circ g_{l(j)} \circ f_j(y)$ and, for $i < l(j)$, $g_{i+1} \circ \ldots \circ g_{l(j)} \circ f_j(s) \neq g_{i+1} \circ \ldots \circ g_{l(j)} \circ f_j(y)$. And denote by $C_{l(j)+1}$ the set of $y \in COL_j$ such that $f_j(s) = f_j(y)$. Fix an integer k, $j \leq k \leq l(j) + 1$, such that with probability at least $(1/2)\epsilon/((\lfloor (\log n)^{1+\delta} \rfloor + 1)(l(j) - j + 2))$ the algorithm A, on input x and $h \in_R H_n$, returns a value $A(x, h) \in C_k$. Such a k must exist by the pigeonhole principle (wlog we assume now that k is known). We distinguish two cases: $k \leq l(j)$ and $k = l(j) + 1$.

Suppose $k \leq l(j)$. Consider the algorithm A' that on input $z = f_j(w)$, where $w \in_R \{0,1\}^j$:

1. runs A to produce x;

2. randomly chooses $g \in_R G_{n,\lfloor (\log n)^{1+\delta} \rfloor}$;

3. for $i = n - \lfloor (\log n)^{1+\delta} \rfloor, \ldots, j - 1, j + 1, \ldots, n$, randomly chooses $\hat{h}_i \in_R \hat{H}_i$;

4. for $i = j, \ldots, k - 1, k + 1, \ldots, l(j)$, randomly chooses $g_i \in_R G_i$;

5. computes $s = \hat{h}_{j+1} \circ \ldots \circ \hat{h}_n \circ f_n(x)$, and randomly chooses $g_k \in_R G_k$ such that $g_k \circ g_{k+1} \circ \ldots \circ g_{l(j)} \circ f_j(s) = g_k \circ g_{k+1} \circ \ldots \circ g_{l(j)}(z)$;

6. constructs $\hat{h}_j = g_j \circ g_{j+1} \circ \ldots \circ g_k \circ \ldots \circ g_{l(j)} \circ f_j$;

7. constructs h as the pair g and the composition $\hat{h}_{n-\lfloor(\log n)^{1+\delta}\rfloor} \circ \ldots \circ \hat{h}_n$

8. gets u by running A on input x and h;

9. outputs $y = \hat{h}_{j+1} \circ \ldots \circ \hat{h}_n(u)$.

Notice that there is a negligible probability that $f_j(s) = z$, in which case we invert z. Thus, assume this is not the case.

Every hash function (but g_k at step 5) is randomly chosen by A'. Let us denote by D the distribution under which such g_k is chosen by A'. If D were the uniform distribution and were independent from the choices of the other hash functions, then the probability of A' returning a colliding u in the set C_k would be the same as for A. The expected number of f values whose inputs are from $\{0,1\}^j$ that collide under the universal$_2$ hash function $g_k \circ \ldots \circ g_{l(j)}$ is at most $2^j/2^{k-1} \leq 2$; and thus it is unlikely that there are more than polynomially many collisions (from the Markov's inequality). Thus, when A gives a collision $u \in C_k$, the value $f \circ \hat{h}_{j+1} \circ \ldots \circ \hat{h}_n(u)$ is one of the f colliding values, and there is a non-negligible probability that this is exactly $z = f(w)$ (in which case we invert z). Unfortunately, D is not the uniform distribution but, as in the A''s algorithm of Theorem 2, it is close enough (for our purposes) to it. The proof of this case is essentially the same of Theorem 2 and, thus, is omitted here.

Now, suppose $k = l(j) + 1$. Here, the hypothesis is that A on input x and $h \in_R H_n$ returns, with probability at least $(1/2)\epsilon/((\lfloor(\log n)^{1+\delta}\rfloor + 1)(l(j) - j + 2))$, a $y \neq x$ such that $f_j \circ \hat{h}_{j+1} \circ \ldots \circ \hat{h}_n(x) = f_j \circ \hat{h}_{j+1} \circ \ldots \circ \hat{h}_n(y)$ and $\hat{h}_{j+1} \circ \ldots \circ \hat{h}_n(x) \neq \hat{h}_{j+1} \circ \ldots \circ \hat{h}_n(y)$. Now, we construct an algorithm A' that, by first running A, computes a pair of colliding and different elements in C_k. And then, by running algorithm A again, exploits the computed collision to try to invert an f value. More formally, the algorithm A' on input $z = f_{j+1}(w)$, where $w \in_R \{0,1\}^{j+1}$:

1. runs A to produce an initial x;

2. "computes a colliding pair $p_1 \neq p_2$ such that $f_j(p_1) = f_j(p_2)$"

   (a) randomly chooses $g \in_R G_{n,\lfloor(\log n)^{1+\delta}\rfloor}$;

   (b) for $i = n - \lfloor(\log n)^{1+\delta}\rfloor, \ldots, n$, randomly chooses $\hat{h}_i \in_R \hat{H}_i$;

   (c) constructs h as the pair g and the composition $\hat{h}_{n-\lfloor(\log n)^{1+\delta}\rfloor} \circ \ldots \circ \hat{h}_n$;

   (d) gets u by running A on input x and h;

   (e) computes the pair p_1, p_2 as $p_1 = \hat{h}_{j+1} \circ \ldots \circ \hat{h}_n(x)$ and $p_2 = \hat{h}_{j+1} \circ \ldots \circ \hat{h}_n(u)$;

3. for $i = j + 2, \ldots, l(j + 1)$, randomly chooses $g_i \in_R G_i$;

4. randomly chooses $g_{j+1} \in_R G_{j+1}$ such that $g_{j+1} \circ \ldots \circ g_{l(j+1)} \circ f_{j+1} \circ \hat{h}_{j+2} \circ \ldots \circ \hat{h}_n(x) = p_1$ and $g_{j+1} \circ \ldots \circ g_{l(j+1)}(z) = p_2$;

5. constructs $\hat{h}_{j+1} = g_{j+1} \circ g_{j+2} \circ \ldots \circ g_{l(j+1)} \circ f_{j+1}$;

6. constructs h as the pair g and the composition $\hat{h}_{n-\lfloor(\log n)^{1+\delta}\rfloor} \circ \ldots \circ \hat{h}_n$;

7. gets v by running A on input x and h;

8. outputs $y = \hat{h}_{j+2} \circ ... \circ \hat{h}_n(v)$.

The probability that the pair p_1, p_2, computed at step 2, is such that $p_1 \neq p_2$ and $f_j(p_1) = f_j(p_1)$ is at least $(1/2)\epsilon/((\lfloor(\log n)^{1+\delta}\rfloor + 1)(l(j) - j + 2))$ (this is a bound on the probability of A outputting an element in C_k). Assume such a pair is successfully obtained at step 2. Let us denote by D the distribution under which g_k is chosen by A' at step 4. All other hash functions (but g_k) are randomly chosen by A'. As for the previous case, if D were the uniform distribution and were independent from the choices of the other hash functions, then the probability of A' returning a colliding v in the set C_k would be the same as for A. The expected number of f_{j+1} values that collide under the universal$_2$ hash function $g_{j+1} \circ ... \circ g_{l(j+1)}$ is at most $2^{j+1}/2^j = 2$, and thus it is unlikely that there are more than polynomially many collisions (from the Markov's inequality). Thus, at step 7 when A returns a collision $v \in C_k$, the value $f_{j+1} \circ \hat{h}_{j+2} \circ ... \circ \hat{h}_n(v)$ is one of the f_{j+1} colliding values, and there is a non-negligible probability that this is exactly $z = f_{j+1}(w)$ (in which case we invert z). Unfortunately, D is not the uniform distribution but, as in the A'''s algorithm of Theorem 2 and in the previous case, it is close enough (for our purposes) to it. The formal proof of this case, based on similar techniques as in the previous proof, will be given in the final paper. ☐

4.2 One-way functions with almost-known preimage-size

In this subsection we show how to construct a small expected preimage-size function if there is a function which has a feasible algorithm that when given an element z in the range, gives a good estimate on the size of the preimage set $f^{-1}(z)$.

Definition 9 *A one-way function has a almost-known preimage-size if there is a polynomial p and a poly-time deterministic algorithm PRE_SIZE such that, on input $z = f(x)$, returns a value*

$$|f^{-1}(f(x))| - p(n) \leq PRE\_SIZE(z) \leq |f^{-1}(f(x))| + p(n)$$

for all $x \in \{0,1\}^n$, except a negligible fraction of them.

A particular case of almost-known preimage-size one-way function is a regular function [GKL]. This is a function where every image of an n-bit input has the same number of preimages of length n, another such function is decoding random linear codes (see [GKL]). Subset sum [IN] is another example of this function. Also a $p(n)$-preimage size function is a particular case of a almost-known preimage-size function. In [IN] the function subset-sum is used directly as a UOWHF, exploiting the instances of subset-sum which compress their argument (from n bits to $l(n) = (1 - \epsilon)n$). On the other hand, the most secure instance of subset-sum is shown to be length-preserving instances (from n bits to $l(n) = n$ bits or $l(n) = n + O(\log n)$); all subset-sum instances are "almost-known preimage-size" and (if one-way) can be used in our scheme.

Let f be a almost-known preimage-size one-way function, and PRE_SIZE be an algorithm that gives an approximation within $p(n)$ to the preimage size of f. Define $i(x) = \lceil \log(PRE\_SIZE(f(x)) + p(|x|)) \rceil$, and define f' as

$$f'(x \diamond g) = f(x) \diamond g \diamond [g(x)]_{i(x)}$$

where $x \in \{0,1\}^n$, $g \in G_{n,n}$ and $[y]_k$ is the first k bits of y.

Along the line of Lemma 5.1 of Impagliazzo, Levin, and Luby [ILL], it is possible to prove the following lemma.

Lemma 3 f' is a one-way function, when $x \in_R \{0,1\}^n$ and $g \in_R G_{n,n}$. Moreover, given randomly chosen $x \in_R \{0,1\}^n$ and $g \in_R G_{n,n}$, the expected number of y, $y \neq x$, that collide $f'(y) = f'(x)$, is $|f^{-1}(f(x))|/2^{i(x)} \leq 1$.

The range of f' may contain elements of different lengths. But, as mentioned in Section 2, it is an easy task to construct an equivalent function with the range of the same length. Thus, as a corollary of Theorem 3 we have the following general result.

Theorem 4 If there is a almost-known preimage-size (or a regular) function then there is a UOWHF, and thus a signature scheme.

5 On Various Notions of Security of Cryptographic Hash

In this section we give a number of definitions of hardness of one-way hash functions or one-way functions, with respect to collision finding. Our motivation is to demonstrate an equivalence among a large set of possible definitions so that any primitive which satisfies one of the conditions will automatically satisfy the other definitions. This will demonstrate the robustness of the definition of CFHF, and may be suggesting a possible way to attack the problem of finding new and less restrictive implementations of CFHF.

Next we define a few classes of functions according to the hardness properties which they satisfy.

We identify the following families of functions; essentially the idea is to classify all possible ways a collision is generated by a family of cryptographic hash functions. We assume that all the functions below are accessible and computable in polynomial-time.

- $\mathcal{H}$ is a Collision-free hash family.

- $\mathcal{F} = \{\mathcal{F}_k\}$ is a collection of pairs of functions such that there is a polynomial p and when $(f_1, f_2) \in_R \mathcal{F}_k$ $Pr[|\{(x,y) : f_1(x) = f_2(y)\}| > 0] > 1/q(k)$. But, for all polynomials q, for all efficient algorithms A, and for all sufficiently large k, $Pr[A(f_1, f_2) = (x,y) : f_1(x) = f_2(y)] < 1/q(k)$ when $f_1, f_2 \in_R \mathcal{F}_k$.

- $\mathcal{G} = \{\mathcal{G}_k\}$ is a collection of pairs of functions such that there is a polynomial q and when $(g_1, g_2) \in_R \mathcal{F}_k$ $Pr[|\{x : g_1(x) = g_2(x)\}| > 0] > 1/q(k)$.

But, for all polynomials p, for all efficient algorithms A, and for all sufficiently large k, $Pr[A(g_1, g_2) = x : g_1(x) = g_2(x)] < 1/p(k)$ when $g_1, g_2 \in_R \mathcal{F}_k$.

- $\mathcal{R}$ is a collection of functions such that there is a polynomial q and when $r \in_R \mathcal{R}$ $Pr[|\{x : r(x) = 0\}| > 0] > 1/q(|x|)$. But, for all polynomials p, for all efficient algorithms A, and for all sufficiently large k, $Pr[A(r) = x : r(x) = 0] < 1/p(k)$ when $r \in_R \mathcal{R}_k$.

- Given a fixed polynomial-time computable function g', $\mathcal{S}$ is a collection of functions such that there is a polynomial q and when $s \in_R \mathcal{S}$ $Pr[|\{x : s(x) = g'(x)\}| > 0] > 1/q(|x|)$. But, for all polynomials p, for all efficient algorithms A, and for all sufficiently large k, $Pr[A(s, g') = x : s(x) = g'(x)] < 1/p(k)$ when $s \in_R \mathcal{S}_k$.

The above functions demonstrate various ways of defining collisions of hash functions, assuming the above functions actually compress the size of their argument. As mentioned is section 2, we can assume that for a given function, range elements with preimages of the same length n, have the same length $l(n)$.

$\mathcal{H}$ is the original CFHF family, which implies that on the same function finding a collision between two different arguments is hard. $\mathcal{F}$ is a family in which a collision between two functions on two different arguments is hard to find, while $\mathcal{G}$ is the family in which for two different functions a collision on the same argument is hard to find. $\mathcal{R}$ is a family in which non-negligible fraction belongs to the kernel and it is hard to find an element in the preimage of the kernel (similarly it can defined with respect to any constant in range, the family $\mathcal{S}$ captures the hardness of finding an argument colliding with a given fixed efficiently computable function. Below we show that collision freeness is robust with respect to the exact notion of collision.

Theorem 5 *The following relations on the function families exist:*

- *The families $\mathcal{F}, \mathcal{G}, \mathcal{R}, \mathcal{S}$ are information theoretically equivalent.*
- *The families $\mathcal{F}, \mathcal{G}, \mathcal{R}, \mathcal{S}$ are information theoretically reducible to $\mathcal{H}$.*

Proof. The following reductions can be observed. $\mathcal{F} \Rightarrow \mathcal{R}$, randomly draw a pair (f_1, f_2) and set the function $r(x \diamond y) = f_1(x) \oplus f_2(y)$. $\mathcal{G} \Rightarrow \mathcal{R}$, randomly draw a pair (g_1, g_2) and set the function $r(x) = g_1(x) \oplus g_2(x)$. In both reductions, the probability of the kernel is polynomially related to the probability of the non-emptiness of the collision set. $\mathcal{R} \Rightarrow \mathcal{S}$, simply draw $r \in_R \mathcal{R}$ and set $s = r \oplus g'$. Similarly, $\mathcal{S} \Rightarrow \mathcal{R}$, simply draw $s \in_R \mathcal{S}$ and set $r = s \oplus g'$. $\mathcal{R} \Rightarrow \mathcal{F}$, draw a random r and set $f_1 = r$ and $f_2 = 0$, and the same reduction $\mathcal{R} \Rightarrow \mathcal{G}$. The probability of the non-emptiness of the kernel in the above reductions is polynomially related as well to the probability of the respected collision set. This concludes the proof of equivalence of $\mathcal{F}, \mathcal{G}, \mathcal{R}$, and $\mathcal{S}$.

Next we show that $\mathcal{H} \Rightarrow \mathcal{R}$. Let $\mathcal{H}_k = \{h : \{0,1\}^k \rightarrow \{0,1\}^{l(k)}\}$. Let $neq(x, y)$, where x, y are both k-bit long strings (otherwise, the function is undefined), a function which gives $1^{l(k)}$ if $x = y$ and $0^{l(k)}$ otherwise. Given $h \in \mathcal{H}_k$, define $r(x \diamond y) = h(x) \oplus h(y) \oplus neq(x, y)$. Notice that the probability that the kernel of r is not empty is polynomially related to the probability of collision in h.

[]

The above theorem justifies the generality of the definition of collision-free functions as the hardest condition among possible function families.

References

[BeM] Bellare M. and S. Micali, *How to Sign Given any Trapdoor Function*, Proceedings of the 20th Annual Symposium on the Theory of Computing, Chicago, Il, 1988, pp. 32-42.

[CW] J. L. Carter and M. N. Wegman, *Universal Classes of Hash Functions*, Journal of Computer and System Sciences 18 (1979), pp. 143-154.

[D] I. B. Damgård, *Collision Free Hash Functions and Public Key Signature Schemes*, Eurocrypt 1987.

[DH] W. Diffie and Hellman, *New Directions in Cryptography*, IEEE Trans. on Information Theory, vol. IT-22, 6 (1976), pp. 644-654.

[E] P. Elias, *Universal Codeword Sets and Representations of the Integers*, IEEE Trans. on Inform. Theory, vol. 21, n. 2, March 1975, pp. 194–203.

[Gi] M. Girault, *Hash-functions using modulo-N Operations*, Eurocrypt, 1987.

[Go] O. Goldreich, *Two Remarks Concerning the GMR Signature Scheme*, Crypto 1986.

[GKL] O. Goldreich, H. Krawczyk, and M. Luby, *On the existence of Pseudorandom Generators*, Proceedings of the 29th Symposium on the Foundation of Computer Science, 1988, pp. 12-24.

[GMRi] S. Goldwasser, S. Micali, and R. Rivest, *A secure digital signature scheme*, Siam Journal on Computing, Vol. 17, 2 (1988), pp. 281-308.

[GMY] S. Goldwasser, S. Micali, and A. C. Yao, *Strong signature schemes*, Proceedings of the 15th Annual Symposium on the Theory of Computing, Boston, MA, 1983, pp. 431-439.

[ILL] R. Impagliazzo, L. Levin, and M. Luby, *Pseudo-Random Generation from One-way Functions*, Proceedings of 21st STOC, May 1989.

[IL] R. Impagliazzo and M. Luby, *One-way Functions are Essential for Complexity Based Cryptography*, Proceedings of the 30th Symposium on the Foundation of Computer Science, 1989.

[IN] R. Impagliazzo and M. Naor, *Efficient Cryptographic Schemes Provably secure as Subset Sum*, Proceedings of the 30th Symposium on the Foundation of Computer Science, 1989.

[La] L. Lamport, *Constructing digital signatures from one-way functions*, SRI intl. CSL-98, October 1979.

[M] R. Merkle, *A Digital Signature based on Conventional Encryption Function*, Crypto 1987, Springer Verlag.

[M1] R. Merkle, *Secrecy, Authentication and Public Key Systems*, Ph.D. Thesis (1982), UMI Research Press, Ann Arbor, Michigan.

[M2] R. Merkle, *One-way Hash Functions and DES*, Crypto 1989.

[MH] R. Merkle and M. Hellman, *Hiding Information and Signature in Trapdoor Knapsack*, IEEE Trans. on Inform. Theory, vol. 24, n. 5, 1978, pp. 525–530.

[NY] M. Naor and M. Yung, *Universal One-way Hash Functions and their Cryptographic Applications*, Proceedings of 21st STOC, May 1989.

[R] M. O. Rabin, *Digital Signatures and Public Key Functions as Intractable as Factoring*, Technical Memo TM-212, Lab. for Computer Science, MIT, 1979.

[Ro] J. Rompel, *One-way Functions are Necessary and Sufficient for Signature*, STOC 90.

[RSA] R. Rivest, A. Shamir, and L. Adleman, *A Method for Obtaining Digital Signature and Public Key Cryptosystems*, Comm. of ACM, 21 (1978), pp. 120-126.

[Y] A. C. Yao, *Theory and Applications of Trapdoor functions*, Proceedings of the 23th Symposium on the Foundation of Computer Science, 1982, pp. 80-91.

Fast Signature Generation with a Fiat Shamir – Like Scheme

H. Ong
Deutsche Bank AG
Stuttgarter Str. 16–24
D – 6236 Eschborn

C.P. Schnorr *
Fachbereich Mathematik / Informatik
Universität Frankfurt
Postfach 111932
D – 6000 Frankfurt/M. 11

Abstract

We propose two improvements to the Fiat Shamir authentication and signature scheme. We reduce the communication of the Fiat Shamir authentication scheme to a single round while preserving the efficiency of the scheme. This also reduces the length of Fiat Shamir signatures. Using secret keys consisting of small integers we reduce the time for signature generation by a factor 3 to 4. We propose a variation of our scheme using class groups that may be secure even if factoring large integers becomes easy.

1   Introduction and Summary

The Fiat–Shamir signature scheme (1986) and the GQ–scheme by Guillou and Quisquater (1988) are designed to reduce the number of modular multiplications that are necessary for generating signatures in the RSA–scheme. Using multicomponent private and public keys Fiat and Shamir generate signatures much faster than with the RSA–scheme. The drawback is that signatures are rather long. They are about t–times longer than RSA–signatures, where t is the round number in the Fiat–Shamir scheme. Using single component keys Guillou and Quisquater obtain signatures of about the same length as in the RSA–scheme but the cost for signature generation is only slightly reduced (by a factor of about 3) compared to the RSA–scheme.

In this paper we propose a new signature scheme and a corresponding authentication scheme that reduces the length of signatures in the Fiat–Shamir scheme to about the length of RSA–signatures. Signature generation with the new scheme is about

---

*This research was performed while the second author visited the Department of Computer Science of the University of Chicago

3 to 4 times faster than with the Fiat–Shamir scheme. The efficiency of the new signature scheme is comparable to that of the discrete logarithm signature scheme by Schnorr (1989): In the new scheme signature generation is somewhat slower, signature verification about 5 times faster than in the discrete logarithm scheme. Signatures, private and public keys are longer in the new scheme.

We present the basic version of the new signature scheme in section 2. This basic version preserves the efficiency of the Fiat–Shamir scheme but reduces the length of signatures. In section 3 we present a variant of the new scheme that generates signatures about 3 to 4 times faster than with the Fiat–Shamir scheme. The authentication scheme that corresponds to the signature scheme is presented in section 4. It is shown to be secure unless computing non trivial 2^t-th roots modulo N is easy. A variation of our scheme using class groups is given in section 5. This variant may be secure even if factoring large integers is easy.

2 A condensed variant of Fiat Shamir signatures

Notation. For $N \in \mathbb{N}$ let $\mathbb{Z}_N$ denote the ring of integers modulo N. The numbers t and k are security parameters, typically $4 \leq t$, $k \leq 20$.

The role of the key authentication center (KAC). The KAC chooses

- random primes p and q such that $p, q \geq 2^{256}$
- a one–way hash function $h : \mathbb{Z}_N \times \mathbb{Z} \to \{0,1\}^{tk}$.
- its own private and public key.

The KAC publishes $N = p \cdot q$, h and its public key.

COMMENTS. The KAC's private key is used for signing the public keys issued by the KAC. The KAC can use any secure public key signature scheme whatsoever for generating this signature.

The user's private and public key. Each user chooses a private key $s = (s_1, \ldots, s_k)$ consisting of random numbers $s_i \in [1, N]$ such that $gcd(s_i, N) = 1$ for $i = 1, \ldots, k$. The corresponding public key $v = (v_1, \ldots, v_k)$ consists of the integers $v_i = s_i^{-2^t} \pmod{N}$ for $i = 1, \ldots, k$.

Registration of users. The KAC checks the identity of a user, prepares an identification string I (containing name, address etc.) and generates a signature S for the pair (I, v) consisting of I and the user's public key v.

Signature generation.
input message $m \in \mathbb{Z}$, private key $s = (s_1, \ldots, s_k)$ and modulus N.

1. *Preprocessing* pick a random $r \in [1, N]$, $x := r^{2^t}$ (mod N).

2. $e = (e_{11}, \ldots, e_{tk}) := h(x, m) \in \{0, 1\}^{tk}$.

3. $y := r \prod_{j=1}^{k} s_j^{\sum_{i=1}^{t} e_{ij} 2^{i-1}}$ (mod N).

Output signature (e, y).

Our signature concept reduces multicomponent signatures of the Fiat Shamir scheme to single components. The efficiency of signature generation is preserved. Step 3 can be performed as follows

$$y := \prod_{e_{t,j}=1} s_j \ (\text{mod } N)$$

$$y := y^2 \prod_{e_{t-i,j}=1} s_j \ (\text{mod } N) \text{ for } i = 1, \ldots, t - 1$$

$$y := y \cdot r \ (\text{mod } N).$$

Step 3 requires at most $kt + t - 1$ modular multiplications; for random e only $t(k+2)/2 - 1$ modular multiplications are required on the average. Step 1 requires t squarings and can be done in a preprocessing stage that is independent of the message m.

Signature verification.
input signature (e, y), message m, $v = (v_1, \ldots, v_k), I, S, N$.

1. check the signature S for (I, v).

2. $z := y^{2^t} \prod_{j=1}^{k} v_j^{\sum_{i=1}^{t} e_{ij} 2^{i-1}}$ (mod N)

3. check that $e = h(z, m)$.

Signature verification can be done using at most $kt + t$ modular multiplications. For random e only $t(k+2)/2 + 1$ modular multiplications are required on the average. Step 2 can be performed as follows:

$$z := y^2 \prod_{e_{t,j}=1} v_j \ (\text{mod } N)$$

$$z := z^2 \prod_{e_{t-i,j}=1} v_j \ (\text{mod } N) \text{ for } i = 1, \ldots, t - 1.$$

Security of signatures. In order to falsify a signature for message m the cryptanalyst has to solve the equation

$$e = h\left(y^{2^t}\prod_j v_j^{\sum_{i=1}^t e_{i,j} 2^{i-1}}\pmod{N},\ m\right),$$

for e and y. No efficient method is known to solve this equation.

3 Fast Signatures

The generation of signatures can be accelerated by choosing secret keys s consisting of small integers $s_1,\ldots,s_k$. The security of this variation of the scheme is based on the assumption that computing 2^t–roots modulo N is difficult. No particular algorithms are known to compute 2^t–roots modulo N given that these 2^t–roots are of order $N^{2^{-t+1}}$.

Let the private key $(s_1,\ldots,s_k)$ consist of random primes $s_1,\ldots,s_k$ in the interval $[1,2^{64}]$. The interval $[1,2^{64}]$ must be large enough so that it is infeasible to find the s_i by exhaustive enumeration. We must have $t \geq 4$ so that $s_j^{2^t}$ is at least of order N^2. We next explain the requirement for the numbers $s_1,\ldots,s_k$ to be prime. For if $s_i = \alpha\cdot\beta$ with $\alpha,\beta \in [1,2^{32}]$ we can find s_i by solving

$$\beta^{-2^t} = v_i\alpha^{2^t}\pmod{N}\quad\text{for }\alpha,\beta\in[1,2^{32}].$$

This can be done using about 2^{32} steps.

For the efficiency of the scheme we suppose that $\sum_j e_{i,j} \leq 8$ for $i = 1,\ldots,t$. Then we have $\prod_{e_{i,j}=1} s_j < 2^{512}$ for all i and computing this product does not require any modular reduction. Consequently step 3 of the procedure for signature generation requires only $2t - 1$ full modular multiplications; the other multiplications are with small numbers. Thus step 3 costs an equivalent of about $2.5t - 1$ full modular multiplications. Step 1 of the procedure for signature generation requires t additional modular multiplications, but these multiplications are done in preprocessing mode independent of the message that is to be signed. The total cost of about $2.5t - 1$ modular multiplications for signature generation compares favourable with the average of $(k/2 + 1)t$ modular multiplications in the original Fiat–Shamir scheme.

4 The authentication scheme and its security

Let the private and public keys s, v be as in the previous sections. In particular we can use the small integer variant for the private key s.

The authentication protocol.
(Prover A proves its identity to verifier B)

1. *Preprocessing.* A picks a random number r with $1 \leq r \leq N$ and computes $x := r^{2^t}(\bmod N)$.

2. *Initiation.* A sends to B its identification string I, its public key v, the KAC's signature S for (I, v) and x.

3. B checks v by verifying the signature S and sends a random string $e \in \{0,1\}^{tk}$ to A.

4. A sends $y := r \prod_j s_j^{\sum_{i=1}^{t} e_{i,j} 2^{i-1}} (\bmod N)$ to B

5. B checks that $x = y^{2^t} \prod_j v_j^{\sum_{i=1}^{t} e_{i,j} 2^{i-1}} (\bmod N)$ and accepts A's proof of identity if this holds.

Obviously if A and B follow the protocol then B always accepts A's proof of identity. We next consider the possibility of cheating for A and B. Let $\tilde{A}$ ($\tilde{B}$, resp.) denote a fraudulent A (B, resp.). $\tilde{A}$ ($\tilde{B}$, resp.) may deviate from the protocol in computing x, y (e, resp.). $\tilde{A}$ does not know the secret s. $\tilde{B}$ spies upon A's method of authentication.

A fraudulent A can cheat by guessing the exam e and sending for an arbitrary $r \in \mathbb{Z}_N$ the crooked proof

$$x := r^{2^t} \prod_j v_j^{\sum_{i=1}^{t} e_{i,j} 2^{i-1}} (\bmod N), \ y := r.$$

The probability of success for this attack is 2^{-tk}.

We prove in the following theorem that this success rate cannot be increased unless we can easily compute some nontrivial 2^t-th root modulo N. For this let $\tilde{A}$ be an interactive, probabilistic Turing machine that is given the fixed values k, t, N. Let RA be the internal random bit string of $\tilde{A}$. Let the success bit $S_{\tilde{A},v}(RA, e)$ be 1 if $\tilde{A}$ succeeds with v, RA, e and 0 otherwise. The success probability $S_{\tilde{A},v}$ of $\tilde{A}$ for v is the average of $S_{\tilde{A},v}(RA, e)$, where RA, e are uniformly distributed. We assume that the time $T_{\tilde{A},v}(RA, e)$ of $\tilde{A}$ for v, RA, e is independent of RA and e, i.e. $T_{\tilde{A},v}(RA, e) = T_{\tilde{A},v}$. This is no restriction since limiting the time to twice the average running time for successful pairs (RA, e) decreases the success rate $S_{\tilde{A},v}$ at most by a factor 2.

Theorem 1. *There is a probabilistic algorithm AL which on input $\tilde{A}, v$ computes a 2^t-root of $\prod_j v_j^{c_i} (\bmod N)$ for some $(c_1, \ldots, c_k) \neq 0$ with $|c_j| < 2^t$ for $j = 1, \ldots, k$. If $S_{\tilde{A},v} > 2^{-tk+1}$ then AL runs in expected time $O(T_{\tilde{A},v}/S_{\tilde{A},v})$.*

Proof. The argument extends Theorem 5 in Feige, Fiat, Shamir (1987). We assume that $T_{\tilde{A},v}$ also covers the time required for B.

Algorithm with input v

1. Pick RA at random. Compute $x = x(\tilde{A}, RA, v)$, i.e. compute x the same way as algorithm $\tilde{A}$ using the coin tossing sequence RA. Pick a random $e \in \{0,1\}^{tk}$. Compute $y = y(\tilde{A}, RA, v, e)$ the same way as algorithm $\tilde{A}$. If $S_{\tilde{A},v}(RA, e) = 1$ then fix RA, retain x, y, e and go to step 2. Otherwise repeat step 1 using a new independent RA.

2. Let u be the number of probes (i.e. passes of step 1) in the computation of RA, x, y, e. Probe up to $4u$ random $\bar{e} \in \{0,1\}^{tk}$ whether $S_{\tilde{A},v}(RA, \bar{e}) = 1$. If some 1 occurs with $\bar{e} \neq e$ then compute the corresponding $\bar{y} = \bar{y}(\tilde{A}, RA, \bar{e}, v)$ and *output* $c_i = \sum_{i=1}^{t}(e_{ij} - \bar{e}_{ij})2^{i-1}$ for $j = 1, \ldots, t$ and $\bar{y}/y \pmod{N}$.

Time analysis. Let $S_{\tilde{A},v} > 2^{-tk+1}$. For fixed $\tilde{A}$ and v let the success bits $S_{\tilde{A},v}(RA, e)$ be arranged in a matrix with rows RA and columns e. A row RA is called *heavy* if the fraction of 1-entries is at least $S_{\tilde{A},v}/2$. At least half of the 1-entries are in heavy rows since the number of 1-entries in non-heavy rows is at most $S_{\tilde{A},v} \cdot \#\text{rows} \cdot \#\text{columns}/2$. Thus the row RA that succeeds in step 1 is heavy with probability at least $1/2$. A heavy row has at least two 1-entries.

We abbreviate $\varepsilon = S_{\tilde{A},v}$. The probability that step 1 probes $i\varepsilon^{-1}$ random RA for some $i \in \mathbb{N}$ without finding an 1-entry is at most $(1 - \varepsilon)^{i/\varepsilon} < 2.7^{-i}$. Thus the average number of probes for the loop of step 1 is

$$\leq \sum_{i=1}^{\infty} i\varepsilon^{-1} 2.7^{-i+1} = O(\varepsilon^{-1}).$$

We have with probability at least $1/2$ that $u \geq \varepsilon^{-1}/2$. The row RA is heavy with probability at least $1/2$. If these two cases happen then step 2 finds a successful $\bar{e}$ with probability $\geq 1-(1-\varepsilon/2)^{2/\varepsilon} > 1-2.7^{-1}$, and we have $e \neq \bar{e}$ with probability $\geq 1/2$. Thus AL terminates after one iteration of steps 1 and 2 with probability

$$\geq \frac{1}{4}(1 - 2.7^{-1})\frac{1}{2} > 0.07.$$

The probability that AL performs exactly i iterations is at most 0.93^{i-1}. Alltogether we see that the average number of probes for AL is at most

$$O\left(5\varepsilon^{-1} \sum_{i=0}^{\infty} 0.93^{i-1} t\right) = O(\varepsilon^{-1}).$$

This proves the claim.

QED

5 A variation of the new scheme using class groups

One can obviously modify the new scheme so that the private and public key components s_i, v_i are elements of an arbitrary finite abelian group G, i.e. we can replace the group $\mathbb{Z}_N^*$ of invertible elements in $\mathbb{Z}_N$ by the group G. The efficiency of signature generation and signature verification relies on the efficiency of the multiplication in G. For the generation of the public key components $v_i = s_i^{-2^t}$ we need an efficient division algorithm in G. The security of the authentication and the signature scheme requires that computing 2^t-th roots in G is difficult.

A particular type of suitable groups are class groups C_Δ of equivalence classes of binary quadratic forms $aX^2 + bXY + CY^2 \in \mathbb{Z}[X,Y]$ with negative discriminant $\Delta = b^2 - 4ac$. The multiplication in C_Δ, which is called *composition*, is only slightly slower than modular multiplication for integers of the order of Δ. All known algorithms for computing 2^t-th roots in C_Δ require knowledge of the group order h_Δ of C_Δ which is called the *class number*.

Class groups C_Δ have the following advantage over the group $\mathbb{Z}_N^*$:

- The problem of computing class numbers h_Δ is harder than the problem of factoring integers N of the order $N \approx |\Delta|$.

- Computing the class number h_Δ is hard no matter whether Δ is prime or composite.

- No trusted authority is required for the generation of Δ, since there is no hidden secret, as is the factorization of the modulus N in the Fiat–Shamir scheme.

For the sake of completeness we give all the details for the operation in class groups.

5.1 Class groups. A polynomial $aX^2 + bXY + cY^2 \in \mathbb{Z}[X,Y]$ is called a *binary quadratic form*, and $\Delta = b^2 - 4ac$ is its *discriminant*. We denote a binary quadratic form $aX^2 + bXY + cY^2$ by (a,b,c). A form for which $a > 0$ and $\Delta < 0$ is called *positive*, and a form is *primitive* if $\gcd(a,b,c) = 1$. Two forms (a,b,c) and (a',b',c') are *equivalent* if there exist $\alpha, \beta, \gamma, \delta \in \mathbb{Z}$ with $\alpha\delta - \beta\gamma = 1$ such that $a'U^2 + b'UV + c'V^2 = aX^2 + bXY + cY^2$, where $U = aX + \gamma Y$, and $V = \beta X + \gamma Y$. Two equivalent forms have the same discriminant.
Now fix some negative integer Δ with $\Delta \equiv 0$ or $1 \bmod 4$. We will often denote a form (a,b,c) of discriminant Δ by (a,b), since c is determined by $\Delta = b^2 - 4ac$. The set of equivalence classes of positive, primitive, binary quadratic forms of discriminant Δ is denoted by C_Δ. The existence of the form $(1,\Delta)$ shows that C_Δ is non–empty.

5.2 Reduction algorithm. Each equivalence class in C_Δ contains precisely one *reduced* form, where a form (a, b, c) is reduced if $|b| \leq a \leq c$ and $b \geq 0$ if $|b| = a$ or if $a = c$.

5.3 Composition algorithm. The set C_Δ is a finite abelian group, the *class group*. The group law, which we will write multiplicatively, is defined as follows. The inverse of (a, b) follows from an application of the reduction algorithm to $(a, -b)$, and the unit element 1_Δ is $(1, 1)$ if Δ is odd, and $(1, 0)$ if Δ is even. To compute $(a_1, b_1) \cdot (a_2, b_2)$, we use the Euclidean algorithm to determine $d = gcd(a_1, a_2, (b_1 + b_2)/2)$, and $r, s, t \in \mathbb{Z}$ such that $d = ra_1 + sa_2 + t(b_1 + b_2)/2$. The product then follows from an application of the reduction algorithm to $(a_1 a_2 / d^2, \ b_2 + 2a_2(s(b_1 - b_2)/2 - tc_2)/d$, where $c_2 = (b_2^2 - \Delta)/(4a_2)$.

5.4 Prime forms. For a prime number p we define the Kronecker symbol $\left(\frac{\Delta}{p}\right)$ by

$$\left(\frac{\Delta}{p}\right) = \begin{cases} 1 & \text{if } \Delta \text{ is a quadratic residue modulo } 4p \text{ and } gcd(\Delta, p) = 1 \\ 0 & \text{if } gcd(\Delta, p) \neq 1 \\ -1 & \text{otherwise.} \end{cases}$$

For a prime p for which $\left(\frac{\Delta}{p}\right) = 1$, we define the *prime form* I_p as the reduced form equivalent to (p, b_p), where $b_p = \min\{b \in \mathbb{N}_{>0} : b^2 \equiv \Delta \bmod 4p\}$.

5.5 Factorization of forms. A form (a, b, c) of discriminant Δ, with $gcd(a, \Delta) = 1$, for which the prime factorization of a is known, can be factored into prime forms in the following way. If $a = \prod_{p \, prime} p^{e_p}$ is the prime factorization of a, then $(a, b) = \prod_{p \, prime} I_p^{s_p e_p}$, where $s_p \in \{-1, +1\}$ satisfies $b \equiv s_p b_p \bmod 2p$, with $I_p = (p, b_p)$ as in 3.4. Notice that the prime form I_p is well-defined because the prime p divides a, $gcd(a, \Delta) = 1$, and $b^2 \equiv \Delta \bmod 4a$.

5.6 Choice of the discriminant and the private and public keys. We can choose $\Delta = -q$ to be the negative of any prime with $q = 3 \bmod 4$ so that q is at least 512 bits long. This particular choice of Δ implies that h_Δ is odd, and thus every class (a, b) in C_Δ has a unique square root.

We can choose the components s_i of the private key $s = (s_1, \ldots, s_k)$ to be prime forms $s_i = I_{p_i}$ with random primes p_i, $2^{63} < p_i < 2^{64}$. We must have $t \geq 3$ so that $p_i^{2^t}$ is much larger than $\sqrt{|\Delta|}$. Given s_i one can easily compute the corresponding public key component $v_i = s_i^{-2^t}$.

Acknowledgement The second author wishes to thank the Department of Computer Science of the University of Chicago for its support during this research. He also wishes to thank A. Shamir for inspiring discussions on this subject.

References

FEIGE, U., FIAT, A. and SHAMIR, A.: *Zero Knowledge Proofs of Identity*. Proceedings of STOC 1987, pp. 210 – 217, and J. Cryptology 1 (1988), pp. 469 – 472.

FIAT, A. and SHAMIR, A.: *How to Prove Yourself: Practical Solutions of Identification and Signature Problems*. Proceedings of Crypto 1986, in Lecture Notes in Computer Science (Ed. A. Odlyzko), Springer Verlag, 263, (1987) pp. 186 – 194.

GOLDWASSER, S., MICALI, S. and RACKOFF, C.: *Knowledge Complexity of Interactive Proof Systems*. Proceedings of STOC 1985, pp. 291 – 304.

GUILLOU, L.C. and QUISQUATER, J.J.: *A Practical Zero–Knowledge Protocol Fitted to Security Microprocessor Minimizing Both Transmission and Memory*. Proceedings of Eurocrypt'88. Lecture Notes in Computer Science, Springer–Verlag (Ed. C. G. Günther) 330 (1988), pp. 123 – 128.

MICALI, S. and SHAMIR, A.: *An Improvement of the Fiat–Shamir Identification and Signature Scheme*. Crypto 1988.

SCHNORR, C.P.: *Efficient Identification and Signatures for Smart Cards*. Proceedings of Crypto'89 (Ed. G. Brassard) Lecture Notes in Computer Science, Springer–Verlag 435, (1990) pp. 239 – 252.

A REMARK ON A SIGNATURE SCHEME WHERE FORGERY CAN BE PROVED

Gerrit Bleumer Birgit Pfitzmann Michael Waidner

Institut für Rechnerentwurf und Fehlertoleranz, Universität Karlsruhe
Postfach 6980, D-7500 Karlsruhe 1, F. R. Germany

I.   INTRODUCTION

A new type of signature scheme, *a signature scheme where forgery by an unexpectedly powerful attacker is provable*, was suggested in [11]: if the signature of an honest participant Alice is forged, she can *prove* this forgery with arbitrarily high probability.

The possibility of proving forgeries does not depend on any unproven assumptions. The impossibility of forgery is based on the existence of pairs of claw-free permutations.

We improve this scheme for the special case that the GMR-generator for pairs of claw-free permutations is used [5]: During the set-up phase, Bob generates a pair (f_0, f_1). Alice's security depends entirely on the sufficiency of Bob's choice. Therefore, in the general case, Bob has to prove to Alice the sufficiency of his choice by a zero-knowledge proof (ZKP). We show that for the GMR-generator, this expensive ZKP can be replaced by the simple condition that the modulus chosen by Bob is odd.

In Section II, we sketch a simplified version of the signature scheme of [11]. Section III contains the necessary notations. The GMR-generator is described in Section IV. In Section V we present our result.

II.   A SIMPLE SIGNATURE SCHEME WHERE FORGERY CAN BE PROVED

The signature scheme where forgery can be proved of [11] is based on the idea of LAMPORT's one-time signatures [3]:

Assume two parties, the signer Alice and the recipient of her signatures, Bob. If Alice has to sign at the most L bits, she chooses a one-way function g and $2 \cdot L$ values $r_{i,0}, r_{i,1}, i = 1,...,L$, randomly from dom($g$), the domain of g. She publishes g and the $2 \cdot L$ images

$$g(r_{1,0})\, g(r_{1,1}) \dots g(r_{L,0})\, g(r_{L,1}).$$

To sign the t-th bit with value $b \in \{0,1\}$, Alice sends the preimage $r_{t,b}$ of $g(r_{t,b})$ to Bob. As usual, if the forger Felix can invert g, Alice's security is lost completely.

The new idea was that in such a case Alice should be able to prove to Bob or a judge Judy that someone has inverted g [11, 10]. For this, function g has to fulfil two conditions:

i. for a fixed value $\sigma > 0$, and for each $x \in \text{dom}(g)$, the value $g(x)$ has at least 2^σ preimages,

ii. g is (computationally) collision-free for Alice, i.e. it is hard for Alice to find a pair (x, y) with $g(x) = g(y)$ and $x \neq y$.

If Felix (or even Bob) forges a signature, at least for one image $g(r_{t,b})$ he computes a new preimage x. Since $r_{t,b}$ was randomly chosen from $\text{dom}(g)$, with probability $(1-2^{-\sigma})$ the pair $(x, r_{t,b})$ is a g-collision. Each g-collision convinces Bob and Judy that condition ii. is violated, i.e. that the signature scheme is broken.

Condition i. guarantees that with high probability Alice can prove a forgery. This is called *Alice's (information theoretical) security*.

Conditon ii. guarantees that it is hard to forge signatures or proofs of forgery. Since Alice can almost always prove a forgery, her security is not influenced by forgeries. Conversely, after a proved forgery, all of Alice's signatures become invalid, i.e. Bob's security depends completely on this condition. Thus the (computational) impossibility of forgery is called *Bob's security*.

Such a function g can be constructed from each pair (f_0, f_1) of claw-free permutations (as defined in [5]): Let $D := \text{dom}(f_i)$. Then define

$$g: \{0,1\}^\sigma \times D \to D$$

$$(\alpha_0, \ldots, \alpha_{\sigma-1}, x) \to f_{\alpha_0}(f_{\alpha_1}(\ldots f_{\alpha_{\sigma-1}}(x)\ldots)). \tag{1}$$

Since g hides the first argument in an unconditional way, g is called *hiding function* in [11].

Finding claws is only proved to be hard for parties who cannot observe the process of generation. Thus instead of Alice, Bob has to generate the pair for Alice.

(Please note that now nothing prevents Bob from inverting g and forging signatures, i.e. nobody but Bob can be sure that a signature was really created by Alice. Therefore, to sign a message for n recipients $\text{Bob}_1, \ldots, \text{Bob}_n$, Alice has to sign the message n times, each time using a different function g_i generated by Bob_i. Alice can prove a forgery by presenting collisions for each function g_i.)

Bob's security is now ensured by Bob himself.

Alice's security depends completely on condition i. Thus Bob must prove the sufficiency of his choice. In general, this can be done by an unconditionally correct zero-knowledge proof [6, 1]. In Section V, we show that for the GMR-generator this expensive proof can be omitted.

Some efficiency improvements are mentioned in [11]. Since they are all based on the function g, they all cause the same problem, thus we only mention that most of the improvements for ordinary one-time signatures described in [7, 9, 8] can be applied. The most important one is MERKLE's tree-authentication for decreasing the length of the public key. To apply this idea, a collision-free hash-function, chosen by Bob, must be used [2], and each collision of the hash-function must be accepted as a proof of forgery.

For a complete description and a formal proof of the security of the signature scheme, see [10, 1].

III. NOTATIONS

For $m \in \mathbb{N}$, $\mathbb{Z}_m$ denotes the ring of all residues modulo m, and $\mathbb{Z}_m^*$ the set of all residues x modulo m with $\gcd(x, m) = 1$. We use the symmetric representation for $\mathbb{Z}_m$, i.e. for odd m, we use the set $\{-(m-1)/2, \ldots, -1, 0, 1, \ldots, (m-1)/2\}$ to represent $\mathbb{Z}_m$.

$(\frac{x}{m})$ denotes the JACOBI-Symbol. QR_m denotes the set $\{x^2 \mid x \in \mathbb{Z}_m^*\}$ of all quadratic residues modulo m, $-QR_m$ the set $\{-z \mid z \in QR_m\}$.

For odd m and $x \in \mathbb{Z}_m$, define the absolute value of x by

$$|x| := \begin{cases} x & \text{if } x \in \{0, \ldots, (m-1)/2\} \\ -x & \text{if } x \in \{-1, \ldots, -(m-1)/2\} \end{cases}$$

For a set M, the symbol $|M|$ denotes the cardinality of M.

IV. THE GMR-GENERATOR FOR PAIRS OF CLAW-FREE PERMUTATIONS

Let $k \in \mathbb{N}$ be the security parameter for the GMR-generator. On input k, the generator randomly selects a number m from the set

$$H_k := \{p \cdot q \mid p, q \text{ prime} \wedge \lfloor \log_2(p) \rfloor = \lfloor \log_2(q) \rfloor = k-1 \wedge p \equiv 3 \bmod 8 \wedge q \equiv 7 \bmod 8\}$$

of BLUM-integers of length $2 \cdot k$ or $2 \cdot k - 1$.

The functions $f_i : D \to D$ are then defined by

$$f_0(x) := |x^2|$$
$$f_1(x) := |4 \cdot x^2|.$$

Their common domain D is given by

$$D := \{x \mid x \in \mathbb{Z}_m^* \wedge (\tfrac{x}{m}) = 1 \wedge x \in \{1, \ldots, \tfrac{m-1}{2}\}\}. \tag{2}$$

Both functions are permutations of D. Finding a claw, i.e. a triple (x_0, x_1, z) with $z = f_0(x_0) = f_1(x_1)$, is as hard as factoring m [5].

In this case, the hiding function $g : \{0,1\}^\sigma \times D \to D$ defined in (1) can be described by

$$g(\alpha, x) = |4^\alpha \cdot x^{2^\sigma}|,$$

where $\alpha = (\alpha_0, \ldots, \alpha_{\sigma-1})$ is interpreted as the integer $\alpha_{\sigma-1} \cdot 2^{\sigma-1} + \ldots + \alpha_1 \cdot 2 + \alpha_0$ (similar to [4], proof by induction on σ).

Finding a g-collision, i.e. a pair $((\alpha, x), (\beta, y))$ with $g(\alpha, x) = g(\beta, y)$ and $(\alpha, x) \neq (\beta, y)$, is as hard as finding a claw.

V. NUMBER OF PREIMAGES OF g

To guarantee that Alice can prove each forgery with probability at least $(1-2^{-\sigma})$, each $z \in \text{im}(g)$, the image of g, must have at least 2^{σ} preimages.

If (f_0, f_1) are permutations of D, e.g. because Bob is honest and uses the GMR-generator, this is satisfied since then the functions $g(\alpha, \cdot)$, $\alpha \in \{0,1\}^{\sigma}$, are permutations of D, too. Thus for each value $z \in D$ and each $\alpha \in \{0,1\}^{\sigma}$, there is exactly one x with $g(\alpha, x) = z$. Thus $|g^{-1}(z)| = 2^{\sigma}$.

In the following, we show that to be convinced of condition i, Alice just has to check that m is odd.

For general odd m, instead of using the domain D defined in (2), we use the domain

$$E := \{ |x^2| \mid x \in \mathbb{Z}_m^* \},$$

from which it is also easy to choose a random element.

Lemma 1. If $m \in H_k$, the domains D and E are equal.

Proof. Let $m = p \cdot q$, where $p \equiv q \equiv 3 \pmod 4$. Hence $\left(\frac{-1}{m}\right) = \left(\frac{-1}{p}\right) \cdot \left(\frac{-1}{q}\right) = (-1) \cdot (-1) = +1$.

$E \subseteq D$: Assume $y := |x^2 (\bmod m)| \in E$. From $\left(\frac{-1}{m}\right) = \left(\frac{x^2}{m}\right) = 1$, it follows that $\left(\frac{|x^2|}{m}\right) = 1$.

Since $|x^2| \in \{1, \dots, \frac{m-1}{2}\}$ by definition, we have $y \in D$.

$D \subseteq E$: Assume $y \in D$, i.e. $\left(\frac{y}{m}\right) = 1$, $y \in \{1, \dots, \frac{m-1}{2}\}$.

If $y \in \text{QR}_m$ then there is an x such that $y = x^2 = |x^2| \pmod m$.

Otherwise $\left(\frac{-y}{p}\right) = \left(\frac{-y}{q}\right) = -1$ holds. In this case there is an x such that $y = -x^2 = |x^2| \pmod m$,

because $\left(\frac{-y}{p}\right) = \left(\frac{-y}{q}\right) = \left(\frac{-1}{p}\right) \cdot \left(\frac{y}{p}\right) = \left(\frac{-1}{q}\right) \cdot \left(\frac{y}{q}\right) = (-1)^2 = 1$ which means $-y \in \text{QR}_m$. $\square$

Lemma 1 says that nothing is changed if Bob is honest, thus Bob's security is not influenced. We only need to consider Alice's security.

Lemma 2. If m is an arbitrary odd integer, then for all $z \in E$, $|g^{-1}(z)| \geq 2^{\sigma}$.

Proof. The proof is in four steps. Each one can be proved by basic algebraic calculations omitted here for shortness.

1st step. The sets $\{+1,-1\}$ and $\text{QR}_m \cup -\text{QR}_m$ are subgroups of $(\mathbb{Z}_m^*, \cdot)$. Consider the quotient group $G := (\text{QR}_m \cup -\text{QR}_m) / \{+1, -1\}$.

Set E is a representation system of G, the element $\pm x^2 \cdot \{+1,-1\} \in G$ is represented by $|x^2| \in E$. Multiplication in E is defined by $|x| \cdot |y| := |x \cdot y|$.

2nd step. For all $\alpha, \beta \in \{0,1\}^{\sigma}$ and $x, y \in G$, let

$$(\alpha, x) * (\beta, y) := (\alpha + \beta \bmod 2^{\sigma}, |x \cdot y \cdot 4^{(\alpha+\beta) \text{ div } 2^{\sigma}}|)$$

Then $(\{0,1\}^{\sigma} \times G, *)$ is an ABELian group.

3rd step. Function g is a group-homomorphism from $(\{0,1\}^{\sigma} \times G, *)$ into G.

Proof: For all $\alpha, \beta \in \{0,1\}^\sigma$ and $x, y \in E$

$$
\begin{aligned}
g((\alpha, x) * g(\beta, y)) &= g(\alpha + \beta \bmod 2^\sigma, \lfloor x \cdot y \cdot 4^{(\alpha + \beta)} \operatorname{div} 2^\sigma \rfloor) \\
&= |4^{\alpha + \beta \bmod 2^\sigma} \cdot (x \cdot y \cdot 4^{(\alpha+\beta)\operatorname{div} 2^\sigma})^{2^\sigma}| \\
&= |4^{\alpha+\beta \bmod 2^\sigma + 2^\sigma \cdot ((\alpha+\beta)\operatorname{div} 2^\sigma)} \cdot (x \cdot y)^{2^\sigma}| \\
&= |4^{\alpha+\beta} \cdot (x \cdot y)^{2^\sigma}| \\
&= \|4^\alpha \cdot x^{2^\sigma}| \cdot |4^\beta \cdot y^{2^\sigma}\| \\
&= |g(\alpha, x) \cdot g(\beta, y)|
\end{aligned}
$$

<u>4th step.</u> Since g is a group-homomorphism, each $z \in \operatorname{im}(g) \subseteq G$ has the same number of preimages. Thus

$$
| g^{-1}(z) | = \frac{| \{0,1\}^\sigma \times G |}{| \operatorname{im}(g) |} = 2^\sigma \cdot \frac{| G |}{| \operatorname{im}(g) |} \geq 2^\sigma. \qquad \square
$$

VI. SUMMARY

If the signature scheme where forgery can be proved of [11] is implemented using the claw-free permutation-pairs of [5], the signer Alice just needs to check whether the modulus m chosen by the recipient Bob is odd. The zero-knowledge proof used in [11] to convince Alice that Bob has generated a suitable m can be omitted.

REFERENCES

[1] Gerrit Bleumer: Vertrauenswürdige Schlüssel für ein Signatursystem, dessen Brechen beweisbar ist; Studienarbeit, Universität Karlsruhe 1990 (in preparation).

[2] Ivan Bjerre Damgård: Collision free hash functions and public key signature schemes; Eurocrypt '87, LNCS 304, Springer-Verlag, Berlin 1988, 203-216.

[3] Whitfield Diffie, Martin E. Hellman: New Directions in Cryptography; IEEE Transactions on Information Theory 22/6 (1976) 644-654.

[4] Oded Goldreich: Two Remarks Concerning the Goldwasser-Micali-Rivest Signature Scheme; Crypto '86, LNCS 263, Springer-Verlag, Berlin 1987, 104-110.

[5] Shafi Goldwasser, Silvio Micali, Ronald L. Rivest: A Digital Signature Scheme Secure Against Adaptive Chosen-Message Attacks; SIAM J. Comput. 17/2 (1988) 281-308.

[6] Oded Goldreich, Silvio Micali, Avi Wigderson: Proofs that Yield Nothing But their Validity and a Methodology of Cryptographic Protocol Design; 27th FOCS, IEEE 1986, 174-187.

[7] Ralph C. Merkle: Protocols for Public Key Cryptosystems; Symposium on Security and Privacy, Oakland 1980, 122-134.

[8] Ralph C. Merkle: A digital signature based on a conventional encryption function; Crypto '87, LNCS 293, Springer-Verlag, Berlin 1988, 369-378.

[9] Ralph C. Merkle: Secrecy, authentication, and public key systems; UMI Research Press 1982.

[10] Birgit Pfitzmann: Für den Unterzeichner sichere digitale Signaturen und ihre Anwendung; Diplomarbeit, Universität Karlsruhe 1989.

[11] Michael Waidner, Birgit Pfitzmann: The Dining Cryptographers in the Disco: Unconditional Sender and Recipient Untraceability with Computationally Secure Serviceability; Universität Karlsruhe 1989; presented at Eurocrypt '89.

Membership Authentication for Hierarchical Multigroups Using the Extended Fiat-Shamir Scheme

Kazuo Ohta   *Tatsuaki Okamoto*   *Kenji Koyama*[†]

NTT Communications and Information Processing Laboratories
Nippon Telegraph and Telephone Corporation
1-2356, Take, Yokosuka-shi, Kanagawa, 238-03 Japan

[†]NTT Basic Research Laboratories
Nippon Telegraph and Telephone Corporation
3-9-11, Midori-cho, Musashino-shi, Tokyo, 180 Japan

Abstract

We propose two membership authentication schemes that allow an authorized user to construct one master secret key for accessing the set of hierarchically ordered groups defined by the user, without releasing any private user information. The key allows the user to prove his membership of his true groups and all lower groups, without revealing his name or true groups. The user can calculate the secret member information needed to access a group from his master secret key, and can convince a verifier using the extended Fiat-Shamir scheme. Each of two proposed schemes can generate the master secret key. To ensure the user's privacy, one uses the blind signature and pseudonym encryption techniques, and the other uses Euclid's algorithm. Because each user stores only one master secret key, memory usage is very efficient. Moreover, verifiers can check membership validity using public information independent of the number of users in an off-line environment. Therefore, our schemes are suitable for smart card applications.

1. Introduction

There are many situations in which a user must prove his authority to others. The easiest and most direct way is to prove his identity. From the standpoint of privacy protection, however, the user often prefers to conceal his identity, that is, to prove his authority as an anonymous user [C1]. When a user is granted service privileges based on his membership of a certain group, for example, a special discount rate is available to members of a group, it is more essential to prove his authority rather than to show his identity. We call this type of authentication, *membership authentication* [C2, KMI1]. This authentication convinces verifiers that the user is a valid member of a certain group without revealing his identity, while user authentication proves the validity of a user by displaying his identity.

When these membership authentication schemes are implemented with smart cards, the following problems have to be considered.

- Efficiency: When a user participates in many groups, he must keep one smart card for each group. This is very inefficient. Thus, these cards should be combined into a single card. In other words, if a secret key represents membership in a group, many secret keys must be replaced with a single secret key.

- Group Isolation: When a user participates in many groups, he may want to conceal group membership so that no third party or group can determine the user's membership in other groups.

- Hierarchy: The range of services available or soon to be available to smart card users is extremely rich and varied. It is obvious that financial institutions and credit companies will want to issue cards with different service ratings. High level cards can access a broadrange of services, while low level cards are restricted to just one or two services. The services or groups are arranged hierarchically, and a user, who is a member of a *superior* group, is automatically a member of all affiliated lower groups.

To implement membership authentication or related services, several schemes have been proposed by [C2], [KMI1], [K] and [AT, MTMA]. However, the schemes of [C2, KMI1] do not treat the hierarchical situation, and do not satisfy the group isolation conditions either. The scheme of [K] is an inefficient authentication in which a high position user must keep excessive secret information. The access control scheme proposed by [AT, MTMA] uses cryptographic key assignment, however, it can not be used directly as an authentication scheme.

In this paper, we propose two membership authentication schemes using an extension of the Fiat-Shamir scheme, which solve the above mentioned problems. Each scheme stores only one master secret key in a card in order to prove various memberships, that is, memory usage is very efficient in the proposed schemes. These schemes allow a user, who occupies one position in a hierarchical structure, to authenticate his membership of any lower position without revealing his identity or original position, that is, the hierarchical property is realized and the group isolation is ensured with our schemes. Moreover, each verifier can check membership validity using public information independent of the number of users in an off-line environment. Therefore, they are suitable for smart card applications.

2. Components

Each scheme has the following components.

- Center - an organization established through the cooperation of various groups. It issues multipurpose smart cards. There is secret information known only to the center. The center can not access the secret information of the various groups.

- Group Administrator - an organization that authenticates a member's identification when he registers with the group. The administrator also maintains a member's database, which stores each member's qualification information; address, salary, age etc., which is private information.

- User - a member of one or several groups.
- Verifier - an entity that checks membership validity. Typical examples of verifiers are terminals that can read various smart cards. These terminals are located in shops where various smart cards are used.

3. Requirements

There are seven requirements for membership authentication for hierarchical multigroups [OkOh].

(1) Completeness - a true user is judged valid by any verifier who uses public information.

(2) Soundness - a false user is not judged valid by verifiers.

(3) Anonymity - identity of user is secret to the verifier and any third party.

(4) Group isolation - information as to which groups a user belongs is secret to everyone except the respective group administrators. Hereafter, we assume there is no conspiracy of group administrators.

(5) Efficiency of verifier - verification is implemented in an off-line environment, that is, a verifier does not have to access the center or group administrators for verification. The amount of information used by a verifier does not depend on the number of users.

(6) Efficiency of user - the amount of information used by a user does not depend on the number of groups the user belongs to.

(7) Group hierarchy - when the set of groups is hierarchical, a user, who is a member of a certain set of groups, is a member of all lower affiliated groups. That is, the card can generate secret membership information corresponding to lower groups from the one piece of secret information that corresponds to the highest groups a user belongs to. It is important to ensure that the card can only generate information about the user's groups.

4. The Proposed Schemes

We propose two schemes that satisfy the above mentioned requirements. The first one is center oriented, and the second one is user oriented. In both schemes, the user's card stores the secret hierarchical
membership information defined as a form similar to those proposed by
Akl et al. in the construction method of hierarchical access key [AT,
MTMA] and by Chaum in the denomination scheme [C3]. A user uses
one piece of secret hierarchical membership information, we call it his
master secret key. We apply both the blind signature technique [C1] and
the pseudonym encryption technique [C4] to generate the master secret
key in the first scheme, and we use Euclid's algorithm to calculate it from
several pieces of secret membership information issued by the group administrators separately in the second scheme. These techniques ensure
group isolation and anonymity. The user generates membership information from the master secret key, and proves that he has the membership
information by using the extended Fiat-Shamir scheme [GQ, OhOk1] or
its symmetric version [O] in both schemes.

4.1 Groups Hierarchical Structure

We assume a set of groups has a structure of partial order (see Fig.1).
The notation ($\bigcirc$) indicates a group, and a line indicates an order relationship. That is, $G_i \geq G_j$ means that group G_i has a higher position
than group G_j. This notation ($\geq$) satisfies the order relationship.

4.2 Center Key Generation and Distribution

The center randomly generates two large prime numbers p and q and
keeps them secret. It also generates public information, such as $n(= p \times q)$,
$a \in Z_n$, where $Z_n = \{0, 1, \cdots, n-1\}$, and b_i, which corresponds to
group G_i ($i = 1, 2, \cdots$) and satisfies both $\gcd(b_i, b_j) = 1$ ($i \neq j$) and

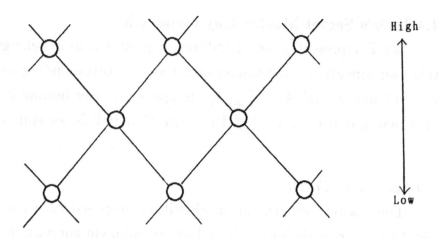

Fig.1 Hierarchical structure of groups

$\gcd(b_i, L) = 1$, where $L = lcm(p - 1, q - 1)$. It moreover calculates

$$e_i = \prod_{j \in \{j \mid G_i \geq G_j\}} b_j,$$

that is, e_i is the product of $\{b_j\}$ whose index j correponds to that of G_i or groups lower than G_i. Finally, the center distributes public keys to users and verifiers.

Only in the second scheme, the center secretly distributes membership information $w_i = a^{1/e_i} \bmod n$, where $1/e_i$ is the inverse element of e_i in $\bmod L$, to the group administrator of G_i.

4.3 Group Administrators Signature Key Generation and Registration

Each group administrator generates a key for a digital signature scheme, and registers the public key in a public information directory. This procedure is necessary in the first scheme.

4.4 User's Secret Master Key Generation

Let Γ represent a set of indexes of groups a user belongs to, and this user already has the master secret key w corresponding to Γ, where $w = a^{\frac{1}{A}} \bmod n$ and $A = \prod_{j \in \Gamma} b_j$. Suppose the user becomes a member of a new group G_i (*i.e.*, $i \notin \Gamma$). Note that if Γ is an empty set then $w = a$.

(1) Scheme 1 (Center Oriented)

The master key generation algorithm proposed here combines the blind signature technique [C1] and the pseudonym encryption technique [C4] in order to protect user's privacy.

First, the user calculates $c = r^{e_i} \cdot w \bmod n$, where r is a random number in Z_n satisfying $\gcd(r, n) = 1$, and sends it to the group administrator of G_i. After the group administrator confirms the user's qualifications, the user receives a digital signature s of c from G_i, and sends (c, s) to the center.

The center checks the validity of (c, s) using the public information of the group administrator G_i. When the check is passed, the center calculates $d = c^{1/e_i} \bmod n$, where $1/e_i$ is the inverse element of e_i in $\bmod L$, and sends it to the user.

Finally, the user calculates a new master secret key w' corresponding to Γ', where $\Gamma' = \Gamma \cup \{j \mid G_i \geq G_j\}$, as follows:

$$B = \prod_{j \in \Gamma \cap \{j \mid G_i \geq G_j\}} b_j,$$

$$w' = (d/r)^B \bmod n.$$

Note that the blind signature technique is used for the communications between users and group administrators to ensure group isolation, and the pseudonym encryption technique is used for communications between users and the center to ensure user anonymity.

(2) Scheme 2 (User Oriented)

Euclid's algorithm is used here in order to generate master secret keys. Since the user calculates his new master secret key by himself, the blind signature and pseudonym encryption techniques are not necessary.

First, the user requests the group administrator G_i to issue secret membership information $w_i = a^{1/e_i} \bmod n$, where $1/e_i$ is the inverse element of e_i in $\bmod L$.

Then, the user calculates a new master secret key w' corresponding to Γ', where $\Gamma' = \Gamma \cup \{j | G_i \geq G_j\}$, from w and w_i in the following way:

Step 1: The user calculates $B = \prod_{j \in \Gamma'} b_j$.

Step 2: He calculates $c = B/A$ and $d = B/e_i$.

Step 3: He calculates α and β satisfying

$$\alpha c + \beta d = 1$$

using Euclid's algorithm. Note that $gcd(c, d) = 1$ holds, since $B = lcm(A, e_i)$ holds.

Step 4: He calculates $w^{\alpha} w_i^{\beta} \bmod n$. (If Γ is an empty set then $w = a$ and $A = 1$.)

Note that since $w_i = a^{1/e_i} \bmod n$, $w' = a^{1/B} \bmod n$ and $w = a^{1/A} \bmod n$ imply $w'^c \equiv w \pmod{n}$ and $w'^d \equiv w_i \pmod{n}$, then $w^{\alpha} w_i^{\beta} \equiv (w'^c)^{\alpha}(w'^d)^{\beta} = w'^{\alpha c + \beta d} \equiv w' \pmod{n}$ holds.

4.5 Users Membership Authentication

When the user attempts to prove his membership of group G_k (i.e., $k \in \Gamma'$), he calculates member information w_k corresponding to G_k from the master secret key w' as follows:

$$B' = \prod_{j \in \{k' | G_k \geq G_{k'}\} \cap \Gamma'} b_j,$$

$$w_k = w'^{B'} \bmod n = a^{1/e_k} \bmod n,$$

where the notation $\overline{S}$ means the complement set of S.

Finally, the user convinces a verifier that he has secret membership information w_k, which satisfies $w_k^{e_k} \equiv a \pmod{n}$, by using the extended Fiat-Shamir scheme or its symmetric version, where e_k is the public information corresponding to group G_k and a is the public information of the system.

5. Discussion

Since each scheme stores only one master secret key in a card, efficiency of user is realized.

During the authentication phase, since the extended Fiat-Shamir scheme [GQ, OhOk1] or its symmetric version [O] are used, completeness, soundness and efficiency of verifier are realized.

Since both the blind signature and pseudonym encryption techniques are applied to generate the master secret key in *Scheme 1*, and Euclid's algorithm is used in *Scheme 2*, group isolation and anonymity properties are ensured.

The group hierarchy property, that the card can *only* generate information $a^{1/e_k} \bmod n$ corresponding to lower groups G_k from the master secret key w', was ensured by the recent result of Everste and van Heyst [EH].

6. Applications

The proposed schemes are applicable to membership authentication *without hierarchy*. Consider the three groups, G_j, G_i and G_i'. G_i and G_i' have the same level, and are subordinated to the higher group G_j where $G_j = G_i \cap G_i'$, $b_j = 1$, and $e_j = e_i \times e_i'$. Our scheme is applicable to this situation providing "$(G_i \cap G_i')$ *authentication*" [KMI2] without modifying any public information.

With the proposed schemes, if new relationships between the highest group and another group are introduced or a new group is added to the highest position in the hierarchical structure, only the public information corresponding to the new group is influenced. However, if lower group sets are restructured, the public information corresponding to all higher groups is influenced. Therefore, extension of the hierarchical structure of groups should be considered in advance.

Membership signature schemes for hierarchical multigroups can also be realized in a similar way using the extended Fiat-Shamir scheme [GQ, OhOk1] or its symmetric version [O].

Our schemes are also applicable to membership authentications in a company's hierarchical organization or in access control services of computer systems, without revealing any private information such as the person's name or true position.

The schemes can also be used, when the set of groups are constructed hierarchically in *reverse order*.

7. Conclusion

We have proposed two membership authentication schemes that generate master secret keys for hierarchical multigroups. In order to generate a master secret key ensuring the user's privacy, one uses the blind signature and pseudonym encryption techniques and the other uses Euclid's algorithm. The schemes satisfy all requirements for membership authentication in hierarchical multigroups. However, the security of the proposed schemes has not yet been confirmed.

Acknowledgement

The authors would like to thank Professor Matsumoto of Yokohama National University for his valuable suggestions.

References

[AT] S.G.Akl and P.D.Taylor, "Cryptographic Solution to a Problem of Access Control in a Hierarchy," ACM Trans. on Computer Systems, 1, 3, pp.239-248 (1983)

[C1] D.Chaum, "Security without Identification: Transaction Systems to Make Big Brother Obsolete," Comm. of the ACM, 28, 10, pp.1030-1044 (1985)

[C2] D.Chaum, "Showing credentials without identification: Signatures transferred between unconditionally unlinkable pseudonyms," Advances in Cryptology, Eurocrypt'85, Springer-Verlag, 1986, pp.241-244

[C3] D.Chaum, "Online Cash Checks," Eurocrypt'89 (1989)

[C4] D.Chaum, "Untraceable electronic mail, return addresses and digital pseudonyms", Comm. of the ACM, 24, 1981, pp.84-88

[EH] J.H.Everste and E. van Heyst, "Which RSA signatures can be computed from some given RSA signatures?", in these proceedings

[GQ] L.C.Guillou and J.J.Quisquater, "A Practical Zero-Knowledge Protocol Fitted to Security Microprocessor Minimizing Both Tranamission and Memory," Eorocrypt'88 (1988)

[K] K.Koyama, "Demonstrating membership of a group using the Shizuya-Koyama-Itoh (SKI) protocol," The 1989 Symposium on Cryptography and Information Security (CIS'89), Gotenba, Japan (1989)

[KMI1] M.Kurosaki, T.Matsumoto and H.Imai, "Simple Methods for Multipurpose Certification," The 1989 Symposium on Cryptography and Information Security (CIS'89), Gotenba, Japan (1989)

[KMI2] M.Kurosaki, T.Matsumoto and H.Imai, "Proving that you belong to at least one of the specified groups," The 1990 Symposium on Cryptography and Information Security (SCIS'90), Hihondaira, Japan (1990)

[MTMA] S.J.Mackinnon, P.D.Taylor, H.Meijer and S.G.Akl, "An Optimal

Algorithm for Assigning Cryptographic Keys to Control Access in a Hierarchy," IEEE Trans. on Computers, 34, 9, pp.797-802 (1985)

[O] K.Ohta, "Efficient Identification and Signature Scheme," Electro. Lett., 24, 2, pp.115-116 (1988)

[OhOk1] K.Ohta and T.Okamoto, "Modification of the Fiat-Shamir Scheme," Crypto'88 (1988)

[OhOk2] K.Ohta and T.Okamoto, "Membership authentication for Hierarchical Multigroups Using Master Secret Information," The 1990 Symposium on Cryptography and Information Security (SCIS'90), Hihondaira,
Japan (1990)

[OkOh] T.Okamoto and K.Ohta, "Membership authentication for Hierarchical Multigroups Using the Extended Fiat-Shamir Scheme," 1989 Autumn Natinal Convention Record, IEICE, Engineering Science, SA-8-5, (Sept. 1989)

Zero-Knowledge Undeniable Signatures
(extended abstract)

David Chaum

Centre for Mathematics and Computer Science
Kruislaan 413 1098 SJ Amsterdam

SUMMARY: Undeniable signature protocols were introduced at Crypto '89 [CA]. The present article contains new undeniable signature protocols, and these are the first that are zero-knowledge.

INTRODUCTION & MOTIVATION

Digital signatures [DH] are easily verified as authentic by anyone using the corresponding public key. This "self-authenticating" property is quite suitable for some uses, such as broadcast of announcements and public-key certificates. But it is unsuitable for many other applications. Self-authentication makes signatures that are somewhat commercially or personally sensitive, for instance, much more valuable to the industrial spy or extortionist.

Thus, self-authentication is too much authentication for many applications. On the other hand, the remaining previously known authentication schemes offer too little authentication. A judge or arbiter cannot use them to resolve disputes as is possible with self authentication. With zero-knowledge "identification" techniques, for example, a judge would not be convinced of anything by a transcript of the interaction, because by definition anyone could generate indistinguishable transcripts. Also with conventional "identify-friend-or-foe" protocols, or any other system where both parties have all relevant secret keys, the cryptography cannot stop either party from producing valid transcripts.

In short, cooperation of the signer should be necessary to convince another party that a particular signature is valid—but a signer, falsely accused of having signed a particular message, should be able to prove his innocence.

Undeniable Signatures

The relatively new technique called "undeniable signatures" [CA] achieves these objectives. An undeniable signature, like a digital signature, is a number issued by a signer that depends on the signer's public key as well as on the message signed. Unlike a digital signature, however, an undeniable signature cannot be verified without cooperation of the signer.

The validity or invalidity of an undeniable signature can be ascertained by conducting a protocol with the signer, assuming the signer participates. If a "confirmation" protocol is used, the cooperating signer gives exponentially-high certainty to the verifier that the signature does correspond to the message and the signer's public key. If instead a "disavowal" protocol is conducted, the signer gives exponentially-high certainty that the signature does not correspond to the message and the signer's public key. In both protocols a cheating signer, even with infinite computing power, has only an exponentially small chance of success and an overwhelming probability of being detected.

Applications

Undeniable signatures are preferable to digital signatures for many upcoming applications.

Consider, for example, the signature a software supplier may issue on its software, allowing customers to check that the software is genuine and unmodified. With undeniable signatures, only paying customers are able to verify the signature, and they are ensured that the supplier remains accountable for the software.

All manner of inter-organizational messages, such as so called EDI, are a natural candidate for signatures that provide for dispute resolution. But self-authentication would greatly increase the illicit salability of such information.

Also for personal transactions, non-repudiation may be an essential component of security for the service provider; but the customer would like to ensure that, for instance, the signatures do not later end up in the newspaper.

Outline

First the underlying cryptography and form of a signature will be presented, which are the same as in [CA]. Then the new confirmation protocol will be described in detail and its security argued. Next the new disavowal protocol is presented followed by sketches of proofs for its properties. Finally some more recent results are discussed.

CRYPTOGRAPHIC SETTING AND SIGNATURES

Consider using the group of known prime order p. All values transmitted between the participants are elements of this group, the multiplicatively denoted group operation is easily computed by all participants, and taking the discrete log in the group is assumed to be computationally infeasible.

One potentially suitable representation is the multiplicative group of the field $GF(2^n)$, where $p = 2^n-1$ is prime. A second is the group of squares modulo prime q, where $q = 2p+1$. (Notice that such choices rule out the Pohlig-Hellman attack on the discrete log [PH].) An attractive variation on the second approach represents group elements by the integers 1 to p; the group operation is the same, except that all results are normalized by taking the additive inverse exactly when this yields a smaller least positive representative.

A suitable group of prime order p and a primitive element g are initially established and made public for use by a set of signers. Consider a particular signer S having a private key x and a corresponding public key g^x. A message m ($\neq 1$) is signed by S to form a signature, denoted z, which should be equal to m^x.

Computing the private key from the public key, assuming only random messages are signed, is the the discrete log problem; forging signatures on random messages is at least as hard as breaking Diffie-Hellman key exchange.

CONFIRMATION PROTOCOL

A verifier V receiving z, which is claimed to be the signature of signer S on message m and thus equal to m^x, can establish the signature's validity using the confirmation protocol of Figure 1.

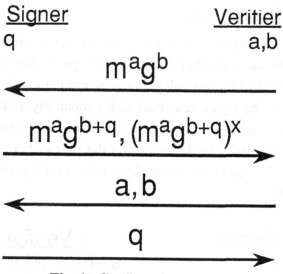

Fig. 1. Confirmation protocol

Each party should initially choose secret random group elements uniformly: S chooses q and V chooses a and b. The first message is formed by V as shown by the first arrow of Figure 1. The second message arrow shows the response of S as a pair of group elements. Next V sends a and b in message 3 so that S can reconstruct the first message. Only once this reconstruction is successful does S send message 4 to reveal q. Finally, by substituting z for m^x, V can reconstruct message 2 and ensure that it was formed properly.

SECURITY OF CONFIRMATION

There are two essential properties:

Theorem 1: The protocol of Figure 1 is zero-knowledge [GMR].

Proof: If V sends a message 3 that should result in a message 4 being sent, V can form the message 2 determined by any random message 4. Any V not sending such a valid message 3 does not receive message 4, but can simulate the message 2 pair as g^y and g^{xy}, by choosing y as a random group element.

Theorem 2: Even with infinite computing power S cannot with probability exceeding p^{-1} provide a valid response for an invalid signature.

Proof: Essentially the same argument as that of [CA] suffices.

DISAVOWAL PROTOCOL

An alleged signer may wish to convince a verifier that a particular message z is not a valid signature corresponding to the signer's public key g^x and message m, i.e. that $z \neq m^x$. To do this, the alleged signer cooperates in an instance of a disavowal protocol. The signer can cheat with probability $1/(k+1)$, where k is a mutually agreed constant and order k operations must be performed by the signer. In practice k might be 1023, for instance, and the protocol could be conducted 2 times for a chance of cheating that is less than one in a million or 10 times to give a chance of only 2^{-100}.

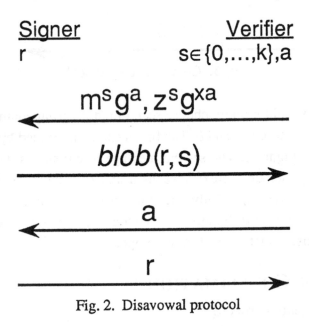

Fig. 2. Disavowal protocol

Consider a single use of the protocol of Figure 2. Initially V chooses an integer s uniformly between 0 and k and chooses a independently and uniformly over the group elements. The first arrow shows how the pair of values sent by V should be formed. Now S can determine the value of s by trial and error. An efficient approach for this raises the first component of the message to the x power and forms a quotient with the second component. The $k+1$ trial quotients can then be computed each by a single multiply from the quotient of the valid signature with z. (Since these quotients are independent of a they can be used for multiple instances of the protocol.) If no s is found, S uses a random value.

Next S sends message 2 containing a blob [BCC] committing to the value of s, but hiding s until the randomly selected r is revealed. (An attractive example is multivalued-blobs based on the discrete log problem that protect the verifier unconditionally, as described in [BCC] §§6.6 and 6.2.2.) Upon receiving the blob as message 2, V can send a. And before finally providing r as the final message, S checks that a can be used to reconstruct the first message.

SECURITY OF DISAVOWAL

Again two things are proved:

Theorem 3: The protocol of Figure 2 is zero-knowledge.

Proof: An interaction in which V sends the correct a, which V can always recognize, is trivially simulated. Any V not supplying an acceptable a only receives a blob, and so the type of zero-knowledge depends on the type of blob.

Theorem 4: Even with infinite computing power S cannot with probability exceeding $1/(k+1)$ provide a valid response for a valid signature.

Proof: if $z = m^x$, a hides s perfectly in the first message. Since the value committed to by the blob cannot be changed, S's best strategy is to guess s.

RECENT WORK

One new result is "convertible" undeniable signatures [BCDP]. These allow the signer to make a single value public that turns all of his undeniable signatures into self-authenticating digital signatures. The signer does not lose the exclusive ability to make signatures and can even selectively convert individual signatures.

The author is aware of some work in preparation:

A signer can "distribute" his undeniable signature signing and/or disavowal abilities among a set of trustees in such a way that a majority of the trustees are necessary and sufficient to perform these functions.

By confirming signatures on random messages in advance, a signer can later simply send, such as by electronic mail, undeniable signatures that the recipient can confirm without further interaction.

The confirmation and disavowal protocols remain zero-knowledge even if multiple instances are conducted in parallel, because of the initial commitment made by the verifier. Another consequence of such "verifier commit" protocols is that it can be made infeasible for covertly cooperating verifiers to be convinced by choosing their single challenge based on coin-flips.

Blobs formed from undeniable signatures can be used to show that the signer can satisfy an agreed predicate. These proofs require only a few messages because blob opening is a parallelizeable confirmation protocol. Such proofs are "undeniable" in the sense that anyone who trusts the randomness of the challenges can later conduct either the confirmation or disavowal protocol with the singer and be convinced whether or not the proof transcript is valid.

CONCLUSION

Undeniable signatures that are Zero-Knowledge can be achieved. They are essentially as efficient in confirmation, and nearly so in disavowal, as other known undeniable signature schemes.

ACKNOWLEDGEMENTS

It is a pleasure to thank the following people for contributing to this paper in one way or another through discussions with the author: Charles Bennett, Jurjen Bos, Joan Boyar, Gilles Brassard, Ivan Damgård, Eugène van Heyst, Tatsuaki Okamoto, and Torben Pedersen.

REFERENCES

[BCC]   Brassard, G., D. Chaum, and C. Crépeau, "Minimum disclosure proofs of knowledge," *Journal of Computer and System Sciences*, vol. 37, 1988, pp. 156–189.

[BCDP] Boyar, J., D. Chaum, I. Damgård, and T. Pedersen, "Convertible undeniable signatures," to be presented at CRYPTO '90.

[CA]   Chaum, D. and H. van Antwerpen, "Undeniable signatures," Advances in Cryptology—CRYPTO '89, Springer-Verlag, 1990, pp. 212–216.

[DH]   Diffie, W. and M.E. Hellman, "New directions in cryptography," *IEEE Transactions on Information Theory*, Vol. IT-22, 1976, pp. 644–654.

[GMR]   Goldwasser, S., S. Micali, and C. Rackoff, "The knowledge complexity of interactive proof-systems," Proceedings, 17th Annual ACM Symposium on the Theory of Computing, May 1985," pp. 291–304.

[PH]   Pohlig, S. and M.E. Hellman, "An improved algorithm for computing logarithms over GF(p) and its cryptographic significance," *IEEE Transactions on Information Theory*, vol. IT-24, 1978, pp. 106–110.

Precautions taken against various potential attacks

in   ISO / IEC   DIS   9796
«Digital signature scheme giving message recovery»

Louis Claude GUILLOU [1]   Jean-Jacques QUISQUATER [2]

with the help of all the experts of ISO/IEC JTC1/SC27/WG20.2 and more specifically

Mike WALKER [3]   Peter LANDROCK [4]   Caroline SHAER [5]

ABSTRACT

This paper describes a «*digital signature scheme giving message recovery*» in order to submit it to the public scrutiny of IACR (the International Association for Cryptologic Research). This scheme is currently prepared by Subcommittee SC27, *Security Techniques*, inside Joint Technical Committee JTC1, *Information Technology*, established by both ISO (the International Organization for Standardization) and IEC (the International Electrotechnical Commission).

The digital signature scheme specified in DIS 9796 does not involve any hash-function. It allows a minimum resource requirement for verification. And it avoids various attacks against the generic algorithms in use.

Definition : An operation (addition, multiplication, power...) modulo n is «**natural**» when, being less than the modulus, the result does not involve the modulo reduction.

— **Attacks by natural products** — The exponential function is the basis of the signature schemes based upon RSA (odd verification exponents), and more generally, based upon exponentials in a ring (including even verification exponents). Under the exponential function, the image of a product of several constants is the product of the images of these constants. A subtle and efficient attack has been recently formulated by Don Coppersmith against annex D of CCITT X509, alias ISO/IEC 9594-8. The attacks by natural products have been definitely excluded in DIS 9796.

— **Attacks by natural powers** — If a natural v-th power is a legitimate argument of the secret function «raising to the power s mod n», then anyone can easily produce the natural v-th root of this argument as a legitimate signature. And even more dangerous, if the verification exponent is even, then signing a natural v-th power may reveal the modulus factorization (cf. Rabin syndrom). In DIS 9796, the natural powers cannot be legitimate arguments to the secret function «raising to the power s mod n».

DIS 9796 is under a 6-month DIS ballot (closed in september 1990) by ISO and IEC Members. This is a major step towards the adoption of an International Standard.

[1] CCETT / EPT, 4, Rue du Clos Courtel, BP 59, F-35512, Cesson Sévigné, France
[2] Philips Research Laboratories, 2, Avenue Van Becelaere, B-1170, Bruxelles, Belgique
[3] Racal Research Ltd, Worton Grange, Reading, Berks RG2 0SB, UK
[4] Department of Mathematics and Computer Science, Aarhus University, Denmark
[5] Racal Research Ltd, Worton Grange, Reading, Berks RG2 0SB, UK

1. Introduction

A digital signature in electronic exchange of information is the counterpart of a handwritten signature in classical mail.

According to the analysis of the ISO experts, two types of digital signature schemes have been clearly identified during preliminary intensive studies.

    — When the verification process needs the message as part of the input, the scheme is named a **signature scheme with appendix**. The elaboration of an appendix involves the use of a hash-function.

    — When the verification process reveals the message together with its specific redundancy (also called "*shadow of the message*"), the scheme is named a **signature scheme giving message recovery**.

DIS 9796 specifies a digital signature scheme giving message recovery for messages of limited length. During the signature process, messages to be signed are padded and extended if necessary. And then an artificial redundancy is added depending upon the message itself. The artificial redundancy is revealed by the verification process. The removal of this artificial redundancy gives message recovery.

The message to be signed need not be in a natural language. It may be any string of bits of limited length. Examples of such messages are cryptographic key materials and the result of hashing another longer message (which is also called the "*imprint of a message*"). Therefore, owing to a hash-function providing the imprint of a message on 128 bits, this digital signature scheme giving message recovery may easily be turned into a digital signature scheme giving imprint recovery, which is a particular case of signature with appendix.

2. Short description of DIS 9796

Any digital signature scheme includes three basic operations : a key production, a signature process and a verification process. The following figures summarize the signature process and the verification process.

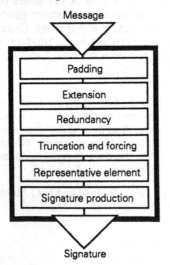

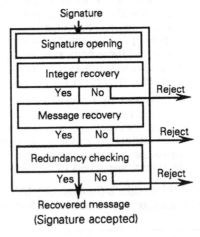

A good implementation of the signature process should physically protect the operations in such a way that there is no direct access to the secret function "raising to the power s modulo n".

2.1. Key production

Each signing entity shall select a positive integer v as its public verification exponent.

NOTE — Values 2 and 3 may have some practical advantages.

Each signing entity shall secretly and randomly select two secret odd prime factors p and q according to the following conditions.
- If v is odd, then $p-1$ and $q-1$ shall be coprime to v.
- If v is even, then $(p-1)/2$ and $(q-1)/2$ shall coprime to v, and moreover, p and q shall not be congruent to each other mod 8.

The public modulus n is the product of the two secret prime factors p and q.

Number k, to be used later on, is the length of the modulus : $2^{k-1} < n < 2^k$.

NOTE — In order to deter modulus factorization, some additional conditions may well be taken into account. These conditions fall outside the scope of this International Standard.

The secret signature exponent is the least positive integer s such that $sv-1$ is a multiple of
- $\text{lcm}(p-1, q-1)$ if v is odd ;
- $\frac{1}{2}\text{lcm}(p-1, q-1)$ if v is even.

NOTE — Some forms of the modulus simplify the modulo reduction and need less storage.
- In the positive forms, after a single most significant bit valued to one, all the bits of the y most significant bytes are valued to zero, up to a quarter of the length of the modulus.
- In the negative forms, all the bits of the y most significant bytes are valued to one, up to a quarter of the length of the modulus.

These forms, where : $1 \le y \le 2x$ and $c < 2^{64x-8y} < 2c$, are
- Form $F_{x, y, +}$: $n = 2^{64x}+c$ of length : $k = 64x+1$ bits ;
- Form $F_{x, y, -}$: $n = 2^{64x}-c$ of length : $k = 64x$ bits.

2.2. Signature process

The message to be signed is a string of bits to be padded by 0 to 7 zeroes to the left of the most significant bit so as to obtain a string of z bytes which codes the padded message MP. Number z multiplied by sixteen shall be less than or equal to number k plus two : $16z \le k+2$. Index r, to be used later on, is the number of padded zeroes plus one. And number t, to be used later on, is the least integer such that a string of $2t$ bytes is at least $k-2$ bits.

Consequently, the message to be signed is the string of the $8z+1-r$ least significant bits of MP ; number z is less than or equal to number t ; and the equality may occur if and only if k is congruent modulo 16 to one of the five values : 14, 15, 0, 1 and 2.

The z bytes of MP are repeated in order and chained to the left, as many times as necessary, until producing a string of t bytes. This result codes the extended message ME. Therefore, for i valued from 1 to t, and j equals $i-1$ (mod z) plus 1 (j is thus valued from 1 to z), the i-th byte of ME equals the j-th byte of MP.

Permutation Π is a HAMMING code with odd parity summarized in the following table.

μ	0	1	2	3	4	5	6	7	8	9	A	B	C	D	E	F
$\Pi(\mu)$	E	3	5	8	9	4	2	F	0	D	B	6	7	A	C	1

The shadow $S(m)$ of any byte m coding the two nibbles μ_1 and μ_2 is defined as $\Pi(\mu_1) \| \Pi(\mu_2)$.

The extended message with redundancy MR is the string of $2t$ bytes defined in the following way : for i valued from 1 to t, the $(2i-1)$-th byte of MR equals the i-th byte of ME, and the $2i$-th byte of MR equals the shadow of the i-th byte of ME, except for the $2z$-th byte of MR which equals the exclusive-or of index r with the shadow of the z-th byte of ME.

The intermediate integer IR is a string of $k-1$ bits where the most significant bit forced to 1 is followed by the $k-2$ least significant bits of the extended message with redundancy MR, except for the least significant byte which is replaced : if $\mu_2 \parallel \mu_1$ are the two least significant nibbles of MR, then the two least significant nibbles of IR equal $\mu_1 \parallel 6$.

The representative element RR is
— IR if v is odd or if the Jacobi symbol of IR with respect to n is $+1$;
— $IR/2$ if v is even and if the Jacobi symbol of IR with respect to n is -1.

The representative element RR is raised to the power s mod n. And signature Σ is either the result or its complement to the modulus, the least one.

2.3. Verification process

The resulting integer IS is the v-th power mod n of signature Σ. And the recovered intermediate integer IR' results from IS by the following decoding.
— If either IS or $(n-IS)$ is congruent to 6 mod 16, then this value is IR'.
— If v is even and if either IS or $(n-IS)$ is congruent to 3 mod 8, then twice this value is IR'.

In all the other cases, and also if IR' does not range from 2^{k-2} to $2^{k-1}-1$, the signature shall be rejected. Consequently, the transformation from IR into IR' is the identity.

The recovered message with redundancy MC is a string of $2t$ bytes obtained by padding 0 to 15 zeroes to the left of the $k-2$ least significant bits of IR', except for the least significant byte which is replaced. If the four least significant nibbles of IR' are $\mu_4 \parallel \mu_3 \parallel \mu_2 \parallel 6$, then the least significant byte of MC equals $\Pi^{-1}(\mu_4) \parallel \mu_2$.

From the $2t$ bytes of MC, t sums are computed. The i-th sum results by exclusive-oring the $2i$-th byte with the shadow of the $(2i-1)$-th byte.

Number z is recovered as the position of the first non-null sum. If the t sums are null, then the signature shall be rejected.

Index r is recovered as the value of the least significant nibble of the first non-null sum. If this nibble is not valued from 1 to 8, then the signature shall be rejected.

The padded message MP' is recovered as the string of the z least significant bytes in odd positions in MC. If the $r-1$ most significant bits of MP' are not all null, then the signature shall be rejected. The message is recovered as the string of the $8z+1-r$ least significant bits of MP'.

Two different methods are proposed for checking redundancy :
— a constructive method,
— and a deductive method.

In the **constructive method**, the signature shall be accepted if and only if the $k-2$ least significant bits of the recovered message with redundancy MC are equal to those of an extended message with redundancy constructed from the recovered padded message MP' according to the operations (extension and redundancy) specified in the signature process.

In the **deductive method**, the signature shall be accepted if and only if the recovered message with redundancy MC satisfies
— checking rule A if z equals t (no extension) ;
— checking rule B if z is less than t (extension).

Checking rule A

> • All the first t–1 sums shall be null.
>
> • The length k of the modulus n shall equal one of the five values : $16t$–2, $16t$–1, $16t$, $16t$+1, $16t$+2.
>
> • In the most significant nibble of the t-th sum, the k+2 (mod 16) least significant bits (0 to 4 bits) shall be null.

Checking rule B

> • All the first t–1 sums shall be null, but one which shall be valued from 1 to 8.
>
> • For i valued to 1, 2, 3... while i is less than t–z, the $(2z+2i-1)$-th and $(2i-1)$-th byte shall be equal.
>
> • In the sixteen most significant bits and in the $(2t-2z-1)$-th byte preceded by its shadow, the k–3 (mod 16) plus one least significant bits (1 to 16 bits) shall be equal.

3. Four precautions taken in DIS 9796

3.1. Elimination of natural powers

THEOREM : A representative element can never be a v-th natural power.
  — A natural power cannot be congruent to 6 mod 16.
  — An even natural power cannot be congruent to 3 mod 8.

PROOF : The following table summarizes the values mod 16 of all the natural powers.

If :	x mod 16 is:	0	1	2	3	4	5	6	7	8	9	A	B	C	D	E	F
Then :	x^2 mod 16 is:	0	1	4	9	0	9	4	1	0	1	4	9	0	9	4	1
	x^3 mod 16 is:	0	1	8	B	0	D	8	7	0	9	8	3	0	5	8	F
	x^4 mod 16 is:	0	1	0	1	0	1	0	1	0	1	0	1	0	1	0	1
		...		...		...		...		...		...		...		...	
And if :	x^{4k} mod 16 is:	0	1	0	1	0	1	0	1	0	1	0	1	0	1	0	1
Then :	x^{4k+1} mod 16 is:	0	1	0	3	0	5	0	7	0	9	0	B	0	D	0	F
	x^{4k+2} mod 16 is:	0	1	0	9	0	9	0	1	0	1	0	9	0	9	0	1
	x^{4k+3} mod 16 is:	0	1	0	B	0	D	0	7	0	9	0	3	0	5	0	F
	x^{4k+4} mod 16 is:	0	1	0	1	0	1	0	1	0	1	0	1	0	1	0	1

Numbers 0 and 1 appear on every line. Numbers 2, 6, A, C and E appear only on the first line. Numbers 3, 5, 7, B, D and F appear on each line corresponding to an odd exponent. Number 4 appears only on the first and second lines (exponent two : natural squares). Number 8 appears only on the first and third lines (exponent three : natural cubes). Number 9 appears on each line corresponding to an exponent not multiple of 4.

Consequently, number 6 appears only on the first line, and numbers 3 and B appear only for odd exponents.
                                                                                    Q.E.D.

3.2. Elimination of shifts and complementations

THEOREM : Shifting or complementing representative elements is not an internal operation.

PROOF : The proof is trivial. In the least significant nibble,
  — a shift replaces 6 by C, 8 or 0 and, when v is even, 3 by 6, C, 8 or 0 ; but when v is even, a shift of one position should be discounted because it reverses the Jacobi symbol.
  — a complementation replaces 6 by A and, when v is even, 3 by C.
                                                                                    Q.E.D.

3.3. Elimination of natural multiplications

THEOREM : There is no constant (except the trivial solution +1) such that the natural product of this constant with a representative element is a representative element.

PROOF : Such a constant shall be congruent to
   1 mod 8 for maintaining a congruency to 6 mod 16 ;
   1 mod 8 for maintaining a congruency to 3 mod 8 ;
   2 mod 8 for transforming a congruency to 3 mod 8 into a congruency to 6 mod 16.
There is no solution for transforming a congruency to 6 mod 16 into a congruency to 3 mod 8.

The first potential constant is $k=2$. It may occur only with an even verification exponent. It is discounted because it reverses the Jacobi symbol with respect to n.

The intermediate integers are less than n, but more than $n/4$. And the representative elements are less than n, but more than $n/8$. Therefore the multiplication modulo n of any representative element by any other potential constant (at least 9) shall involve a modulo reduction. **Q.E.D.**

3.4. Forcing Jacobi symbols to +1 when v is even

When the public verification exponent is even, the Rabin syndrom is removed by restricting moduli to Williams numbers and by forcing to +1 Jacobi symbols of representative elements. This precaution is strengthened by introducing redundancy and by eliminating natural powers.

According to Fermat, if p is an odd prime, then for any integer x from 0 to $p-1$, x^{p+1} and x^2 are congruent to each other modulo p. Therefore, «raising to the power $(p+1)/2$ in the field GF(p)» transforms any element x either into itself if x is a quadratic residue of GF(p), or into its complement to p if x is a quadratic non-residue of GF(p). Moreover if prime p is congruent to 3 mod 4, then $(p+1)/2$ is even, and «raising any quadratic residue x to the power $(p+1)/4$ in GF(p)» computes a square root of x in field GF(p). The Legendre symbol of any integer x with respect to any prime p is obtained by raising x to the power $(p-1)/2$ in field GF(p).

In DIS 9796, both p and q are congruent to 3 mod 4. Therefore, the secret signature exponent corresponding to $v=2$ is $s=(n-p-q+5)/8$. «Raising to the power $(n-p-q+5)/8$ in ring Zn» is indeed equivalent to «raising to the power $(p+1)/4$ in field GF(p)» and «raising to the power $(q+1)/4$ in field GF(q)» before reconstructing the global result from the two partial results.

A Williams integer (also called a Blum integer) is defined as the product of two primes p and q both congruent to 3 mod 4, but not congruent to each other mod 8. In fact, one prime factor is congruent to 3 mod 8, and the other one to 7 mod 8. Therefore Williams integers are congruent to 5 mod 8 (7x3 = 21 = 5 + 2x8).

Number 2 is a quadratic non-residue in field GF(p) if prime p is congruent to 3 mod 8, but a quadratic residue in field GF(q) if prime q is congruent to 7 mod 8. The Jacobi symbol of 2 with respect to any Williams integer n is -1, and multiplying by 2 in ring Zn reverses the Jacobi symbol of integers coprime to n.

The public verification exponent v may be written as v' times 2^e where v' is odd. And let us name s' the secret signature exponent corresponding to the public verification exponent v'. The «global process» is defined as the composition of «raising successively to the powers s and v in Zn», which is also equivalent to composing the successive operations in Zn : «raising e times to the power $(n-p-q+5)/8$» followed by «raising to the power s'», and then «raising to the power v'» followed by «squaring e times». The composition of «raising to the power s'» and «raising to the power v'» is the identity. Therefore the «global process» is equivalent to «raising e times to the power $(n-p-q+5)/8$» and then «squaring e times», which is also «raising e times to the even power $(n-p-q+5)/4$ ».

The «central operation» is defined as «raising to the even power $(n-p-q+5)/4$ in Zn», which is also equivalent to «raising to the even power $(p+1)/2$ in $GF(p)$» and «raising to the even power $(q+1)/2$ in $GF(q)$» before reconstructing the global result from the two partial results. In both fields $GF(p)$ and $GF(q)$, the central operation, as well as the «global process», reverses the sign of quadratic non-residues, but is the identity for all the other numbers. The result of the central operation (as well as the result of the «global process») is always a quadratic residue in both fields $GF(p)$ and $GF(q)$, except for multiples of either p or q.

The Jacobi symbol with respect to n is the product of the Legendre symbols with respect to p and q.

Let us first take any integer x ranging from 0 to $n-1$ and having -1 as Jacobi symbol with respect to n. Then the «global process» results in integer y having $+1$ as Jacobi symbol with respect to n. On one hand, x^2 and y^2 are equal mod n. Therefore n divides x^2-y^2 which is $(x-y)$ times $(x+y)$. On the other hand, x has -1 as Jacobi symbol while y and $-y$ have $+1$. Therefore n divides neither $x-y$ nor $x+y$. And either p or q divides $x-y$. Therefore a prime factor is easily computed as the greater common divider of n and $x-y$. Applying the «global process» to any argument having -1 as Jacobi symbol reveals the factorization.

Let us take now any integer z ranging from 0 to $n-1$, and having $+1$ as Jacobi symbol with respect to n. Then the «global process» results in either z if z is a quadratic residue in both fields $GF(p)$ and $GF(q)$, or $n-z$ if z is a quadratic non-residue in both fields $GF(p)$ and $GF(q)$. Applying the «global process» to any argument having $+1$ as Jacobi symbol does not reveal the factorization.

4. Conclusion

During the signature process, artificial redundancy is added to the messages to be signed so as to obtain the corresponding «representative elements» which are the only legitimate arguments to the secret function «raising to the power s mod n».

Four reasons in favour of DIS 9796

• Shifting or complementing representative elements does not result into other ones.
• The natural product of any representative element by any constant other than 1 is not a representative element.
• Natural v-th powers are not representative elements.
• If the public verification exponent v is even, then moduli are restricted to Williams integers and the Jacobi symbol of the representative elements with respect to n is forced to $+1$.

In DIS 9796 as opposed to previous versions of 9796, permutation Π plays no direct part in protecting against shifts, complementations, natural multiplications and natural powers. This protection is **totally ensured** by constructing the intermediate integers as strings of $(k-1)$ bits where the most significant bit is forced to 1 and the least significant nibble is forced to 6.

Permutation Π is used for increasing the distance between strings of bits which code representative elements and for avoiding long strings of constant bits in these representative elements. These simple requirements are fulfilled by a HAMMING code with odd parity.

Annex A :   Illustrative example

A.1. Key production

The public verification exponent is $v=3$. Therefore the secret prime factors p and q shall both be congruent to 2 mod 3.

$p =$	BA09106C	754EB6FE	BBC21479	9FF1B8DE
	1B4CBB7A	7A782B15	7C1BC152	90A1A3AB
$q =$	16046EB39	E03BEAB6	21D03C08	B8AE6B66
	CFF955B6	4B4F48B7	EE152A32	6BF8CB25

The public modulus n is here of the form : $2^{512} + c$, with $2c > 2^{128} > c$, coded over 513 bits.

$n =$	100000000	00000000	00000000	00000000
	BBA2D15D	BB303C8A	21C5EBBC	BAE52B71
	25087920	DD7CDF35	8EA119FD	66FB0640
	12EC8CE6	92F0A0B8	E8321B04	1ACD40B7

The secret signature exponent is $s=(n-p-q+3)/6$.

$s =$	2AAAAAAA	AAAAAAAA	AAAAAAAA	AAAAAAAA
	C9F0783A	49DD5F6C	5AF651F4	C9D0DC92
	81C96A3F	16A85F95	72D7CC3F	2D0F25A9
	DBF1149E	4CDC3227	3FAADD3F	DA5DCDA7

A.2. Length of various variables

Because number k is 513, intermediate integers IR and IR', signatures Σ and resulting integers IS are strings of 512 bits. Messages to be signed are strings of 1 to 256 bits. Number z is valued from 1 to 32. Padded messages MP and MP' are strings of 1 to 32 bytes. Number t is 32. Extended messages ME are strings of 32 bytes. Messages with redundancy MR and MC are strings of 64 bytes.

A.3. Signature process

The message is : C BBAA 9988 7766 5544 3322 1100. Its length is 100 bits. After padding four zeroes to the left, the padded message MP is a string of 13 bytes. Therefore $z=13$ and $r=5$.

$MP =$	0C	BBAA9988	77665544	33221100

The extended message ME results by repeating the successive bytes of MP, in order and concatenated to the left, until obtaining a string of 32 bytes.

$ME =$	55443322	11000CBB	AA998877	66554433
	2211000C	BBAA9988	77665544	33221100

The extended message with redundancy MR is a string of 64 bytes where the 32 bytes of ME are interleaved with 32 bytes of redundancy. The message border is signalled by a break (let us compare E20C to E70C) in the redundancy.

$MR =$	44559944	88335522	3311EE00	E70C66BB
	BBAADD99	0088FF77	22664455	99448833
	55223311	EE00E20C	66BBBBAA	DD990088
	FF772266	44559944	88335522	3311EE00

The intermediate integer IR results from MR by truncating to 511 bits, by padding to the left one bit valued to 1 and by replacing the least significant byte $\mu_2 \parallel \mu_1 = 00$ by $\mu_1 \parallel 6 = 06$.

$IR =$	C4559944	88335522	3311EE00	E70C66BB
	BBAADD99	0088FF77	22664455	99448833
	55223311	EE00E20C	66BBBBAA	DD990088
	FF772266	44559944	88335522	3311EE06

The representative element RR equals here IR because v is odd. And the signature Σ results by raising RR to the power s mod n and keeping here the complement to n.

$\Sigma =$	309F873D	8DED8379	490F6097	EAAFDABC
	137D3EBF	D8F25AB5	F138D56A	719CDC52
	6BDD022E	A65DABAB	920A8101	3A85D092
	E04D3E42	1CAAB717	C90D89EA	45A8D23A

A.4. Verification process

Signature Σ is indeed less than $n/2$. The resulting integer IS is obtained by cubing Σ mod n.

$IS =$	3BAA66BB	77CCAADD	CCEE11FF	18F39944
	FFF7F3C4	BAA73D12	FF5FA767	21A0A33D
	CFE6460E	EF7BFD29	27E55E52	896205B7
	13756A80	4E9B0774	5FFEC5E1	E7BB52B1

Because IS is congruent to 1 mod 16, the intermediate integer IR' (a string of 512 bits where the most significant bit is valued to 1 and the least significant nibble is valued to 6) equals $n-IS$.

$IR' =$	C4559944	88335522	3311EE00	E70C66BB
	BBAADD99	0088FF77	22664455	99448833
	55223311	EE00E20C	66BBBBAA	DD990088
	FF772266	44559944	88335522	3311EE06

The recovered message with redundancy MC is here a string of 64 bytes equal to IR', except for the most significant bit which is forced to 0 and the least significant byte which is replaced. The four least significant nibbles of IR' equal $\mu_4 \| \mu_3 \| \mu_2 \| 6$ = EE06. Therefore the least significant byte of MC equals $\Pi^{-1}(\mu_4) \| \mu_2 = 00$ because $\Pi(0)$ is E.

$MC =$	44559944	88335522	3311EE00	E70C66BB
	BBAADD99	0088FF77	22664455	99448833
	55223311	EE00E20C	66BBBBAA	DD990088
	FF772266	44559944	88335522	3311EE00

The first non-null sum is the 13-th sum which is 5. Therefore $z=13$ and $r=5$, and the length of the recovered message is 100 bits ($8z+1-r = 8\times13+1-5 = 100$). In the 25-th byte (0C), the 4 most significant bits are null ($r-1=4$, $2z-1=25$). Then the padded message MP' is recovered.

$MP' =$	0C	BBAA9988	77665544	33221100

And the message itself is recovered as : C BBAA 9988 7766 5544 3322 1100.

The signature is accepted because the recovered message with redundancy MC is a string of bytes satisfying checking rule B.

Redundancy : The first 31 sums are null, except for the 13-th sum which is 5.
Extension : For i valued from 1 to 18 ($t-z-1=32-13-1=18$), the $(2i+25)$-th and $(2i-1)$-th bytes are equal.
Truncation : In the sixteen most significant bits (4455) and in the 37-th byte preceded by its shadow ($2t-2z-1 = 64-26-1 = 37$; $S(55) = 44$), the 15 least significant bits ($k-2=15$ mod 16) are equal.

Software Run-Time Protection: A Cryptographic Issue

Josep Domingo-Ferrer
Departament d'Informàtica
Universitat Autònoma de Barcelona
08193 Bellaterra, Catalonia-Spain

Abstract. *A new method is featured which solves the software integrity problem by properly coding rather than enciphering. Adopting the lengthy and expensive solution which consists of having the whole program signed/encrypted by an authority would require full decryption and secure storage for the whole program before execution, whereas one signed instruction, pipe-lined decoding-executing, and secure recording of a few of the last read instructions suffice in our case. A general use of the proposed system could practically prevent any viral attack with minimum authority operation.*

0. Introduction

Our goal is *program integrity for the user*, i. e. ensuring that, given an image code, any instruction insertion, deletion or modification *before or during* execution, will cause execution to stop. This requires that the image be stored under a suitable structure, which can be almost completely worked out by the same user who wrote the program. A one-way *function F* [Diff76], such that in general $F(X \oplus Y) <> F(X) \oplus F(Y)$ (where $\oplus$ denotes addition modulo 2 on the binary representations of the operands), and a *public-key signature scheme* must be agreed upon before implementing the method. The signature consists of a private transformation D, *exclusively owned by an authority*, and a publicly registered inverse transformation E. Also, a *normalized instruction format* must be defined. Suppose an algorithm A consisting of machine-code executable instructions $i_1, i_2, ..., i_n$. Assume that i_n *is not a branch instruction* (it can be for instance an **END** or a **RET** instruction). Call I_j the instruction resulting from padding i_j to a fixed length and adding a redundance pattern to i_j.

1. User Preparation Phase

In order for the user to turn a program he has written into a *trusted program*, he first normalizes it into a sequence $I_1, ..., I_n$, where n is the number of instructions in the program. Then he replaces each I_j with a *trace* T_j. The traces are computed in a *reverse order*, from T_n to T_1. In this way, the sequential program $I_1, ..., I_n$ looks like

$$
\begin{aligned}
T_1 &= F(T_2) \oplus I_2 \\
T_2 &= F(T_3) \oplus I_3 \\
&\cdots \\
T_{n-1} &= F(T_n) \oplus I_n \\
T_n &= F(I_n)
\end{aligned}
\tag{1}
$$

I_1 does not appear in the sequence (1): it will be dealt with in section 2. Once the structure for a sequential program has been designed, we must solve the forward unconditional, forward conditional and subroutine branchings in order to be able to treat any program having no backward branches. *Although for clarity we will present I_k as located in a sequential trace T_{k-1}, this need not be true*, as it will become evident. For the same reason, in the rest of the paper we will also sometimes write the traces following a branch as sequential ones. A *forward unconditional branch* at instruction I_k to instruction I_j is translated as

$$
\begin{aligned}
\cdots \quad T_{k-1} &= F(T_k) \oplus I_k \\
T_k &= F(T_j) \oplus I_j \\
T_{k+1} &= \cdots \\
&\cdots \\
T_j &= \cdots
\end{aligned}
\tag{2}
$$

When I_k is a *forward conditional branch* to instruction I_j, the following traces are computed (also in an index decreasing order)

$$
\begin{aligned}
\cdots \quad T_{k-1} &= F(T_k) \oplus I_k \\
T_k &= F(T_{k'}) \oplus F(T_{k+1}) \oplus I_{k+1} \\
T_{k'} &= F(T_j) \oplus I_j \\
T_{k+1} &= \cdots \\
&\cdots \\
T_j &= \cdots
\end{aligned}
\tag{3}
$$

For a *branch to subroutine* (machine-code subroutine) we must also guard against the right subroutine being replaced at run-time; so, assuming that the instructions $I^\wedge_1, ..., I^\wedge_m$ of the subroutine are already encoded as $T^\wedge_0, T^\wedge_1, ..., T^\wedge_m$, we retrieve $F(T^\wedge_i) \oplus I^\wedge_i$ from $T^\wedge_0 = D(F(T^\wedge_i) \oplus I^\wedge_i)$ (see section 2 about the heading trace $T^\wedge_0$) and include it in the calling program as follows

$$
\begin{aligned}
T_{k\text{-}1} &= F(T_k) \oplus I_k \\
T_k &= F(T_{k'}) \oplus F(T_{k+1}) \oplus I_{k+1} \\
T_{k'} &= F(T^\wedge_i) \oplus I^\wedge_i \\
T_{k+1} &= ...
\end{aligned}
\tag{4}
$$

If I_k is a branch to I_j with $j < k$, the branch trace structures proposed so far cannot be used to compute T_k (for a backward unconditional branch) or $T_{k'}$ (for a backward conditional branch), since a trace T_j is needed which has not yet been computed and depends on T_k (resp. $T_{k'}$). So a *backward unconditional branch* at instruction I_k to instruction I_j is translated as

$$
\begin{aligned}
T_{j\text{-}1} &= F(T^-_j) \oplus F(T_j) \oplus I_j \\
T^-_j &= F(T_j) \oplus T^-(j) \\
T_j &= ... \\
&\quad ... \\
T_{k\text{-}1} &= F(T_k) \oplus I_k \\
T_k &= F(T^-(j)) \oplus I_j \\
T_{k+1} &= ...
\end{aligned}
\tag{5}
$$

Finally, the trace structure for a *backward conditional branch* is straightforward

$$
\begin{aligned}
T_{j\text{-}1} &= F(T^-_j) \oplus F(T_j) \oplus I_j \\
T^-_j &= F(T_j) \oplus T^-(j) \\
T_j &= ... \\
&\quad ... \\
T_{k\text{-}1} &= F(T_k) \oplus I_k \\
T_k &= F(T_{k'}) \oplus F(T_{k+1}) \oplus I_{k+1} \\
T_{k'} &= F(T^-(j)) \oplus I_j \\
T_{k+1} &= ...
\end{aligned}
\tag{6}
$$

Both in (5) and (6), $T^-(j)$ has been computed by applying a one-to-one function to j.

2. Authority Endorsement Phase

After user trace computation, the authority owning the private transformation D endorses the trace sequence by computing a closing trace $T_0 = D(F(T_1) \oplus I_1)$. Notice that the missing instruction I_1 appears now in the trace sequence, and that *the whole program need not be supplied to the authority, but just $F(T_1) \oplus I_1$.*

3. Program Execution with Controlled Instruction Flow

<u>Theorem 1 (Correctness)</u>. The program $i_1, i_2, ..., i_n$ can be retrieved and executed from its corresponding trace sequence $T_0, T_1, T_2, ..., T_n$.

<u>Proof (sketch)</u>. We have six cases: (a) sequential instruction blocks, (b) forward unconditional branchings, (c) forward conditional branchings, (d) subroutine branchings, (e) backward unconditional branchings, and (f) backward conditional branchings. Due to lack of space, we only prove the first case here. The run-time setting used consists of a coprocessor p', whose task is retrieving the instructions I_k and forwarding them to a usual processor p; it is assumed that p' and p are pipe-lined. The path between p and p' must be a secure one, so that it is advisable that both processor and coprocessor be encapsulated in a single chip (with a hybrid circuit [Ebel86], this is achieved at low redesign cost).

Now, for a **sequential instruction block**, operation at cycle k is: T_k is being read, T_{k-1} is available in a coprocessor internal register, T_{k-2} is being *evaluated* by p' and I_{k-2} is being executed by p (actually the i_{k-2} stripped from I_{k-2} is executed, after redundance checking). Evaluating a trace T_m means to retrieve the instruction contained in the trace (I_{m+1} for a sequential trace, I_j for a branch trace to T_j, see section 1). Then, following this scheme, after reading T_0 and T_1 during cycles 0 and 1, at cycle 2 T_2 is read and p' evaluates T_0 by computing $F(T_1) \oplus E(T_0) = I_1$. It must be pointed out this computation is feasible because of the transformation E being easy and public and T_1 being available to p'. Thus instruction I_1 has been retrieved, *if its redundance pattern is all right*, of course. Now suppose that at cycle $k-1$ I_1 through I_{k-2} have been retrieved and the program has been executed till I_{k-3}. Then at cycle k, I_{k-2} is executed and I_{k-1} is retrieved from T_{k-2} by computing $F(T_{k-1}) \oplus T_{k-2} = I_{k-1}$. Again this is possible because of T_{k-1} being available at cycle k and F being public and easily computable. The result follows by induction. Execution stops at cycle $n+2$ after executing I_n, which means that only two overhead cycles have been introduced (the first read at cycle 0 is unavoidable even for a conventional execution, see diagram 1). As for T_n, this trace is only used during evaluation of T_{n-1} for I_n.

p								
EXECUTE	*	*	*	I_1	...	I_{n-2}	I_{n-1}	I_n

p'								
EVALUATE	*	*	T_0	T_1	...	T_{n-2}	T_{n-1}	T_n

p'								
READ	T_0	T_1	T_2	T_3	...	T_n	*	*

| CYCLE | 0 | 1 | 2 | 3 | ... | n | $n+1$ | $n+2$ |

DIAGRAM 1. Sequential Block. $n+2$ Usable Cycles for n Instructions.

4. Run-Time Integrity

<u>Theorem 2 (Run-Time Integrity)</u>. If a program $i_1, ..., i_n$ is stored as $T_0, ..., T_n$, and is evaluated as described in section 3, any instruction substitution, deletion or insertion before or during execution will be detected at run-time, thus causing the processor to stop executing *before* the substituted, deleted or inserted instruction(s). Moreover, only the last five read traces must be kept in the internal secure memory of the processor (they are kept even in case of interrupt).

<u>Proof (Sketch)</u>. Since the arithmetic link between two consecutively executed instructions in the sequential case (I_k and I_{k+1}) is essentially the same as in a forward unconditional branching (I_k and I_j), both cases can be reduced to a single one. Thus five cases must be considered for the proof: 1) sequential instruction blocks and forward unconditional branchings, 2) forward conditional branchings, 3) subroutine branchings, 4) backward unconditional branchings, and 5) backward conditional branchings. Because of space reasons, we will only develop the proof for a **sequential block**, which uses only the last read trace (no need for all five last ones); some additions to the main idea are used for

the other cases. First consider that an intruder attempts a **substitution**, by replacing I_k with I_k^*; it follows from (1) that he can then choose either to maintain T_{k-1} or to modify it. If he tries the first thing, he must find a T_k^* s. t. $F(T_k^*) \oplus I_k^* = T_{k-1}$, but this is unfeasible given the unidirectionality of F. Consequently T_{k-1} is changed to T_{k-1}^*; now if nothing more is done it will not be possible for the processor p' to retrieve I_{k-1} by evaluation of T_{k-2} (the resulting garbage is not likely going to be a valid instruction because of the redundance field). Thus a change in I_k causes p to stop after execution of I_{k-2}, which is a good behaviour. On the other hand if we recompute $T_{k-2}^* = F(T_{k-1}^*) \oplus I_{k-1}$ then p' will not be able to recover I_{k-2} and execution will stop earlier. Eventually if we proceed the backward recomputation, we see by induction that T_1 will be replaced by T_1^*, and this will be detected since T_0 cannot be replaced (it is signed by the authority), and it will not be possible to retrieve a valid I_1 by evaluation of T_0 at cycle 2. Thus a modification of I_k enforces a modification of T_{k-1}, due to the unidirectionality of F. For this change to remain undetected, backward recomputation of traces should be made, but this is stopped by the signature in T_0; in any case, a defective instruction is *never* executed. If a change of a subset of instructions $I_{k1}, ..., I_{ks}$ is attempted a similar argument can be used because this also implies changing some traces. As for **deletion** of a trace T_k from a sequential flow, it also breaks the natural arithmetic link because it amounts to substituting T_{k+1} for T_k, and we have shown that substitutions are detected. Finally, **insertion** of a new trace T_{k*} between T_k and T_{k+1} is neither feasible: we can compute $T_{k*} = F(T_{k+1}) \oplus I_{k*}$ so as to link with T_{k+1}, but this is useless for we cannot link T_k and T_{k*} without changing the former or having a garbage I_{k+1} in T_k, which will be both detected, as shown above.■

5. Applications and Conclusion

Our system allows branches and guarantees full integrity while requiring secure storage only for the last five read traces; also decoding and execution are pipe-lined. The work performed by the authority in our scheme is rather small, so that *a great deal of users may share a single authority*, which simplifies most of applications. For example, imagine a large software company, where all programmers write and prepare their programs as specified in sections 1 and 2 in order to protect against computer viruses. Then a single authority can be used to endorse every program. Finally, our proposal requires that only $F(T_1) \oplus I_1$ be supplied to the authority for endorsement; the preparation phase can be carried out by the user himself, so that *the authority need not know what is being signed,* but just a user's valid identification (software privacy).

References

[Diff76] W. Diffie and M. E. Hellman, "New Directions in Cryptography", *IEEE Transactions on Information Theory*, vol. IT-22, pp. 644-654, Nov. 1976.

[Ebel86] G. H. Ebel, "Hybrid Circuits: Thick and Thin Film", in *Handbook of Modern Microelectronics and Electrical Engineering*, ed. C. Belove, Wiley, New York, 1986.

[Park89] G. Parkin and B. Wichmann, "Intelligent Modules", in *The Protection of Computer Software - Its Technology and Applications*, ed. D. Grover, Cambridge University Press, Cambridge, 1989.

[RiAD78] R. L. Rivest, L. Adleman and M. L. Dertouzos, "On Data Banks and Privacy Homomorphisms", in *Foundations of Secure Computation*, ed. R. A. DeMillo et. al., Academic Press, New York, 1978.

An identity-based identification scheme based on discrete logarithms modulo a composite number

Marc Girault

Service d'Etudes communes des Postes et Télécommunications (SEPT)
42 rue des Coutures, Caen, France

Abstract. *We first describe a modification of Schnorr's identification scheme, in which the modulus is composite (instead of prime). This modification has some similarity with Brickell-McCurley's one, presented at the same conference. Then, by establishing a new set-up, we derive the first identity-based identification scheme based on discrete logarithms. More precisely, it is based on discrete logarithm modulo a composite number, a problem known to be harder than factorization problem. This scheme has interesting and somewhat paradoxical features. In particular, any user can choose his own secret, and, provided the parameters have convenient sizes, even the trusted center is unable to retrieve it from the public key (contrary to any identity-based scheme known until now).*

1. Introduction

At CRYPTO'90 conference, Schnorr presented a new identification scheme [Sc], the security of which is based on the difficulty of discrete logarithm problem modulo a prime.

The set-up of Schnorr's scheme is as follows. A trusted center (or KAC : "Key Authentication Center") generates two primes p and f such that f divides p-1, and an integer b of order f modulo p (i.e. f is the smallest integer such that $b^f = 1$ [mod p]). The integers p, f and b are published by the center. Each user chooses a secret s, smaller than f, and his public key is $P = b^{-s}$ (mod p). Now, the center signs a message composed of user's identity I, his public key P and other parameters such as validity dates. This signed message will be called certificate, in accordance with ISO/CCITT vocabulary. Any signature scheme (including the one derived from this identification scheme) can be used to produce this certificate.

Compared to other identification schemes, Schnorr's one has some advantages but also some drawbacks. One of these drawbacks is the necessity for the center to produce certificates, and for the verifier to check them. This drawback does not exist in identity-based schemes [Sh], in which a user's public key is nothing but his identity I.

Conversely, one important drawback of identity-based schemes is the fact that not only a user cannot choose his secret key, but this key is calculated by the center and can be calculated again by it at any moment during its period of validity.

In this paper, we present the first identity-based identification scheme in which every user can choose himself his secret key, the center being unable to retrieve it from the public key. To achieve this goal, we first design a Schnorr-like identification scheme using a composite modulus instead of a prime one (section 3), then we modify the set-up of the scheme in order to render it identity-based (section 4). Beforehand, we briefly recall what Schnorr's identification protocol is (section 2).

Because we are limited in space, only the security of the set-up of these schemes will be discussed. The security of the protocols themselves (in particular their zero-knowledgeness) will not be addressed.

Before starting, we inform the reader that a very closely related paper has been presented at ESORICS'90 conference [GP] in October 1990.

2. Schnorr's identification scheme

The set-up of Schnorr's scheme has been described in the introduction. The typical sizes of the parameters are 512 bits for p, and 140 bits for f.

When Alice wants to prove to Bob she is Alice, the two partners use the following protocol :

1) Alice picks a random integer r in the interval $[0, f-1]$, calculates $x = b^r \pmod{p}$ and sends it to Bob along with her certificate.

2) Bob checks the certificate, picks a random integer c in the interval $[0, 2^t-1]$ (where, typically, t lies between 20 and 70) and sends it to Alice.

3) Alice calculates $y = r + sc \pmod{f}$ and sends it to Bob.

4) Bob checks that $b^y P^c = x \pmod{p}$.

It can be proven that a fraudor (somebody who wants to impersonate Alice but does not know s), has only one chance over 2^t not to be detected, and that this protocol is zero-knowledge provided discrete logarithm modulo a prime is a hard problem. Some optimizations features of the scheme, though important, are not considered here for short.

3. A Schnorr-like scheme with a composite modulus

Description

We now describe a modification of Schnorr's scheme. Let n be the product of two primes p and q such that $p = 2fp' + 1$ and $q = 2fq' + 1$, where f, p' and q' are distinct primes. In the basic version, f is 200 bit-long, p' and q' are 300-bit long, so n is a 1000-bit integer. In a variant, f is 140-bit long, p' and q' are 210-bit long, so n is 700-bit long (this should be the minimal length of n). The difference of security between both versions will be discussed later.

Let b be an integer of order f both modulo p and modulo q. Therefore the order of b modulo n is f and we have : $b^f = 1 \pmod{n}$. Note that all these parameters can be easily generated by the center (we omit the details). Now, the integers n, f and b are made public, whilst p and q are kept secret.

The rest of the scheme is exactly the same as in Schnorr's scheme except that we replace every occurrence of *p* by *n*. To be explicit, each user chooses a secret s, smaller than f, and his public key is $P = b^{-s} \pmod{n}$. The certificate is calculated by the authority as shown in the introduction and the identification protocol is as follows :

1) Alice picks a random integer r in the interval [0, f-1], calculates $x = b^r \pmod{n}$ and sends it to Bob along with her certificate.

2) Bob checks the certificate, picks a random integer c in the interval [0, 2^t-1] and sends it to Alice.

3) Alice calculates $y = r + sc \pmod{f}$ and sends it to Bob.

4) Bob checks that $b^y P^c = x \pmod{n}$.

Note that Schnorr introduced the parameter f, a small divisor of p-1, only in order to optimize the performances of this scheme. The security of his scheme is not compromised if f is equal to p-1, since p is public. On the contrary, choosing f as a very small divisor of p-1 and q-1 is crucial for the security of our modification, since p and q are secret.

Security of the scheme

Generally speaking, the security of the scheme lies on the difficulty of computing a discrete logarithm modulo a composite number. This problem is known to be harder than factorization problem. More precisely, factorization problem can be reduced to general discrete logarithm problem in probabilistic polynomial time [W]. Recently, Schrift and Shamir [SS] proved that almost all bits of the discrete logarithm modulo a Blum integer were individually secure (and the right half of them simultaneously secure) provided factorization of Blum integers is hard.

Now, the moduli we use are Blum integers, but of a very special form. Moreover, some side information on factors of n is revealed, since a divisor f of (p-1)(q-1), the Euler function of n, is made public, as well as an integer b of order f modulo n. The security of the scheme therefore lies on a very specific assumption, namely the difficulty of finding s, given b^{-s} (mod n) and all this knowledge about n.

However, as far as we know, even factoring n seems to be hard. The only attack we found, apart from existing factorization algorithms, costs $2^{|p'+q'|/|f|}$ operations, where $|u|$ denotes the number of bits of u. This is why we chose $|p'+q'|/|f|$ at least equal to 2^{70}.

Note that, in the basic version, factoring n is not enough to retrieving s from P, since the enemy still has to compute discrete logarithms modulo p and q, which are 500-bit primes. In fact, as long as factoring (our special) n *or* discrete logarithm modulo (our special) p is hard, then our scheme is secure. This is an interesting similarity with Brickell and McCurley's scheme [BC], reported in these proceedings. This last scheme also is a modification of Schnorr's scheme, designed so as to preserve its security even if either factorization or discrete logarithm modulo a prime (but not both) became no more infeasible.

Of course, the property that the scheme remains secure even if n is factorized does not subsist if n is a 700-bit number (and p, q 350-bit numbers) since existing algorithms may compute discrete logarithms for primes of this size. Nonetheless, if this property is not specifically required, then these sizes of parameters can be chosen.

4. A "paradoxical" identity-based scheme

Description

We now come to our final scheme. Let n, f, b be as in the previous section, e be a public exponent coprime with p-1 and q-1 (where, typically, the length of e lies between 20 and 70 bits) and d be the inverse of e modulo l.c.m. (p-1, q-1). The integers n, f, b and e are published by the authority, whilst p, q and d are kept secret.

Alice chooses her secret key s as before, calculates b^{-s} (mod n) and gives it to the center. Then, the center calculates $P = I^{-d}b^{-s}$ (mod n), and P will be Alice's public key . This public key is somewhat particular in that it depends both on user's identity I and on his secret s. Note that I and P are connected by the equation : $P^eIh^s = 1$ (mod n), where $h = b^e$ (mod n). A similar set-up can be found in [TO].

The protocol is as follows :

1) Alice picks a random integer r in the interval [0, f-1], calculates $x = h^r$ (mod n) and sends it to Bob along with her certificate.

2) Bob checks the certificate, picks a random integer c in the interval [0, e-1] and sends it to Alice.

3) Alice calculates $y = r + sc$ (mod f) and sends it to Bob.

4) Bob checks that $h^y(P^eI)^c = x$ (mod n).

Security of the scheme

Two distinct questions have to be discussed. First, what is the level of difficulty to impersonate a user (by finding his secret s) ? Second, what is the level of difficulty to impersonate the trusted center (by finding his secret d) ?

In non-identity-based schemes, as Schnorr's scheme, the second question is irrelevant since the answer only depends on the signature scheme which is used to produce certificates. And this signature scheme may be completely independent of the identification scheme itself.

In usual identity-based schemes, the two questions become only one, because s is calculated from I using d, and finding d seems to be the only way to find s.

In our scheme, the two questions are separate since, as already mentioned, d is not sufficient to calculate s. Factoring n is enough to impersonate the trusted center whilst calculating a discrete logarithm modulo n (a harder problem, see previous section) is required to impersonate a user.

Other topic

The set-up of this scheme also provides material for identity-based key-exchange. The whole package has been exposed in [GP].

Acknowledgements

Many thanks to Brigitte Vallée, who improved the security of the scheme. Also thanks to K. Ohta, C. Schnorr and G. Robin, who listened to me exposing it.

Bibliography

[BC] E. F. Brickell and K. S. McCurley, "An interactive identification scheme based on discrete logarithms and factoring", EUROCRYPT'90, these proceedings.

[GP] M. Girault and J. C. Paillès, "An identity-based scheme providing zero-knowledge authentication and authenticated key-exchange", Proc. of ESORICS'90, 24-26 oct. 90.

[Sc] C. P. Schnorr, "Efficient identification and signatures for smart cards", Advances in Cryptology, Proc. of CRYPTO'89, Springer-Verlag.

[Sh] A. Shamir, "Identity-based cryptosystems and signature schemes", Advances in Cryptology, Proc. of CRYPTO'84, LNCS 196, Springer-Verlag, 1985, pp.47-53.

[SS] A. Schrift and A. Shamir, "The discrete log is very discreet", presented at 1st Oberwolfach Conference on Cryptography, 24-30 sept. 89.

[TO] K. Tanaka and E. Okamoto, "Key distribution system using ID-related information directory suitable for mail systems", Proc. of SECURICOM'90, pp.115-122.

[W] H. Woll, "Reductions among number theoretic problems", Information and Computation 72, pp.167-179, 1987.

A NOISY CLOCK-CONTROLLED SHIFT REGISTER CRYPTANALYSIS CONCEPT BASED ON SEQUENCE COMPARISON APPROACH

Jovan Dj. Golić

Miodrag J. Mihaljević

Institute of Applied Mathematics and Electronics, Belgrade
Faculty of Electrical Engineering, University of Belgrade
Bulevar Revolucije 73, 11001 Beograd, Yugoslavia

Abstract: A statistical cryptanalysis method for the initial state reconstruction of a noisy clock-controlled shift register using the noisy output sequence only, is proposed. The method is based on the sequence comparison approach.

1. PROBLEM STATEMENT

A review of clock-controlled shift registers is presented in [1]. A statistical model of the clock-controlled shift register structure, which is under consideration in this correspondence, is shown in Fig.1. For simplicity, we assume that the shift register whose output is correlated with the generator output is one-two clocked.

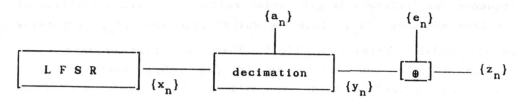

Fig.1. A model of the clock-controlled shift register structure.

A binary sequence $\{x_n\}$ is the output of a linear feedback shift register (LFSR) with characteristic polynomial $f(X)=\Sigma_{\ell=0}^{L}c_{L-\ell}X^{\ell}$, $c_0=1$, and $X_0=[x_{-\ell}]_{\ell=1}^{L}$ is the LFSR initial state. For example, a decimation sequence $\{a_n\}$ is the output of another binary shift register. The decimation box output is defined by $y_n = x_{f(n)}$, $f(n) = n + \Sigma_{j=1}^{n} a_j$, $n=0,1,2,\ldots$.

In the statistical model, $\{a_n\}$ is regarded as a realization of the sequence of i.i.d. binary variables $\{A_n\}$ such that $Pr(A_n=1) = 0.5$ for every n. A binary sequence $\{e_n\}$ is a realization of a sequence of i.i.d. binary variables $\{E_n\}$ such that $Pr(E_n=1) = p < 0.5$ for every n, where p is the cross-correlation parameter, which may involve the plaintext statistics as well, [2]. Finally, a binary sequence $\{z_n\}$ is defined by

$$z_n = x_{f(n)} \oplus e_n \quad , \quad f(n) = n + \Sigma_{j=1}^{n} a_j \quad , \quad n=0,1,2,\ldots \ . \qquad (1)$$

In this correspondence, the problem of the initial state ($X_0 = [x_{-\ell}]_{\ell=1}^{L}$) reconstruction when $f(X)$, p, and a segment $\{z_n\}_{n=1}^{N}$ are known, is considered.

2. INITIAL STATE RECONSTRUCTION

A correlation attack [2] is based on the Hamming distance between two binary sequences of the same length. Obviously, the same statistical approach can not be applied here. However, suppose we defined a suitable distance measure d between two binary sequences of different length, which reflects the transformation of the LFSR sequence $\{x_n\}$ into the output sequence $\{z_n\}$ according to the model displayed in Fig.1. Then, we could proceed along essentially the same lines as in [2], thus establishing a statistical procedure which we call a generalized correlation attack.

Due to the assumed statistical model, each X_0 gives rise to a conditional probability distribution on the set of all binary sequences $\{z_n\}_{n=1}^{N}$. We thus have a pattern recognition system with 2^L classes corresponding to all the initial states of the LFSR. Given an observed segment $\{z_n\}_{n=1}^{N}$, an optimal decision strategy (yielding the minimum probability of decission error) is to decide on the initial state with maximum posterior probability. When the LFSR is regularly clocked, as in [2], it is optimal to decide on the initial state $\hat{X}_0$ such that the Hamming distance between $\{z_n\}_{n=1}^{N}$ and $\{\hat{x}_n\}_{n=1}^{N}$ is minimum (a sufficient statistics). However, when the LFSR is clocked irregularly it is not clear how to find an optimum decision rule. Anyway, given an appropriate distance measure, we can define a minimum distance decision procedure which may be close to optimal.

Let $\{\hat{x}_n\}_{n=1}^{M}$ be an LFSR sequence corresponding to the initial state $\hat{X}_0$ (typically, $M \cong 3N/2$). Let d be the distance between $\{\hat{x}_n\}_{n=1}^{M}$ and $\{z_n\}_{n=1}^{N}$. Two cases-hypotheses are possible:

H_0 : the observed sequence $\{z_n\}_{n=1}^{N}$ is produced by $\hat{X}_0$;

H_1 : the observed sequence $\{z_n\}_{n=1}^{N}$ is not produced by $\hat{X}_0$.

Consequently, d is a realization of a random variable D with two possible probability distributions (averaged over the ensemble of all the initial states): $\{Pr(D|H_0)\}$ and $\{Pr(D|H_1)\}$. How to determine or estimate these distributions will be discussed in the next Section. Suppose that they are known. Note that they depend on N , assuming that $M = M(N)$. First determine the threshold t and length N so as to achieve the given probabilities of "the missing event" P_m and "the false alarm" P_f . As in [2], P_m is chosen close to zero (f.e., 10^{-3}) and P_f is picked very close to zero, $P_f \cong 2^{-L}$, so that the expected number of false alarms is very small ($\cong 1$). Then, the decision procedure goes through the following steps, for every possible initial state $\hat{X}_0$:

Step 1: generate $\{\hat{x}_n\}_{n=1}^{M}$.

Step 2: calculate the distance d between $\{\hat{x}_n\}_{n=1}^{M}$ and $\{z_n\}_{n=1}^{N}$.

Step 3: according to the threshold t accept H_0 or H_1 .

The output of the procedure is the set of the most probable candidates for the true initial state. The computational complexity is proportional to the number of possible initial states (for example, 2^L).

3.   A DISTANCE MEASURE AND RELEVANT PROBABILITY DISTRIBUTIONS

A distance measure should be defined so that it enables statistical discrimination between the two cases: first, when $\{\hat{x}_n\}_{n=1}^M$ and $\{z_n\}_{n=1}^N$ are picked at random, uniformly and independently, and second, when $\{z_n\}_{n=1}^N$ is obtained from $\{\hat{x}_n\}_{n=1}^M$, according to the model in Fig.1, that is, by the deletion of some bits subject to the decimation constraints and by the complementation of the remaining ones, with probability p. This problem is a special case of the comparison problem between two sequences when one sequence is obtained from the other by symbol substitution, deletion, and insertion, which is extensively studied in the literature. For example, the sequence matching problem is considered in coding theory (see [3], for example) and text processing (see [5], for example). A review of the sequence matching techniques and applications is presented in [4].

According to [4], one of the widely used distances is the Levenshtein distance [3]. Let the edit operations that transform one sequence into another be substitution, deletion, and insertion. Then, the Levenshtein distance between two sequences is defined as the minimum number of edit operations required to transform one sequence into the other. The various extensions of the basic Levenshtein distance are proposed in the literature. For our problem, the Constrained Levenshtein Distance (CLD) concept [7] is relevant, because the constraints are inherent to the decimation function (see relation (1)). In [5], [6], an efficient algorithm for the constrained Levenshtein distance computation is proposed when the constraints relate to the total number of deletions, insertions, and substitutions, respectively.

We define CLD^*, the distance measure between $\{\hat{x}_n\}_{n=1}^M$ and $\{z_n\}_{n=1}^N$ as the minimum number of deletions and complementations required to obtain $\{z_n\}_{n=1}^N$ from $\{\hat{x}_n\}_{n=1}^M$ subject to the assumed constraint on the number of consecutive deletions. Whether this distance is a sufficient statistics remains an open question, but it is reasonable to believe that this is approximately the case.

With the CLD^* so defined, a problem is to determine the probability distributions $\{Pr(D|H_0)\}$ and $\{Pr(D|H_1)\}$. According to the literature the problem appears very difficult. One approach is a nonparametric estimation .

Another problem is to define a procedure for efficient

computation of the defined distance measure. Following the main ideas from [7], a novel dynamic programming algorithm can be derived that computes the desired distance measure CLD^* in the following way.

The Constrained Levenshtein Distance (CLD^*) Computation Procedure:

1. Input: binary sequences $\{\hat{x}_n\}_{n=1}^M$ and $\{z_n\}_{n=1}^N$.

2. Initialization: $d(k,0) = k$, $k = 0, 1, \ldots, M-N$,

   $$d(0,\ell) = d(0,\ell-1) + (\hat{x}_\ell \oplus z_\ell) , \quad \ell = 1, 2, \ldots, N .$$

3. Recursive calculation for $M > N$:

   $$d(k,\ell) = \min\{ \ d(k-1,\ell-1) + (\hat{x}_{k+\ell-1} \oplus z_\ell) + 1 \ , \ d(k,\ell-1) + (\hat{x}_{k+\ell} \oplus z_\ell) \ \} .$$

   $$\ell = 1, 2, \ldots, N \quad , \quad k = \max\{1, \ M-2N+\ell\}, \ldots, M-N.$$

4. Output: the CLD^* between $\{\hat{x}_n\}_{n=1}^M$ and $\{z_n\}_{n=1}^N$: $d^* = d(M-N, N)$.

The computational complexity of the procedure is quadratic $O(N(M-N))$.

Note that an arbitrary number of initial deletions is allowed, since the length M of $\{\hat{x}_n\}_{n=1}^M$ that actually produced $\{z_n\}_{n=1}^N$ is not known (therefore, one can assume that $M = 2N+1$).

4. REFERENCES

[1] D.Gollman, W.G.Chambers, "Clock-controlled shift registers: A review", IEEE Journal on Selected Areas in Communications, vol. SAC-7, May 1989., pp.525-533.

[2] T.Siegenthaler, "Decrypting a class of stream ciphers using ciphertext only", IEEE Trans. Comput. vol. C-34, Jan. 1985, pp.81-85.

[3] A.Levenshtein, "Binary codes capable of correcting deletions, insertions, and reversals", Sov. Phy. Dokl., vol.10, pp.707-710, 1966.

[4] D.Sankoff, J.B.Kruskal, *Time Warps, String Edits and Macromolecules: The Theory and Practice of Sequence Comparison*. Reading, MA: Addison-Wesley, 1983.

[5] B.J.Oommen, "Recognition of noisy subsequences using constrained edit distance", IEEE Trans. Pattern Analysis Mach. Intell., vol. PAMI-9, Sep. 1987., pp.676-685.

[6] B.J.Oommen, "Correction to 'Recognition of noisy subsequences using constrained edit distance'", IEEE Trans. Pattern Analysis Mach. Intell., vol. PAMI-10, Nov. 1988., pp.983-984.

[7] B.J.Oommen, "Constrained string editing", Inform. Sci., vol.40, 1986., pp.267-284.

The MD4 Message Digest Algorithm

Burton S. Kaliski Jr.
RSA Data Security Inc.
Redwood City, CA

Abstract. *MD4 is a new, fast message digest algorithm. It inputs a message of any length and outputs a digest of 128 bits. It is conjectured that it is computationally infeasible to find two messages with the same digest or a message with a prespecified digest. MD4 processes 1.45M bytes/s on a SUN Sparc station, 70K bytes/s on a DEC MicroVax II, and 32K bytes/s on a 20MHz 80286. MD4 is also quite compact. MD4 is being placed in the public domain for review and possible adoption as a standard.*

Details of the MD4 message digest algorithm will be presented by Ronald L. Rivest at CRYPTO '90 and will appear in the proceedings of that conference.

A remark on the efficiency of identification schemes

Mike Burmester

RHBNC - University of London

Egham, Surrey TW20 OEX

U.K.

Abstract

The efficiency parameters of identification schemes (memory size, communication cost, computational complexity) are based on given security levels and should allow for the 'worst-case' probability of error (forgery). We consider instances of the schemes in [OO88] and [Sch90] for which the efficiency is not as good as claimed.

Introduction. Ohta-Okamoto presented [OO88] a modification of the Fiat-Shamir [FS86] identification scheme which claims to reduce the probability of error (forgery) from 2^{-kt} to L^{-kt} (for suitable L). Here k is the number of secret information integers, t the number of iterations and L the exponent (for the Fiat-Shamir scheme $L = 2$). We shall see that this is not always true, e.g., there are instances for which this probability is 2^{-kt} and indeed 2^{-t}, if we use the argument in [BD89]. In particular, for the parallel implementation the probability of error can be $1/2$. Similar instances occur with the scheme in [Sch90].

The schemes that we consider are based on interactive proof systems. A formal setting for such systems is given in [GMR89,FFS88]. Let A be the prover, B the verifier and (A, B) an interactive proof of membership in a language $\mathcal{L}$. For every dishonest prover $\tilde{A}$ there is a probability that B will accept when the input $x \notin \mathcal{L}$. The probability of error of (A, B) is the largest such probability, taken over all $\tilde{A}$. This is negligible when the proof (A, B) is sound [GMR89]. The probability of error for proofs of knowledge [FFS88] is defined in a similar way.

The Ohta-Okamoto scheme. Let $n = pq$, p and q distinct odd primes, $L \geq 2$, and $x = (I; n, L)$, $1 < I < n$, be the input. The prover A proves that there exists (or that it knows) an S such that $I = S^L \bmod n$. The protocol has four steps which are repeated $t = O(\log n)$ times. In Step 1, A sends B the number $X = R^L \bmod n$, R random in Z_n. In Step 2, B sends A a random query $E \in Z_L$ and in Step 3, A replies with $Y = R \cdot S^E \bmod n$. Finally in Step 4, B verifies that $Y^L \equiv X \cdot I^E \pmod{n}$. B accepts (the proof of A) if the verification is valid for all t iterations.

We shall show that the probability of error can be as large as 2^{-t}. Suppose that $L = 2L_1$, L_1 odd, and that $I, S_1 \in Z_n^*$ are such that $I = S_1^{L_1} \bmod n$ with S_1 a quadratic non-residue mod n. Then I is a non-residue and does not have an L-th root mod n. Let $\tilde{A}$ be a dishonest prover which guesses the parity of the queries randomly, with uniform distribution. In Step 1, $\tilde{A}$ sends $X = R^L \bmod n$ if the guessed parity is even, and $X = R^L \cdot I^{-1} \bmod n$ if the guessed parity is odd. In Step 3, $\tilde{A}$ sends $Y = R \cdot S_1^{E/2} \bmod n$ if the (actual) query E is even, and $Y = R \cdot S_1^{(E-1)/2} \bmod n$ if E is odd. Then B will accept when $\tilde{A}$ has guessed the parity correctly. Indeed for E even, $X \cdot I^E \equiv R^L \cdot S_1^{L_1 E} \equiv (R \cdot S_1^{E/2})^L \equiv Y^L \pmod{n}$ and for E odd, $X \cdot I^E \equiv R^L \cdot I^{-1} \cdot S_1^{L_1 E} \equiv R^L \cdot S_1^{L_1(E-1)} \equiv (R \cdot S_1^{(E-1)/2})^L \equiv Y^L \pmod{n}$. So B will accept with probability $1/2$ for each iteration. Therefore the probability of error for the proof (A, B) is at least 2^{-t}.

A similar example can be used with proofs of knowledge. For 'unrestricted input' proofs the input I has to be an L-th root. Again we take $L = 2L_1$, only this time L_1 need not be odd and S_1 is a quadratic residue. The dishonest prover $\tilde{A}$ is given on its knowledge tape S_1 but not $\sqrt{S_1} \bmod n$ (the soundness condition for proofs of knowledge [FFS88] does not restrict the contents of the knowledge tape of $\tilde{A}$: we assume that it is hard to compute $\sqrt{S_1} \bmod n$, given S_1). So $\tilde{A}$ does not *know* an L-th root of I. As before, if $\tilde{A}$ guesses the parities then B will accept with probability 2^{-t}.

This argument can be easily extended to other values of L which have a common factor with $p-1$ or $q-1$. An illustration of a more general case for which n is a product of three primes and L is a prime is given in [BD89].

In [OO88, p.241] it is argued that the probability of cheating is $1/L$ when $t = 1$ and L is the product of distinct primes with $(L, p-1) = L$, provided that there is no probabilistic polynomial time algorithm for factoring. This is not true for our example. For us, with such $L > 2$, L even, the probability of error is $1/2$, and there is no reason why factoring should be any easier (*e.g.*, when S_1 is a quadratic non-residue, for proofs of membership, or when $\tilde{A}$ has S_1 on its knowledge tape, for proofs of knowledge).

In conclusion, the probability of error (forgery) lies between L^{-kt} and 2^{-t}, depending on L. Even though this is negligible when $t = \Theta(\log n)$, the larger value must be taken into account when considering the efficiency parameters of the scheme. We get the lowest probability (and hence the best efficiency) when L is a prime number [GQ88] which is large (non-constant, polynomial in $\log n$), provided that the input is of the 'proper' form and that $Y \notin Z_n^*$, for proofs of membership, or $Y \neq 0$, for proofs of knowledge [BD89].

The Schnorr scheme. Let p, q be odd primes with $q \mid p-1$, $\alpha \in Z_p$ have order q, $L = 2^l$, and $x = (v; \alpha, p, q, L)$, $v \in Z_p^*$, be the input. The prover A proves that it knows an s such that $v = \alpha^{-s} \bmod p$. Again the protocol has four steps. In Step 1, A sends $z = \alpha^r \bmod p$, r random in $[1 : p-1]$, in Step 2, B sends the random query $e \in Z_L$, and in Step 3, A replies with $y = r + se \pmod{q}$. In Step 4, B checks that $z = \alpha^y v^e \bmod p$ and accepts if equality holds.

For this protocol the probability of error is $1/2$. Indeed let γ be a primitive element of Z_p and $\alpha = \gamma^{p-1/q} \bmod p$, $\beta = \gamma^{p-1/2q} \bmod p$, and $v = \beta^{-s} \bmod p$, s odd. Then $v \neq \alpha^i \bmod p$ for all i (v has even order) and there is no s such that $v = \alpha^{-s} \bmod p$. $\tilde{A}$ is a dishonest prover which is given s on its knowledge tape. As before $\tilde{A}$ guesses the parity of the query and sends either $z = \alpha^r \bmod p$ or $z = \alpha^r v \bmod p$ in Step 1. In Step 3, $\tilde{A}$ sends

$y = r + se/2 \pmod{q}$ if e is even, and $y = r + s(e-1)/2 \pmod{q}$ otherwise. Again B will accept when $\tilde{A}$ has guessed the parity correctly. So the probability of error is $1/2$.

In [Sch90, Proposition 2.1] it is argued that if the probability of error ε is greater than 2^{-l+2} then $\log_\alpha v$ can be computed in time $O(\varepsilon^{-1})$ with constant, positive probability. For us, when $l > 3$, this is not true since $\varepsilon = 1/2$ and $\log_\alpha v$ does not exist.

To prevent this situation (of 'proving' knowledge of logarithms which do not exist) the verifier must check in the protocol that $v^q \equiv 1 \pmod{p}$. Then $\log_\alpha v$ always exists. Of course this is only possible when q is 'public'. The example described above also applies to the Brickell-McCurley identification scheme [BrMcC90] as presented at Eurocrypt'90. This scheme has now been adjusted so that the prover first proves to a Key Issuing Authority that $\log_\alpha v$ exists.

Acknowledgement. The author wishes to thank Yvo Desmedt for helpful discussions and Kevin McCurley for a remark about the Schnorr identification scheme.

References

[BrMcC90] E.F. Brickell, and K.S. McCurley. An interactive identification scheme based on discrete logarithms and factoring. Presented at Eurocrypt'90.

[BD89] M. Burmester, and Y. Desmedt. Remarks on the soundness of proofs. *Electronics Letters*, **25**(22), 1989, pp.1509–1511.

[FFS88] U. Feige, A. Fiat, and A. Shamir. Zero knowledge proofs of identity. *Journal of Cryptology*, **1**(2), 1988, pp. 77–94.

[GQ88] L.C Guillou and J.-J. Quisquater. A practical zero-knowledge protocol fitted to security microprocessor minimizing both transmission and memory. *Proceedings of Eurocrypt'88*, Lecture Notes in Computer Science 330, Springer-Verlag, Berlin, pp. 123–128.

[FS86] A. Fiat, and A. Shamir. How to prove yourself: Practical solutions to identification and signature problems. *Proceedings of Crypto'86*, Lecture Notes in Computer Science 206, Springer-Verlag, New York, pp. 186–194.

[GMR89] S. Goldwasser, S. Micali, and C. Rackoff. The knowledge complexity of interactive proof systems. *Siam J. Comput.*, **18**(1), 1989, pp. 186–208.

[OO88] K. Ohta, and T. Okamoto. A modification of the Fiat-Shamir scheme. *Proceedings of Crypto'88*, Lecture Notes in Computer Science 403, Springer-Verlag, New York, pp. 232–243.

[Sch90] C.P. Schnorr. Efficient Identifications and Signatures for Smart Cards. *Proceedings of Crypto'89*, Lecture Notes in Computer Science 435, Springer-Verlag, New York, pp. 239–251.

ON AN IMPLEMENTATION OF THE
MOHAN-ADIGA ALGORITHM

Gisela Meister

GAO, Gesellschaft für Automation und Organisation mbH
Euckenstr.12, D - 8000 München, FR Germany

ABSTRACT. Several asymmetric cryptographic systems such as the RSA system [6] require modular exponentiation of large integers. This paper discusses a modular routine described in [2], which is suited for smart cards. It is based on the Mohan-Adiga algorithm [5]. This algorithm is comparatively fast, if the leading half of the bits of the modulus is 1. It will be shown that this restriction has some severe implications on the number of suitable primes and on the security of the system. If one decrements the number of leading 1's then the security level of the system is increased while the speed is decreased.

1 INTRODUCTION

Software implementations of Public Key Algorithms based on modular exponentation are at present not very suited for say smart cards. This is due to the amount of time and RAM it takes to encipher a block or produce a digital signature for a document using a 512 bit modulus. In light of recent advances in factoring [4] it can be assumed that the length of the modulus and the execution time of the algorithm will be increased. This makes the need for a fast modular exponentiation method even more important. One approach to avoid the rather time-consuming bitwise reduction, often used in modular exponentiation [3], is the Mohan-Adiga algorithm which uses integer multiplication instead. Several authors ([1],[2],[5]) suggest that the speed be increased by choosing the modulus m to satisfy

$$(1.1) \qquad 2^{LM} - 2^{LM/2} \leq m < 2^{LM} ,$$

where LM denotes the length of the modulus in bits. This means that the leading half of the bits of m is 1. We assume LM to be a multiple of 4.

One problem is of course to find suitable primes p and q which satisfy not only condition (1.1) but also those in [6]. In this paper it will be shown that, for a given prime p of length LM/2, there is at most one prime q of length LM/2+1 such that the product m=p*q satisfies (1.1).

2 CONSTRUCTION OF THE MODULUS

Knobloch [2] uses the following approach to construct a modulus satisfying condition (1.1).

Let p be an integer satisfying the following inequality

(2.1) $2^{LM/2-1} \leq p < 2^{LM/2}$,

and let

(2.2) $q_0 = [(2^{LM} - 2^{LM/2} -1)/p] +1$,

where [x] denotes the integer part of the real number x. Then

(2.3) $2^{LM/2+1} > q_0 \geq 2^{LM/2}$ and

(2.4) $2^{LM} > p * q_0 \geq 2^{LM} - 2^{LM/2}$.

This means that $m = p*q_0$ satisfies condition (1.1).

In the prime number theorem [3, pp 366] it is proved that a given integer p is prime with probability

(2.5) $P(p) = 1 / \ln 2^{LM/2}$.

In [2] it is stated that there exists a small intervall around q_0 which contains 'sufficiently many' primes q such that the product p*q satisfies condition (2.4).This is however not the case. For integers q with $q < q_0$ or $q > q_0+1$ do not satisfy (2.4) as can easily be seen by substitution of q_0 by q_0-1 or q_0+2, respectively.

Hence there is at most one candidate for q. This is q_0, if q_0 is odd, and q_0+1, if q_0 is even. If neither q_0 nor q_0+1 are prime, the process has to start again.

In the sequel let f(p) denote the odd candidate. It is straightforward to show by means of the prime number theorem that the conditional probability that f(p) is prime, is bounded by

(2.8) $P(f(p)|p) < 1 / \ln 2^{LM/2+1}$.

So the number of suitable pairs is

(2.9) $P(p) * P(f(p)|p) * 2^{(LM/2)}$

           $< \quad ANZ = \quad 2^{LM/2} / (\ln 2^{LM/2} * \ln 2^{LM/2})$.

If (p,q) does not have to satisfy (2.4), we obtain

(2.10) $P(p) * P(q) * 2^{LM/2} * 2^{LM/2} = ANZ * 2^{LM/2}$.

In ([1],[2],[5]) it is stated that there are enough pairs of primes (p,q) such that the product m=p*q satisfies (1.1). Note however that the number ANZ of pairs which satisfy (2.4) is equal to a quarter of the number of pairs with modulus of length LM/2 satisfying $2^{LM/4-1} < p < 2^{LM/4}$ and $2^{LM/4+1} > q_0 > 2^{LM/4}$.
The number of pairs is

(2.11) $2^{LM/2} / \ln(2^{LM/4})^2 = ANZ * (LM/2)^2 / (LM/4)^2 = 4\ ANZ$.

3   SECURITY RISKS

Since there is a functional relation between p and f(p), we can obtain p as the unique solution of the functional equation

(3.1) $f(p) * p = m$

and the unique root of the function

(3.2) $g(p) = m - p * f(p)$.

For instance, equation (3.2) is easily solved if p satisfies

$2^{LM/2} - 2^{LM/4} \leq p < 2^{LM/2}$ or $2^{LM/4} - 2^{LM/16} \leq p < 2^{LM/4}$.

Say $p = 2^{LM/2} - x$ with $x < 2^{LM/4}$. Then $f(p) = 2^{LM/2} + x$ and (3.2) reduces to the solution of the quadratic equation

$$(3.3) \qquad 0 = g(p) = m + 2^{LM/2+1} * p - p^2.$$

Therefore the approach suggested in [2] seems to be not feasible.

4 MODIFICATION

If we modify (1.1) to become

$$(4.1) \qquad 2^{LM} > m \geq 2^{LM} - 2^{LM*a} = L_a , \quad 3/4 \leq a < 1$$

as suggested in [7], then the security of the system increases while the speed decreases. To be precise, we define

$$(4.2) \qquad q_0 = [(L_a / p)] + 1,$$

then the interval I_a, in which each q together with p satisfies (4.1), is

$$(4.3) \qquad I_a = [q_0, q_0 + 2^{LM*a-LM/2+1}].$$

According to [3, p 366] the number of its primes is roughly

$$(4.4) \qquad |I_a| / \ln 2^{LM/2+1},$$

and the probability that q in I_a is prime is equal to

$$(4.5) \qquad 1 / \ln 2^{LM/2}.$$

With increasing a the number ANZA of pairs, satisfying (4.1) increases by $2^{LM*(a-1/2)}$. It is

$$(4.6) \qquad ANZA = ANZ * |I_a|.$$

The closer a is to 1, the slower the Mohan-Adiga algorithm will be. So the optimal factor a depends on the security level as well as on the required speed of the algorithm.

BIBLIOGRAPHY

[1]   W.Fumy and A.Pfau, *On the Complexity of Asymmetric Smart Card Authentication*, Proceedings Smart Card 2000, 1989.

[2]   H.-J.Knobloch, *A Smart Card Implementation of the Fiat Shamir-Identificaton Scheme*, in: Advances in Cryptology-Proceedings of Eurocrypt'88, Lecture Notes in Computer Science 330, Springer-Verlag 1988, 87-95.

[3]   E.Knuth, *The Art of Computerprogramming*, vol.2, Seminumerical Algorithms, Addison-Wesley Publishing Company, Reading, 1980.

[4]   A.K.Lenstra and A.K.Manasse, *Factoring with two large primes*, Abstracts of Eurocrypt '90.

[5]   S.B.Mohan and B.S.Adiga, *Fast Algorithms for Implementing RSA Public Key Cryptosystem*, Electronic Letters (1985) vol. 21, no. 7, 761.

[6]   R.Rivest, A.Shamir and A.Adleman, *A Method for Obtaining Digital Signatures and Public Key Cryptosystems*, Commun.ACM (1978) 120-126.

[7]   C.Guillou and J.-J.Quisquater, *Precautions Taken Against Various Potential Attacks in ISO/IEC DP 9796*, Abstracts of Eurocrypt'90.

Vol. 437: D. Kumar (Ed.), Current Trends in SNePS – Semantic Network Processing System. Proceedings, 1989. VII, 162 pages. 1990. (Subseries LNAI).

Vol. 438: D. H. Norrie, H.-W. Six (Eds.), Computer Assisted Learning – ICCAL '90. Proceedings, 1990. VII, 467 pages. 1990.

Vol. 439: P. Gorny, M. Tauber (Eds.), Visualization in Human-Computer Interaction. Proceedings, 1988. VI, 274 pages. 1990.

Vol. 440: E.Börger, H. Kleine Büning, M. M. Richter (Eds.), CSL '89. Proceedings, 1989. VI, 437 pages. 1990.

Vol. 441: T. Ito, R. H. Halstead, Jr. (Eds.), Parallel Lisp: Languages and Systems. Proceedings, 1989. XII, 364 pages. 1990.

Vol. 442: M. Main, A. Melton, M. Mislove, D. Schmidt (Eds.), Mathematical Foundations of Programming Semantics. Proceedings, 1989. VI, 439 pages. 1990.

Vol. 443: M. S. Paterson (Ed.), Automata, Languages and Programming. Proceedings, 1990. IX, 781 pages. 1990.

Vol. 444: S. Ramani, R. Chandrasekar, K.S.R. Anjaneyulu (Eds.), Knowledge Based Computer Systems. Proceedings, 1989. X, 546 pages. 1990. (Subseries LNAI).

Vol. 445: A. J. M. van Gasteren, On the Shape of Mathematical Arguments. VIII, 181 pages. 1990.

Vol. 446: L. Plümer, Termination Proofs for Logic Programs. VIII, 142 pages. 1990. (Subseries LNAI).

Vol. 447: J. R. Gilbert, R. Karlsson (Eds.), SWAT 90. 2nd Scandinavian Workshop on Algorithm Theory. Proceedings, 1990. VI, 417 pages. 1990.

Vol. 448: B. Simons, A. Spector (Eds.), Fault-Tolerant Distributed Computing. VI, 298 pages. 1990.

Vol. 449: M. E. Stickel (Ed.), 10th International Conference on Automated Deduction. Proceedings, 1990. XVI, 688 pages. 1990. (Subseries LNAI).

Vol. 450: T. Asano, T. Ibaraki, H. Imai, T. Nishizeki (Eds.), Algorithms. Proceedings, 1990. VIII, 479 pages. 1990.

Vol. 451: V. Mařík, O. Štěpánková, Z. Zdráhal (Eds.), Artificial Intelligence in Higher Education. Proceedings, 1989. IX, 247 pages. 1990. (Subseries LNAI).

Vol. 452: B. Rovan (Ed.), Mathematical Foundations of Computer Science 1990. Proceedings, 1990. VIII, 544 pages. 1990.

Vol. 453: J. Seberry, J. Pieprzyk (Eds.), Advances in Cryptology – AUSCRYPT '90. Proceedings, 1990. IX, 462 pages. 1990.

Vol. 454: V. Diekert, Combinatorics on Traces. XII, 165 pages. 1990.

Vol. 455: C. A. Floudas, P. M. Pardalos, A Collection of Test Problems for Constrained Global Optimization Algorithms. XIV, 180 pages. 1990.

Vol. 456: P. Deransart, J. Maluszyński (Eds.), Programming Language Implementation and Logic Programming. Proceedings, 1990. VIII, 401 pages. 1990.

Vol. 457: H. Burkhart (Ed.), CONPAR '90 – VAPP IV. Proceedings, 1990. XIV, 900 pages. 1990.

Vol. 458: J. C. M. Baeten, J. W. Klop (Eds.), CONCUR '90. Proceedings, 1990. VII, 537 pages. 1990.

Vol. 459: R. Studer (Ed.), Natural Language and Logic. Proceedings, 1989. VII, 252 pages. 1990. (Subseries LNAI).

Vol. 460: J. Uhl, H.A. Schmid, A Systematic Catalogue of Reusable Abstract Data Types. XII, 344 pages. 1990.

Vol. 461: P. Deransart, M. Jourdan (Eds.), Attribute Grammars and their Applications. Proceedings, 1990. VIII, 358 pages. 1990.

Vol. 462: G. Gottlob, W. Nejdl (Eds.), Expert Systems in Engineering. Proceedings, 1990. IX, 260 pages. 1990. (Subseries LNAI).

Vol. 463: H. Kirchner, W. Wechler (Eds.), Algebraic and Logic Programming. Proceedings, 1990. VII, 386 pages. 1990.

Vol. 464: J. Dassow, J. Kelemen (Eds.), Aspects and Prospects of Theoretical Computer Science. Proceedings, 1990. VI, 298 pages. 1990.

Vol. 465: A. Fuhrmann, M. Morreau (Eds.), The Logic of Theory Change. Proceedings, 1989. X, 334 pages. 1991. (Subseries LNAI).

Vol. 466: A. Blaser (Ed.), Database Systems of the 90s. Proceedings, 1990. VIII, 334 pages. 1990.

Vol. 467: F. Long (Ed.), Software Engineering Environments. Proceedings, 1989. VI, 313 pages. 1990.

Vol. 468: S. G. Akl, F. Fiala, W. W. Koczkodaj (Eds.), Advances in Computing and Information – ICCI '90. Proceedings, 1990. VII, 529 pages. 1990.

Vol. 469: I. Guessarian (Ed.), Semantics of Systems of Concurrent Processes. Proceedings, 1990. V, 456 pages. 1990.

Vol. 470: S. Abiteboul, P.C. Kanellakis (Eds.), ICDT '90. Proceedings, 1990. VII, 528 pages. 1990.

Vol. 471: B. C. Ooi, Efficient Query Processing in Geographic Information Systems. VIII, 208 pages. 1990.

Vol. 472: K. V. Nori, C. E. Veni Madhavan (Eds.), Foundations of Software Technology and Theoretical Computer Science. Proceedings, 1990. X, 420 pages. 1990.

Vol. 473: I. B. Damgård (Ed.), Advances in Cryptology – EUROCRYPT '90. Proceedings, 1990. VIII, 500 pages. 1991.

Printed in the United States
By Bookmasters

Printed in the United States
By Bookmasters